DATE DUE FOR RETURN

MYCOTOXINS

MYCOTOXINS

edited by

I.F.H. PURCHASE

B.V.SC., M.R.C.V.S., PH.D., M.R.C. PATH.,

Head, Experimental Pathology Unit
Central Toxicology Laboratory
Imperial Chemical Industries Limited
Alderley Park, Near Macclesfield, Cheshire
Great Britain

ELSEVIER SCIENTIFIC PUBLISHING COMPANY
AMSTERDAM — OXFORD — NEW YORK 1974

ELSEVIER SCIENTIFIC PUBLISHING COMPANY
335 Jan van Galenstraat
P.O. Box 211, Amsterdam, The Netherlands

AMERICAN ELSEVIER PUBLISHING COMPANY, INC.
52 Vanderbilt Avenue
New York, New York 10017

Library of congress card number: 74-83314
ISBN 0-444-41254-9
With 111 illustrations and 57 tables

Printed in The Netherlands

CONTRIBUTORS TO THIS BOOK

B. Altenkirk — National Research Institute for Nutritional Diseases, Tiervlei, C.P. (Republic of South Africa)

L.G. Atherton — Atlantic Regional Laboratory, National Research Council for Canada, Halifax, Nova Scotia (Canada)

Steven D. Aust — Department of Biochemistry, Michigan State University, East Lansing, Mich. (U.S.A.)

Michael R. Boyd — Department of Pharmacology, School of Medicine, Vanderbilt University, Nashville, Tenn. 37232 (U.S.A.)

D. Brewer — Atlantic Regional Laboratory, National Research Council for Canada, Halifax, Nova Scotia (Canada)

W.H. Butler — MRC Toxicology Unit, Medical Research Council Laboratories, Woodmansterne Road, Carshalton, Surrey (Great Britain)

Clyde M. Christensen —Department of Plant Pathology, University of Minnesota, Minneapolis, Minn. (U.S.A.)

Robert M. Eppley — Division of Chemistry and Physics, Bureau of Foods, Food and Drug Administration, U.S. Department of Health, Education and Welfare, Washington, D.C. (U.S.A.)

Makoto Enomoto — Department of Pathology, St. Marianna University School of Medicine, Kawasaki (Japan)

J. Harwig — Food Research Laboratories, Health Protection Branch, Department of National Health and Welfare, Tunney's Pasture, Ottawa, Ontario, K1A OL2 (Canada)

Hiroshi Iizuka — Institute of Applied Microbiology, University of Tokyo, Tokyo (Japan)

A.Z. Joffe — Laboratory of Mycology and Mycotoxicology, Department of Botany, The Hebrew University of Jerusalem (Israel)

Palle Krogh — Institute of Hygiene and Microbiology, Royal Veterinary and Agricultural University, Copenhagen (Denmark)

W.F.O. Marasas — Plant Protection Research Institute, Pretoria (Republic of South Africa)

C.J. Mirocha — Department of Plant Pathology, University of Minnesota, Minneapolis, Minn. (U.S.A.)

Paul M. Newberne — Department of Nutrition and Food Science, Massachusetts Institute of Technology, Cambridge, Mass. (U.S.A.)

K. Ohtsubo — The Institute of Medical Science, The University of Tokyo, Shirokanedai, Minato-ku, Tokyo-108 (Japan)

I.F.H. Purchase — Central Toxicology Laboratory, Imperial Chemical Industries Ltd., Alderley Park, Macclesfield, Cheshire (Great Britain)

Joseph V. Rodricks — Division of Chemistry and Physics, Bureau of Foods, Food and Drug Administration, U.S. Department of Health, Education and Welfare, Washington, D.C. (U.S.A.)

M. Saito — The Institute of Medical Science, The University of Tokyo, Shirokanedai, Minato-ku, Tokyo-108 (Japan)

P.M. Scott — Health Protection Branch, Health and Welfare Canada, Ottawa, Ontario K1A OL2 (Canada)

Eugene B. Smalley — Department of Plant Pathology, University of Wisconsin, Madison, Wis. 53607 (U.S.A.)

Frank M. Strong — Department of Biochemistry, University of Wisconsin, Madison, Wis. 53607 (U.S.A.)

A. Taylor — Atlantic Regional Laboratory, National Research Council for Canada, Halifax, Nova Scotia (Canada)

CONTRIBUTORS (*Continued*)

Ikuko Ueno — Department of Carcinogenesis and Cancer Susceptibility, Institute of Medical Science, University of Tokyo, Tokyo (Japan)

Yoshio Ueno — Department of Chemical Microbiology, Faculty of Pharmaceutical Sciences, Tokyo University of Science, Ichigaya, Tokyo (Japan)

J.J. van der Watt — Division of Toxicology, National Research Institute for Nutritional Diseases, South African Medical Research Council, P.O. Box 70, Tiervlei, C.P. 7503 (Republic of South Africa)

S.J. van Rensburg — National Research Institute for Nutritional Diseases, Tiervlei, C.P. (Republic of South Africa)

Benjamin J. Wilson — Center in Environmental Toxicology, Department of Biochemistry, Vanderbilt University, Nashville, Tenn. 37232 (U.S.A.)

PREFACE

Mycotoxins and the diseases caused by them were relatively obscure in the scientific literature until the discovery of aflatoxin in the early 1960s. Since then there has been a growing interest in them with a resulting rapid increase in the number of publications describing the mycological, chemical, toxicological and epidemiological aspects of mycotoxins and the isolation of many new toxins. The field is now bewilderingly complex with a great variation in the detail and extent of the knowledge of different mycotoxins. This has not detracted from the great importance attached to a full knowledge of all aspects of mycotoxicoses based on the potent activity, the widespread occurrence and the real or potential hazard to human health of mycotoxins. This book has been compiled as a handbook for students, or for those faced with possible mycotoxicoses in the field and to a lesser extent for research workers working in this and related fields. For this reason the chapters have been arranged, as far as possible, with a common format. The chapters have been written around a particular group of toxins or fungi or diseases as the information warrants.

In a book of this length it is impossible to cover all aspects of all mycotoxicoses. There are particular subjects which fall just outside the field of mycotoxins. An example of this is celery dermatitis. This disease has been known for many years as a dermatitis affecting the hands of workers handling celery. For many years it was thought to be caused by celery oil but in 1961 Birmingham and co-workers (Arch. Dermatol., 83 (1961) 73) attributed the disease to contact with the pink rot fungus *Sclerotinia sclerotiorum*. This fungus attacks healthy celery plants and in the infected tissue phototoxic furanocoumarins (mainly 4,5',8-methylpsoralen and 8-methoxypsoralen) are present. Contact of these compounds with the skin in the presence of light at a wavelength of 3650 Å produces dermatitis in 48 hours.

Other diseases have not been included. The range of diseases known as *Aspergillus toxicoses* occurring in chickens and other animals have been known for many years. The knowledge on these diseases is, however, not well developed. Another area not covered is the recently described association between spina bifida in man in the United Kingdom and damaged potatoes.

Other subjects may also have been excluded, but the major mycotoxicoses, about which some detail is available, are described in this text.

Alderley Park,   I.F.H. Purchase
July 1974

CONTENTS

Chapter 20 — MYCOTOXIC NEPHROPATHY 419
by Palle Krogh

Chapter 1

AFLATOXIN

W.H. BUTLER

MRC Toxicology Unit, Medical Research Council Laboratories, Woodmansterne Road, Carshalton, Surrey (Great Britain)

Outbreaks of disease attributable to the fungus

In the latter part of 1959, the s.s. Rossetti out of Brazil landed a cargo of groundnut meal in the United Kingdom. This meal was incorporated as a protein supplement into the feed of various farm animals and had an immediate effect on turkey poults in the South East of the country. In the early months of 1960, extensive losses of turkeys were reported from a hitherto unknown disease characterised by rapid deterioration of the condition, subcutaneous haemorrhages and death. At postmortem the livers of these animals were pale in colour, fatty and showed extensive necrosis and biliary proliferation. This disease syndrome was called "Turkey X Disease" [1]. Concurrent with the loss of turkey poults, large-scale losses of partridges, pheasants and ducklings were also reported [2]. The common factor in this disease was shown to be the incorporation of the batches of groundnut meal which had been imported from Brazil. The disease syndrome seen in the turkeys appeared as a new disease and a search for known toxins was unsuccessful [1]. Investigation of the meal and extraction of the toxin was carried out at the Central Veterinary Laboratory, Weybridge, where the toxicity of the isolates was tested on 1-day-old ducklings. The isolation procedures, and characterisation of the compounds found, called aflatoxins by an Interdepartmental Working Party, are described later (see page 3).

Later in 1960, outbreaks of disease occurred in pigs, also apparently caused by an unknown toxin [3]. Similarly a disease of unknown aetiology was reported in calves [4]. In both cases the aetiology was traced to the incorporation of the Brazilian groundnut meal into the rations of the livestock. During 1960, disease attributed to the imported Brazilian groundnut meal involved turkeys, pheasants, partridges, ducklings, calves and pigs. Allcroft and Carnaghan [5] reported that duckling livers sent from East Africa indicated that a disease similar to that investigated at Weybridge occurred there. Coincidentally in 1960 attention was drawn to an increasing incidence of trout hepatomas in hatchery-reared trout in the United States. Following the reports of investigation into the cause of "Turkey X Disease" in the

United Kingdom it was established that the same aetiologic agent was present in the cottonseed meal [6] used as feed which was later shown to be contaminated with aflatoxin [7].

Although the biological activity of the aflatoxins was not recognised until 1960—61 there is evidence that these compounds had produced disease prior to that time. Paget [8] described in guinea pigs an "exudative hepatitis" of unknown aetiology which is pathologically similar to the disease reported by Patterson *et al.* [9] in guinea pigs fed toxic groundnut meal. Schoental [10] fed the diet used by Paget [8] to rats and induced hepatic carcinoma. Le Breton *et al.* [11] also described a high incidence of hepatic carcinoma in a colony of rats, which was later attributed to aflatoxin. Salmon and Newberne [12], investigating the mechanism of hepatic carcinoma induced by choline deficiency, found that the tumours were always associated with methanol-extracted peanut meal. Newberne *et al.* [13] have subsequently shown that choline deficiency alone in the absence of aflatoxin does not induce hepatic carcinoma.

Newberne *et al.* [14] reported an outbreak of liver disease (Hepatitis X) in kennel-reared dogs in South Eastern United States, which was shown to be related to a commercial diet but the toxic principle was not isolated. The pattern of liver necrosis and biliary proliferation, seen in the dogs, has been reproduced experimentally by aflatoxin [15]. During 1952 a disease of pigs also occurred in South Eastern United States in which, again, the most striking feature was acute liver disease [16]. This disease was related to the ingestion of mouldy corn from which *Aspergillus flavus* and *Penicillium rubrum* were isolated. Culture extracts of both these fungi were shown to be toxic producing massive liver damage and haemorrhage. The *P. rubrum* toxins were shown to be the more toxic and hence received somewhat greater attention than those of *A. flavus.*

There have been many reports since 1935 of high incidence of hepatomas in hatchery-raised trout from Italy, France, Japan and the United States. Although it is not possible to attribute this directly to aflatoxins it would appear that these compounds are the most likely causative agents.

It is, therefore, apparent that the problems associated with aflatoxin-contaminated foodstuffs were present prior to the recognition of the aflatoxins. This evidence is only presumptive in that the foodstuffs are those that are known to support aflatoxin production and the disease states resemble those induced experimentally. There is, however, some direct evidence for the presence of aflatoxin as Keppler and De Iongh [17] isolated aflatoxin from a 40-year-old sample of groundnuts.

On completion of the studies which demonstrated that groundnut meal was contaminated with a toxic compound, the Central Veterinary Laboratory at Weybridge put aside a large consignment of the toxic meal, designated "Rossetti", for experimental use. Lancaster *et al.* [18] reported that rats fed the toxic groundnut meal developed hepatocarcinoma. In 1963 Butler and Barnes [19] reported a series of feeding experiments with "Rossetti" meal, assayed for aflatoxin, which showed that aflatoxin is an extremely potent carcinogen for the rat. This was confirmed later using crystalline aflatoxins [20]. It is this biological activity of the aflatoxins which has promoted large programmes throughout the world designed to assess the risk to man and reduce the potential hazard. As a result of the early demonstration that the aflatoxins were toxic to a wide range of both farm [21] and laboratory [22] animals, steps were taken to ensure that the level of contamination was reduced so that field cases of disease are now rare.

Isolation of toxic metabolites of the fungus

Asplin and Carnaghan [2] demonstrated that a hot methanol extract of toxic peanut meal, when partially purified by extraction into chloroform, produced a proliferation of bile duct epithelium in 1-day-old ducklings comparable to that seen when the peanut meal itself was fed. By paper chromatographic methods Sargent et al. [23] isolated crystalline, toxic material which had an R_F of 0.7, and a blue fluorescence in UV light. Although Austwick and Ayerst [24] demonstrated only the presence of dead fungal hyphae within the cotyledons of the "Rossetti" meal, Sargent et al. [23] isolated cultures of A. flavus from toxic groundnuts grown in Uganda. The isolates grown on Czapek's solution/agar produced a similar blue fluorescent material with an R_F of 0.7 which was toxic to 1-day-old ducklings. On thin-layer chromatography (TLC) Nesbitt et al. [25] resolved two fluorescent spots, one with an R_F of 0.6 and the other of lower R_F with a green fluorescence. The toxic compounds were given the trivial name of aflatoxin and the spots were designated aflatoxin B and aflatoxin G. The blue fluorescent aflatoxin had an empirical formula of $C_{17}H_{12}O_6$ and the green aflatoxin G, $C_{17}H_{12}O_7$. Smith and McKernan [26] isolated 12 fluorescent fractions of which 5 were toxic to the 1-day-old duckling. Independently Van der Zijden et al. [27] and De Iongh et al. [28] also isolated 2 toxic fluorescent compounds with similar empirical formulae. Hartley et al. [29] isolated 4 fluorescent compounds on thin-layer chromatography using chloroform : methanol (98:2) which were designated aflatoxin B_1, B_2 and G_1 and G_2 in order of decreasing R_F. The infrared and UV absorption spectra were similar, indicating related structures. B_2 and G_2 were established as dihydro-derivatives of B_1 and G_1.

The structures of the aflatoxins were elucidated by Asao et al. [30] and are given in Fig. 1. The chemical characteristics of the aflatoxins are given in Table I. The total synthesis of racemic aflatoxin B_1 has been reported by Büchi et al. [31,32].

Allcroft and Carnaghan [33,34] found that cows when fed toxic groundnut meal excreted a toxic factor in the milk which had the same biological effect on 1-day-old ducklings as aflatoxin B_1. De Iongh et al. [35] showed, by TLC, that this toxic factor, designated "milk toxin", was a blue fluorescent compound with an R_F lower than that of aflatoxin B_1. This toxin, aflatoxin M_1, was shown to occur when A. flavus was grown in culture and was also

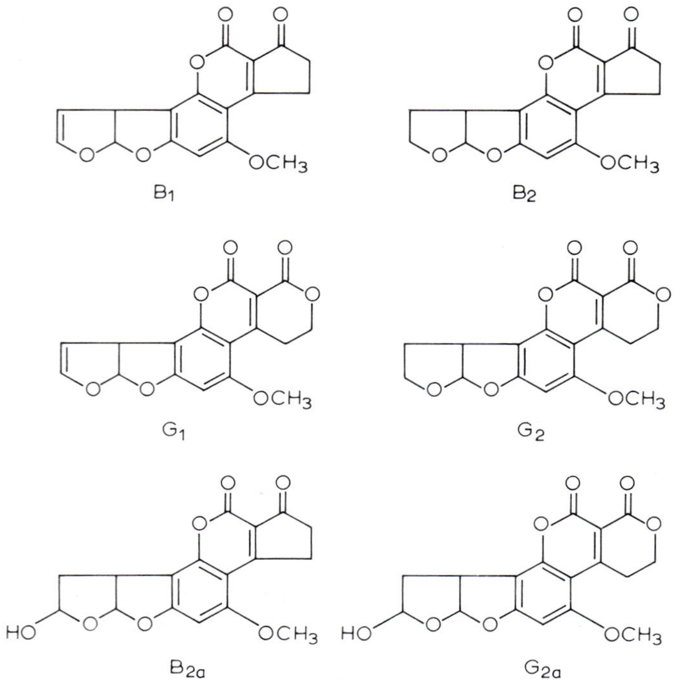

FIG. 1

TABLE I

CHEMICAL PROPERTIES OF THE AFLATOXINS

Aflatoxin	Molecular formula	Molecular weight	Melting point	362—363 nm absorption (ϵ)	Fluorescence emission (nm)
B_1	$C_{17}H_{12}O_6$	312	268—269	21,800	425
B_2	$C_{17}H_{14}O_6$	314	286—289	23,400	425
G_1	$C_{17}H_{12}O_7$	328	244—246	16,100	450
G_2	$C_{17}H_{14}O_7$	330	237—240	21,000	450
M_1	$C_{17}H_{12}O_7$	328	299	19,000 (357 nm)	425
M_2	$C_{17}H_{14}O_7$	330	293	21,000 (357 nm)	—
GM_1	$C_{17}H_{12}O_8$	344	276	12,000 (358 nm)	—
B_{2a}	$C_{17}H_{14}O_7$	330	240	20,400	—
G_{2a}	$C_{17}H_{14}O_8$	346	190	18,000	—
Aflatoxicol	$C_{17}H_{16}O_6$	314	230—234	14,100	425

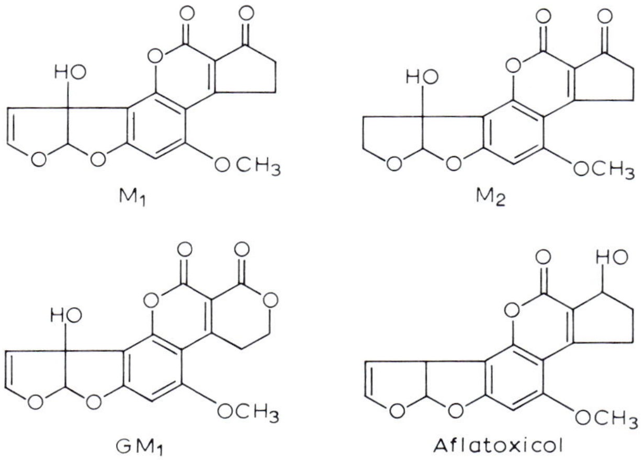

M₁ M₂ GM₁ Aflatoxicol

FIG. 2

found in the liver, kidneys and urine of sheep [36] and in the livers of rats when given aflatoxin B₁ [37]. Holzapfel *et al.* [38] isolated 2 components from the urine of sheep fed mixed aflatoxins which were similar to the milk toxin and designated aflatoxin M_1 and M_2. Aflatoxin M_1 was shown to be a hydroxylated aflatoxin B_1. The chemical structure is given in Fig. 2 and the chemical characteristics are listed in Table I. Aflatoxin GM_1 is a similarly hydroxylated derivative of aflatoxin G_1. Two additional aflatoxins have been isolated from cultures of *A. flavus* [39] which were shown to be 2-hydroxy derivatives of aflatoxin B_1 and G_1 and were named aflatoxin B_{2a} and aflatoxin G_{2a} whose structures are shown in Fig. 1. These two compounds are not toxic to 1-day-old ducklings.

Biosynthesis

The aflatoxins are a group of secondary metabolites which are formed after the logarithmic phase of growth of the fungus. The biosynthetic pathways have not been described in detail for the aflatoxins. Adye and Mateles [40] studied the incorporation of ^{14}C-labelled methionine, phenylalanine, acetate and tryptophan and demonstrated an incorporation of label into aflatoxins. In a detailed study of the distribution of ^{14}C derived from acetate and methionine it has been demonstrated that the ring carbons are labelled from the acetate and methoxy carbon from the methionine [41]. This method of labelling has been used to prepare labelled aflatoxins for biological use [42].

Holker and Underwood [43] studied the synthesis of aflatoxins and suggested that they were derived from sterigmatocystin which is a toxic secondary metabolite of *A. versicolor*. Biollaz *et al.* [41] considered that either there is a common synthetic pathway for aflatoxin and sterigmatocystin or that the sterigmatocystin is a precursor of the aflatoxin and concluded that there is a common pathway for the two metabolites. For a detailed discussion of the synthesis of the secondary fungal metabolites see Detroy *et al.* [44]

Environmental and laboratory conditions favouring production of toxic metabolites

The *A. flavus* group of fungi is a constituent of the microflora of the air and soil throughout the world, and contributes to the deterioration of many foodstuffs [45]. However, it was soon recognised that *A. flavus* does not invade groundnuts to any significant extent prior to harvest [24]. Aflatoxin has been found as a natural contaminant in many types of food including peanut, cottonseed meal, corn, cassava, rice, soya beans, wheat, sorghum and barley [46]. Although *A. flavus* has a worldwide distribution various factors restrict the areas in which natural contamination with aflatoxin may occur. *A. flavus* and *A. parasiticus* will under certain conditions produce aflatoxin. However, all strains of these fungi are not good producers of toxins. Hesseltine *et al.* [47] showed that there were strain differences with strain 1999 producing the most aflatoxin B_1 on sorghum, peanuts and rice, while strain 3000 produced the most on wheat and corn. Stubblefield *et al.* [48] showed that strain 3145 produced the largest amounts of aflatoxin B_1 on either oats, groats or oat hulls. For any given strain of fungus the substrate influences the amount of aflatoxin produced. Although aflatoxin has been isolated from many different foodstuffs there is a wide variation in the ability of the substrate to sustain the fungus in the production of aflatoxin [49]. Groundnuts have been demonstrated consistently as being a good substrate for aflatoxin production, but it is of interest for the field control of the problem that Rao and Tulpule [50] found one variety (US 26) when inoculated with toxin-producing strains of *A. flavus* did not support production of aflatoxin.

Austwick and Ayerst [24], demonstrated that the most important factor in the growth and production of aflatoxin by *A. flavus* was the relative humidity surrounding a natural substrate. At a humidity at less than 70% very few fungi will grow on stored foods but with a relative humidity of 85% and a moisture content of 30% within the groundnut *A. flavus* is able to produce aflatoxin. After drying to a moisture content of 15% and a 50% relative humidity, little aflatoxin is produced [49]. As a result of this, rapid dehydration of the groundnuts prior to storage will greatly reduce aflatoxin contamination of the stored material. As much of the groundnut crop is dried in the sun, any rain or humid weather will result in an increased moisture content and possibility of aflatoxin contamination. The optimum temperature for growth of *A. flavus* is 36—38° with a range of 8—46°, while the optimum temperature for aflatoxin production by *A. flavus* has been shown to be 25°. The relative amounts of aflatoxin B_1 and G_1 produced is also modified by temperature. Rabie and Smalley [51] reported a temperature optimum of 24° for aflatoxin B_1 and an optimum for G_1 of 30°. Sorenson *et al.* [52] showed that when aflatoxins B_1 and G_1 were produced on rice an increasing temperature resulted in relatively more B_1 than G_1 being produced.

Austwick and Ayerst [24] showed that from any batch of groundnuts only a relatively small number of the groundnuts were contaminated with aflatoxin. As a result it is widely accepted that intact kernels are rarely invaded by *A. flavus* but when the kernel is damaged, either by termites or direct mechanical damage, it is more frequently contaminated by *A. flavus* and also frequently contains aflatoxin.

Aflatoxin is produced when *A. flavus* is grown in a liquid medium but the yields are less than on natural substrates. The aeration, pH and temperature of the medium influences both the final yield of aflatoxin and the relative amount of B_1 and G_1. In a series of large-scale preparations of aflatoxins undertaken by Sargent at Microbiological Research Establishment, Porton Down, G_1 was consistently the most abundant, but from batch to batch there was variation in the relative proportions of B_1 and G_1 [53]. For a detailed summary of the conditions for production of aflatoxin see Detroy *et al.* [44].

Analysis for toxic metabolites

Toxic principles of mouldy groundnut meal were demonstrated by Sargent *et al.* [23]. The meal was extracted with methanol, partially purified in chloroform and then purified by paper chromatography in butanol—acetic acid. A single blue fluorescent spot was seen in UV light. Subsequently, many different solvent systems have been suggested for the initial extraction of the toxins. Different products such as cottonseed, cocoa or raw milk require modifications to the solvent system to remove naturally occurring pigments and other contaminants (see refs. 54 and 44 for detailed results).

Following the primary extraction, the products were further purified by partition between methanol and petroleum ether in a separatory funnel [35]. Other workers have suggested modifications to the system [54]. Finally TLC provides the best separation of the individual aflatoxins. Van der Zijden *et al.* [27], using chloroform—1% ethanol separated crystalline material by TLC on Kieselgel G. Using chloroform : methanol De Iongh *et al.* [35] resolved the 4 major aflatoxins and classified them according to their R_F's and the colour. Subsequently different solvent systems have been used (see ref. 54). Following isolation of aflatoxins, quantitative estimation relied upon visual comparison of the UV fluorescence with known standards and serial dilution to extinction. As further information was obtained with pure aflatoxins more accurate methods of measurement were evolved depending upon fluorodensitometric measurements on the eluate of TLC plates [55].

Aflatoxin M_1 and M_2 have been extracted from milk by acetone : chloroform : water [56] or aqueous acetone followed by solvent partition with chloroform after defatting [57]. Aflatoxin M was then resolved on TLC, using chloroform—methanol.

Bioassay system

The original bioassay system was described by Sargent *et al.* [23] and was based on the observation that 1-day-old Khaki Campbell ducklings were extremely sensitive to toxic groundnut meal [2]. The chloroform—methanol extract in propylene glycol was given by a tube to the lower end of the oesophagus on 5 consecutive days and the survivors were killed at day 8. The animals which died prior to day 8 showed massive necrosis of the liver and the survivors a biliary proliferation which was similar to that seen when ducklings were fed toxic meal. The degree of biliary proliferation was scored +, ++, +++, ++++ (Figs. 3 and 4). This test is reproducible and still forms the basis of the duckling bioassay. Butler [58] studied effects of single doses of aflatoxin B_1 and demonstrated that the biliary proliferation was very rapid, reaching a peak at 3 days and then regressed. As little as 1.5 μg B_1 produced a detectable proliferation but there is considerable variation in response of the ducklings to dose levels up to 15 μg per duckling. Although the different aflatoxins have varying LD_{50} values for the Khaki Campbell duckling, the lesions induced are histologically similar. A similar lesion is also produced by dimethylnitrosamine and cycasin but not by ethionine, carbon tetrachloride or thioacetamide.

Fish of many species are very susceptible to the aflatoxins. Abedi and McKinley [59], suggested the use of Zebra fish larvae (*Brachydanio rerio*) and demonstrated a sensitivity at a level of 1 μg/ml B_1.

The chick embryo has also been used [60]. The aflatoxins were injected into the yolk or air cell prior to incubation and the mortality occurring over the 21-day incubation period was assessed. Although the system is reasonably sensitive (0.025 μg is an LD_{50}) reproducibility between laboratories has limited the usefulness of the test.

Various tissue culture systems have been proposed. Legator and Withrow [61] showed that as little as 0.01 μg B_1 could produce a significant mitotic inhibition of cultured embryonic lung cells. Daniel [62] suggested an assay system based on inhibition of growth of cultured rat fibroblasts. She demonstrated that 0.06 μg/ml caused 50% inhibition of growth. There have been a few studies on micro-organisms which have demonstrated that bacteria are relatively insensitive to the direct action of the aflatoxins [63]. However, Legator *et al.* [64] demonstrated that the system of induction of bacteriophage in lysogenic bacteria was sensitive at concentrations of 0.06 μg/ml.

All bioassay systems suffer from the defect of non-specificity and in themselves will not conclusively demonstrate or assay aflatoxin. It would appear essential that if, following the isolation and chemical characterization, further evidence is required, a biological assay should be performed on the extracted aflatoxin. The choice of system depends somewhat upon the facilities available within the laboratory but the duckling test would still appear to be the system of choice.

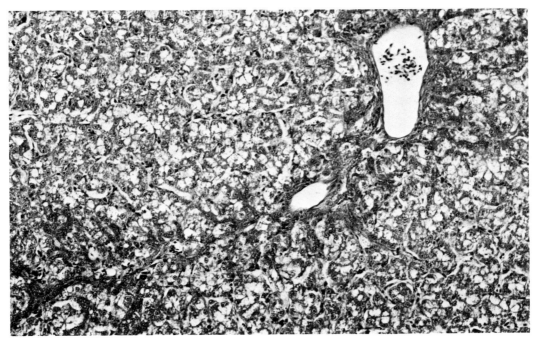

Fig. 3. Duckling liver 3 days after a single dose of aflatoxin B_1 showing biliary proliferation scored 1+. H. and E. × 180.

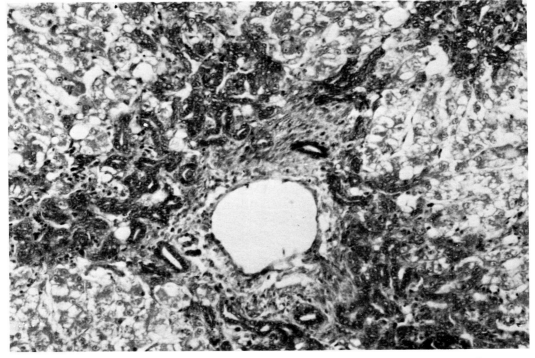

Fig. 4. Duckling liver 3 days after a single dose of aflatoxin B_1 showing biliary proliferation scored +++. H. and E. × 250.

Natural occurrence

The original outbreak of "Turkey X Disease" was the result of contamination of groundnuts grown in Brazil [1]. Contaminated peanut products have now been found in all producing countries. It was soon recognized that other foods could also be contaminated with aflatoxin, including cottonseed meal [65]. In investigations of trout hepatomas in the United States, Sinnhuber *et al.* [7], demonstrated that rations containing cottonseed meal and peanut meal were responsible for the outbreaks of disease, and demonstrated the presence of aflatoxin. A field survey of local foods in the Philippines [66] found aflatoxin in a variety of peanut products, cassava, cocoa, corn, rice and other foods. Similar distribution of aflatoxin in food is found in other parts of the world. Although rice is a cereal of very widespread usage, the levels of aflatoxin appear to be less than in other foods with the possible exception of parboiled rice. High levels of contamination (1.7 mg/kg) have been reported in cassava from Uganda, which was associated with a case of acute liver disease in a boy [67]. In field samples there is always very wide variation in the level of aflatoxin present in food [68].

Aspergillus flavus is able to synthesise the aflatoxins when grown on a wide variety of substrates. However, the yields of the different aflatoxins vary with the substrate (see refs. 47 and 44 for summary). In contrast to natural contamination where aflatoxin B_1 is frequently the largest component (for example Rossetti meal which caused "Turkey X Disease" contained 10 parts per million (ppm) B_1 and negligible G_1) when grown in culture aflatoxin G_1 may be the most abundant.

Biological activity in experimental systems

As has been referred to in the introduction to this chapter aflatoxin was recognised following the outbreak of acute disease in farm animals. Following this, the acute toxicity of the aflatoxins was studied in conventional laboratory animals which demonstrated that the lesion described in the 1954, by Paget [8], as "exudative hepatitis" in guinea pigs, was indeed attributable to aflatoxin. This naturally occurring disease has now been reproduced under experimental conditions, both by feeding groundnut meal and by dosing with purified aflatoxins. There is a wide range of susceptibility from the highly susceptible 1-day-old duckling, and guinea pig, to the more resistant sheep, mouse and adult turkey.

Turkeys and ducklings

In both field cases and experimental birds dosed with aflatoxin turkey poults have convulsions and die in opisthotonus with multiple subcutaneous haemorrhages. The consistent pathological lesion seen is that of the liver which is large and mottled; on histological examination widespread periportal zone necrosis, haemorrhage and some increase of fat is seen. At dose levels which allow the birds to survive for some weeks the livers become fibrotic [69]. Magwood *et al.* [70], have demonstrated that if poults are fed low levels of aflatoxin for some months the birds acquire a degree of tolerance to much higher levels of the toxin. The livers of these animals showed some increase of fibrosis and bile duct proliferation.

Ducklings

One-day-old ducklings are extremely susceptible to the acute effects of aflatoxin with an LD_{50} for aflatoxin B_1 of 0.335 mg/kg and G_1 of 0.784 mg/kg [53]. Prior to the characterization of aflatoxin the duckling was utilised in its bioassay. Acutely the birds failed to grow and there were subcutaneous haemorrhages. In the liver, which is the major organ affected, an increase of fat, and a periportal zone necrosis are seen and if the animals have survived for a few days, an extensive biliary proliferation [2]. At sublethal doses of aflatoxin the bioassay system depended upon an assessment of the degree of biliary proliferation. A single dose of aflatoxin B_1 induces a rapidly developing biliary proliferation reaching a maximum by 3 days, which then regresses, so that by 14 days only very slight biliary proliferation is still present [58]. One-year survivors of an LD_{50} failed to show any persistent hepatic abnormality. Although this lesion formed the basis for the bioassay system, as described above, it was found that following a single dose of aflatoxin B_1 there was only a poor correlation between the dose and biliary proliferation with considerable variation between the groups of animals. This lesion is not specific to aflatoxin and has been described following dimethylnitrosamine and cycasin, although the naturally occurring alkaloids such as retrorsine do not produce this lesion.

Although a single dose of aflatoxin does not lead to longterm hepatic changes, Carnaghan [71] has demonstrated that feeding aflatoxin at levels of 0.03 ppm for 14 months results in the induction of hepatic carcinoma.

Chickens

Chickens are considerably less susceptible than ducklings and turkey poults to the acute toxic action of aflatoxin. Asplin and Carnaghan have described feeding experiments in Rhode Island Red chickens with toxic groundnut meal. Very few of the chickens died and at dose levels of 0.5 ppm there was some retardation in growth of the chicken; during the first 3 weeks of feeding the livers were enlarged and fatty. By 6 weeks there was some fibrosis and biliary proliferation. However, these experiments were not carried on to demonstrate whether aflatoxin is a carcinogen for the chicken. Other strains of chicken are less susceptible to the toxin. A dose level of 1.6 ppm aflatoxin had little effect on growth or on the laying ability of Arbor-Acres hybrid hens [72].

Pigs

Loosmore and Harding [3] described field outbreaks in pigs due to the same contaminated meal that caused the "Turkey X Disease" and this has been reproduced experimentally. Histologically there is a distortion of lobular pattern of the liver by dissecting fibrosis and some biliary proliferation. By 26 weeks there is considerable fibrosis of the liver with the formation of small hyperplastic nodules. The other striking feature seen in these livers is the considerable variation in parenchymal nuclear size with the formation of megalocytes. The acute LD_{50} of aflatoxin B_1 in weanling pigs is 0.62 mg/kg, while doses of 1 to 2 mg/kg resulted in death in 18—24 h. The main lesions seen were again confined to the liver and consisted of centrilobular zone haemorrhagic necrosis. The gall bladders of these animals were oedematous with multiple petechial haemorrhages, similar to that described for the dog (see below). In the animals which survived a few days the normal lobular pattern of the liver was accentuated by biliary proliferation [73].

Cattle

As with the pig, field cases of poisoning due to contaminated Brazilian groundnut meal were reported following the outbreak of "Turkey X Disease" [4]. This outbreak occurred in calves 3 to 9 months old that had been fed, for at least 6 weeks, a compounded feed containing 15% Brazilian groundnut meal, in which the aflatoxin content had not been assayed. The livers of the animals were fibrotic with biliary proliferation but also showed lesions similar to those of veno-occlusive disease, which have been described following ragwort (*Senecio jacobea*) poisoning. At the same time Clegg and Bryson [74] reported an outbreak of a similar disease in cattle, which resulted in 4 out of 16 cattle dying. Histologically the liver showed severe diffuse fibrosis and extensive bile duct proliferation and again veno-occlusive disease. Allcroft and Lewis [75] demonstrated essentially the same pathological changes in the liver in calves and older cattle fed a diet containing 2 ppm aflatoxin. As with the field cases the adult cattle are more resistant than calves to the acute toxicity of the aflatoxins. However, there is a marked effect upon the milk yield of the cows. The veno-occlusive lesion in the bovine is distinctive and has, as yet, not been described in other species.

Sheep

There have been no field cases reported of aflatoxicosis in sheep. However, large doses of the toxins have induced hepatic lesions. In the Central Veterinary Laboratory in Weybridge 2 rams which were being used for the collection of metabolites of aflatoxin died, following very large doses. These sheep had a centrilobular necrosis with some fatty change. Armbrecht *et al.* [76], however, have demonstrated an LD_{50} in the order of 1 mg/kg, a figure considerably lower than that found at Weybridge. In a long-term feeding trial sheep were maintained for 3 years on a diet containing 20% of a toxic groundnut meal which resulted in a daily intake 1.75 mg of aflatoxin B_1. Only 1 animal surviving 3.5 years from the start of the experiment had hepatic carcinoma. A further animal killed 5 years after the initiation of the experiment and a ewe lamb born from one of the experimental ewes developed chondromas from the ethnoid bone. At the present moment it is very difficult to know whether these neoplasms are related to the experimental conditions [77].

Dogs

Prior to the recognition of the problems associated with the aflatoxins Newberne *et al.* [14], described a disease in dogs called "hepatitis X", which appeared to results from a noninfective agent associated with commercial dog food. This disease can now be reproduced experimentally by either feeding dogs with food naturally contaminated with aflatoxin or by dosing with purified aflatoxin [15]. The LD_{50} is 0.5—1.00 mg/kg with the main damage, again, being found in the liver and consisting of a centrilobular zone necrosis, congestion and increase of fat. By 7 days there is a significant biliary proliferation. As with the pig the gall bladder is severely affected with a haemorrhagic oedema of the mucosa. Long-term studies have not been reported with the dog.

Cat

Adult mixed-breed cats are sensitive to acute single doses of aflatoxin with an LD_{50} 0.55 mg/kg, which results in a periportal cell necrosis, and in those animals surviving beyond 72 h, biliary hyperplasia [78].

Rabbits

Little work has been carried out with rabbits but the LD_{50} is 0.3 mg/kg, which results in a midzonal area of hepatic necrosis and as in other species, a bile duct hyperplasia. There has been no long-term feeding experiments reported in the rabbit [78].

Guinea pig

Guinea pigs are very susceptible to aflatoxin as demonstrated by Patterson *et al.* [9] who reported the experimental production of a disease in guinea pigs which was similar to the "exudative hepatitis" described by Paget [8] in 1954. The feeding of toxic groundnut meals assayed for aflatoxin has been described in detail by Butler and Barnes [19]. We described the development of biliary proliferation, dilatation of periportal lymphatics and a scattered individual parenchymal cell necrosis at a dose level of 0.35 to 0.40 ppm aflatoxin. Most animals were dead by 27 weeks, but one survived at 44 weeks, at which time the liver was coarsely granular with many bizarre hyperplastic nodules, fibrosis and biliary proliferation. This lesion could not be classified as carcinoma. In a further experiment 6 guinea pigs survived from between 106—160 weeks and in one, killed at 127 weeks at a dietary level of 0.15 ppm aflatoxin, a single anaplastic hepatocellular carcinoma was found. There is no other report of hepatic carcinoma in the guinea pig induced by aflatoxin. The acute LD_{50} is 1.40 mg/kg with most animals dying between 1 and 3 days [79]. As with most other species the liver was the major organ affected with the production of a centrilobular zone hepatic necrosis, which, in animals living for more than 48 h was associated with extensive biliary proliferation. This lesion slowly resolved so that by between 4 and 6 weeks the liver architecture was normal. Although the liver was the organ most consistently and severely affected by single doses of aflatoxin

B_1, necrosis of the proximal tubules of the kidney is seen. Haemorrhage of the zona reticularis of the adrenal cortex was a consistent finding. Small petechial haemorrhages were found in the spleen, pancreas, and lung, and the small gut was often found with altered blood, although no large areas of mucosal ulceration could be found.

Mouse

There is little detailed work reported on the acute effects of aflatoxin in the mouse. The LD_{50} has been estimated at 9 mg/kg in a Swiss strain mouse [78], but there is possibly considerably strain variation in this. Extensive feeding trials undertaken by Newberne have failed to demonstrate a significant increase in liver tumours in the adult mouse [78]. Vesselinovitch *et al.* [80] have reported the increased incidence of hepatomas when aflatoxin B_1 was administered to infant mice. In contrast to this pulmonary tumours have been induced in strain A adult mice, given repeated injections of aflatoxinn B_1 [81].

Trout

As has been mentioned previously, in the 1960s the incidence of hepatomas in trout in the United States reached epidemic proportions and it was established that this was associated with cottonseed meal in the feed. After the aflatoxins had been isolated in the United Kingdom it was soon recognised that the cottonseed meal fed to the trout was contaminated with aflatoxin. Subsequently, it was demonstrated that very low doses of crystalline aflatoxin added to the trout diet resulted in metastasising hepatocellular carcinoma [7]. When the aflatoxin was compared to the other carcinogens in the trout it was found that the trout is extremely susceptible to very low levels of many known hepatic carcinogens. Recently Wales and Sinnhuber [82], have reported the induction of hepatomas in Sockeye salmon.

Hamsters

The acute LD_{50} of aflatoxin B_1 for hamsters is 10.2 mg/kg but there is little information as to the pathological changes induced [83]. Feeding experiments have been inconclusive as, although the animals failed to grow at a level of 2 ppm aflatoxin, there was not a significant difference in the lesions induced in dosed and control animals [84]. Aflatoxin has also been shown to be teratogenic to the hamster [85].

Monkeys

Various species of primates have received increasing attention in the study of the problems associated with aflatoxin. Madhavan and his colleagues have reported that at doses of 500 µg of mixed crystalline aflatoxin given for 18 days followed by a dose of 1 mg per day, deaths occurred on days 32 and 34. At higher dose levels the animals died earlier. Histologically the livers showed fatty infiltration, biliary proliferation and portal fibrosis. It was further demonstrated that the Rhesus monkey fed a low protein diet was more susceptible to the acute toxicity of aflatoxin [86,87]. Cuthbertson *et al.* [88] reported the effects of feeding toxic groundnut meal to Cynomologus monkeys. At a level of 5 ppm aflatoxin in the diet the animals rapidly developed liver cell damage, biliary proliferation and died. At lower dietary levels (1.8 ppm aflatoxin) some animals survived 3 years. In one of these there was a coarse nodular cirrhosis, while another monkey exhibited irregulatity of parenchymal nuclear size. Deo *et al.* [89] also described the acute and chronic toxicity of aflatoxin in the Rhesus monkey and noted that at high dose levels an extensive haemorrhagic necrosis to the liver was produced and when the dose was reduced, a prominent biliary proliferation with large bizarre hyperchromatic parenchymal cells was found at 5 months. No carcinomas of the liver were reported and the effects of protein deficiency described by Madhavan [87] were not reproduced. Similar findings of the acute reaction have been reported by Alpert *et al.* [90] and Svoboda *et al.* [91]. As part of the investigation of the possible hazards to man in Thailand, Bourgeois *et al.* [92] described the acute toxicity of aflatoxin B_1 in the Macaque. They reported that all animals which were receiving between 13.5 and 40.5 mg/kg aflatoxin B_1 died within 149 h. The most consistent clinical observation was vomiting. At postmortem centrilobular hepatic necrosis was found with some biliary proliferation and also a massive increase of fat. This increase of fat was also demonstrated in the heart and kidneys and cerebral oedema was reported. The authors considered that the clinicopathological syndrome was similar to that of Reye's syndrome in children and suggested that the human syndrome might be associated with the ingestion of aflatoxin.

Rats

Rats have been the most extensively studied of all laboratory animals. The LD_{50} for mature male rats of the Porton-Wistar strain is 7.2 mg/kg and for females 17.9 mg/kg [93]. There are strain differences in the susceptibility of rats with inbred Fischer rats being considerably more susceptible than the Porton-Wistar. The lesion induced in both the male and female and in different strains is the same and consists of a periportal zone necrosis and biliary proliferation. The necrosis develops slowly over 3 days, and is accompanied by marked biliary proliferation. There is only slow recovery of this lesion. The injury induced by aflatoxin differs in this respect from that of either carbon tetrachloride or dimethylnitrosamine and is possibly related to the observations of Rogers and Newberne [94] that aflatoxin inhibits parenchymal cell mitosis in a normal liver and also following partial hepatectomy [95].

Aflatoxin G_1 is also acutely toxic to the rat but the distribution of lesions seen is somewhat different to that following aflatoxin B_1. Again the liver is affected and a periportal necrosis is seen but this lesion is not as consistent or dramatic as that seen following B_1. In most animals there is recovery with slight residual biliary proliferation, whilst in the severely affected animals hyperplastic nodules, persistent bile duct proliferation and cholangio-fibrosis may be seen at 4 weeks. The most consistent lesions occur in the kidneys, where there is necrosis of the tubular epithelium of the inner zone of the cortex, and in the adrenals, where necrosis involves the zona reticularis and the zona fasciculata [96].

At the time that the original outbreaks of "Turkey X Disease" were being investigated it was shown that the same batches of groundnut meal when fed to rats resulted in the induction of hepatic carcinoma [18]. The development of the hepatic lesion which is induced by feeding meals assayed for aflatoxin was described by Butler and Barnes [19] in which it was shown that the lesions were similar to those produced by other carcinogens such as ethionine, dimethylnitrosamine and 4-dimethylaminoazobenzene. The main difference in the lesions induced by aflatoxin was the absence of fibrosis and cholangiofibrosis in the stages preceding the diagnosis of hepatic carcinoma. When purified aflatoxins were available it was confirmed that the active principle in the groundnut meal was aflatoxin [20]. These feeding trials demonstrated that aflatoxin was an extremely potent carcinogen, as in the male, 0.5 ppm aflatoxin induced 100% incidence of carcinoma. A summary of the feeding results is given in Table II. When purified aflatoxin B_1 was fed to an inbred strain of Fischer rats, levels as low as 0.015 ppm in the diet induced a high incidence of hepatic carcinoma. In the same strain a dose of 40 μg/day for 10 days was sufficient to induce hepatic carcinoma diagnosed 82 weeks later [97]. In the experiments undertaken at Carshalton using Porton-Wistar rats, feeding 5 ppm aflatoxin B_1 in the form of contaminated meal for short periods also resulted in the induction of hepatic carcinoma [98] (Figs. 5 and 6). A summary of these experiments is given in Table II. There is an obvious strain difference in the susceptibility of rats to aflatoxin in that the inbred Fischer rats are more susceptible than the random-bred Porton-Wistar rats. Aflatoxin G_1 is also hepatocarcinogenic to the rat [99] but there is less evidence as to the other minor aflatoxins. Aflatoxin M, fed as contaminated milk to either Porton-Wistar or inbred Fischer rats has not induced hepatic carcinoma [100].

In the original series of experiments investigating the carcinogenic action of the Rossetti meal, as well as hepatic carcinoma there was an increased incidence of neoplasms at other sites, and it is now recognised that the aflatoxins are able to induce carcinomas of the oesophagus, glandular stomach and colon and possibly the duodenum. Carcinoma of the kidney may also be induced. In an experiment investigating the carcinogenicity of aflatoxin B_1 and G_1 [99] we reported an increased incidence of epithelial kidney tumours in the groups of rats receiving aflatoxin G_1 but not B_1. In both groups hepatic carcinoma was found. However, in experiments reported by Epstein et al. [101], there was a significant difference in incidence of renal carcinoma induced in certain strains of rats. It is interesting that in Porton-Wistar rats, derived from the colony at MRC Laboratories, Carshalton, aflatoxin G_1 is able to induce renal carcinoma. This is also the same strain of rat in which it was possible to demonstrate that G_1 is a very potent renal toxin

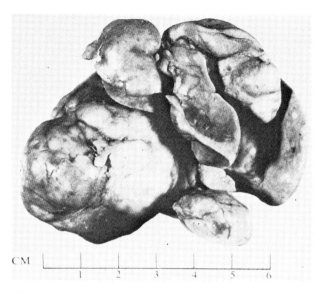

Fig. 5. Liver from a male rat fed 5 ppm aflatoxin B_1 for 6 weeks followed by 47 weeks normal diet showing multiple large hepatocarcinomas.

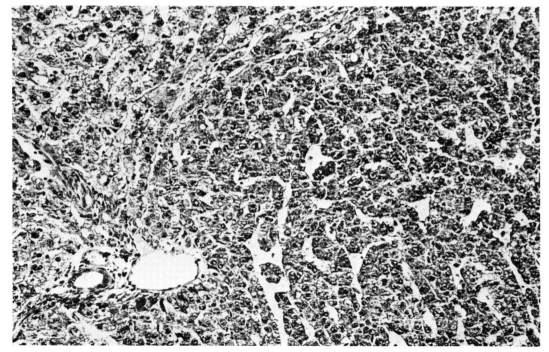

Fig. 6. Edge of hepatocellular carcinoma from a male rat fed 5 ppm aflatoxin B_1 for 6 weeks followed by 47 weeks normal diet. This neoplasm metastasised to the lungs and was maintained through multiple generations of transplantation. H. and E. × 180.

TABLE II

SUMMARY OF FEEDING EXPERIMENTS

Sex	% toxic meal	Aflatoxin B_1 (ppm in diet)	Duration, wk. (av.)	Normal diet, wk. (av.)	Liver tumours
Contaminated meal (Rossetti)[a]					
Porton-Wistar rats					
Male	50	5	36		5/6
Female	10	1	75		5/6
Female	5	0.5	88		26/33
Male	5	0.5	82		25/25
Female	1	0.1	91		5/30
Male	1	0.1	100		22/44
Male	50	5	9	54	6/6
Male	50	5	6	83	12/19
Male	50	5	3	93	3/20
Male	50	5	1	97	0/13
Fischer inbred rats[b]					
Male	50	5	6	45	20/20
Purified aflatoxin B_1[c]					
Male	1	1	41		18/22
Female		1	64		4/4
Male		0.3	52		6/20
Female		0.3	70		11/11
Male		0.015	68		12/12
Female		0.015	82		13/13
Male		40 µg/day	10 days	82	4/24
Female		40 µg/day	10 days	82	0/23

[a] Ref. 98.
[b] W.H. Butler unpublished.
[c] Ref. 97.

in high doses [96]. At present it is not possible to say which of the various aflatoxins is responsible for inducing the carcinoma of the glandular stomach. The carcinomas of the colon have been induced by aflatoxin B_1 and the incidence of this may be modified by a vitamin A deficient diet [102].

In all the experiments which have been reported of hepatic carcinogenesis with the aflatoxins there is no evidence of a significant cirrhosis of the liver, which would indicate that cirrhosis is not a necessary precursor of hepatic neoplasia. Newberne *et al.* [13] demonstrated that although dietary induced cirrhosis by itself did not result in liver tumour induction, when aflatoxin was superimposed upon the cirrhosis, there was an increased incidence of hepatic carcinoma. These observations have led to the conclusion that the occurrence of neoplasm of the liver in rats following prolonged choline deficiency was the result of contamination of the diet by aflatoxin. The carcinogenic action of aflatoxin may also be drastically reduced by hypophysectomy [103] or by treating the animals with diethylstilboestrol

[104]. McLean and Marshall [105] demonstrated that phenobarbitone fed to rats in the drinking water, in order to induce the mixed-function oxidases, afforded a certain degree of protection to the liver from the carcinogenic action of aflatoxin.

Tissue culture
        Tissue culture systems have been used extensively both in the bioassay of aflatoxin as in the study of its mechanism of action. The first report was that of Juhász and Greczi [106], who demonstrated that extracts of mouldy groundnut meal were toxic to calf kidney cells. Shortly after this observation Daniel [62] demonstrated that the growth of the culture of rat fibroblasts was inhibited by mixed aflatoxins at levels of 0.025 μg/ml with a concentration of 0.06 μg/ml causing 50% inhibition of growth. She suggested that this could form a suitable assay system. Legator and Withrow [61] reported an inhibition of mitosis in embryonic lung cultures at levels as low as 0.01 μg/ml. Since these early reports aflatoxins have been shown to be toxic to human embryo liver cells resulting in inhibition of RNA and DNA synthesis at levels of 2 μg/ml aflatoxin [107,108]. The mechanism of action of aflatoxin has been studied *in vitro* using HeLa cells, which has demonstrated a similar mode of action to that occurring *in vivo* [109,110].

Metabolism and mode of action of the aflatoxins

        Different species vary in their susceptibility to acute poisoning by aflatoxin with LD_{50} values ranging from 0.3 to 17.9 mg/kg (Table III). Mature animals tend to be less susceptible than neonates. These observations may indicate differing pathways of metabolism and clearance of the aflatoxin. The first metabolite to be isolated was aflatoxin M extracted from milk of cows and sheep. Aflatoxin M_1 (see Fig. 1 for structure) is hydroxylated aflatoxin B_1. This metabolite has been isolated from urine of many species

TABLE III

ACUTE TOXICITY AFLATOXIN B_1

Species	LD_{50} mg/kg	Reference
Duckling	0.335	53
Rabbit	0.3	78
Cat	0.55	78
Pig	0.62	73
Dog	0.5—1	15
Guinea pig	1.40	79
Sheep	1.0	76
Rat male	7.2	93
female	17.9	93
Mouse	9	78
Hamster	10.2	83
Chick embryo	0.025 μg/embryo	60

including man [66]. Aflatoxin M_1 has been isolated from rat liver and has been demonstrated following *in vitro* metabolism of B_1 by liver preparations from a wide range of species [111]. Aflatoxin M_1 has the same acute LD_{50} on ducklings as B_1 but there is no evidence that it is a carcinogen for the rat. The compound has been shown to induce hepatocarcinoma in the trout [7] and occasional subcutaneous sarcoma following injection [112]. Similarly hydroxylated derivatives of B_2 and G_1 have been described. Patterson and Roberts [113] showed that aflatoxin B_1 was metabolised *in vitro* by liver microsome preparations and demonstrated that this was NADPH dependent. Patterson has grouped the species into fast, intermediate and slow metabolisers of the aflatoxin. These are grouped according to their ability to metabolise an LD_{50} dose of aflatoxin. In the first group including rabbit, duckling and guinea pig an LD_{50} dose is metabolised in under 12 min. In the intermediate group of chick, mouse, pig and sheep this time ranges from 1 h to 4 h and in the slow group, of which the sole example is the rat, the time is 0.8 to 2.6 days [114].

As well as the metabolism to the 4-hydroxy derivative of aflatoxin B_1 it became apparent that some livers were also able to form aflatoxin hemiacetals *in vitro*. Both the aflatoxin B_1 and G_1 were metabolised *in vitro* by liver microsomes in an NADPH-dependent system to form the 2-hydroxy derivatives known as aflatoxin B_{2a} and G_{2a} (Fig. 2) [115]. These compounds are non-toxic and were originally described by Dutton and Heathcote [39] from cultures of *A. flavus*. Rabbit, duckling, guinea pig, chick and mouse liver microsomes are able to carry out this hydroxylation.

A further form of metabolism may be demonstrated in mouse, rat and rabbit microsomes which are able to *O*-demethylate the aflatoxins, resulting in the formation of aflatoxin P_1 (Fig. 2). Other evidence for this pathway is given by the *in vivo* metabolism of ^{14}C methoxy-labelled aflatoxin [116]. The avian species of chick, duck, quail and turkey and also the rabbit have the alternative pathway for aflatoxin metabolism [113]. Also in a microsomal $NADPH_2$-dependent system the aflatoxins are metabolised to the corresponding cyclopentonols known as aflatoxicol (Fig.2).

The formation by microsomal systems of an unidentified toxic metabolite has been described by Garner *et al.* [117]. This was demonstrated by the toxicity to *Salmonella typhimurium* TA-1530. It has been suggested that this toxic metabolite is an epoxide of aflatoxin, but this has not yet been characterised.

In studying the clearance of aflatoxin, Butler and Clifford [37] were unable to find extractable fluorescent aflatoxin in rat liver after 24 h. While Lijinsky, Lee and Gallagher [118] using tritiated aflatoxin demonstrated the persistence of label within the liver protein for up to 1 month. Indeed it is possible to obtain a positive autoradiograph at that time (Butler, unpublished). Wogan *et al.* [116] investigated the metabolism of a ^{14}C-labelled aflatoxin B_1 and demonstrated that when the label was in the methoxy group about 25% of the label was expired as CO_2 over a 24-h period with a peak at 2 h. In contrast, when aflatoxin was ring-labelled 60% of the activity appeared in the faeces as a result of biliary excretion. The residual radioactivity in the liver is very similar following both types of labelling and the liver was more heavily labelled than any other tissue which correlates well with the observation that the liver is the main target site for the aflatoxin. The excretion and tissue distribution of [^{14}C]aflatoxin B$_1$ was also studied in the mouse and although the pattern of labelling and route of excretion are similar to the ring-labelled compound, the mouse appears to excrete the aflatoxin more efficiently than the rat.

It is abundantly evident that the aflatoxins are metabolised in biological systems. However, it is less clear whether the aflatoxins acquire metabolism prior to exerting either their acute toxic or carcinogenic action. The acute toxicity of the aflatoxins may either be enhanced or reduced. The toxicity is enhanced by feeding a low-protein diet [87] or by hypophysectomy [119], in which cases it is possible that the rate of metabolism of the toxin is reduced. However, in contrast to this both feeding a choline-deficient diet [120], which reduces the activity of certain of the oxidases, and administration of DDT [121], which induces certain of the mixed-function oxidases, afford some protection. There are similar inconsistencies in the response to the carcinogenic action. The carcinogenic response may be reduced by hypophysectomy [103], treatment with phenobarbitone [122], diethylstilboestrol [104] or by feeding a low-protein diet [87]. In contrast a low-choline diet appears to result in an enhanced susceptibility [13].

Other factors have been shown to enhance the carcinogenicity of the aflatoxins. Using the trout as a test system Sinnhuber et al. [123] demonstrated that gossypol and 3-methylcoumarin given to trout previously fed with aflatoxinn B$_1$ doubled the final incidence of hepatic tumours. In more detailed investigations of the diets containing cottonseed, which is frequently used as trout food Lee et al. [124] demonstrated that cyclopropenoid fatty acid, malvalic acid and sterculic acid all enhanced the production of hepatic tumours in the trout.

The mechanism by which aflatoxin produces the acute toxicity has been studied mainly in the rat and as yet no complete scheme can be proposed which will explain the mode of action. Aflatoxin B$_1$ binds to DNA altering the absorption spectrum, producing a shift in absorption maximum from 363 nm to 366—368 nm [125]. Using tritiated aflatoxin B$_1$ Lijinsky et al. [118], showed that in vivo DNA, RNA and protein were labelled. A concurrent autoradiographic study demonstrated both nuclear and cytoplasmic labelling throughout the liver lobule [126]. DNA synthesis is inhibited by aflatoxin in both the normal liver [94], and following partial hepatectomy [127]. Both groups demonstrated marked inhibition of mitosis which, in part, may be due to the inhibition of DNA synthesis, but also a further factor, possibly associated with the spindle must be involved.

Alterations in RNA metabolism are among the earliest demonstrable effects of aflatoxin in the rat liver. The inhibition of incorporation of RNA precursors has been reprted by many groups [125, 128, 129]. Following this Gelboin et al. [130] demonstrated that the activity of RNA polymerase in isolated rat liver nuclei was inhibited by B$_1$. However, it is uncertain that the inhibition is a result of the interaction of B_1 with DNA or a direct inhibition of the polymerase enzyme.

The consistent morphological observation has been that of nucleolar segregation [131,132] (Fig. 7). This change, which may also be induced by compounds such as actinomycin D [133], has been correlated with the inhibition of the RNA polymerase [134], the inhibition forming the basis of the active toxicity. However, Butler [131,135] found no zonal distribution of the nucleolar segregation or correlation between the cytoplasmic changes of disruption of the rough endoplasmic reticulum (RER) and the nucleolar segregation. In a further study comparing the effects of B_1 on the RNA polymerases and nuclear morphology in choline-deficient animals, there was no good correlation between the two observations[136]. At present the role of the inhibition of nucleic acid metabolism in the production of the hepatic necrosis is not understood. One of the earliest biochemical effects to be described was that of the inhibition of the incorporation of labelled leucine into the protein of liver slices [137]. This was confirmed by Clifford and Rees [129,139] but *in vivo* no such inhibition could be demonstrated [118]; however, Wogan and Friedman [140] and later Clifford and Rees [139], demonstrated that B_1 inhibited the cortisone induction of rat liver tryptophan pyrrollase which results from *de novo* synthesis of the enzyme protein. Pong and Wogan [141] also demonstrated that the induction of zoxazolamine hydroxylase by benzpyrene is inhibited by B_1. In contrast to this, Butler [135] described a considerable increase of smooth endoplasmic reticulum in hepatocytes of rats treated with aflatoxin B_1. The mechanism of this proliferation which appears to be common to many forms of acute liver injury is not understood.

While the correlation between the nuclear changes and nucleic acid synthesis have received much attention aflatoxin B_1 also has dramatic cytoplasmic effects. In the work discussed above it has been suggested that aflatoxin inhibits protein synthesis *via* the nucleus. However, one of the earliest and most consistent change seen in disruption of the cytoplasmic RER [131] (Fig. 8) and breakdown of polysome profiles [142]. This has been studied by Williams and Rabin [143], who demonstrated that B_1 has a direct action on polysome binding to microsomal membrane and hence produce the observed degranulation of the RER.

It is apparent that aflatoxin has many sites of action both in the cytoplasm and the nucleus. At present it is not possible to identify any single site of action as the prime target from which the events leading to cell death stem. A more probable explanation would be that the compound has multiple sites of action, the combination of which may result in irreversible damage.

Risks to man

Based on the premise that man may be susceptible to the aflatoxins, much work has been carried out to reduce the levels of human exposure. There is

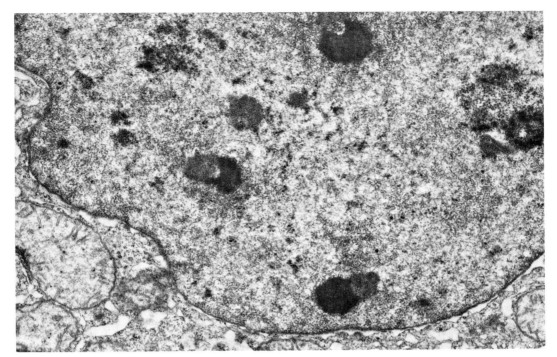

Fig. 7. Electron micrograph of a hepatic parenchymal cell nucleus 6 h after treatment with aflatoxin G_1 showing segregation of nucleoli. × 16 500.

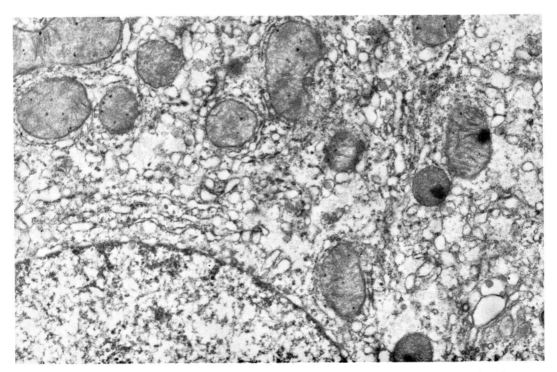

Fig. 8. Electron micrograph of the cytoplasm of a hepatic parenchymal cell showing disruption of the rough endoplasmic reticulum. × 19 000.

considerable species variation in the acute toxicity. The published range of the LD_{50} is given in Table III. Although an accurate LD_{50} for the different strains of non-human primates has not been published, it appears such species are susceptible. It is not known if man is as acutely susceptible as the guinea pig, duckling or monkey, or relatively resistant as the mouse and rat. The aflatoxins may induce hepatocarcinomas in the rat, ferret, duck and trout and possibly also in the guinea pig and monkey and it would seem reasonable to assume that man is also susceptible. There is, however, no direct evidence of this.

The evidence of acute toxicity in man is slender. Serck-Hansen [67] reported the death of a boy in Uganda, from acute liver damage, and it was found that the cassava eaten by the boy was heavily contaminated with aflatoxin. In Thailand the aetiology of Reye's syndrome has been linked with the ingestion of aflatoxin. Reye's syndrome, in which there is an encephalopathy and fatty infiltration of the liver and kidney, has been reproduced experimentally in the Macaque monkey by feeding aflatoxin [92].

The evidence suggests that the aflatoxins play a role in the induction of the human neoplasia which has a well defined geographical distribution with high incidence areas in the Sub-Saharal areas of Africa and South East Asia. Oettle [144] formulated the mycotoxin hypothesis for the aetiology of this disease. He suggested that such factors as viral hepatitis, parasitic infestation, haemosiderosis, kwashiorkor or the ingestion of toxic pyrrollizidine alkaloids failed to account for the distribution of this neoplasm. It is known that man is exposed to the aflatoxins. Alpert *et al.* [145] found that 40% food samples tested in Uganda contained aflatoxin with 15% containing more than 1 ppm. Levels of 0.015 ppm may induce carcinoma in the rat with continuous ingestion. In Kenya, Swaziland, Thailand, and Mozambique, where the incidence of hepatic carcinoma is high, aflatoxins are present in human food. In the Philippines, Campbell [66] analysed the urine and faeces of people who had eaten peanut butter known to be contaminated with aflatoxin and found aflatoxin M_1 in 24 h urine samples. He noted that an ingestion of 10—15 μg aflatoxin B_1 resulted in detectable M_1 in the urine. There is, therefore, ample evidence that man is exposed to the aflatoxins in those areas of the world in which hepatic carcinoma is prevalent. This exposure results directly from the problems of harvesting and food storage in developing countries; therefore, in order to reduce the body burden of aflatoxin these practices must be improved. It is also in these countries where the greatest problems of protein malnutrition occur. In recognition of these problems the WHO recommend that protein supplement foods should not contain more than 300 ppb aflatoxin in the knowledge that aflatoxin is the most effective carcinogen known for some species.

References

1 W.P. Blount, Turkeys, 9 (1961) 52.
2 F.D. Asplin and R.B.A. Carnaghan, Vet.Rec., 73 (1961) 1215.

3   R.M. Loosmore and J.D.J. Harding, Vet. Rec., 73 (1961) 1362.
4   R.M. Loosmore and L.M. Markson, Vet. Rec., 73 (1961) 813.
5   R. Allcroft and R.B.A. Carnaghan, Chem. and Ind., (1963) 50.
6   H. Wolf and E.W. Jackson, Science, 142 (1963) 676.
7   R.O. Sinnhuber and J.H. Wales, J. Natl. Cancer Inst., 41 (1968) 711.
8   G.E. Paget, J. Pathol. Bacteriol., 67 (1954) 393.
9   J.J. Paterson, J.C. Crook, A. Shand, G. Lewis and R. Allcroft, Vet. Rec., 74 (1962) 639.
10 R. Schoental, Brit. J. Cancer, 15 (1961) 812.
11 E. Le Breton, C. Frayssinnet and J. Boy, Compt. Rend., 255 (1962) 784.
12 W.D. Salmon and P.M. Newberne, Cancer Res., 23 (1963) 571.
13 P.M. Newberne, D.H. Harrington and G.N. Wogan, Lab. Invest., 15 (1966) 962.
14 J.W. Newberne, W.S. Bailey and H.R. Seibold, J. Am. Vet. Med. Ass., 127 (1955) 59.
15 P.M. Newberne, R. Russo and G. N. Wogan, Pathol Vet., 3 (1966) 331.
16 J.E. Burnside, W.L. Sippel, J. Forgacs, W.T. Carll, M.B. Atwood and E.R. Doll, Am. J. Vet. Res., 18 (1957) 817.
17 J.G. Keppler and H. de Iongh, Fd Cosmet. Toxicol., 2 (1964) 675.
18 M.C. Lancaster, F.P. Jenkins and J.M. Philp, Nature, 192 (1961) 1095.
19 W.H. Butler and J.M. Barnes, Brit. J. Cancer, 17 (1963) 699.
20 J.M. Barnes and W.H. Butler, Nature, 202 (1964) 1016.
21 R. Allcroft, Aflatoxicosis in farm animals, in L.A. Goldblatt (Ed.), Aflatoxin, Academic Press, New York, 1969, p. 237.
22 W.H. Butler, Aflatoxicosis in laboratory animals, in L.A. Goldblatt (Ed.), Aflatoxin, Academic Press, New York, 1969, p. 223.
23 K. Sargeant, J. O'Kelly, R.B.A. Carnaghan and R. Allcroft, Vet. Rec., 73 (1961) 1219.
24 P.K.C. Austwick and G. Ayerst, Chem. and Ind., (1963) 55.
25 B.F. Nesbett, J. O'Kelly, K. Sargeant and A. Sheridan, Nature, 195 (1962) 1060.
26 R.H. Smith and W. McKernan, Nature, 195 (1962) 1301.
27 A.S.M. van der Zijden, W.A.A.B. Koelensmid, J. Boldingh, C.B. Barrett, W.O. Ord and J. Philp, Nature, 195 (1962) 1060.
28 H. de Iongh, R.K. Beerthius, R.O. Vles, C.B. Barrett and W.O. Ord, Biochim. Biophys. Acta, 65 (1962) 549.
29 R.D. Hartley, B.F. Nesbitt and J. O'Kelly, Nature, 198 (1963) 1056.
30 T. Asao, G. Büchi, D.M. Abdel-Kader, S.B. Chang, E.L. Wick and G.N. Wogan, J. Am. Chem. Soc., 98 (1965) 882.
31 G. Büchi, E.M. Foulkes, M. Kurono and G.F. Mitchell, J. Am. Chem. Soc., 88 (1966) 4534.
32 G. Büchi, D.M. Foulkes, M. Kurono, G.F. Michell and R.S. Schneider, J. Am. Chem. Soc., 89 (1967) 6745.
33 R. Allcroft, and R.B.A. Carnaghan, Vet. Rec., 74 (1962) 863.
34 R. Allcroft and R.B.A. Carnaghan, Vet. Rec., 75 (1963) 259.
35 H. de Iongh, R.O. Vles and J.G. van Pelt, Nature, 202 (1964) 466.
36 R. Allcroft, H. Rogers, G. Lewis, J. Nabney and P.E. Best, Nature, 209 (1966) 154.
37 W.H. Butler and J.I. Clifford, Nature, 206 (1965) 1045.
38 C.W. Holzapfel, P.S. Steyn and I.F.H. Purchase, Tetrahedron Lett., 25 (1966) 2799.
39 M.F. Dutton and J.G. Heathcote, Chem. and Ind., (1968) 418.
40 J. Adye and R.I. Mateles, Biochim. Biophys. Acta, 86 (1964) 418.
41 M. Biollaz, G. Buchi and G. Milne, J. Am. Chem. Soc., 90 (1968) 5017.
42 G.N. Wogan, G.S. Edwards and R.C. Shank, Cancer Res., 27 (1967) 1729.
43 J.S.E. Holker and J.G. Underwood, Chem. and Ind., (1964) 1865.
44 R.W. Detroy, E.B. Lillehoj and A. Ciegler, Aflatoxin and related compounds, in A. Ciegler, S. Kadis and S.J. Ajl (Eds.), Microbial Toxins, Vol. 6, Academic Press, New York 1971, pp. 4—178.
45 C.M. Christensen, Botan. Rev., 23 (1957) 108.
46 G.N. Wogan, Federation Proc., 17 (1968) 932.

47 C.W. Hesseltine, O.L. Shotwell, M. Smith, J.J. Ellis and R.D. Stubblefield, Bacteriol. Rev., (1966) 795.

48 R.D. Stubblefield, O.L. Shotwell, C.W. Hesseltine, M.L. Smith and H.H. Hall, Appl. Microbiol., 15 (1967) 186.

49 U.L. Diener and N.D. Daivis, Aflatoxin formation by *Aspergillus flavus*, in L.L. Goldblatt (Ed.), Aflatoxin, Academic Press, New York, 1969, pp. 13—54.

50 K.S. Rao and P.J. Tulpule, Nature, 214 (1967) 738.

51 C.J. Rabie and E.B. Smalley, Symposium Mycotoxins Foodstuffs, Agr. Aspects, Pretoria, South Africa (1965) p. 18.

52 W.G. Sorenson, C.W. Hesseltine and O.L. Shotwell, Mycopathol. Mycol. Appl., 33 (1967) 49.

53 W. Lijinsky and W.H. Butler, Proc. Soc. Exptl. Biol., 123 (1966) 151.

54 W.A. Pons and L.A. Goldblatt, Physicochemical assay of aflatoxins, in L.A. Goldblatt (Ed.), Aflatoxin, Academic Press, New York, 1969, pp. 77—106.

55 J. Nabney and B.F. Nesbitt, Analyst, 90 (1965) 155.

56 I.F.H. Purchase and M. Steyn, J. Assoc. Offic. Anal. Chem., 50 (1967) 363.

57 B.A. Roberts and R. Allcroft, Fd. Cosmet. Toxicol., 6 (1968) 339.

58 W.H. Butler, J. Pathol Bacteriol., 88 (1964) 189.

59 Z.H. Abedi and W.P. McKinley, J. Assoc. Offic. Anal. Chem., 51 (1968) 902.

60 M.J. Verrett, J.P. Marliac and J. McLaughlin, J. Assoc. Offic. Anal. Chem., 47 (1964) 1003.

61 M.S. Legator and A. Withrow, J. Assoc. Offic. Anal. Chem., 47 (1964) 1007.

62 M.R. Daniel, Brit. J. Exptl. Pathol., 46 (1965) 183.

63 M.S. Legator, Biological assay for aflatoxins, in L.A. Goldblatt (Ed.), Aflatoxin, Academic Press, New York, 1969, pp. 107—149.

64 M.S. Legator, Bacteriol. Rev., 30 (1966) 471.

65 R.M. Loosmore, R. Allcoft, E.A. Tutton and R.B.A. Carnaghan, Vet. Rec., 76 (1964) 64.

66 T.C. Campbell, J.P. Caedo, J. Bulatoa-Jayme, L. Salomat and R.W. Engel, Nature, 227 (1970) 403.

67 A. Serck-Hanssen, Arch. Environ. Health, 20 (1970) 729.

68 R.C. Shank, G.N. Wogan, J.B. Gibson and A. Mondasuta, Fd Cosmet. Toxicol., 10 (1972) 61.

69 W.G. Siller and D.C. Ostler, Vet. Rec., 73 (1961) 134.

70 S.E. Magwood, E. Annau and A.H. Cover, Canad. J.Comp.Med., 30 (1966) 17.

71 R.B.A. Carnagham. Nature, 208 (1965) 308.

72 F.H. Kratzer, D. Bandy, M. Wiley and A.N. Booth, Proc. Soc. Exptl. Biol., 131 (1969) 1281.

73 P.M. Newberne, U.S. Dept. Interior, Fish and Wildlife Service Report 70, p. 130.

74 F.G. Clegg and H. Bryson, Vet. Rec., 74 (1962) 992.

75 R. Allcroft and G. Lewis, Vet. Rec., 75 (1963) 487.

76 B.H. Armbrecht, W.T. Shalkop, L.D. Rollins, A.E. Pohland and L. Stoloff, Nature, 225 (1970) 1062.

77 G. Lewis, L.M. Markson and R. Allcroft, Vet. Rec., 80 (1967) 312.

78 P.M. Newberne and W.H. Butler, Cancer Res., 29 (1969) 236.

79 W.H. Butler, J. Pathol. Bacteriol., 91 (1966) 277.

80 S.D. Vesselinovitch, N. Mihailovich, G.N. Wogan, L.S. Lombard and K.V.N. Rao, Cancer Res., 32 (1972) 2289.

81 R. Wieder, G.N. Wogan and M.B. Shimkin, F. Natl. Cancer Inst., 40 (1968) 1195.

82 J.H. Wales and R.O. Sinnhuber, J. Natl. Cancer Inst., 48 (1972) 1529.

83 G.N. Wogan, Bacteriol. Rev., 30 (1966) 460.

84 F.C. Chesterman and A. Pomerance, Brit. J. Cancer, 19 (1965) 802

85 J.A. Di Paolo, J. Elis and H. Erwin, Nature, 215 (1967) 638.

86 T.V. Madhavan, P.G. Tulpule and C. Gopalan, Arch.Pathol., 79 (1965) 466.

87 T.V. Madhavan, S.K. Rao and P.G. Tulpule, Ind. J. Med. Res., 53 (1965) 984.

88 W.F.J. Cuthbertson, A.C.Laursen and D.A.H. Pratt, Brit. J. Nutr., 21 (1967) 893.
89 M.G. Deo, Y. Dayel and V. Ramalingaswami, J.Pathol., 101 (1970) 47.
90 E. Alpert and A. Serck-Hanssen, Arch. Environ. Health, 20 (1970) 723.
91 D.J. Svoboda, J.K. Reddy and C. Liu, Arch. Pathol., 91 (1971) 452.
92 C.H. Bourgeois, R.C. Shank, R.A. Grossman, D.O. Johnsen, W.L. Woodling and P. Chandavimol, Lab. Invest., 24 (1971) 206.
93 W.H. Butler, Brit. J. Cancer, 18 (1964) 756.
94 A.E. Rogers and P.M. Newberne, Cancer Res., 27 (1967) 855.
95 C. Frayssinet, C. Lafarge, A.M. De Recondo and E. Le Breton, Compt. Rend., 259 (1964) 2143.
96 W.H. Butler and W. Lijinsky, J. Pathol., 102 (1970) 109.
97 G.N. Wogan and P.M. Newberne, Cancer Res., 27 (1968) 2370.
98 W.H. Butler and J.M. Barnes, Fd Cosmet. Toxicol., 6 (1968) 135.
99 W.H. Butler, M. Greenblatt and W. Lijinsky, Cancer Res., 29 (1969) 2206.
100 W.H. Butler, Pure Appl. Chem., 35 (1973) 217.
101 S.M. Epstein, B. Bartus and E. Farber, Cancer Res., 29 (1969) 1045.
102 P.M. Newberne and A.E. Rogers, J. Natl. Cancer Inst., 50 (1973) 439.
103 C.M. Goodall and W.H. Butler, Intern. J. Cancer, 4 (1969) 422.
104 P.M. Newberne and G. Williams, Arch. Environ. Health, 19 (1969) 489.
105 A.E.M. McLean and A. Marshall, Brit. J. Exptl. Pathol., 52 (1971) 322.
106 S. Juhász and E. Gréczi, Nature, 203 (1964) 861.
107 A.J. Zuckerman, K.N. Tsiquaye and F. Fulton, Brit. J. Exptl. Pathol., 48 (1967) 20.
108 A.J. Zuckerman, K.R. Rees, D.R. Inman and I.A. Robb, Brit. J. Exptl. Pathol., 49 (1968) 33.
109 E.H. Harley, K.R. Rees and A. Cohan, Biochem. J., 114 (1969) 289.
110 L. Garvican, F. Cajone and K.R. Rees, Chem.-Biol.Interactions, 7 (1973) 39.
111 D.S.P. Patterson and R. Allcroft, Fd Cosmet. Toxicol., 8 (1970) 43.
112 I.F.H. Purchase and L.J. Vorster, S. Afr. Med. J., 42 (1968) 219.
113 D.S.P. Patterson and B.A. Roberts, Biochem. Pharmacol., 20 (1971) 3377.
114 D.S.P. Patterson and R. Allcroft, Fd Cosmet. Toxicol., 8 (1970) 43.
115 D.S.P. Patterson and B.A. Roberts, Fd Cosmet. Toxicol., 8 (1970) 527.
116 G.N. Wogan, G.S. Edwards and R.C. Shank, Cancer Res., 27 (1967) 1729.
117 R.C. Garner, E.C. Miller and J.A. Miller, Cancer Res., 32 (1972) 2058.
118 W. Lijinsky, K.Y. Lee and C.H Gallagher, Cancer Res., 30 (1970) 2280.
119 C.M. Goodall, N. Z. Med. J., 67 (1968) 32.
120 A.E. Rogers and P.M. Newberne, Nature, 229 (1971) 62.
121 A.E.M. McLean and E.K. McLean, Proc. Nutr. Soc., 26 (1966) XIII.
122 A.E.M. McLean and A. Marshall, Brit. J. Exptl. Pathol., 52 (1971) 323.
123 R.O. Sinnhuber, D.J. Lee, J.H. Wales and J.L. Ayres, J. Natl. Cancer Inst., 41 (1968) 1293.
124 B.J. Lee, J.H. Wales and R.O. Sinnhuber, Cancer Res., 31 (1971) 960.
125 M.B. Sporn, W.C. Dingman, H.L. Phelps and G.N. Wogan, Science, 151 (1966) 1539.
126 W.H. Butler, Liver injury induced by aflatoxin, in H. Popper and F. Schaffner (Eds.), Progress in Liver Diseases, Vol. 3, Grune and Stratton, New York 1970, pp. 408-418.
127 A.M. De Recondo, C. Frayssinet, C. Lafarge and E. Le Breton, Compt. Rend., 261 (1965) 1409.
128 C. Lafarge, C. Frayssinet and R. Simard, Compt. Rend., 263 (1966) 1011.
129 J.I. Clifford and K.R. Rees, Nature, 209 (1966) 312.
130 H.V. Gelboin, J.S. Wortham, R.G. Wilson, M.A. Friedman and G.N. Wogan, Science, 154 (1966) 1205.
131 W.H. Butler, Am. J. Pathol., 49 (1966) 113.
132 D. Svoboda, H.J. Grady and J. Higginson, Am. J. Pathol., 49 (1966) 1023.
133 E.A. Smuckler and E.P. Benditt, Lab. Invest., 14 (1965) 1699.

134 R.S. Pong and G.N. Wogan, Cancer Res., 30 (1970) 294.
135 W.H. Butler, Chem.-Biol. Interactions, 4 (1971) 49.
136 W.H. Butler, and G.E. Neal, Cancer Res., 33 (1973) 2878.
137 R.H. Smith, Biochem. J., 88 (1963) 50 P.
138 R.C. Shank and G.N. Wogan, Toxicol. Appl. Pharmacol., 9 (1966) 468.
139 J.I. Clifford and K.R. Rees, Biochem. J. 102 (1967) 65.
140 G.N. Wogan and M.A. Friedman, Arch. Biochem. Biophys., 128 (1968) 509.
141 R.S. Pong and S.N. Wogan, Federation Proc., 25 (1966) 662.
142 A.K. Roy, Biochim. Biophys. Acta, 169 (1968) 206.
143 D.J. Williams and B.R. Rabin, FEBS Letters, 4 (1969) 103.
144 A.G. Oettle, S. Afr. Med. J., 39 (1965) 817.
145 M. E. Alpert, M.S.R. Hutt, G.N. Wogan and C.S. Davidson, Cancer, 28 (1971) 253.

Chapter 2

PITHOMYCES CHARTARUM: A FUNGAL PARAMETER IN THE AETIOLOGY OF SOME DISEASES OF DOMESTIC ANIMALS

L.G. ATHERTON, D. BREWER and A. TAYLOR

Atlantic Regional Laboratory, National Research Council for Canada, Halifax, Nova Scotia (Canada)

Under optimum conditions, lambs fed lush pasture gain 350—400 g live weight [1] per day. Such growth is, of course, the basis for the use of the cow and the sheep as farm animals pre-eminently suitable for meat production. Such weight gains are not commonly achieved under practical commercial farming conditions and there are many reasons for this inefficient production. Some of the reasons can be attributed to lack of understanding of newer agricultural techniques by farmers, and conversely, to the marginal commercial advantage of agricultural practices elaborated in the ivory towers of experimental research stations. It is, however, well-known [2] that poor growth may be observed on pasture that is apparently highly nutritious. In a number of cases, sluggish growth has been found to be due to the absence in the diet of trace amounts of vitamins or minerals [3] but there are many instances where such an explanation is untenable. An example of this phenomenon is a clutch of diseases reported in New Zealand, Australia, South Africa and Texas. The diseases are known by several names, *e.g.*, "facial eczema" in New Zealand and Australia, which do not indicate the pathology of the conditions. Clinically, the diseases present as dehydration and photosensitisation and post-mortem examination usually reveals the condition to be of some duration. However, whilst these extreme pathological signs are of considerable importance to individual farmers, it is doubtful that these outbreaks are the most important source of loss so far as world meat production is concerned. Rather, the large numbers of icteric or post-icteric, emaciated carcasses, brought to light in the slaughter house, constitute a serious reduction in world production of high quality edible protein.

Facial eczema has been studied in New Zealand for about 70 years [4]; it occurs mainly in the North Island, north of latitude 40°S, but even in the latitudes where it is common, its occurrence is, fortunately, patchy. For example, a farm at Manetuke, near Gisbourne, was purchased by the New Zealand Department of Agriculture because facial eczema occurred there almost every year, whilst the condition is rarely seen at the Experimental Farm at Ruakura, near Hamilton, 200 miles away to the west. There is considerable variability in the severity of the incidence of the disease from one year to another, but its occurrence is observed only in late summer and autumn, even in years when large numbers of animals are affected. Based on many years experience, the opinion is widely held that outbreaks of facial eczema occur after a period of rain and warm weather, and for some time, meteorological forecasts were used, with limited success, to warn farmers of conditions likely to result in the appearance of facial eczema in their flocks. Naturally, the possibility of a fungal vector in the aetiology of the syndrome was considered early in the investigations but no isolation of a toxin-producing organism was achieved [5]. It was concluded that the condition was initiated by some change in the metabolism of pasture plants, possibly resulting in the production, by the plants, of a toxic metabolite. Research on this theory was, therefore, commenced and White [7] was able to purify material from toxic, cut, dried grass about $2 \cdot 10^6$ times. During this work, White [7] observed that in the early stages of purification of the toxic material, the poisonous fractions were always associated with crystalline, non-toxic, compounds whose chemical and physical properties suggested them to be related to the enniatins [6]; that is, to typical fungal metabolites [7]. Shortly afterwards, Thornton [8], who was engaged in the study of the mycological flora of rye-grass pastures, noticed that the fungal population in areas prone to facial eczema differed from that of areas where the disease rarely occurred. He isolated 5 or 6 species from the former pastures but none appeared to be toxic until he selected a high-sporing sector from a laboratory culture and Te Punga and MacKinnon [135] found that when this was fed to lambs, the animals developed all the symptoms of facial eczema. The fungus concerned was *Pithomyces chartarum* and this result stimulated the need to develop methods of study of the metabolism and toxicology of its metabolites in farm animals and of the growth and metabolism of populations of fungi in the field. These subjects are discussed in the subsequent sections of this chapter. The chapter is concluded with a critical summary of the methods of control of the disease that are based on the belief that *P. chartarum* plays a role in the aetiology of facial eczema.

Taxonomy and morphology of *Pithomyces chartarum*

The identity of the organism associated with facial eczema was originally determined as *Sporidesmium bakeri* Sydow [9,10]. Subsequently, Ellis [11], in a study of the genus Pithomyces, re-classified *S. bakeri* as *Pithomyces chartarum* (Berk. and Curt.) M.B. Ellis. In the intervening years, the organism had been called *Scheleobrachea echinulata* (Speg.) Hughes [12] and *Piricauda chartarum* (Berk. and Curt.) R.T. Moore [13]. The fungus had also been described previously as *Sporidesmium chartarum* Berk. and Curt. [14] and *Sporidesmium echinulatum* Speg. [15].

Hughes [16], Ellis [11] and Dingley [17] have given descriptions of this organism as it appears on natural substrata. The superficial mycelium on plant debris is composed of a compact network of branched and anastomosed, lightly pigmented hyphae. The conidia are borne in clusters on this network. The conidiophores are short, $2.5-10 \cdot 2-3.5 \ \mu$, hyaline and

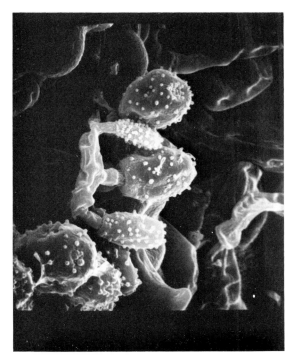

Fig. 1. Scanning electron micrograph of *P. chartarum* (× 1505). Material fixed in formalin—ethanol—water, dehydrated in an ethanol series and gold coated.

thin-walled and they are borne laterally on the hyphae. Each conidiophore bears a single conidium, produced as the blown-out end. Initially, the spores are lightly pigmented, strongly echinulate and non-septate but, during maturation, septa are formed and the outer wall thickens with a resultant masking of the echinulations. Mature conidia are broadly oval, 18—29 · 10—17 μg, with 3 to 4 (usually 3) transverse septa and the middle cells are usually divided by longitudinal septa. The conidia are brown to dark brown, verruculose to echinulate (Fig. 1) and may be constricted at the septa. The spores are liberated by a fracturing of the wall of the conidiophore which appears as a frill at the base of the spore (Fig. 2). It is this frill of tissue that serves to distinguish the spores of *P. chartarum* from superficially similar spores of other fungi, such as Stemphylium. The conidia of *P. chartarum* are very similar to those of *P. maydicus* but the latter usually have only two transverse septa, rarely three, and are generally smaller, 12—20 · 6—12 μ [11].

Ellis [11], Dingley [17] and Marasas and Schumann [18] have described the organism in culture. We have found, as also reported by Dingley [17], that there is a great variation in the appearance of different isolates in culture. In general, on 2% (w/v) malt agar at 25°, the cultures grow relatively rapidly, producing woolly or floccose colonies, which are

Fig. 2. Photomicrograph of spores of *P. chartarum* (× 1400).

generally light grey to olive grey in the centre and often have white margins. The conidiophores are often in groups and the conidia are thus produced in clusters. In other cultures, the conidia are produced relatively evenly over the surface of the culture but vary in number from one isolate to another. The densely floccose cultures are generally low sporing. We have found that sectoring does occur in cultures of some of the isolates when grown on 2% malt agar. Dingley [17] reported that sectoring of the cultures was extremely common on potato dextrose agar, especially at temperatures below 20° and above 24°. Since the number of the nuclei in the mycelium and the conidiophores was erratic and hyphal anastomosis often occurred, she suggested that the organism occurred in the field and often in culture as a heterokaryon and thus genetic recombination, rather than mutation, could explain the frequent sectoring in culture.

Dingley [17] also found that cultures from single spore isolates, when grown at 32°, could be classified as *P. maydicus* on the basis of spore septation. Cultures of the same isolates grown at below 24° could be classified as *P. chartarum*. She examined herbarium material deposited at the Commonwealth Mycological Institute, Kew, England, and found a few conidia typical of *P. maydicus* in most preparations of *P. chartarum*. The reverse situation was found to exist also in specimens of *P. maydicus*. Considering this, she felt that the present morphological characters employed to differentiate the species, namely, spore size and septation, were not sufficiently constant under a variety of conditions to classify naturally infected material collected at different times of the year. This point has also to be considered when spores collected in a spore trap are examined.

The spores that have been examined of isolates of *P. chartarum* and *P. maydicus* have crystalline coats of spicules of cyclodepsipeptides [19,20] (Fig. 3). Isolates of *P. chartarum* from Australia, England, New Zealand and the U.S.A., including spores from field collections, have been examined and 4 depsipeptides present (sporidesmolides I, II, III and pithomycolide) have been characterised [21]. Sporidesmolides I and II have been shown to be on the spore surface [19]. In the case of *P. maydicus*, it has also been shown that the spicules on the spore surface are depsipeptides, but they are of

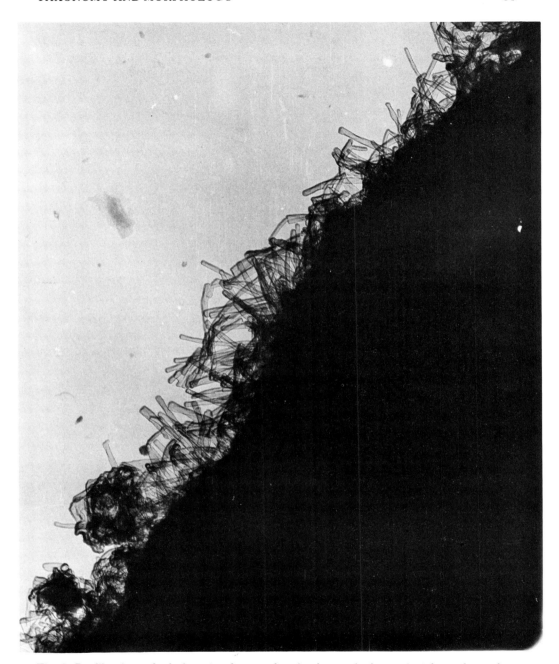

Fig. 3. Profile view of whole natural spore showing how spicules project from the surface (× 32000).

different composition to those of *P. chartarum;* one of these compounds — sporidesmolide IV — has been isolated and characterised [22]. It is possible that these natural products will be useful in the identification of particular isolates when there is serious doubt whether they are *P. chartarum* or *P. maydicus;* however, at present, too few isolates of *P. maydicus* have been examined.

Ecology of *Pithomyces chartarum*

Geographical distribution

Pithomyces chartarum has been reported from Ghana, Malaya, Mauritius, Zambia, Malawi, Jamaica, Philippines, Sierra Leone, Rhodesia, Sudan, U.S.A., and Queensland, Australia by Ellis [11]; from Victoria [23, 24], Queensland [25], New South Wales [26] and Western Australia [27] in Australia; from New Zealand [9]; from India [28,29]; from Guinea [30]; from Texas [31] and Illinois [32]; from Honduras [33]; from Panama [34]; from Jamaica [35]; from Ontario [36], Saskatchewan [37] and Nova Scotia in Canada [38]; from South Africa [18]; from England [39,40]; from Japan [41]; Brazil (private communication) and Italy [15]. In the herbarium of the Department of Agriculture, Ottawa, Canada, there are specimens of the fungus from France, Kansas and Ohio, U.S.A., and Ontario, Canada.

Habitats of *Pithomyces chartarum*

Though commonly associated with plant debris, in particular of grass species, the fungus has been recorded from a wide variety of sources, namely: plant leaves and stems [9,11,16,18,26], inflorescence [30], roots [32] and seed [18,27,31]; from soil [33,36], air [11,25,35], plywood [28], paper [15], and human foodstuffs [41]. In the herbarium of the Department of Agriculture, Ottawa, Canada, there are specimens that were isolated from soil, air and sewage sludge.

Ecology of *Pithomyces chartarum* on pasture

The fungus colonizes and sporulates on the debris in the sward. It has also been found on the cut edges of leaves and on dead sheathing leaf bases attached to grass plants.

The anatomy of spore production by *P. chartarum* and its habitat in the pasture made the collection of its spores mechanically feasible, thus giving some hope to precise estimates of the population densities of the fungus in the field. Thus to appreciate the discussion of populations of this fungus in later parts of this chapter, the techniques of measuring spore populations are reviewed.

Methods of determining populations of fungal spores

Several techniques are used to determine population densities of fungal spores. In most studies of fungal populations in soil, water or sediment, a dilution series is made and small samples from the various dilutions are incubated in an agar medium. The common medium for fungal growth is one containing rose bengal which limits colony size and has a bacteriostatic effect. Disadvantages of estimates of spore density by this method are the selectivity of the medium which may be unfavourable to some types of fungi known to occur in the population, as well as the selectivity of the incubation temperature. Some species on a plate may inhibit the growth of another species present while the spores of some fungi may be inactive. Another important factor in such estimates is the determination of the nature of the fungal propagule. This involves study of the plates to determine which colonies arise from spores. Sometimes soil plates are made by putting a small amount of soil from a sample into a petri dish before pouring the agar. This usually gives a higher number of species for a particular sample, but, in general, both dilution and soil plates give a similar picture of the population [42]. Again, analysis of the origin of a colony must be carried out.

To gain information about the presence of spores of particular species, a simple washing and counting method can be employed [43]. After the herbage is cut and washed, the filtrate from the washings is examined in a counting chamber and the number of spores expressed as spore numbers/ gram of dried herbage. Thus a minimum value is obtained, the percentage recovery of the spores from the grass being unknown.

The simplest method of estimating airborne fungal flora is the gravity settling culture plate; a petri dish of sterile medium exposed to the air, incubated and observed. No quantitative evaluation of the population can be made due to inherent inaccuracies. Only particles large enough to be pulled by gravity, or impacted by air turbulence, onto the collecting surface are sampled, while those that are too small or too light to settle out quickly are missed. Also, variations in wind, humidity, rainfall and temperature destroy uniformity of results. Long exposures needed to counter-balance these factors tend to cause drying with consequent poor growth.

In an effort to overcome some of these difficulties, Andersen [44] designed an apparatus called a sequential impaction cascade sieve volumetric air sampler. This is essentially a series of 6 plates, one beneath the other, each covered by sieving with a mesh of diminishing size and connected to a vacuum source that delivers one cubic foot of air per min to the plates. Sayer et al. [45] found that this method gave quantitatively reproducible results for measuring spore density in air with many more and different genera isolated than were collected by gravity settling culture plates. Although there was some overlapping of particle size between the stages, the meshes worked fairly well with slippage generally no more than one stage. The disadvantages associated with this system are lack of orientation to the wind preventing isokinetic air sampling; the possibility of contamination when loading and unloading the plates; only the viable spores are counted; and there is no determination of the percentage of spores that are not impacted onto the plates by the volume of air put through the apparatus.

Some studies have been made comparing plates and slides which are covered with a sticky substance [46] to trap the spores. Generally higher counts of the spore load of the air are obtained from the slides, although this was not true for the genus Cladosporium [47—50]. When Pady *et al.* [46] germinated the spores on the slides, the viable count was approximately the same for both slides and plates.

If the air contains a large number of spores during the sampling time, the sample is difficult to count with any degree of accuracy. To overcome this difficulty and also to give more precise information about the time of occurrence of various spores in the air (the main object of many of these studies), various types of slide sampler have been constructed. The simplest [51] described as a 24-h slide spore collector contained 24 glycerol-coated microscope slides that were rotated past an orifice so that 1 sq. inch of each slide was exposed per hour. The apparatus was mounted in an air-tight box with a fan in the lid. The fan, run by a small battery, drew air through the box at a constant rate calculated to give an efficiency of 70% or more. Spores of all sizes, from 5—140 μ, were collected.

Another type of volumetric spore trap was developed by Hirst [52]. Earlier, May [53] had built a cascade impactor that was made for separating the particles in aerosol by size by impacting a sample of air drawn through jets of decreasing width onto the sticky surface of microscope slides. Hirst used this impactor to discover the suction rate that would give the greatest efficiency of impaction so that in his experiments the air stream entering the trap would bear a representative spore load. Ideally, with the orifice facing the wind-stream, the suction rate should equal the external wind speed. As this is not possible due to fluctuations in wind speed over a sampling period, the most useful suction rate (10 l per min) was determined. Hirst's trap consists of an impactor unit (Fig. 4), a wind vane mounting, and the motor housing which provides the suction. The wind vane directs the orifice (14 × 2 mm) of the impactor into the wind, while the slide is moved past the orifice at a constant rate giving a continuous trace of the day's spores. The orifice was made as small as possible with the same bore throughout to avoid spore deposition on its walls — a drawback of the cascade impactor. The deposition of spores on the exposed surface of the slide was found to be fairly uniform, while the smallest interval of time at which the spore content could be reliably estimated was 2 h.

A further refinement of this system was made by Pady [54]. He built a similar trap with a timing mechanism which deposits spores in 24, 1-h bands on a single microscope slide.

Gregory [55] designed a light portable volumetric trap in which the spores are impacted on a slide by an air current maintained at a steady rate by a hand-operated sliding vane pump, attached to a low vacuum reservoir with a gauge. This trap collected about 75% of the spores in the air sample,

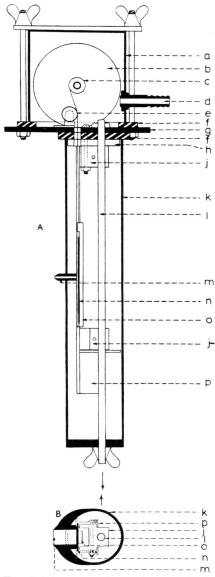

Fig. 4. Diagrammatic sections of the impactor unit (one-third of actual size). A, vertical section; B, horizontal section at the level of the orifice. Parts of the air-tight shell and the slide, where cut by the plane of the section, are shown in solid black, except the rubber seals which are shaded. a, air tight top cover; b, clock; c, thread drum; d, outlet tube; e, guide roller; f, rubber seal; g, mounting plate; h, positioning disk; j, spacing piece; k, removable orifice cylinder; l, tie rod; m, orifice; n, slide; o, slide carrier; p, slide runners of 1 section brass.

References p. 65

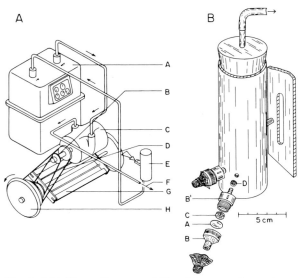

Fig. 5. (A) Flow diagram of Brook's spore trap. A, gas meter; B, odometer; C, pump; D, pump drive; E, slide holder, impactor unit; F, direction of air flow; G, agitator; H, wheels of trap. (B) Exploded diagram of spore-trap. A, Millipore filter disk; B,B', top and bottom parts of filter holder; C, backing plate; D, rubber grommet.

and, as with the Hirst trap, retained the smallest fungal spores with high efficiency. Besides the advantage of being able to move the trap easily, time of sampling is readily shortened, a definite help in counting when spore density is high. This is also the main disadvantage; for, if the density is low, there is the labour of operating the pump a considerable time.

To estimate the fungal population of a pasture, further modifications were made by Brook [56] to Gregory's trap. The trap (Fig. 5A) is mounted on wheels between which are rotating blades that throw the spores up so that they are sucked into the impactor. The air current through the impactor is maintained by the sliding vane pump being attached to the wheels. A gas meter between the impactor and the pump measures the flow of air and maintains an even current. The efficiency of such a trap is unknown, but comparative studies of kinds and densities of spores can be made. This trap was further changed (Fig. 5B) [2] by replacing the slide for collecting the sample with two millipore filters. This allows for direct observation of one filter by mounting it in immersion oil and the cultivation of spores from the other by using it to make a dilution series.

Seasonal changes of populations of Pithomyces chartarum

Estimates as to the numbers of *P. chartarum* present in pastures have been made by a washing technique and also by spore trapping methods. Thornton and Sinclair [43], using a simple washing technique, found that high spore counts were favoured by large amounts of dead and damaged plant materials and warm moist conditions. They sampled the pastures and counted the number of spores at approximately weekly intervals. In normal pastures, the spore count increased to a peak in March. They found that they could advance and intensify the peak numbers of spores by regularly cutting the pasture and leaving the debris to accumulate. The increase in the supply of nutrients and the tendency to maintain humid conditions within the pasture would thus favour the development of the fungus.

Smith *et al.* [57] carried out an intensive study of the seasonal variation in spore numbers of *P. chartarum*. This study was carried out in the Waikato region of New Zealand during 1960 and 1961. They employed the mobile spore trap designed by Brook [56], as it was more adaptable to rapid collection of data than was the washing technique of Thornton and Sinclair [43]. However, some measurements of the spore loads on the herbage were made by the latter method. Concurrent with the spore counts, meteorological data were recorded and, from their results, they were able to relate peaks in spore numbers with certain meteorological conditions. Two distinct periods of different types of conditions appeared to be required for high counts. The first was a few days of warm moist weather needed for sporulation of the organism. Following this, 2 or 3 days of dry, sunny weather were required. Smith *et al.* [57] note that high spore counts do not reflect rapid sporulation of the organism but are the result of the presence of large numbers of readily detached spores on the herbage. The surveys of 1960 and 1961 showed a general trend for peaks of spore counts at Ruakura and in the Waikato region to occur at the beginning of March and again at the end of April (Fig. 6A). In 1960, there was also a peak at the beginning of February. Grass minimum temperatures appeared to have a pronounced effect upon

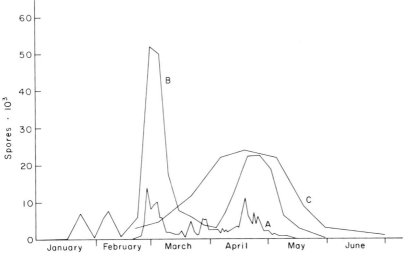

Fig. 6. Spores/ft³ of air in 1961. A, in the Waikato region of New Zealand [57]; B, Dargaville, Auckland province [58]; C, spore numbers measured by washing technique on irrigated pastures, Rosedale, Victoria, Australia [60].

References p. 65

the spore count and the variation in these temperatures, together with the variation in relative humidity, produced marked fluctuations in the numbers of spores during peak periods.

Brook [58], working with the mobile spore trap on farm pastures near Dargaville and plots of perennial rye grass in Auckland, reported a composite picture of the seasonal development of *P. chartarum*. He felt that during September and October, the temperatures were not favourable to the development of *P. chartarum* and that the debris was colonized by other saprophytes. Although in December, the temperatures were sufficiently high for strong growth of the fungus, there was a delay in the rise in activity probably due to the competition from earlier established saprophytes. During January and February, though the temperatures were high and there was an accumulation of debris, there were only occasional peaks of activity, dependent entirely on falls of rain (Fig. 6B). He determined that the optimum temperature for growth and sporulation was 24° and that a moisture level at or near 100% relative humidity was required for maximum sporulation. A period of 24 h at 100% relative humidity was required for maximum infection of leaves from spores. He also found that, in the field at a mean incubation temperature of 18°, the fungus could complete the cycle from infection to sporulation in 3 days and that 4 days after the rain commenced, there were large numbers of new conidia on the leaves [59]. The warm, rainy weather in late February and early March encouraged rapid growth and sporulation of *P. chartarum*. This weather also initiated the autumn flush of grass growth; thus, after the accumulated debris of the summer was gone and because there was little dead material being accumulated, the activity of the fungus declined and remained low until April, when there was sufficient debris again. After this second seasonal climax, there was a decline to a low winter level.

Fisher [60] in Rosedale, Victoria, Australia, using a Hirst spore trap, found that during April there was a significant increase in the numbers of *P. chartarum* spores in the air. Using the "washing-technique", she also studied the development of the fungus under flood irrigation. From February 21 until April 18, there was a steady increase in the number of spores per g of herbage, after which there was a steady decline to approximately the February 21 level at the beginning of June (Fig. 6C). The decline continued until a negligible level was reached on September 25. The elimination of the moisture variable may be the reason for the regularity of the curve of spore production. However, she suggested the alternative that the results may have been just a reflection of more even distribution of spores throughout the sampling area.

More recently, di Menna and Parle [61], during an examination of the moulds of leaves of perennial rye-grass and white clover and litter from a rye-grass dominant pasture by dilution platings of the washings of the samples, found that *P. chartarum* was not isolated very frequently. However, it was isolated when expected from leaves of rye-grass and white clover, even though it constituted only a minor part of the total mould flora. At a sampling when the washed spore count of the organism was 850000 per gram wet weight, only 13 of 141 isolates were *P. chartarum*.

Metabolism of *Pithomyces chartarum*

(1) Growth and nutrition of P. chartarum strain C

Ross [62] showed that this laboratory isolate of the fungus grows well on a defined medium (g/l: potassium dihydrogen phosphate, 1; potassium chloride, 0.5; magnesium sulphate heptahydrate, 0.5; ferrous sulphate heptahydrate, 0.01; sodium nitrate, 2; and glucose, 30). Growth was increased by about 30% by adding trace metals (ppm: $FeCl_6$ 6 H_2O, 2.6; $ZnSO_4$ 7 H_2O, 0.4; $CuSO_4$ 5 H_2O, 0.1; $MnCl_2$ 4 H_2O, 0.05; and $Na_2 MoO_4$ 2 H_2O, 0.04). Growth of the fungus was also supported by fructose, xylose, mannose, sucrose and starch, but only cellobiose increased it (by 15%) as compared to glucose. Ross also determined the effect on growth of a number of nitrogen sources added at a concentration of 0.0167% N. Of those tested, only asparagine and L-glutamic acid increased growth by 17.5 and 31% respectively. Closely similar results were obtained by Done *et al.* [63] (Fig. 7) in studies which used an extract from carrots and potatoes as the basal medium. By analysis of the medium constituents, these authors showed that glutamine was rapidly metabolised by the mould and that glutamine added

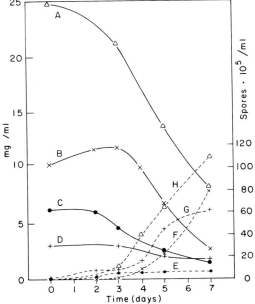

Fig. 7. Utilization of medium constituents and production of sporidesmin and sporidesmolides as a function of culture age. A, dry weight of media; B, sugar; C, nitrogen (× 10); D, ash; G, mycelium dry weight; H, spore count; F, sporidesmolides (× 100); E, sporidesmin (× 1000).

to the medium before autoclaving increased growth by 27%. Later experiments [19] showed that DL-valine and DL-isoleucine also increased growth by 15% and 7% respectively, and, in addition, the residual amino acid recovered from the growth medium was largely of the D configuration. It emerged from these experiments that, provided the carbon : nitrogen ratio was kept at about 20, growth was proportional to the richness of the medium. Extremely heavy growth could be achieved on such materials as bran, whole wheat and rye-grain.

All of the experimental results given in the previous paragraph were obtained by growing the fungus in surface culture on liquid or agar medium. Much greater growth rates are obtained in submerged culture, either in shaken flasks [64,65] or in stirred fermenters [66]. In general, about twice the weight of fungal tissue was obtained in 4 days' growth in submerged culture as was harvested after 7 days' growth in surface culture in otherwise strictly comparable conditions. After 4 days, lysis of mycelium occurred and the weight of fungal tissue harvested decreased. During the growth phase, the changes in carbohydrate and nitrogen concentrations were similar to those described above (Fig. 7) for surface cultures except their depletion was accelerated. In all fermentations, the pH of the culture medium changed from about 5 initially to about 7 after 7 days' growth.

(2) Sporulation of P. chartarum

In studies of the effect of temperature, pH, carbohydrate source and concentration of certain cations on the number of spores produced by *P. chartarum* in culture, Ross concluded [67] that these parameters were not important. However, it was shown that the amino acids, L-asparagine, L-valine, L-leucine and L-alanine, when added to give a final nitrogen concentration of 0.017%, increased the number of spores produced as compared to sodium nitrate by factors of 5.2, 2.8, 1.25 and 4.3 respectively. Thus sporulation is dictated to some degree by the nitrogen source. Dingley *et al.* [64] pointed out that the utilization of the nitrogen in the medium to produce 1 g of fungal tissue was 50% greater in cultures that produced appreciable numbers of spores than in non-sporing cultures. A possible reason for this observation was provided [19] when it was shown that about 3.53% of the wet weight of spores of *P. chartarum* was depsipeptide; thus the nitrogen of sporidesmolide I [68] represented 0.3% of the total spore weight. In biosynthetic experiments using uniformly labelled [14C]valine, 27% of the 14C added to the culture was recovered in the depsipeptide fraction [69] largely associated with the spores. It is, therefore, clear that the synthesis of this spicular coat competes in a significant way with protein synthesis in this organism. By contrast, lipid synthesis, as estimated by its total weight, is smaller in the spore than in the mycelium [70,71], the lipid extracted being 1.39% of the dry matter of spores as compared to 4.5% in the fungal tissue as a whole. It was possible to distinguish (Table I) between the total fatty acid composition of the spore — which closely resembled the

TABLE I

COMPARISON OF THE FATTY ACIDS IN THE LIPID FROM MYCELIUM, SPORES AND SPORE SURFACES OF *PITHOMYCES CHARTARUM*

Source of lipid	C_{12}	C_{14}	C_{15}^{br}	$C_{16}^{2=}$	$C_{16}^{1=}$	C_{16}	$C_{17}^{1=}$	C_{17}	$C_{18}^{3=}$	$C_{18}^{2=}$	$C_{18}^{1=}$	C_{18}	$C_{20}^{1=}$	C_{20}	C_{22}
Spore surface	0.3	0.8	0.4	0	1.1	26.9	0.5	0.7	0.4	33.6	25.8	7.6	0	0.6	0.7
Spores	?	0.5	0	0.2	0.6	27.9	?	0.5	0.4	51.5	12.9	4.4	0	0.6	?
Mycelium	0.4	0.2	0	—	0.7	29.5	—	0	1.1	41.3	15.0	8.1	—	—	—

C, number of carbon atoms; subscripts indicate number of carbon atoms in chain; superscripts indicate the number of double bonds in the chain; br, branched. A dash means that the analysis was not recorded and a query (?) shows that only a trace of the fatty acid was found.

References p. 65

fatty acid composition of fungal tissue as a whole — and the fatty acid compositon of the lipophilic material on the spore surface [19]. This lipophilic material comprised about 25% of the total spore lipid and hence about 0.32% of the dry weight of the spore. Thus a maximum of 1% of the spore dry weight is available for metabolic purposes having lipid as an energy source, a remarkably low figure in an organ designed for survival.

(3) Production of toxic metabolites by P. chartarum

In the laboratory, the production of sporidesmin and its chemical relation by *P. chartarum* is very low; maximum yields being about 3 mg/l in liquid surface or submerged cultures (Table II) [63,64]. In surface culture, sporidesmin is produced at the onset of the rapid phase of growth and it is distributed equally between the fungal tissues and the culture medium. After about the third day of incubation, the concentration changes little in surface culture, but in shaken culture, the concentration rapidly declines after the fourth day when mycelial lysis occurs. A statistical analysis of almost 100 batches of *P. chartarum* strain C revealed that there was a significant correlation between the yield of sporidesmin and the initial dry weight of the potato—carrot medium used. This led to the adoption of a fermentation technique used in the early days of penicillin production, *i.e.*, the culture of the fungus on semi-solid media. It was found that appreciably higher yields of sporidesmin, *i.e.* calculated on the basis of grams of sporidesmin produced/100 g dry weight of medium, could be achieved by cultivation of the organism on bran, wheat or rye, and the latter was selected for routine production because of its better mechanical properties. However, good yields of sporidesmin were produced by Dr. Thornton by growing the organism on wheat.

Very small changes in sporidesmin production were obtained by changing the initial pH of the medium, in the range 4.5—8, or the temperature of incubation in the range 18—28°, though Ross and Thornton [72] found generally lower yields at higher temperatures. These workers also gained some evidence that the addition of L-asparagine to the medium increased the yield of sporidesmin.

The relationship of sporidesmin production to the ability of the organism to sporulate is one that has been studied by several workers. The

TABLE II

PRODUCTION OF SPORIDESMINS BY *PITHOMYCES CHARTARUM* ISOLATE C, ON RYE-GRAIN MEDIUM

Metabolite	Yield (mg/100 g dry medium)
Sporidesmin	12.9
Sporidesmin B	1.3
Sporidesmin E	1.3
Sporidesmin G	0.1
Sporidesmin H	0.1

question has practical importance because it is relatively easy to determine spore numbers in the field and thus if sporidesmin production was proportional to the density of spores on pasture, a forecast of toxicity could be made with accuracy. Recently, data on this question have been reported [65]. A statistical analysis of these data is difficult because only very poor polynomial least mean square fits on the results are possible. The linear fit has an index of determination of 0.57 and a positive gradient with an intercept on the Y axis of about $1.4 \cdot 10^8$ spores/l ($1.12 \cdot 10^6$ spores/g of culture media). It is, therefore, highly probable that isolates will be obtained from the field that sporulate profusely on laboratory media but do not produce sporidesmin. It also follows from these data that a slightly better chance than 1 in 2 exists that an isolate that produces more spores will also produce more sporidesmin. Thus a very high sporing isolate is more likely to produce sporidesmin than one that produces few spores. However, two groups of workers have reported the production of sporidesmin in small amounts by isolates that did not sporulate in laboratory culture [64,65].

(4) Comparison of different isolates of P. chartarum

The ability of different isolates of this fungus to produce sporidesmin is reviewed in the previous paragraph. No work has been done on the proportion of the different toxic metabolites produced by isolate C, that are biosynthesised by wild isolates. This is an important lacuna in our knowledge, since it is possible that the toxicity of a pasture would be slight if, for example, sporidesmin B was the principal metabolite, but great if sporidesmin E was produced in quantity. Wild isolates from Australia, New Zealand and the United States all produced depsipeptides and those from Australia and New Zealand in laboratory culture synthesised the different components of the depsipeptide mixture in the same proportion. There was good correlation between the quantity of depsipeptide produced and the number of spores in the culture, even though, as stated in the previous paragraph, the isolates varied widely in their ability to sporulate. Thus work on the concentration of depsipeptides in toxic cut pasture is indicative of the density of the spore population. These facts formed the basis for the beaker-test for toxicity [73].

A little work has been done [62,64,72] on the nutrition of wild isolates of *P. chartarum* but the relevance of this work to the nutrition of the fungus in its natural habitat is hard to assess since the choice of laboratory media probably exerts selection. The point was illustrated by Dingley *et al.* [64] who reported changes in sporidesmin production after each sequential subcultivation of a wild isolate. In the example given, production decreased to about 30% of the original after about 8 subcultures. In other unpublished work, small increases in sporidesmin production were observed in sequential subcultivation of the wild isolates in the laboratory.

(5) Biosynthesis of sporidesmins

When the structure of sporidesmin was proposed, it was clear [74] that it was probably biosynthesised by the condensation of one mole of L-tryptophan with one of L-alanine. Some evidence that this suggestion is true was obtained by Towers and Wright [75] who grew *P. chartarum* in the presence of DL-[3-^{14}C]tryptophan, L-[U-^{14}C]alanine, DL-[1-^{14}C]serine and DL-[*Me*-^{14}C]methionine. The sporidesmin produced was not characterised, nor were its physical properties, other than its partition coefficients in the water [76] and water—methanol chromatographic systems, measured, but the radioactive zones were toxic to guinea pigs [77] and rats [78].

Speculatively, the geometry and substitution of the indoline-pyroline ring junction suggests that the tri-cyclic ring system may be formed by the addition of the lone pair of electrons of the amino group of tryptophan to an epoxy function at the 2,3 position of the indoline system as shown in the scheme [79]:

The cis-fused bicyclic system is found in sporidesmin B and it is perhaps pertinent that chetomin [80], a related metabolite from *Chaetomium cochliodes*, contains a fragment where ring closure has not taken place because the indoline nitrogen atom is alkylated. If these suggestions are true, it may also follow that the secondary hydroxyl group in sporidesmin is introduced after ring closure. These speculations, depending on the known structure of related metabolites, may also be extended to the possible modes of formation of the sulphur bridge. Thus the metabolite sporidesmin F [81] indicates that the sulphur function may be added to either of the carbon atoms of the dioxopiperazine ring that are α to the carbonyl groups. The organism is capable of using a wide range of sulphur compounds as its source of the sulphur atoms of the bridge [82] and it is therefore possible that the bioreagent for the introduction of this entity is a derivative of dihydrogen disulphide. Some evidence that this is the case was obtained when the trithio- and tetrathio-metabolites [83,84] were isolated from cultures of the fungus and when it was found that sporidesmin was smoothly converted to these metabolites by treatment with dihydrogen disulphide [85].

In all considerations of the biosynthesis of sporidesmin and its related metabolites, one fact stands pre-eminent: the very low weights of the metabolites produced in unit volume of culture medium. Progress in the field is, therefore, dependent on finding techniques to improve the yield which is today the same as it was 15 years ago. It is possible that UV irradiation will yield isolates that produce more sporidesmin [65] but no production studies have been reported.

Analysis of sporidesmins
The principal physical properties of the biologically active sporidesmins

TABLE III

PHYSICAL CONSTANTS OF SPORIDESMIN AND ITS RELATED TOXIC METAB-
OLITES

Sporidesmin B,   (R = Cl, R' = H, R'' = OH, n = 2); m.p. 183°; $[M]_D$ = $-$ 78°,
                   (c = 0.78, EtOH), +55° (c = 0.75, CHCl$_3$); λ_{max}(Et$_2$O) 218, 256,
                   307 nm (log. ϵ 4.50, 4.08, 3.41); τ (CDCl$_3$) 2.93 (H), 4.6 (H), 6.14
                   (3 H), 6.19 (3 H), 6.47, 6.76 (3 H), 6.97 (3 H), 7.97 (3 H), 6.8, 7.2
                   (2 H, J_{AB} = 16.1 Hz). Sporidesmin B acetate m.p. 93—114°.

Sporidesmin H,   (R = R' = H, R'' = Cl, n = 2, orientation of methoxyl groups unknown),
                   m.p. 140—150°; λ_{max} (MeOH) 216, 252, 290 nm (log. ϵ 4.37, 4.05,
                   3.81); τ (CDCl$_3$) 3.02, 3.56 (2H, J_{AB} = 8 Hz), 4.60 (H), 6.15 (3 H),
                   6.24 (3 H), 6.68 (3 H), 6.95 (3 H), 7.96 (3 H), 6.70, 7.20 (2 H,
                   J_{CD} = 16 Hz).

Sporidesmin,   (R' = R'' = OH, R = Cl, n = 2), m.p. 179°, benzene solvate m.p.
                   110—120°; $[M]_D^{23}$ = $-$ 185° (c = 1.1, MeOH), +38° (c = 1.4, CHCl$_3$);
                   λ_{max} (Et$_2$O) 218.5, 254, 302 nm (log. ϵ 4.60, 4.12, 3.45); τ (CDCl$_3$)
                   2.92 (H), 4.70 (H), 5.16 (H), 5.42 (H), 6.13 (3 H), 6.18 (3 H), 6.60
                   (H), 6.70 (3 H), 6.93 (3 H), 7.97 (3 H); cell dimensions: a 9.64,
                   b 10.58, c 23.88 A. Diacetate, m.p. 170—171°; $[M]_D^{22}$ + 43.4°
                   (c = 0.51, CHCl$_3$).

Sporidesmin E,   (R' = R'' = OH, R = Cl, n = 3), isolated as its etherate; m.p. 180—185°,
                   $[M]_D^{20}$ = $-$666° (c = 0.064, CHCl$_3$); λ_{max} (MeOH) 217, 252, 295 nm
                   (log. ϵ 4.52, 4.22, 3.50); τ (CDCl$_3$, 37°) 2.92 and 2.93 (H), 4.59 and
                   4.71 (H), 5.38 and 5.48 (H), 6.10, 6.13 and 6.17 (6H), 6.50, 6.67,
                   6.86 and 6.98 (6 H), 8.00 and 8.05 (3 H), signals due to ether omitted;
                   diacetate, m.p. 112—115°; $[M]_D^{20}$ = $-$695°, (c = 0.11, CHCl$_3$). Treat-
                   ment with methyl iodide and sodium borohydride gives sporidesmin
                   D also a metabolite, m.p. 105—107° (ethanolate); $[M]_D^{23}$ = + 35°
                   (c = 0.1, CHCl$_3$).

Sporidesmin G,   (R' = R'' = OH, R = Cl, n = 4), $[M]_D^{23}$ = $-$482° (c = 0.065, MeOH);
                   λ_{max} (MeOH) 219, 252, 300 nm (log. ϵ 4.64, 4.16, 3.78); τ (CDCl$_3$)
                   2.82 (H), 4.90 (H), 5.35 (H), 6.10 (3 H), 6.16 (3 H), 6.60 (3 H),
                   6.93 (3 H), 8.03 (3 H); cell dimensions: a 15.160, b 21.369, c 8.978 Å;
                   D_m 1.40 g $\cdot$ cm^{-3}.

are given in Table III. These data are the basis of all analytical procedures as
they characterise the toxic metabolites. Ideally, the toxins should be isolated
from material derived from natural sources and their physical properties
determined before one embarks on an assay procedure since none of the
assays that have been used to date are unique for the sporidesmins (see, *e.g.*,
Leaver [86]). This onerous task has rarely been attempted, it being assumed

References p. 65

that the biological or chemical effect observed was due exclusively to the poisons.

Chromatography of sporidesmins

Synge and White reported [76] that sporidesmin and sporidesmin B could be separated (R_F 0.5, 0.21 respectively) by paper chromatography using water as the eluting solvent and Russell [87] applied two-dimensional chromatography on Whatman No. 4 paper (light petroleum as the eluting solvent in one direction; acetic acid:benzene (1:49) in the other) to separate sporidesmin from other constituents of toxic grass. The sporidesmins can be separated by thin-layer chromatography [88]. Using commercially available silica gel plates, the R_F values obtained when 5 different solvents were used for development are given in Table IV. It is clear from Table IV that all 4 toxic metabolites can be separated and unpublished data show that they can also be separated from sporidesmin H. The method can be increased in scale and the separations are as good when the toxic materials in quantities of 100 mg are applied as a streak on thick layer plates of dimensions 100 × 20 cm.

All of the sporidesmins can be detected in μg amounts on paper or thin-layer chromatograms by reflected short wave ultraviolet light. Alternatively, if the chromatograms are sprayed with neutral aqueous silver nitrate (5%), quantities of sporidesmins greater than 0.1 μg can be detected. A slight intensification of such spots is achieved by overspraying with 0.1 *N* sodium arsenite. The silver nitrate reacts only slowly with S-methyl derivatives of the sporidesmins and it is useful to keep chromatograms sprayed with this reagent overnight, particularly when the material chromatographed

TABLE IV

R_F VALUES FOR SPORIDESMIN AND ITS RELATED METABOLITES

Compound	Eluting solvent				
	Benzene— ethyl acetate (4:1)	CHCl$_3$	Benzene— ether— acetic acid (70:30:1)	Hexane— *t*-butyl alcohol (9:1)	CHCl$_3$— MeOH (19:1)
Dethiosporidesmin	0.32	0.06	0.31	0.17	0.53
Sporidesmin	0.38	0.09	0.39	0.17	0.57
Sporidesmin E	0.36	0.07	0.39	0.24	0.54
Sporidesmin G	0.21	0.03	0.25	0.24	0.49
Sporidesmin B	0.39	0.11	0.41	0.23	0.59
Sporidesmin D	0.19	0.04	0.24	0.01	0.55

The R_F values are for chromatography of the compounds on commercially available Merck silica gel 20 × 20 cm plates, thickness 0.25 mm, type F$_{254}$.

is from a biological source. The ability of the disulphides to catalyse the reaction of iodine with azide ion (see below) can also be used to detect this component of the mixture of toxic metabolites on paper chromatograms [87]. The chromatograms are first sprayed with iodine and then, after some minutes, with a solution of starch when sporidesmin, sporidesmin B and sporidesmin H appear as white spots on a blue background.

Quantitative estimations of sporidesmins

(a) *Chemically*. 3 methods have been used. The best of the methods depends on the reaction of the alkaline decomposition products of sporidesmins with iodine [63]. Titration of the residual iodine and a comparison of the titre with a duplicate sample not treated with alkali gave a precise estimate of the sporidesmin present; 1 ml of 0.01 N sodium arsenite being equivalent to 0.7 mg of sporidesmin. The method suffers from two disadvantages: relatively large quantities of sporidesmin are required, and, in crude materials from biological sources, the presence of interfering substances often leads to spuriously high results [89]. The second method that has been used for sporidesmin assay is the ability of sulphur compounds to catalyse the reaction of iodine with azide ion:

$$2\,N_3^- + I_2 = 3\,N_2 + 2\,I^-$$

Sporidesmin is a more efficient catalyst of this reaction than other sulphur compounds, *e.g.*, cystine [90]; about 30 moles of iodine are reduced by azide for each mole of sporidesmin present. The method is thus very sensitive; 1 ml of 0.01 N sodium arsenite being equivalent to 0.055 mg of sporidesmin. Further, the catalyst concentration can be determined by manometric estimation of the nitrogen evolved, about 4 μl of nitrogen is obtained per μg of sporidesmin at normal temperature and pressure [91]. The value of this sensitive assay procedure is increased since it has been shown that many kinds of cells do not seriously interfere with the estimation, provided that their cell membranes are intact [91]. The method, however, suffers from several disadvantages. There has been a range of values quoted for the volume of standard iodine equivalent to unit weight of the catalyst and all workers have stressed the importance of time-consuming standardisation procedures. Many other compounds that are present in extracts from plant and animal tissues will reduce iodine and, in addition, some compounds that do not react with iodine will, nevertheless, catalyse its reaction with azide. It suffers from an additional disadvantage when it is to be used as a measure of toxicity because the most toxic trisulphide, sporidesmin E, is a very inefficient catalyst of the reaction. Hence values obtained by titration of plant or animal extracts can be very misleading as far as their toxicity is concerned.

The third method, that of adding radioactive sporidesmin to an unknown sample and determining the decreased specific radioactivity on the re-isolated toxin, has been widely used and is potentially an excellent technique. However, as mentioned above, radioactive sporidesmin of known specific activity is a compound that has not yet been obtained; most authors being content to elute bands from chromatograms, determine their toxicity and assume that the material is pure sporidesmin [25,82,92]. However, fully characterised sporidesmin E of known specific radioactivity has now been prepared [85] and is available for biological work.

(b) Biologically. Sporidesmin may be determined by measuring its inhibition of growth of *Bacillus subtilis* [93]. The toxin can be assayed using this organism either by standardising the inoculum and the time of incubation of the cultures containing known concentrations of the poison or by determining the lag of a known number of cells in the presence of the toxin in comparison with a control. With each assay, standard solutions of sporidesmin must be used as positive controls. The minimum dose required to inhibit growth for 18 h is 80 μg/ml, and in the range 10—50 μg/ml, the lag phase is increased linearly by 75 min/10 μg. The method is very reproducible but suffers from the disadvantage that relatively large quantities of sporidesmin are required (Table V).

An assay requiring similar quantities of material was used very effectively during work on the isolation of sporidesmin from grass, and later from cultures of *P. chartarum*. The material suspected to be toxic was distributed on grass and/or powdered milk from solutions in organic solvents. The solvents were evaporated and the material fed to young guinea pigs [7,76,94] for 3 weeks, followed by a 4th week on a normal diet. After this time, the animals were slaughtered and their livers examined macroscopically (and sometimes histologically) for liver lesions. It was found that 2 mg of sporidesmin per kg body weight produced severe lesions. The assay was very reliable but also time-consuming.

A more rapid, though less precise method, also used before the isolation of sporidesmin, was the rabbit eye corneal opacity test [63]. Here the extract was dispersed at a concentration of 1.2 mg/ml in 1% "Tween 80" solution and 0.05 ml was applied to the eye-ball of New Zealand white rabbits on each of 2 successive days. When 10 μg of sporidesmin was applied in this way, congestion of the scleral vessels with oedema and inflammation of the conjunctival membranes was observed 5—7 days after instillation.

The growth of mammalian cells in culture is extremely sensitive to minute amounts of sporidesmin and sporidesmin E. Indeed, the latter compound is probably one of the most toxic compounds to mammalian cells in tissue culture known. These facts can be used as the basis of what is probably the most satisfactory method of assay of the toxic metabolites of *P. chartarum*. Cells of a sensitive line, *e.g.*, HeLa, Chang appendix cells, *etc.*, are dispensed into 125 × 6 mm pyrex screw-cap test tubes and are incubated in nutrient medium until the cells have adhered to the glass and started to grow. After about 24 h incubation, and before a continuous layer of cells has been established, the material suspected to be toxic is dissolved in ethanol and a dilution series prepared with phosphate buffered saline. Each dilution is then added to a culture of cells. An estimate of the concentration of sporidesmin can be obtained by any of the methods described above and thus the dilution series arranged in which a concentration of sporidesmin of about 3 ng/ml is located in about the centre of the dilution series. The cultures are then incubated at 37° for 3—4 days when they are examined. A sharp "end-point" is almost always observed, the cells below this dilution having grown to cover the entire surface of the culture vessel and the cells incubated with concentrations greater than the end-point showing pyknosis and no growth. As shown in Fig. 8, this end-point can also be demonstrated by the more objective criterion of the concentration of protein in the cultures [63,95,96].

The sensitivity of different cell lines varies considerably but several groups of workers have obtained values for the end-point for sporidesmin in the range 0.4—3 ng/ml. This variability is probably due to small differences in the time of addition of the sporidesmin to such variable constituents of the medium as calf serum and, in some cases, to different amounts of the various sporidesmins in the samples assayed. When these factors are taken into account, the agreement between the various groups of workers is remarkable. The toxicities of 10 derivatives of sporidesmin are given in Table V. These results were obtained in this laboratory and have been selected only because they are comparable and fairly complete. It is clear that sporidesmin and sporidesmin E are considerably more toxic than their closest chemical relations. This high potency not only enables the assay to be very sensitive, it also confers a good deal of specificity since only a few Colombian amphibian poisons, chemically unrelated [97], are known to have toxicity to cell lines at these concentrations. Thus the method has found great use in determina-

TABLE V

GROWTH INHIBITION OF *BACILLUS SUBTILIS* AND HeLa EPITHELIAL CELLS BY SPORIDESMIN AND ITS DERIVATIVES

Metabolite or derivative	*Bacillus subtilis* min. inhibitory conc. (μg/ml)	HeLa cells least toxic dose (ng/ml)
Sporidesmin	80	1.0
Sporidesmin monoacetate	—	4.0
Sporidesmin diacetate	120	6.0
Sporidesmin B	400	3.0
Sporidesmin C acetate	>1000	—
Sporidesmin D	—	>1000
Sporidesmin E	10	0.04
Sporidesmin E diacetate	—	0.8
Sporidesmin G	—	4.0
Sporidesmin H	—	10

A dash in the Table indicates that the analysis was not done.

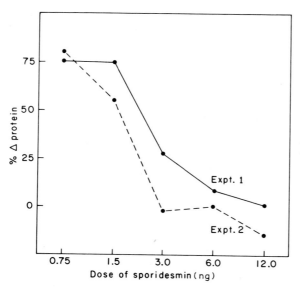

Fig. 8. Response of HeLa cells to sporidesmin doses, by colorimetric measurements of cell protein formed. Doses were added in 1-ml amounts of medium to 24-h cultures which contained initially 50 μg cell protein/ml medium. Measurements of protein formed were made on the 6th day. Control (no dose) cultures contained 178 μg (Expt. 1) and 68 μg (Expt. 2) cell protein. %△ protein = protein formed as a percentage of that of controls.

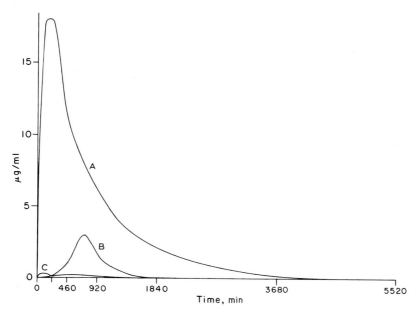

Fig. 9. Excretion of sporidesmin after a single oral dose; A, concentration of sporidesmin in bile; B, in urine; C, in serum.

tions of toxicity of isolates of fungi from pasture (see above) and in following the excretion of sporidesmin after the poison had been administered to experimental animals.

Since disposable tubes, media and cell lines have become commercially available the assay is easy to perform and most of its disadvantages have disappeared.

Excretion of sporidesmin in mammals

The excretion by sheep of a single oral dose of 1 mg of sporidesmin per kg body weight is shown in Fig. 9. The curves in this figure are least mean square fits of the means of the data given by Mortimer and Stanbridge [98] and all have an index of fit >0.5. Using Hodgett's value [99] for the total volume of plasma as 46.7 ml/kg, the results show that 30 min after dosing, 2% of the dose is present in the serum. Between 2 and 4 h after dosing, using the value of 80 ml of bile excreted/h [100], 2.86 mg of sporidesmin, or 14% of the dose, were excreted in bile. These must be regarded as extraordinary figures, showing an amazing ability of the animal to scavenge the rumen contents for this poison and then to effect a remarkably efficient absorption of sporidesmin across the hepatocyte monolayer. The sporidesmin in the bile was isolated and shown to be unconjugated. This work was done before the facile oxidation-reduction of the disulphide groups was known [101] hence sporidesmin may be excreted in the bile as the mono- or di-sodium salt (for example) of the dithiol. One would expect oxidation during its isolation from bulked samples of bile. The maximum concentration of sporidesmin appearing in the urine was found 12 h after dosing (Fig. 9), when it was 10 times that found in serum. Evidently, the normal physiological function of the convoluted tubules was not markedly affected, the void volume of urine being only 0.5% of that secreted by the glomeruli. Although the relative severity of lesions observed in the urinary system and the biliary tree are approximately proportional to the level of the poison found in these excretory ducts, these concentrations are more than 100 times those required to induce cytotoxic effects in epithelial cells in culture.

Very similar results were reported by Towers who studied the distribution of ^{35}S-labelled 'sporidesmin' in female guinea pigs [77]. After 4 days, 16—18% of the administered radioactivity had been excreted in the urine and 22—25% in the faeces. Of the tissues examined, liver, kidney and whole blood showed the highest radioactive levels and, in the latter case, about 90% of the radioactivity was associated with erythrocytes. As Towers also found [78] that 56% of the radioactivity in blood from dosed rats was associated with the red cell fraction, there is the possibility that the level of sporidesmin in serum given in Fig. 9 should be increased by a factor of at least 2 to give the total blood level.

In rats, 42% of the ^{35}S present in the urine after the animals had been dosed with ^{35}S-sporidesmin was precipitated with basic lead acetate at pH 8, suggesting that the ^{35}S was present as a conjugate. This result supports those previously reported by Slater *et al.* [102] who showed that urine glucuronides increased by more than 100% 3 days after sporidesmin administration, despite a marked reduction in the volume of urine excreted. The proportion of sporidesmin apparently excreted as glucuronide in the faeces of dosed rats was similar to that found in their urine.

Thus sporidesmin given orally to sheep, guinea pigs or rats is rapidly absorbed and equally rapidly transported across the hepatocyte monolayer from whence it is excreted in the bile and finally in the faeces. It appears unlikely that is is re-absorbed from the duodenum or small intestine (but see Worker [103,104]).

Toxicity of sporidesmin in mammals

(1) The response of mammals to different doses of sporidesmin

It was apparent at the earliest stages of our investigations [105] that the response of sheep to sporidesmin was acutely dependent on the dose. At 0.1 mg/kg, no signs of intoxication were seen, whilst at 1.0 mg/kg more than 90% of the dosed animals died. In Table VI, the response of sheep to sporidesmin given by various routes and at different dose levels within the range stated above is summarized. In general, many field outbreaks of facial eczema are more severe than the intoxication produced by dosing 0.5 mg/kg, whilst a dose of 1.0 mg/kg produces a response more severe than usually observed in the field [106].

The susceptibility of other mammalian species to sporidesmin varies greatly. Milk-fed calves [107] are slightly less susceptible, a dose of about 3 mg/kg having an effect somewhat similar to that of 1 mg/kg in sheep. Rab-

TABLE VI

SUMMARY OF RESPONSES BY SHEEP TO DIFFERENT DOSES AND ROUTES OF ADMINISTRATION OF SPORIDESMIN

Route	Total spdm. administered (mg/kg)	Number of doses	Interval bwt. doses (h)	Diet	Clinical % photosens.	Effect % died
Oral	0.3	1		G	0	0
Oral	0.3	3	24, 48	G	0	0
Oral	0.9	1		G	83	0
Oral	0.9	3	24	G	83	17 (10)
Intra-abomasum	0.4	2	48	G	50	0
Intra-rumen	0.7	2	12	P	100	0
Oral	0.8	3	24	P	100	10
Subcutaneous	0.4	3	48	G	0	100 (10)
Intravenous	1.0	1		G	0	100 (4)

Values in brackets in the mortality column indicate the longevity of the animals that died. Abbreviations: G, fed indoors on water and dried grass; P, fed on pasture; spdm., sporidesmin.

bits are slightly more susceptible (1—2 mg/kg) [108], and guinea pigs slightly less (2—4 mg/kg). Rats are much more resistant to the poison (20—30 mg/kg) [108] and mice even more so (200—300 mg/kg) [108]. Few other species have been examined. Mortimer reported high mortality in 1-week-old chicks at a dose level of 5—10 mg/kg [108] and some epidemiological studies in man in districts of New Zealand where high spore burdens are common have been reported, but the results were negative [109].

The response of the different sexes to sporidesmin intoxication has only been reported in the case of rats and mice where it has been shown [108,110] that females are much more sensitive; at the same dose rate, male mice survived 4.2 days longer than females and the body weight loss of males was 7.8 g less.

(2) Clinical signs of sporidesmin intoxication in sheep

The first sign of intoxication in penned sheep was a marked reduction in the food consumed. Anorexia developed until, at the 4th day, the consumption of the controls was 4 times that of the dosed. At this time too, diarrhoea was observed in many sheep and this symptom, of course, was the first seen in grazing sheep. Generally speaking, the severity of both these signs was proportional to the dose. After the 4th day, the appetite of the dosed sheep improved as did their demeanour except in animals dosed at 3 mg/kg where diarrhoea and dehydration increased with time until the animals succumbed between the 4th and 8th day after dosing. At 0.3 mg/kg, the animals did not lose weight as compared to the controls, but at doses greater than 0.5 mg/kg, the average body weight of the dosed sheep after 28 days was almost 8 kg less than the control.

10 days after dosing, it was evident that many animals suffered great discomfort in sunlight and sought shade whenever possible. Inflamed, oedematous swellings rapidly developed on the lips, face, eyelids and vulva. The animals were restless, had characteristic drooping ears, and frequently self-inflicted trauma aggravated and complicated the lesions which resembled very closely those seen in field cases of facial eczema. The incidence and time of onset of photosensitization are shown in Table VII.

(3) Morbid anatomy of sporidesmin intoxication

The morbid anatomy of the intoxication process was determined by post-mortem examination of groups of 8 sheep which were killed at intervals from the daty of dosing. 2 days after dosing, the neck of the gall bladder and the cystic duct were oedematous and petechial haemorrhages were infrequently observed in the submucosa of the gall bladder and beneath the epithelium of the urinary bladder. These changes increased in severity through the intoxication period. After 4 days, discolouration of the liver was more obvious and the extrahepatic and some of the larger intrahepatic bile ducts were oedematous. At 7 days, these changes had intensified and some

TABLE VII

MORTALITY OF SHEEP AND THE INCIDENCE OF PHOTOSENSITISATION IN ANIMALS DOSED WITH SPORIDESMIN AT FOUR DIFFERENT RATES

Dose (mg/kg)	Days after administration											
	4		7		10		14		21		24	
	P	M	P	M	P	M	P	M	P	M	P	M
0.3	0	0	0	0	0	0	0	0	0	0	0	0
0.5	0	0	0	0	25	0	33	0	66	0	66	8
1.0	0	0	2.5	2.5	14	0	83	0	81	12.5	—	75
3.0	0	28	0	71	—	100	—	—	—	—	—	—

P, % photosensitized; M, % mortality.

of the bile ducts were plugged. The livers were often fatty and patches of fibrosis were seen. At this stage, the first examples of "white" bile were seen. Icterus was first observed at 10 days; some livers were stained with bile and many intrahepatic ducts were completely obstructed with plugs. At 14 days, 75% of the carcasses were icteric. The liver surfaces were uneven and, in places, shrunken below the surrounding areas; fibrosis was much more marked. Oedema of the hepatic ducts was less noticeable and fibrosis in these ducts was apparent. Colourless bile was observed in 40% of the gall bladders. The ureters were now affected causing distension of the renal pelves with urine. Petechiation and ulceration were sometimes seen in the renal pelves. The adrenal glands were 28% heavier than in the normal sheep, the hypertrophy appeared to be mostly cortical; occasionally, cortical haemorrhages were seen. Between this time (14 days) and the time when the animal succumbed to the intoxication process, these lesions developed progressively. Little evidence was found of liver degeneration. At 24—28 days after dosing the animals were grossly emaciated with little body fat and were also severely dehydrated.

This morbid anatomical chain of events is similar in many respects in intoxicated guinea pigs, rabbits and calves, but in mice and rats, which, in general, require doses that are at least an order of magnitude greater, the pathological changes that occur in the liver are mild and appear to repair rapidly. The difference is probably pharmaceutical since administration of small doses of sporidesmin to mice and rats over a long period induces typical icterus due to inhibition of bile flow and other pathological changes analogous to those reported in the sheep [108]. However, it is important to appreciate that rats and mice succumb to sporidesmin intoxication without these bizarre changes in liver function [102]. 24 h after administration of 5 mg/kg to rats, the animals suffered from diarrhoea due probably to systematically initiated inflammation of the intestine. After 2 days, pronounced adrenal cortical hyperplasia was observed and the level of corticosteroids excreted in the urine was about 3 times normal, the steroids included a low concentration of 11-deoxycorticosterones. A general inflammatory response was perhaps most strikingly demonstrated by following the clearance of Evan's blue after intradermal administration of $2\,\mu g$ of sporidesmin [102,111]. As a result of this increased capillary permeability, the most permanent morbid changes observed after sporidesmin intoxication in mice and rats were ascites and pleural effusions, which associated with the inevitable haemoconcentration, were the major factors causing the death of the animals.

(4) Histology

Details of the pathological changes at a cellular level that occur 2, 4, 7, 10, 14, 21 and 28 days after a single oral dose of sporidesmin to sheep have been reported [112]. Typical lesions in the liver, extrahepatic bile ducts and gall bladder throughout the intoxication process are illustrated in 34 well-reproduced photomicrographs in Mortimer's paper. In addition, some 6 photomicrographs of kidney, ureter and bladder tissue are reproduced as well as a histochemical slide of adrenal cortex taken 21 days after dosing. Obviously, this wealth of detail cannot be reproduced here and the following is a simplified précis of Mortimer's description in which some of the salient facts about the developing lesions are described.

In the early stages of the intoxication (2—4 days), the principal changes in the parenchymal cells of the liver were vacuolation especially adjacent to the portal tracts and the presence of large quantities of lipid in the cytoplasm surrounding, though not in, the vacuoles. In the portal areas, oedematous, inflammatory changes centered round the bile ducts and their associated blood and lymphatic vessels were observed. The severity of the bile duct lesions was proportional to their size. The larger portal tracts contained an inflammatory exudate in the earlier specimens and later, the tunica media of the arterioles of these tracts was necrotic, and similar vascular fibrinoid was noted in the accompanying veins. The cystic ducts were the most severely affected extrahepatic ducts, the epithelium of which was degenerate and necrotic. In the gall bladder, similar, though less severe lesions, were seen; the layers of bundles of the muscle wall were forced apart by a fibrinous exudate, those nearest to the lumen were the most affected and these were sometimes degenerate and contained pyknotic nuclei. Occasionally, some distension of the pancreatic ducts was seen. In about half the animals investigated, degenerative changes in the epithelial cells of the collecting tubules and papillary ducts of the renal medulla were seen. Later vacuolation and the presence of lipid in the cells were observed. The epithelium of the urinary bladder was degenerate, sometimes necrotic and occasionally absent. As in the gall bladder, the muscular layers were separated by oedema and some sub-epithelial capillary loops were ruptured. In 75% of the animals, degenerative cells were found in the adrenal zona reticularis and polymorphonuclear phagocytes were active in most.

In the period 4—10 days after dosing, the vacuolation and lipidosis of the parenchymal cells had largely resolved and their glycogen content returned to normal. Increased numbers of enlarged Kupffer cells were seen at 10 days which sometimes contained bile pigments. In the portal tracts, processes were in train which eventually led to the occlusion of the bile ducts, and the fibroblastic repair developing in the vascular vessels caused varying degrees of occlusion, usually more advanced in the portal veins. These lesions were eccentric, occupying that part of the lumen closest to the bile duct(s) of the tract. The active zone of proliferation was immediately within the internal elastic membrane.

After 14 days, a second phase of deterioration in the parenchymal cells was observed; marked vacuolation and loss of cell integrity, decreased glycogen and stainable lipid, randomly distributed with occasional spicules of material thought to be steroidal in nature, were seen. Large foci of necrosis were present related to the vascular occlusive lesions in the neighbouring portal tracts. In addition, many mitotic figures were seen in the parenchymal cells. The lesions in the portal tracts were severe, sometimes showing extensive necrosis. The epithelium of the original small bile ducts was, at 10 days, abnormal; the swollen acidophilic cytoplasm of these cells reduced the size or completely occluded the lumen of the bile ductules. Occasional thrombi protruding into the portal veins were first seen at this stage of the intoxication. In 25% of the sheep examined 14 days after dosing, lesions were found in the small intestine, ulcerative in nature; the arterioles and veins associated with the lesions having undergone fibrinoid necrosis of the tunica media and thickening of the tunica intima. The lesions in the urinary system were severe; no lipid was found in the cells of the collecting tubules and papillary ducts and some mitotic figures were present.

Tissues were also examined 21 and 24—28 days after administration of sporidesmin. At these times, numerous secondary changes due to renal, cortical and hepatic dysfunction were apparent.

In sheep treated at lower dose rates (0.5 mg/kg), great variability in response was seen, *i.e.*, lesions of diminished severity were not observed in all sheep but rather, the variation was from one animal to another. A few were almost normal and one was as severely affected as the group dosed at 1 mg/kg. Some sheep were examined that had been dosed at 3 mg/kg; here no tissues were taken after the 8th day since all the animals had died by this time. In these animals dosed at 3 mg/kg, lesions, similar to those described above that were seen in the first few days of the intoxication, were observed, but, in general, there was much more necrosis. In one animal, lesions were found in the small intestine below the sphincter of Oddi. As no lesions of this type were reported in the other animals of this group, it is strange that the intestinal submucosa responds with such variability to high concentrations of sporidesmin (see above).

Some histological aspects of the pulmonary lesions in mice [108] and rats [102,110] have also been reported.

(5) Changes in liver function following sporidesmin intoxication

(a) Bromosulphthalein clearance tests

When bromosulphthalein was administered intravenously at a level of 5 mg/kg, the dye was removed from the blood of normal sheep in 20 min [113]. 10 to 14 days after a dose of 1 mg/kg of sporidesmin, the ability of the animal to remove the dye from its plasma was greatly impaired; for 20 min after injection of the dye, the plasma still retained 50% of the bromo-sulphthalein [113]. In the earlier stages of the intoxication process, great variability in the clearance of bromosulphthalein from the blood of different sheep was observed. In some, 50—70% retention of the dye was found 20 min after intravenous injection, 2—7 days after administration of sporides-min. It has been suggested that these results are a further manifestation of the two phases of the intoxication process.

(b) Bile secretion in sporidesmin intoxicated animals

Two studies of the effect of sporidesmin on bile flow in sheep have been reported [86,114]. Leaver [86] collected bile from a gall bladder fistula having ligated the cystic duct close to its junction with the hepatic ducts. He showed that 3 bile ducts joined the cystic duct, draining the dorsal lobe of the liver, between the ligature and the gall bladder. He collected about 12 ml of bile per h, or about 17% of the total bile flow [115], thus the bulk of the bile secreted drained into the duodenum and maintained the enterohepatic circulation of the bile. One to 2 days after dosing at 1 mg/kg, the daily bile flow fell to 25% of normal; the bile was pale yellow and contained less than 10% of the bilirubin present in the bile before intoxica-tion. 4 days after dosing, the flow of bile increased to 50—60% of normal, it was green in colour and contained the normal quantity of bilirubin. Finally after 8 days, the secretion from the fistula was colourless and contained no bilirubin. Presumably, this fluid was an inflammatory exudate secreted by the gall bladder wall equivalent to the white bile reported by Mortimer and Taylor [105]. Very similar results were obtained by Mortimer and Stan-bridge [114] who ligated the cystic duct at the neck of the gall bladder and the common bile duct below its junction with the hepatic and cystic ducts. An external fistula was then installed between the cystic duct and the com-mon bile duct, the latter being ligated anterior to the entry of the canula. In Fig. 10, the two experimental bile collecting systems are shown. The bile collected by Mortimer and Stanbridge thus did not contain any secretions from the gall bladder and this organ was unable to affect the composition or concentration of bile constituents. Thus, as in Leaver's experiments, the bile flowing into the duodenum was abnormal. Mortimer and Stanbridge showed that there was a small reduction in the rate of bile secretion in the 24 h after dosing 1 mg/kg. Thereafter, the flow rate decreased rapidly until after 4 days little secretion occurred. In some sheep, the flow resumed after 6 days but terminated again at about the 12th day after dosing. It may be concluded that the secretion of bile from the liver as a whole behaves similarly to its

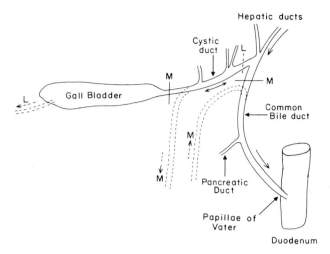

Fig. 10. Diagram of experimental bile excretion collection devices; L, entry and exit fistulas in the system used by Leaver [86]; M, fistulas used by Mortimer and Stanbridge [114].

secretion from the dorsal lobe. On one point of considerable biochemical interest the two groups of workers are not agreed. Thus Mortimer and Stanbridge [114] found no evidence of reduced excretion of bilirubin in the bile of their intoxicated sheep provided that some bile was secreted. They conclude that mechanisms of bilirubin conjugation and transfer across the hepatocyte are not impaired in the preobstructive phase of the disease. By contrast, Leaver [86] reported rapid reduction in the concentration of bilirubin in bile 6 h after dosing whilst appreciable bile flow (>30%) was recorded. Similar results were obtained in rats by Slater and Griffiths [116]. It is possible that these different results are a reflection of experimental physiological situations which are incomparable but if the primary cholangitis is a result of the excretion of sporidesmin (Fig. 9) the reason for the delayed cholestasis reported by Mortimer and Stanbridge is obscure. Nevertheless, these experiments have shed much light on the early hepatic results of sporidesmin toxicity. Thus, in the first few hours after dosing, sporidesmin is excreted in concentrations much greater than those required to induce pyknosis in isolated clones of epithelial cells (but not in continuous layers of such cells [63]); shortly thereafter, a reduction in bile flow occurs and possibly some alteration in the excretory mechanism for bilirubin. Later, these metabolic functions of hepatic cells recover somewhat, but the fibroblastic stenosis of the biliary tree, proceeding in the meantime, prevents a return to normal bile secretion.

(6) Biochemical changes in blood and liver constituents after sporidesmin intoxication

Blood

At the dose levels studied, no changes in the morphology or numbers of erythrocytes have been observed throughout the course of the intoxication in sheep [113]. During experiments with mice, where sporidesmin was fed in daily small doses, a progressive anaemia was observed until at the 43rd day of the experiment the blood of many animals had less than $4 \cdot 10^6$ erythrocytes/ml [108]. In sheep, although no changes were observed [113] in platelet counts and coagulation times during the period of intoxication studied, clot retraction was noticeably less after the 7th day. Of the serum proteins, the albumin concentration was steadily reduced from the onset of the poisoning and γ-globulin levels were markedly increased after the sheep became photosensitised. In the rat [102], a fall in serum proteins has been reported, the extent of which was dependent on the dose of sporidesmin. However, the most striking change in the serum proteins that occurred in jaundiced sheep was the presence of an abnormal lipoprotein that was easily separated by centrifugation at $100\,000$ g and $d = 1.21$. In normal sheep serum, very little lipoprotein is separated under these conditions. The protein was thermolabile, *i.e.*, when the serum was frozen and then thawed, the clear fresh serum became grossly turbid, and centrifugation, at low speed, resulted in a layer of lipophilic material on the surface [117]. It is not known whether this unusual lipoprotein is analogous to that reported by Rimington *et al.* [110], in electrophoretograms of serum from rats dosed with sporidesmin. Other serum proteins, especially those generally regarded as intracellular enzymes, are also present in high concentration in the serum of icteric sheep. The one studied most was glutamic-oxaloacetic transaminase, chiefly because of highly interesting changes in this serum constituent 2—5 days after intoxication [117]. At this time, the levels of this enzyme in individual sheep were very variable, there being normal levels in some animals and greatly elevated ones in others. No explanation for this variability is available. Like sheep that are photosensitive, all animals dosed at 3 mg/kg had very high levels of this enzyme [117].

The observations described above concerning the thermolability of icteric serum from sporidesmin-dosed animals led us to examine the lipophilic phase from lipoproteins that had been thermally denatured. This lipid gave strong positive tests for steroids, and it therefore seemed probable that determination of this group of compounds in intoxicated sheep might be indicative of the severity of the condition. The lipid from the thermolabile lipoproteins also contained high concentrations of bilirubin and phosphorus. In passing, it is noted that icteric sheep with facial eczema also provided serum samples that contained thermolabile lipoproteins of similar composition to those isolated from sera collected from animals with the experimental disease [106]. When this work commenced, however, the normal values for serum steroids, lipid phosphorus and bilirubin in sheep were not known. It was found that methods for the determination of bilirubin and lipid phosphorus in human pathological sera could easily be applied to sheep serum. However, the determination of steroids in sheep serum was difficult because their concentration was low and they contained a high proportion of those compounds which give high values in the Libermann-Burchard colour reaction [117]. These problems were overcome to some extent by using Henly's [118] modification of the Zlatkis method [119] where, *e.g.*, cholestra-5,7-diene-3β-ol gives the same extinction coefficient in the colour reaction as cholesterol. These investigations allowed us to show that the serum steroids were elevated from about the 10th day after intoxication. It was also shown that the lipid phosphorus and bilirubin levels behaved similarly. In addition, some of the animals that were dosed at 3 mg/kg exhibited a reversible bilirubinaemia shortly after the administration of the toxin. The proportion of conjugated bilirubin found in the serum in the obstructive phase of the disease was not determined, but in cases of facial eczema [106] more than 60% of the bilirubin was conjugated, the proportion being independent of the bilirubin concentration in serum. Although the concentration of serum steroids does not increase until the obstructive phase of the intoxication, analysis of the serum steroids revealed important changes in composition at an earlier stage [117,120]. Approximately 7 days after dosing, there was a small increase in the proportion of steroids present in the serum that were not esterified by long-chain fatty acids, and a large increase in the levels of cholic and deoxycholic acids. As the poisoning develops, the proportion of unesterified steroid and cholic acid increases but the deoxycholic acid values decrease [120]. These bile acids are probably present as their taurine esters. Somewhat similar changes occur in the sporidesmin-poisoned rabbit [121] but also with important differences. In sheep, opalescent serum was never seen in the early stages of the condition but in rabbits, serum collected more than 48 h after dosing was frequently opalescent and this has been shown to be due to increased triglyceride and steroid ester components of the serum [121]. Serum cholesterol levels have also been reported in the sporidesmin-intoxicated rat, but consistent increases were not observed even in animals that were seriously ill; the standard deviation of the results from the intoxicated animals was, however, very much greater than normal [102].

Liver

The presence of lipophilic bodies in frozen sections of the livers of sheep 48 h after administration of sporidesmin led Mortimer *et al.* [122] to extract the lipids present in such livers and to compare them with similar extracts taken from normal sheep liver. They found that the lipid extracted from unit weight of liver tissue of the dosed sheep was greater than that from the controls throughout the experimental period, but the greatest increase was found in the period between 2 and 7 days after dosing — in excellent agreement with the histological results reported above. The lipid accumulating in the liver at this time was not phospholipid or cholesterol, but was partly accounted for by the steroid ester fraction. The phenomenon was later studied in detail [123] and it was found that the triglyceride fraction was principally responsible for the lipid accumulation. Very similar results have also been reported in the case of rabbit liver [121].

This accumulation of triglyceride in hepatic cells is a phenomenon widely recognised as a consequence of chemical hepatotoxicity [124,125], although few substances exert such a toxic effect at such low concentrations. It has been suggested that one effect of carbon tetrachloride on hepatic cells is to inhibit the synthesis of the protein that carries the triglyceride synthesised in the liver. Many compounds, including sporidesmin, which have an epidithiodioxopiperazine group, are specific inhibitors of the synthesis of viral RNA by host cells [101]. The concentration of these compounds which exerts this inhibitory effect is of the same order as that of sporidesmin which inhibits protein synthesis in isolated clones of, for example, HeLa cells [63]. The synthesis of virus by host cells is a rapid process and so is the synthesis of the carrier protein of triglycerides by hepatic cells [126,127]. It is, there- fore, attractive to envisage an inhibition, by sporidesmin, of the synthesis of RNA in the hepatic cells that codes for the synthesis of the carrier proteins. However, no experimental evidence is available on this point. The difference in the results of serum lipid analysis in the sheep and rabbit also indicate an effect of sporidesmin on primary energy production. Thus no accumulation of triglyceride occurs in sheep serum, presumably because in ruminants lower molecular weight fatty acids, produced by the rumen microflora, serve as an energy source. Such an accumulation does occur in the rabbit where volatile fatty acids are of much less importance. It can, therefore, be deduced that the transport of glycerides across hepatic cell membranes in rabbits is inhibited, both from the portal veins and to the bile canaliculi. Unfortunately, no analyses of portal venous blood for changes in volatile fatty acid levels after sporidesmin intoxication have been reported.

Control of facial eczema

The facts presented in the preceding sections of this chapter establish several similarities between sporidesmin intoxication and the disease facial eczema. However, like most theories, that of the role of *Pithomyces char- tarum* in the disease stands or falls on its utility; in particular, its use to stimulate further work designed to find methods of control of the disease. Assuming that the theory is true, is it possible to inhibit the growth of the fungus in the field and, having done so, is the incidence of facial eczema reduced? Three different ways have been attempted and all show some degree of promise.

For many years, experience has shown that the disease occurs in late summer and autumn, particularly after a period of warm rainfall. Thus, many years ago, a weather forecasting system was inaugurated which warned farmers in districts where the disease was common of the imminence of rain, thus allowing them to herd their flocks and feed them with fodder known to

be safe [5,128]. These forecasts were supplemented after 1960 with estimates of the density of populations of spores of *P. chartarum* that were
determined throughout the North Island of New Zealand, and, in some
districts, these efforts greatly improved the forecasting of dangerous periods
[129]. However, these methods assume a relationship between spore numbers and toxicity; that the fungus will produce the same 'spectrum' of toxic
metabolites on all substrates and that a small increase in spore numbers is
likely to progress to a dense population. These assumptions render predic-
tion hazardous and it is not surprising that the forecasts have had less than
unanimous response from farmers.

A second approach, of great theoretical interest, is the selective control
of the growth of *P. chartarum* in the field. This is attractive in a number of
ways because the dose—response studies of sporidesmin intoxication show
that quite a small decrease in dose results in a large decrease in the severity
of the disease. Thus a screening programme was initiated early in the
research on *P. chartarum* for substances which inhibited its growth and one
or two active compounds (*e.g.* captan) were tested in the field but without
any decrease in the population densities of the fungus. Some success was
later achieved [130] by spraying pastures with emulsions of C_{10}—C_{12} fatty
acids, but in 1966 on the suggestion of a veterinarian at Gisbourne (Mr. M.W.
Howe) the common anthelmintic 2-(thiazolyl-2)benzimidazole ("thiabendazole") was tried with immediate success. It was shown [131] that application of this compound at 2 lb/acre not only induced a drastic reduction in
the population density of *P. chartarum* spores in the field, but also that
sheep grazing the treated pastures did not develop facial eczema. This discovery has resulted in a great deal of work designed to find the optimum
conditions of application of the fungicide and also whether small changes in
its structure result in compounds with better properties. One important
property of thiabendazole is its rapid binding to spores of the fungus [132].
This not only enables the fungicide to be applied sparingly but also is
probably responsible for its prolonged effect. It is now known that 140
g/hectare gives appreciable protection to stock for extended periods. The
effect of spraying the fungicide is also rapid and it has been found that 24 h
after spraying the pasture is safe. Due to the prolonged effect of thiabendazole, the importance of forecasting dangerous periods of the season is
much reduced but the main disadvantage of the technique is one of cost. It
is, however, likely that further experience and the inflation of world meat
prices will reduce the importance of this factor considerably.

The third control method stems from an observation of Hove and
Wright [133] that a high protein diet had some protective effect on rats
challenged with sporidesmin. The application of this result to sheep dosed
with sporidesmin and to sheep grazing pasture infested with *P. chartarum* has
not been published except in the annual reports of the Ruakura Agricultural
Research Center [134]. In general, it has been found that high-protein sheep
pellets give more protection than white clover-rye grass pasture which in turn
is better than low-protein maize silage or chou moellier. It is clear that this

discovery is important, especially if it results in work designed to find pasture plants which tend to inhibit the growth of the fungus, and in addition, provide a diet to the grazing animal which confers on that animal some degree of resistance to the disease.

The success of these experiments to control facial eczema establishes not only a role for *P. chartarum* in diseases of grazing animals but also a new area of agricultural research, since the part played by the fungal flora in the complex sequence of biological events which are involved in the conversion of pasture plants to useful food and fibre is now seen to be important. It seems likely that facial eczema is the first and perhaps most bizarre example of fungal inhibiting mechanisms which plague the efficient production of farm animals.

Acknowledgements

We thank our colleagues, Miss C. Bird, Drs. F.J. Simpson and R.G. Stevenson, who have read the typescript and made many helpful suggestions. The following have been especially helpful by sending us unpublished information: J.M. Dingley, E.E. Fisher, J.D. Smith, H.P. Upadhyay, C.O. Gourley, W.F.O. Marasas, S.J. Hughes, M. Pantidou, L.E. Henrikkson, L. Holm and M. Przybylski. We also thank Dr. M. Corlett for electron scanning microscopy, Mr. W.S. Bertaud for Fig. 3 and Dr. J.M. Hirst for Fig. 4. We wish to thank Dr. J.A. Parmelee, Curator, Mycological Herbarium, Plant Research Institute, Ottawa, for the loan of specimens.

References

1 C.R.W. Spedding, Sheep Production and Grazing Management, Baillière, Tindall and Cassell, London, (1970) 140.
2 D. Brewer, F.W. Calder, T.M. MacIntyre and A. Taylor, J. Agric. Sci., Camb., 76 (1971) 465.
3 H.R. Marston, Proc. Roy. Soc., B, 137 (1950) 18.
4 J.G. Gilruth, N.Z. Dept. Agric. Ann. Rept., 16 (1908) 189.
5 J. Filmer, N.Z. J. Agric., 97 (1958) 202.
6 Pl. A. Plattner and U. Nager, Helv. Chim. Acta, 31 (1948) 665, 2192, 2203.
7 E.P. White, N.Z. J. Agric. Res., 1 (1958) 433, 859.
8 R.H. Thornton and J.C. Percival, Nature, 183 (1959) 63.
9 J.C. Percival and R.H. Thornton, Nature, 182 (1958) 1095.
10 P. Sydow, Ann. Mycol., Berl., 12 (1914) 195.
11 M.B. Ellis, Mycol. Papers, 76 (1960) 1.
12 S.J. Hughes, Can. J. Bot., 36 (1958) 727.
13 R.T. Moore, Rhodora, 61 (1959) 87.
14 M.J. Berkeley, Grevillea, 3 (1874) 50.
15 P.A. Saccardo, Michelia, 1 (1879) 478.
16 S.J. Hughes, Mycol. Papers, 50 (1953) 1.
17 J.M. Dingley, N.Z. J. Agr. Res., 5 (1962) 49.
18 W.F.O. Marasas and I.H. Schumann, Bothalia, 10 (1972) 509.
19 W.S. Bertaud, I.M. Morice, D.W. Russell and A. Taylor, J. Gen. Microbiol., 32 (1963) 385.

20 E. Bishop, H. Griffiths, D.W. Russell, V. Ward and R.N. Gartside, J. Gen. Microbiol., 38 (1965) 289.
21 A. Taylor, Advan. Appl. Microbiol., 12 (1970) 189.
22 E. Bishop and D.W. Russell, J. Chem. Soc. (C), (1967) 634.
23 B.S. Janes, Nature, 184 (1959) 1327.
24 D.E. Hore, Aust. Vet. J., 36 (1960) 172.
25 R.G. Rees, Aust. J. Bot., 12 (1964) 185.
26 R.J. Lamb and J.F. Brown, Trans. Brit. Mycol. Soc., 55 (1970) 383.
27 W.A. Shipten and S.C. Chambers, Aust. J. Exptl. Agric. Anim. Husb., 6 (1966) 432.
28 S.A. Narayanan, Sydowia, 16 (1962) 77.
29 V.P. Sahni, Mycopathol. Mycol. Appl., 29 (1966) 226.
30 J. Kranz, Sydowia, 19 (1965) 92.
31 R.A. Taber, R.E. Pettit, W.A. Taber and J.W. Dollahite, Mycologia, 60 (1968) 727.
32 S. Nemec, Mycopathol. Mycol. Appl., 41 (1970) 331.
33 R.D. Goos, Mycologia, 55 (1963) 142.
34 E.F. Morris, Mycologia, 48 (1956) 728.
35 D.S. Meredith, Ann. Appl. Biol., 50 (1962) 577.
36 G.L. Barron, The Genera of Hyphomycetes from soil, Williams and Wilkins, Baltimore, 1968.
37 J.D. Smith, personal communication.
38 C. Gourley, personal communication.
39 M.E. Lacey and P.H. Gregory, Nature, 193 (1962) 85.
40 R.G. Pawsey, Trans. Brit. Mycol. Soc., 47 (1964) 357.
41 S.I. Udagawa, M. Inchoe and H. Kurata, in M. Hertsberg (Ed.) Proc. First U.S.–Japan Conference on Toxic Microorganisms, Unnumbered Publ. U.S. Dept. of Interior and U.S.N.R. Panels on Toxic Microorganisms, Washington, D.C., 1970, p. 174.
42 J.H. Warcup, Trans. Brit. Mycol. Soc., 35 (1957) 248.
43 R.H. Thornton and D.P. Sinclair, N.Z. J. Agric. Res., 3 (1960) 300.
44 A.A. Anderson, J. Bacteriol., 76 (1958) 471.
45 J.W. Sayer, B.S. Dudley and J. Ghosseiri, J. Allergy, 44 (1969) 214.
46 S.M. Pady, C.L. Kramer and B.J. Wiley, Mycologia, 54 (1962) 168.
47 C.L. Kramer, S.M. Pady and B.J. Wiley, Mycologica, 55 (1963) 380.
48 D. Frey and E.B. Durie, Mycopathol. Mycol. Appl., 16 (1962) 295.
49 G.F. Harsh and S.E. Allen, J. Allergy, 16 (1945) 125.
50 A.M. Targow and O.A. Plunkett, Ann. Allergy, 9 (1951) 428.
51 J.D. Panzer, E.C. Tullis and E.P. van Arsdel, Phytopathology, 47 (1957) 512.
52 J.M. Hirst, Ann. Appl. Biol., 39 (1952) 257.
53 R.K. May, J. Sci. Instrum., 22 (1945) 187.
54 S.M. Pady, Phytopathology, 49 (1959) 757.
55 P.H. Gregory, Trans. Brit. Mycol. Soc., 37 (1954) 390.
56 P.J. Brook, N.Z. J. Agric. Res., 2 (1959) 690.
57 J.D. Smith, W.E. Crawley and F.T. Lees, N.Z. J. Agric. Res., 4 (1961) 538.
58 P.J. Brook, N.Z. J. Agric. Res., 6 (1963) 147.
59 P.J. Brook, N.Z. J. Agric. Res., 7 (1964) 87.
60 E.E. Fisher, J. Dept. Agric., Victoria, February (1962) 10.
61 M.E. di Menna and J.N. Parle, N.Z. J. Agric. Res., 13 (1970) 51.
62 D.J. Ross., N.Z. J. Sci., 3 (1960) 15.
63 J. Done, P.H. Mortimer, A. Taylor and D.W. Russell, J. Gen. Microbiol., 26 (1961) 207.
64 J.M. Dingley, J. Done, A. Taylor and D.W. Russell, J. Gen. Microbiol., 29 (1962) 127.
65 M.E. di Menna, J. Campbell and P.H. Mortimer, J. Gen. Microbiol., 61 (1970) 87.
66 D.W. Russell, R.J. Sturgeon and V. Ward, J. Gen. Microbiol., 36 (1964) 289.
67 D.J. Ross, N.Z. J. Sci., 3 (1960) 441.
68 D.W. Russell, J. Chem. Soc., (1962) 753.
69 G.W. Butler, D.W. Russell and R.T.J. Clarke, Biochim. Biophys. Acta, 58 (1962) 507.

70 L. Hartman, J.C. Hawke, I.M. Morice and F.B. Shorland, Biochem. J. 75 (1960) 274.

71 L. Hartman, I.M. Morice and F.B. Shorland, Biochem. J., 82 (1962) 76.

72 D.J. Ross and R.H. Thornton, N.Z. J. Sci., 5 (1962) 165.

73 D.D. Perrin, N.Z. J. Agric. Res., 2 (1959) 266.

74 R. Hodges, J.W. Ronaldson, A. Taylor and E.P. White, Chem. Ind., (1963) 42.

75 N.R. Towers and D.E. Wright, N.Z. J. Agric. Res., 12 (1969) 275.

76 R.L.M. Synge and E.P. White, N.Z. J. Agric. Res., 3 (1960) 907.

77 N.R. Towers, N.Z. J. Agric. Res., 13 (1970) 182.

78 N.R. Towers, N.Z. J. Agric. Res., 13 (1970) 428.

79 J.D. Bu'Lock and A.P. Ryles, Chem. Commun., (1970) 1404.

80 S. Safe and A. Taylor, J. Chem. Soc., (Perkin I), (1972) 472.

81 W.D. Jamieson, R. Rahman and A. Taylor, J. Chem. Soc. (C), (1969) 1564.

82 P.J. Brook and R.E.F. Matthews, N.Z. J. Sci., 3 (1960) 591.

83 R. Rahman, S. Safe and A. Taylor, J. Chem. Soc. (C), (1969) 1665.

84 E. Francis, R. Rahman, S. Safe and A. Taylor, J. Chem. Soc. (Perkin I), (1972) 470.

85 S. Safe and A. Taylor, J. Chem. Soc. (C), (1970) 432.

86 D.D. Leaver, Res. Vet. Sci., 9 (1968) 255.

87 G.R. Russell, Nature, 186 (1960) 788.

88 R. Rahman, S. Safe and A. Taylor, J. Chromatog., 53 (1970) 592.

89 N.T. Clare, Microbiological Aspects of Facial Eczema, N.Z. Dept. Sci. and Ind. Res., Information Series No. 37, 1964, p. 15.

90 W.E. Dahl and H.L. Pardue, Anal. Chem., 37 (1965) 1382.

91 D. Brewer and A. Taylor, Can. J. Microbiol., 13 (1967) 1577.

92 J. Marbrook, N.Z. J. Agric. Res., 7 (1964) 596.

93 D. Brewer, D.E. Hannah and A. Taylor, Can. J. Microbiol., 12 (1966) 1187.

94 D.D. Perrin, N.Z. J. Sci. Technol., A38 (1957) 669.

95 L.B. Fastier, N.Z. J. Agric. Res., 4 (1961) 72.

96 P.H. Mortimer and B.S. Collins, Res. Vet. Sci., 9 (1968) 136.

97 J.W. Daly, D.M. Jerina and B. Witkop, Experientia, 28 (1972) 1129.

98 P.H. Mortimer and T.A. Stanbridge, J. Comp. Pathol., 78 (1968) 505.

99 V.E. Hodgetts, Aust. J. Exptl. Biol. Med. Sci., 39 (1961) 187.

100 T.A. Stanbridge and P.H. Mortimer, J. Comp. Pathol., 78 (1968) 499.

101 P.W. Trown, Biochem. Biophys. Res. Commun., 33 (1968) 402.

102 T.F. Slater, U.D. Sträuli and B. Sawyer, Res. Vet. Sci., 5 (1964) 450.

103 N.A. Worker, Nature, 185 (1960) 909.

104 N.A. Worker and D.C. Dodd, N.Z. J. Agric. Res., 3 (1960) 712.

105 P.H. Mortimer and A. Taylor, Res. Vet. Sci., 3 (1962) 147.

106 J. Done, P.H. Mortimer and A. Taylor, Res. Vet. Sci., 1 (1960) 76.

107 P.H. Mortimer, Personal communication.

108 P.H. Mortimer, N.Z. J. Agric. Res., 13 (1970) 437.

109 H.J.H. Hiddlestone, N.Z. Med. J., 60 (1961) 24.

110 C. Rimington, T.F. Slater, W.G. Spector, U. Sträuli and D.A. Willoughby, Nature, 194 (1962) 1152.

111 W.G. Spector, J. Pathol. Bacteriol, 63 (1951) 93.

112 P.H. Mortimer, Res. Vet. Sci., 4 (1963) 166.

113 P.H. Mortimer, Res. Vet. Sci., 3 (1962) 269.

114 P.H. Mortimer and T.A. Stanbridge, J. Comp. Pathol., 79 (1969) 267.

115 I.W. Caple and T.J. Heath, J. Comp. Pathol., 81 (1971) 411.

116 T.F. Slater and D.B. Griffiths, Biochem. J., 88 (1963) 60p.

117 J. Done, P.H. Mortimer and A. Taylor, Res. Vet. Sci., 3 (1962) 161.

118 A.A. Henly, Analyst, 82 (1957) 286.

119 A. Zlatkis, B. Zak and A.J. Boyle, J. Lab. Clin. Med., 41 (1953) 486.

120 J.A. Peters and P.H. Mortimer, Res. Vet. Sci., 11 (1970) 183.

121 J.A. Peters, Nature, 210 (1966) 610.

122 P.H. Mortimer, A. Taylor and F.B. Shorland, Nature, 194 (1962) 550.

123 J.A. Peters and L.M. Smith, Biochem. J., 92 (1964) 379.
124 M.C. Schotz and R.O. Recknagel, Biochim. Biophys. Acta 41 (1960) 151.
125 P.M. Harris and D.S. Robinson, Biochem. J., 80 (1961) 352.
126 V.P. Dole, J. Clin. Invest., 35 (1956) 150.
127 V.P. Dole, A.T. James, J.P.W. Webb, M.A. Rizack and M.F. Sturman, J. Clin. Invest., 38 (1959) 1544.
128 K.J. Mitchell, T.O. Walshie and N.G. Robertson, N.Z. J. Agric. Res., 2 (1959) 584.
129 P.J. Brook and G.V. Mutch, N.Z. J. Agric. Res., 7 (1964) 138.
130 R.H. Thornton and W.B. Taylor, N.Z. J. Agric. Res., 6 (1963) 329.
131 D.P. Sinclair, N.Z. J. Agric., 114 (1967) 23.
132 F.J. Stutzenberger and J.N. Parle, J. Gen. Microbiol., 73 (1972) 85.
133 E.L. Hove and D.E. Wright, Life Sci., 8 (1969) 545.
134 Annual Report of Research Division, N.Z. Dept. of Agriculture., 1970—1971, p. 16, 64.
135 W.A. Te Punga and M.M. McKinnon (personal communication).

Chapter 3

CLAVICEPS PURPUREA — ERGOTISM

S.J. VAN RENSBURG and B. ALTENKIRK

National Research Institute for Nutritional Diseases, Tiervlei, C.P. (Republic of South Africa)

Knowledge of the unique properties of ergot is so old that complete understanding of historical studies is plagued by semantic difficulties. Thus, without any degree of certainty, scholars have claimed that the ancient Chinese knew and used ergot for obstetrical purposes some 5000 years ago. The ancient Greek physician Hippokrates (460—357 B. C.) recommended coarsely powdered barley flour cooked in water to further childbirth. Since barley alone could not be expected to have much effect, it is assumed that the barley was ergotized. Furthermore, the drug Hippokrates invariably mentioned when dealing with the uterus or uterine conditions was "melanthium". The exact meaning of the word "melanthium" is controversial, since the only descriptions of the drug were "arising out of grain" and "to sort out of grain".

Ancient Romans were acutely aware of blighted and spoiled grain and from 700 B. C. observed a religious wine festival to the god Robigus who guarded the planted grain fields. The festival was celebrated on April 24 because this was the time that some kernels started to turn black. Julius Caesar ascribed epidemics to "spoiled" grains. The Roman naturalist Plinius who perished in Pompeii during the eruption in 79 A. D. left a reasonably accurate morphological description of ergot, together with environmental conditions for its development and some associations with toxicity.

There is little doubt that ancient Arabs knew ergot and recognized it as a powerful poison at least 1000 years ago. They described it as occurring in two forms, one black with the shape of small horns and the other with a stalk-like shape like a clove which was blackish tinged with yellow.

The identification of widespread epidemics in ancient times is surprisingly difficult due to the vague terminology used in describing symptoms and the prevalence of infectious diseases. The earliest description of an epidemic which could conceivably have been caused by ergot occurred amongst the Spartans when they were at war with the Athenians in the year 430 B. C. In the year 857 on the lower Rhine in Germany an epidemic characterized by necrosis of the limbs and death was recorded. Good descriptions of the outbreak of fire plague in Paris during 945 are strongly suggestive of ergotism. Innumerable epidemics throughout Europe but mainly in France were reported during the ensuing 600 years. Often they were accompanied by war and famine, up to 40 000 people succumbing under conditions of dreadful suffering. Somewhat bizarre descriptions usually mention the screams of pain, the unbearable stench of rotting flesh, detachment of limbs and death or even, at times, miraculous cures. Desperate survivors at times even resorted to cannibalism.

The influential position of the church on the Middle Ages resulted, as a natural consequence, in many invocations to the Saints for help. St. Anthony, the founder of monastic life, emerged as the most important. His remains are buried in a country church at Dauphiné and many afflicted pilgrims who had journeyed there reported cures. So it was that a wealthy noble named Gastron vowed before St. Anthony that if his son was cured he would donate all his wealth to the church to aid other victims. The donation led to the foundation of a hospital brotherhood which was a major advance in the evolution of hospital care. To this day, the disease is commonly known as St. Anthony's Fire, where the holy fire (post-ischaemic inflammation) consumes the limbs which then turn black (necrosis, gangrene) before detachment.

Strangely there are no accurate descriptions which prove the existence of ergot in Europe, let alone its prevalence, during the great epidemics of the Middle Ages. Although it was mentioned and undoubtedly used as a herbal medicine in Europe during the 16th century, the first known published illustration dates to the middle of the 17th century. Some women consumed three sclerotia to increase contractions during childbirth, yet only an isolated few had vague suspicions about its association with bread poisoning. Not until the Parisian lawyer and physician, Denis Dodart, became interested in botany and presented a classical paper to the French Royal Academy of Science in 1673, did the aetiology of "holy fire" become firmly established. Nevertheless, 100 years were to pass before there was any significant reduction of epidemics.

The precise botanical identification of ergot was controversial for two centuries after Dodart incriminated it aetiologically. Prior to Dodart's paper, some considered it a plant separate and foreign to rye, although the belief that it arose as a result of metamorphosis of a kernel of grain was generally accepted. Excessive rain and sunshine (when mildew growth was plentiful) overnourished the grain resulting in individual kernels which hypertrophied and changed colour — from dark, almost black to purple, brown or straw coloured. Each hypertrophied "kernel", which was actually a single sclerotium, was a slightly curved structure 1.5 to 6 cm in length which bore a striking resemblance to a cockspur, which is *ergot* in French. Slowly evidence accumulated that ergot was a fungus, with much confusion being caused by its three stages until Tulasne elucidated the basic elements of the life cycle in 1853. The various stages were now classified as a single fungus and named *Claviceps purpurea*. It was soon realized that ergot sclerotia were not always due to *Claviceps purpurea* alone and today about 50 other species

of Claviceps have been documented. Similarly, rye was not
and many hundreds of species of the grass family, which su
essential cereal and forage crops, have been incriminated.

The isolation and identificatien of the constituents o
cupied chemists for more than two centuries and the task is au y
plete. Initially, the object was to see if there was any difference between the
chemical composition of ergot and rye. Ergot was shown to contain 25 to
35% oils, and no gluten as did rye. Various minerals, a violet, a yellowish-red
and a yellow pigment were isolated more than 150 years ago, yet the first
clinically useful active extract, prepared by purifying an aqueous extract
with alcohol, only came into use 50 years later. Slowly evidence for the
presence of alkaloids accumulated culminating in Tannet's isolation of pure
but inactive "ergotinine" crystals in 1875, which was only recently shown to
be ergocristinine. He also isolated an active substance which would not crystallize, and termed it ergotinine amorphe. It was only in 1906 that Kraft, a
Swiss country pharmacist, showed that the amorphous substance was the
hydrate of the crystalline ergotinine, and thus at last a pure active compound, which Kraft named hydroergotinine, was known which was a model
for future chemists. At about the turn of the century several other important
constituents of ergot including pigments, ergochrysin, secalonic acid and
ergoxanthein, a glycoside, ergothioneine and even acetylcholine were isolated. Most eventful of all was the year 1918 when the brilliant young
Arthur Stoll isolated crystals of ergotamine, a highly active drug which,
because of its constant activity, revolutionized the treatment of post-partum
haemorrhage.

These and other aspects of ergotism have been the subject of several
reviews [1—7].

Outbreaks of disease attributable to the fungus

The exceedingly distressing and variable symptoms of ergotism lead to
rather bizarre early descriptions of the disease, somewhat biased by the fear
of holy fire (gangrene) and bewitchment (nervous symptoms). Accurate and
meticulous descriptions were written during the later Middle Ages, and then
accuracy was verified by comparison with modern outbreaks and isolated
instances of ergotism, usually as a result of attempts to induce abortion.

Necrotic or gangrenous symptoms seemed to predominate during the
first 38 epidemics described during the Middle Ages. Symptoms began with
lassitude, sometimes a cold or prickling sensation in the limbs, followed by
severe muscular pains, particularly in the calf. Intellect was dulled though
the appetite and pulse remained normal during the early stages. Within a few
weeks the affected limbs became swollen and inflamed; this change was
accompanied by violent burning pains and sensations of intense heat alter-
nated with those of icy coldness. Gradually the affected parts became
numbed and the pains sometimes ceased suddenly. The skin was cold, livid
and wrinkled, but sometimes became covered with red or violet vesicles.

Unaffected parts of the skin, in particular the face and whites of the eyes were often yellow, indicating that the patient was suffering from jaundice. Feet and legs were more commonly affected and eventually turned a charcoal black. The gangrenous part shrank and became mummified. Gradually the lesion spread upwards and sometimes moist gangrene intervened. The severity of peripheral necrosis varied from the mere shedding of nails and the loss of fingers or toes to the loss of all four limbs. Most frequently the feet were affected. In severe cases violent pains for 24 h were the only premonitory signs followed by the rapid onset of gangrene.

Convulsive ergotism was characterized by predominant symptoms of the involvement of the nervous system. Usually there was no gangrene but occasionally mixed outbreaks have been described where both gangrene and nervous symptoms occur in the same or different patients. The nervous form has been given numerous names — sometimes implicating the cause or region — most frequently associated with a prominent symptom such as formication (described as a tingling sensation as if ants or mice were crawling under the skin). In addition, itching, numbness of hands and feet, twitching, muscular cramps and sustained spasms and convulsions were described. The terrible pain associated with this syndrome was often described in a graphic manner. Typically the fists were clenched, hands were in acute flexion and the feet were extended. Depending on the involvement of extensors or flexors, the body was "rolled in a ball" or extended as in tetanus. Sudden acute convulsions followed by extended periods of relaxed drowsiness caused confusion with epilepsy. Between convulsions a voracious appetite and insomnia were frequent.

Patients sometimes succumbed after a few convulsions within hours of the onset of symptoms, but usually they died after some weeks. The only significant post mortem findings were bleeding and softening of the brain and lesions of the posterior horns of the spinal cord.

In severe non-fatal cases the disease lasted up to 2 months and full mental recovery was rare. The patients appeared to become increasingly sensitive to ergot and relapses were frequent. Mortality varied between 11 and 60%. Later authors described a milder form in which all the inhabitants of a village could be affected during an epidemic. Mild formication, numbness, fatigue, giddiness, chest discomfort and sometimes mild diarrhoea and vomiting lasted for several weeks.

Although ergot potentiates uterine contractions in the term sensitized uterus, there is no evidence that it is truly abortifacient. The isolated instances of abortion during gangrenous epidemics are probably no more frequent than the incidence following any serious disease and certainly there is no evidence that the chronic convulsive type ever produced abortion. Similarly attempts to procure abortion are frustratingly unsuccessful, not rarely leading to increased dosage with tragic consequences.

Many patients gave birth to living children at term and the disease was never communicated to breastfed infants. During an epidemic weaning may be followed by an acute onset. In convulsive ergotism particularly there seems little doubt that children generally are more susceptible. No conclusive evidence of any sex differences in the susceptibility of children or adults exists. Both forms of ergotism do produce amenorrhea but there is little evidence that lactation is significantly suppressed in the human.

Early epidemics cannot be identified with great certainty since descriptions of symptoms were rather vague and indeed unequivocal proof of the existence of ergot only dates back some 500 years. In the Middle Ages, between the year 837 up to 1347, some 50 epidemics were recorded in central Europe. They occurred mainly in France and all were of the gangrenous type.

Either the chronicles were quiescent or possibly few epidemics occurred during the 14th, 15th and 16th centuries. Only vague references to the disease were made during this period, some in the form of frescoes in chapels. The recrudescence of ergotism was characterized by the emergence of the convulsive form — 65 epidemics of this type occurred from 1581 to 1889. Of these, 29 were in Germany and the remainder occurred in Russia, Sweden, Italy, Finland, Holland, England, Switzerland, Norway, Hungary and the United States. The gangrenous form again occurred in France and although there were mixed epidemics, none in this country were of the purely convulsive type. Great epidemics raged in France and Germany, particularly from 1770 to 1780; up to 8000 people are said to have died in a single district during such epidemics. The outbreaks generated much scientific and government interest; Dodart's clear exposition of the aetiology a century before had not gained universal acceptance but was now confirmed. The great famine of 1770 which precipitated the epidemics emphasized the importance of supplementary crops and consequently potatoes and maize became staples in the diet in addition to ryebread. The ultimate result was that no further great epidemics occurred in civilized countries.

Well marked outbreaks and smaller epidemics continued during the 19th century throughout Europe, starting with the severe winter of 1812—13 which defeated Napoleon in Russia. The last gangrenous epidemic occurred in southern France in 1855 but the convulsive form persisted in other countries. The last German epidemic of considerable extent was predicted in the local paper during a wet summer and when the first cases appeared contaminated grain was confiscated and the millers were fined. Outbreaks continued particularly in northern and eastern Europe well into the 20th century, the last being recorded in Russia and England in 1928. Cases of ergotism are now generally only seen after attempts to induce abortion or after drug overdosage, although history has shown that it may well reappear after long quiescent periods. Idiopathic peripheral limb gangrene is still a well recognized clinical entity in Central Africa and has been observed in pregnant women after the ingestion of herbal concoctions of unknown composition to facilitate labour [8,9]. Some patients show syndromes with striking clinical similarities to ergotism.

The persistence of ergotism for centuries after its cause was known is understandable. Ignorance and obstinacy played a part — one man ate a glassful of ergot to prove its innocuousness. He died the next day. On the other hand, severe famine and dire necessity precipitated most big epidemics. Up to three-quarters of the grain harvested consisted of impurities, of which ergot was often the major contaminant. Rye was usually the first harvest of the season and at times it was the sole food available, being processed into bread, dumplings, pancake, porridge, soup *etc.* Much evidence exists that ergot sclerotia were more toxic when fresh than when stored. In times of need, the grain that fell out during harvesting was milled immediately and used locally. This portion was particularly rich in ergot and thus ergotism became a disease primarily of agricultural labourers. As early as 1597 the Marbury Medical faculty recognized the increased susceptibility of malnourished individuals to convulsive ergotism. Dairy products, particularly scarce in the affected parts of Germany, were advocated as remedies. Some evidence exists that vitamin A will reduce spinal lesions induced by ergot.

Enormous amounts of ergot were apparently consumed during the great gangrenous epidemics in France. Most reports indicate that about a quarter of the grain consisted of ergot. Undoubtedly the individual susceptibility to particularly the gangrenous form varied greatly, and whole families were not usually affected as was the case in the convulsive form. Recent outbreaks have proved that death from gangrene can follow the consumption of about 100 g of ergot over a few days. During the last Russian epidemics it was found that disease occurred when the flour contained 1% ergot, and death from the convulsive form occurred after the use of flour containing 7% ergot. A 2% contamination is adequate to cause epidemics. A quantity 10 times less produced mild cases in Manchester in 1927.

Serious attempts by governments to control the disease started in 1722 when the Prussian government exchanged good grain for ergotized rye. Legislation to prevent the occurrence of epidemics was introduced notably in Saxony in 1764 — several years before the great epidemic in Europe of 1770—72. The French government, with the aid of the clergy, attempted to educate the population. With the gradual decrease of home milling during the 19th century, legislative control was facilitated. Larger sclerotia were removed by sieving, the smaller oil-rich sclerotia were removed by stirring the grain in water. The most effective purification method is to increase the density of the water with, *e.g.*, salt — a fact only realized in the 20th century.

Chemical and physical properties of the toxic metabolites

The constituents of ergot are many; each small sclerotium contains some 10 different groups of substances with a total of over 100 compounds. Groups of compounds found in ergot include estolides, many diverse amines, amino acids, glucans whose main chain consists of 1—3 linked β-D-glucopyranosyl units, pigments, enzymes and fatty acids. Some of the more unusual compounds include 2,3-dihydroxybenzoic acid, guanine propionate, 12-hydroxystearic acid, (+)-threo 9,10-dihydroxystearic acid, D,L-α-aminoheptanoic acid, paspalin, paspalicin, clavicipitic acid, indoleisopropionic acid and ergothioneine. The most important substances in toxicology and medicine isolated from ergot are the alkaloids. The discovery of ergotamine by Stoll in 1918 ushered in the era of modern ergot alkaloid chemistry [10].

Structurally ergot alkaloids are classified either as derivatives of lysergic acid or as clavine alkaloids.

Type I — Derivatives of lysergic acid

The lysergic acid derivatives are of the acid amide type and are subdivided into: (*a*) Simple lysergic acid amides which include ergine and ergometrine and (*b*) Derivatives of lysergic acid of the peptide type.

The peptide type of derivatives are further subdivided into: (*i*) the ergotamine group; (*ii*) the ergotoxine group; (*iii*) the ergoxine group. The latter subdivision is based on the hydrolysis products given by these alkaloids. Upon mild alkaline hydrolysis the ergot alkaloids of the peptide type decompose to give lysergic acid, 2 amino acids (one of which is always L-proline and the other either L-phenylalanine, L-leucine, L-isoleucine or L-valine), an α-keto acid (pyruvic acid, dimethylpyruvic acid or α-ketobutyric acid) and one equivalent of ammonia. The ergotamine group of alkaloids yields pyruvic acid while dimethylpyruvic acid is the α-keto acid produced by hydrolysis of the ergotoxine group. The ergoxine group yields α-keto butyric acid on hydrolysis.

Type I(a)

The simple derivatives of lysergic acid are derivatives of D-isolysergic acid and D-isolysergic acid (II). These are C-8 atom epimers. Both contain the indole group built into a tetracyclic ring system named ergoline (III) [11] (Fig. 1). The D-lysergic acid and the D-isolysergic acid are readily interconvertible in alkaline medium. They have been isolated from ergot and are also obtained by rigorous alkaline hydrolysis of type I ergot alkaloids. The first synthesis of lysergic acid was accomplished in 1954, [12]. The stereochemistry and absolute configuration of both isomers is known.

Fresh ergot from the sclerotia of *Claviceps purpurea* and saprophytic cultures of *C. paspali* contain the pharmacologically, highly active D-lysergic acid alkaloids. These are all levorotatory or only slightly dextrorotatory and their names end in *-ine*. Isolysergic acid is the characteristic component of those strongly dextrorotatory alkaloids of weak activity whose names end in *-inine* (see Table I). The unstable 8,9-double bond structural isomer of lysergic acid has been reported from submerged cultures of a Portuguese *C. paspali* strain [13].

Ergine (IV) (lysergic acid amide) (see Fig. 2) and erginine (isolysergic acid amide) are the main constituents of the alkaloid mixture from ergot of *Paspalum distichum* L. They have also been isolated as the main components, together with other alkaloids of the clavine group, from seeds of *Rivea corymbosa* (L.) Hall. f. and *Ipomoea tricolor* Cav. [14]. These seeds were

TABLE I

ALKALOIDS OF LYSERGIC AND ISOLYSERGIC ACIDS

	m.p. °C	$(\alpha)_D^{20}$	$(\alpha)_{5461}^{20}$	Ref.
Ergine $(C_{16}H_{17}ON_3)$	242	—	+15	81
Erginine	132–234	+480	+608 (pyridine)	19
Ergocornine $(C_{31}H_{39}O_5N_5)$	182–184	−188	−226	
Ergocristine	228	+409	+624	82
Ergocristinine $(C_{35}H_{39}O_5N_5)$	160–175	−183	−217	19
α-Ergocryptine	226	+366	+460	
α-Ergocryptinine $(C_{32}H_{41}O_5N_5)$	212–214	−190	−226	
β-Ergocryptine	240–242	+480	+508	20
β-Ergocryptinine $(C_{32}H_{41}O_5N_5)$	173	−174	—	
Ergometrine (Ergonovine)	220	+424	—	
	162 (from ethylacetate)	+41 (ethanol)	+60 (ethanol)	83–86
	212 (from acetone)			
Ergometrinine $(C_{19}H_{23}O_5N_3)$	196	+414	+520	87
Ergosecaline	—	a 298°	—	
Ergosecalinine $(C_{24}H_{28}O_4N_4)$	217	−183	−220	18
Ergosine $(C_{30}H_{37}O_5N_5)$	220–230	+420	+522	
Ergosinine	228	−169	—	21
Ergostine $(C_{34}H_{37}O_5N_5)$	204–208	+357		
Ergostinine	215–216	−160	−192	10
Ergotamine	180 (acetone)	+369	+462	
Ergotaminine $(C_{33}H_{35}O_5N_2)$	241–243 (benzene)	+32 (pyridine)	+368 (pyridine)	4
Lysergic acid	238	—		
Isolysergic acid	218	+17 (pyridine)		4
Lysergide $(C_{20}H_{25}N_3O)$	80–85			
Lysergic acid methylcarbinolamide $(C_{18}H_{21}O_2N_3)$	135	+29 (DMF)	—	81
Lysergic acid-L-valinemethylester $(C_{22}H_{27}O_3N_3)$	80–85	−103	—	
Isolysergic acid-L-valinemethylester	167	+421	—	88
Methylergometrine (Methylergonovine) $(C_{20}H_{25}N_3O_2)$	172	−45 (pyridine)	—	

a $(\alpha)_D^{18}$

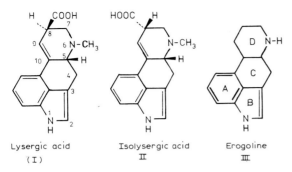

Lysergic acid   Isolysergic acid   Erogoline
   (I)   II   III

Fig. 1. General formulae.

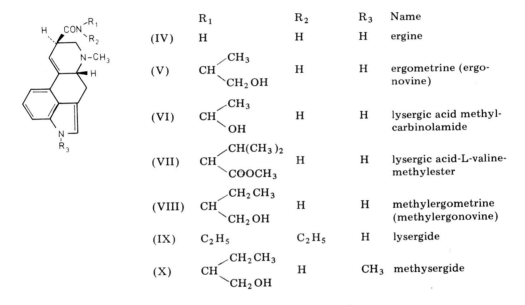

		R₁	R₂	R₃	Name
	(IV)	H	H	H	ergine
	(V)	CH⟨CH₃/CH₂OH	H	H	ergometrine (ergonovine)
	(VI)	CH⟨CH₃/OH	H	H	lysergic acid methyl-carbinolamide
	(VII)	CH⟨CH(CH₃)₂/COOCH₃	H	H	lysergic acid-L-valine-methylester
	(VIII)	CH⟨CH₂CH₃/CH₂OH	H	H	methylergometrine (methylergonovine)
	(IX)	C₂H₅	C₂H₅	H	lysergide
	(X)	CH⟨CH₂CH₃/CH₂OH	H	CH₃	methysergide

Fig. 2. Simple lysergic acid derivatives (Type Ia).

used centuries ago by Central American Indians as the magic drug called
"Ololinqui". The occurrence of lysergic acid alkaloids in the plant family of
Convolvulaceae is a completely unexpected phytochemical discovery as these
alkaloids had been found only in the lower fungi of the genus Claviceps.

Ergometrine (V) (lysergic acid L-2-propanolamide), the specific uterotonic ergot alkaloid was discovered in 1935 in four different laboratories
almost simultaneously and described under four different names. Although
the names ergometrine and ergobasine have remained in use in Europe, ergo-
novine was adopted in the United States of America as the official name.
The structural elucidation was reported by Jacobs and Craig [15] and the
synthesis of ergometrine effected by Stoll and Hofmann [16].

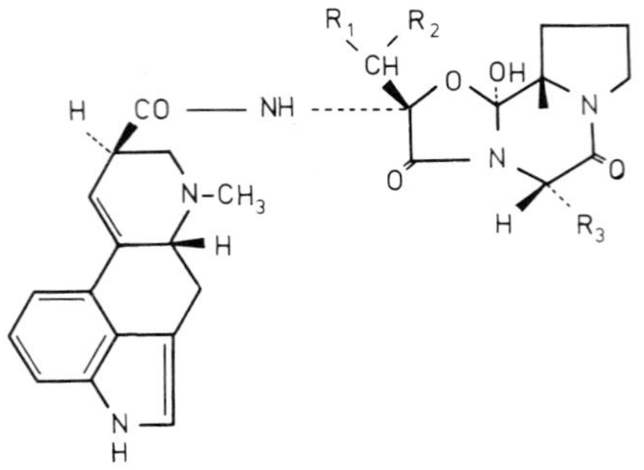

	R_1	R_2	R_3	Name
(XI)	H	H	$CH_2C_6H_5$	ergotamine
(XII)	H	H	$CH_2CH(CH_3)_2$	ergosine
(XIII)	H	H	$CH(CH_3)_2$	ergosecaline
(XIV)	CH_3	CH_3	$CH_2C_6H_5$	ergocristine
(XV)	CH_3	CH_3	$CH_2CH(CH_3)_2$	α-ergocryptine
(XVI)	CH_3	CH_3	$CH(CH_3)_2$	ergocornine
(XVII)	CH_3	CH_3	$CH(CH_3)CH_2CH_3$	β-ergocryptine
(XVIII)	H	CH_3	$CH_2C_6H_5$	ergostine

Fig. 3. Peptide derivatives of lysergic acid.

Ergometrinine (ergonovinine), the C-8 epimer of ergometrine, occurs only in small quantity in ergot together with the isomeric alkaloid.

Lysergic acid methylcarbinolamide (VI) was isolated together with ergine and erginine from saprophytic cultures of *Claviceps paspali*. It decomposes easily in a weak acid solution to form ergine and acetaldehyde.

In many ergot drugs traces of D-lysergic acid-L-valine methyl ester (VII) can be detected.

Synthetic derivatives of great pharmaceutical importance include methylergonovine (VIII) (Methergine ®), lysergide (IX) (lysergic acid diethylamide, LSD) and methysergide (X) (1-methyl-lysergic acid butanolamide).

Type I (b)

(*i*) *Ergotamine group.* These ergot alkaloids are characterised by the pyruvic acid liberated on alkaline hydrolysis. The two main pairs of alkaloids belonging to this group are ergotamine (XI), ergotaminine and ergosine (XII), ergosinine (see Fig. 3). Ergosecaline (XIII), whose structure is not entirely elucidated, gives lysergic acid, ammonia, pyruvic acid and valine as cleavage products and thus can be tentatively grouped in this section.

Ergotamine (XI) isolated from Swiss ergot was the first chemically homogeneous and fully active ergot alkaloid to find widespread medical application [17]. Ergotaminine, the isolysergic acid isomer of ergotamine, is not easily dissolved in most solvents, and as a result, it crystallizes rapidly from the equilibrium set up between ergotamine and ergotaminine in hydroxyl-containing solvents. As is the case with most of the isolysergic acid derivatives, ergotaminine is pharmacologically inactive.

Ergosine (XII) and ergosinine, which were isolated from Iberian ergot, have so far found no medicinal application [18].

(*ii*) *Ergotoxine group.* The 4 known pairs of alkaloids in this group each give dimethylpyruvic acid on hydrolysis.

Ergocristine (XIV), which was isolated from Iberian ergot, is a constituent of the alkaloid mixture known as ergotoxine [19]. The other constituents are α-ergocryptine (XV), the main alkaloid in ergot of Japanese and South American wild grasses, and ergocornine (XVI). Recently β-ergocryptine (XVII), isomeric with α-ergocryptine has been reported [20]. These two isomers differ in the peptide portion of the alkaloid. α-Ergocryptine yields L-leucine upon hydrolysis, and β-ergocryptine L-isoleucine.

The original 3 ergotoxine alkaloids, ergocristine, ergocornine and α-ergocryptine, which crystallize as a 1:1:1 mixture, and which for many years were thought to be a single compound named ergotoxine, find medical application as the hydrogenated derivatives under the name Hydergin ® for the treatment of peripheral and cerebral vascular disorders and of essential hypertension.

(*iii*) *Ergoxine group.* Only a single pair of ergot peptide alkaloids is known to produce α-keto butyric acid on hydrolysis. Ergostine (XVIII) is found in very small quantities as a minor alkaloid in Swiss ergot [21]. A total synthesis of ergostine and its isomer ergostinine has been developed.

The total synthesis of the peptide portion of the peptide alkaloids has been achieved. This has enabled alkaloids to be synthesised that have not yet been found in nature [22].

Pharmacologically inactive acid-alkaloids are produced by the peptide alkaloids and the corresponding dihydro-derivatives in acid solution [23]. The peptide moiety is affected in an unknown way. After intense irradiation with UV light one molecule of water is added to the Δ [9,10] double bond of lysergic acid-derived alkaloids to give lumi-derivatives.

Type II — The clavine alkaloids

These differ from classical lysergic acid derivatives in that the carboxyl group of the lysergic acid has been reduced to hydroxymethyl or methyl group. The clavine alkaloids occur mainly in sclerotia of Claviceps species that parasitize wild grasses in the Far East and in Africa. Saprophytic cultures of these strains also can produce clavines. Trace amounts are found in sclerotia and saprophytic cultures of *C. purpurea* and *C. paspali*. Clavine-type alkaloids also occur in the genera of the family Convolvulaceae and in certain fungi. The chemistry of clavine alkaloids has been comprehensively described by Hofmann [24]. Table II gives a list of the known clavine alkaloids and some of their characteristics. Fig. 4 shows their structures.

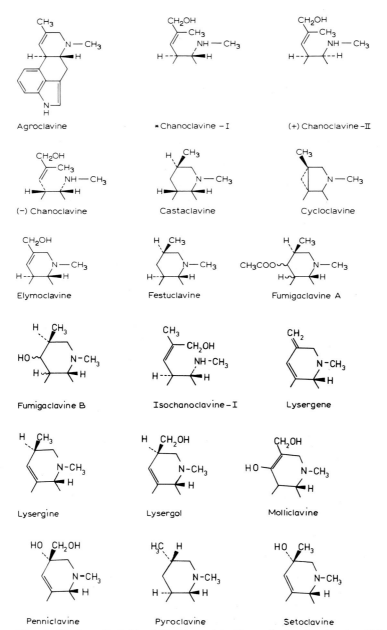

Fig. 4. Clavine alkaloids. *Starting with Chanoclavine-I only ring D is shown.

TABLE II

CLAVINE ALKALOIDS

Name	m.p. °C	$(\alpha)^{20}_D$	Formula	Keller's col. reaction	Ref.
Agroclavine	206	−151	$C_{16}H_{18}N_2$	violet-blue	89
Chanoclavine-I (Secaclavine)	222	−240 (pyr)	$C_{16}H_{20}N_2O$	violet-blue	90, 91
(−) Chanoclavine-II	174	−332 (pyr)	$C_{16}H_{20}N_2O$		92
Costaclavine	182	+44 (pyr)	$C_{16}H_{20}N_2$		90
Cycloclavine	165−166	+63	$C_{16}H_{16}N_2$		93
Dihydrolysergol-I			$C_{16}H_{20}N_2O$		43
Elymoclavine	249	−109 (EtOH)	$C_{16}H_{18}N_2O$	violet-blue	94
Elymoclavine-O-β-D-fructoside			$C_{22}H_{28}N_2O_6$		95
Festuclavine	242−244	−69	$C_{16}H_{20}N_2$	violet-blue	96
Fumigaclavine A	85	−57	$C_{18}H_{22}N_2O_2$		97
Fumigaclavine B	245	−113	$C_{16}H_{20}N_2O$		97
Isochanoclavine-I	181	−216 (pyr)	$C_{16}H_{20}N_2O$		92
Isolysergol			$C_{16}H_{18}N_2O$		98
Isopenniclavine	163−165	+146 (pyr)	$C_{16}H_{18}N_2O_2$	green	91
Isosetoclavine	234−237	+107 (pyr)	$C_{16}H_{18}N_2O$	green	91
Lysergene	247−249	+407	$C_{16}H_{16}N_2$	green	99
Lysergine	286	+65 (pyr)	$C_{16}H_{18}N_2O_2$	violet-blue	99
Lysergol	249−250	+50 (pyr)	$C_{16}H_{18}N_2O$	blue	99, 14
Molliclavine	253	+30 (pyr)[a]	$C_{16}H_{18}N_2O_2$	green	100
Norsetoclavine			$C_{15}H_{16}N_2O$		101
Penniclavine	222	+153 (pyr)	$C_{16}H_{18}N_2O_2$	green	39
Pyroclavine	204	−90 (pyr)	$C_{16}H_{20}N_2$	violet-blue	90
Setoclavine	229−234	+174	$C_{16}H_{18}N_2O$	green	91

a $(\alpha)^{17}_D$

Clavine alkaloids are pharmaceutically unimportant but current research is progressing concerning their therapeutic application.

Biosynthesis of ergot alkaloids

By the use of isotope techniques and saprophytic cultures the biosynthesis of the ergoline skeleton has been elucidated. The ergoline ring system in ergot fungi and higher plants is constructed from L-tryptophan (XIX) and mevalonic acid (XX). The *N*-methyl group of the ergot alkaloids is derived from methionine *via* a transmethylation reaction. The connection of the isoprene unit with the tryptophan occurs at the 4-position of the indole nucleus by way of 4-dimethylallyltryptophan (XXI), which compound has recently been isolated from saprophytic cultures of Claviceps [25] (see Fig. 5). The precursors of tryptophan in ergot are indole, anthranilic acid and indolylpyruvic acid.

Pigments of ergot

The ergot of *Claviceps purpurea* growing on rye contains a mixture of pigments from which the first crystalline product was isolated in 1877. Only

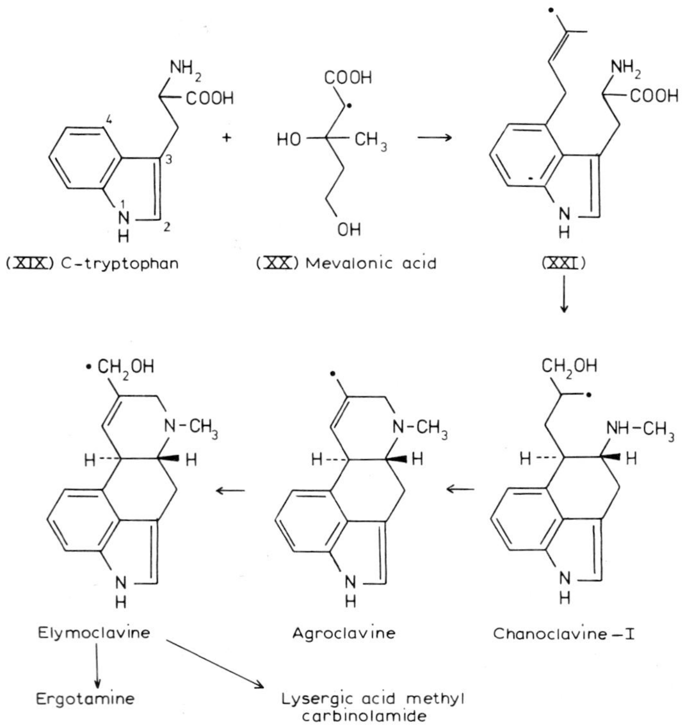

Fig. 5. Biosynthetic pathways to ergot alkaloids.

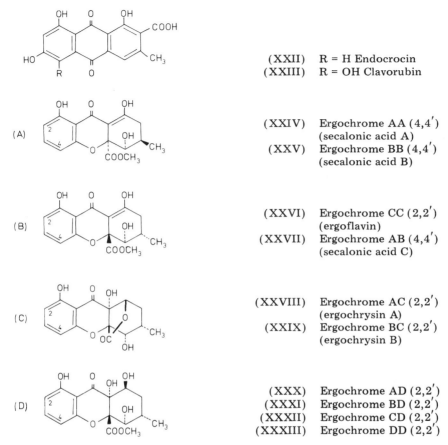

(XXII)   R = H Endocrocin
(XXIII)   R = OH Clavorubin

(XXIV)   Ergochrome AA (4,4')
          (secalonic acid A)
(XXV)   Ergochrome BB (4,4')
          (secalonic acid B)

(XXVI)   Ergochrome CC (2,2')
          (ergoflavin)
(XXVII)   Ergochrome AB (4,4')
          (secalonic acid C)

(XXVIII) Ergochrome AC (2,2')
          (ergochrysin A)
(XXIX)   Ergochrome BC (2,2')
          (ergochrysin B)

(XXX)   Ergochrome AD (2,2')
(XXXI)   Ergochrome BD (2,2')
(XXXII)   Ergochrome CD (2,2')
(XXXIII) Ergochrome DD (2,2')

Fig. 6. Pigments of ergot.

in recent years has the resolution of the mixture into its components permitted structural elucidation to be carried out.

Endocrocin (XXII) and clavorulin (XXIII) are hydroxyanthraquinone carboxylic acids. The majority of ergot pigments are weakly acid, pale yellow substances that are 2,2'- or 4,4'-dimers of the 4 monomeric xanthones designated by the letters A, B, C and D in Fig. 6. Franck *et al.* designated the pigments as ergochromes and by using two letters and two numbers it is possible to characterize the dimers and the type of linkage [26,27]. For some pigments trivial names are retained.

Ergoflavin (XXVI), the principal pigment from ergot, was crystallized in pure form by Freeborn [28] in 1912. The complete structure was elucidated in 1963 [29,30]. The structural elucidation of the other ergochromes has been mainly achieved by research groups in Canada, England and especially in Germany.

With the aid of saprophytic cultures the biogenesis of these natural

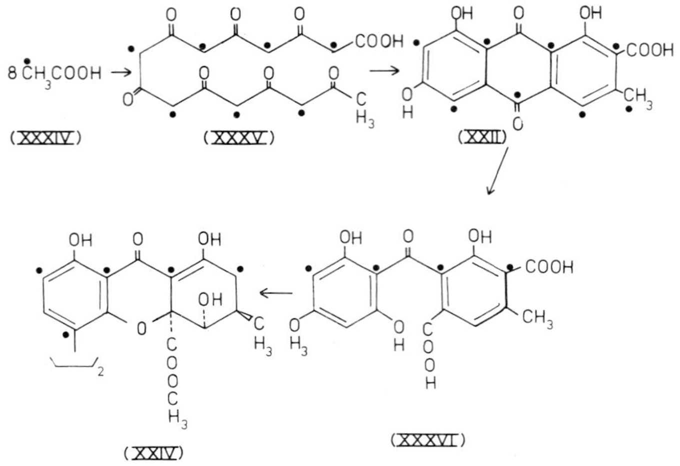

Fig. 7. Biosynthesis of ergochromes.

products has been elucidated. Biosynthetically ergochromes arise by cleavage of an anthraquinone precursor (see Fig. 7). Support for this hypothesis
has been provided by Franck *et al.* who showed that secalonic acid C
(XXVII) produced in the presence of [2-^{14}C]acetate had the expected distribution of radioactivity [31]. More recently Franck *et al.* have shown that
ergochromes produced in the presence of [1-^{14}C]acetate, [2-^{14}C]acetate,
or [2-^{3}H$_2$]acetate have labelled patterns consistent with their derivation *via*
endocrocin [32—34]. A possible biogenetic pathway of ergochrome AA
(4,4') (secalonic acid A) (XXIV) starting from acetic acid (XXXIV) *via* the
hypothetical heptaketopalmitic acid (XXXV) and endocrocin (XXII). The
postulated benzophenonecarboxylic acid (XXXVI) is probably formed by
oxidative ring fission of the anthrone or its hydroperoxide rather than in the
anthraquinione (XXII) itself (see Fig. 7).

Environmental and laboratory conditions favouring the growth of ergot

Centuries before ergot was known to be a fungus, it was noted by
numerous early writers that ergot was plentiful during unusually wet seasons.
Some even mentioned that it occurred when mildew was plentiful. As early
as 1539, Brock in Germany noted the importance of hot spells alternating
with sudden rains. Local factors are important since certain valleys or even
fields were more prone to be ergotized. Moist soils encouraged ergot growth
and some outbreaks were associated with marshy country. In France, however, light sandy soils were incriminated. Since moisture could affect temperature, the interaction of these factors caused some confusion.

In Dodart's classical exposition of the aetiology of ergotism, he mentioned that rainy springs followed by warmer than usual summers favoured
ergot growth. Climatic conditions before the great epidemic starting in 1770

were unusual. The winter was not continuously severe yet the spring was late. In June during the flowering of rye there was much cold and mist followed by heavy rains. The maturing rye kernels were then subjected to heat waves and drought. Unusual climatic conditions have favoured epidemics indirectly by causing a scarcity of food. A large proportion of the crop was lost following frost in May prior to the last Russian epidemic in 1927. The wet and cold summer extended the flowering period of the surviving rye, rendering it more liable to infection.

The hitherto inexplicable effect of climatic conditions became obvious only when Tulasne elucidated the three basic stages of the life cycle of Claviceps in 1853. The small banana-shaped, brittle, blackish ergot is the sclerotium or resting stage of *Claviceps purpurea*. During harvesting many sclerotia fall to the ground where they slowly absorb moisture and by the next spring commence to germinate. At this stage the presence of moisture is important. Many tiny mushrooms appear to grow out of a single sclerotium. The head of each "mushroom" produces numerous submicroscopic filamentous ascospores — up to a million ascospores may be produced in this way by a single sclerotium. These spores must reach the pistil of the flowering rye to develop further. Dry windy weather will facilitate the dissemination of spores. Climatic conditions prolonging the flowering period will increase the chances of infection. The greater susceptibility of rye is based on the fact that rye, unlike other cereals, depends largely on cross-fertilization and opens its glumes in order to receive pollen from other plants.

Damp weather or dew causes moisture to accumulate at the base of the pistil and it is there that the ascospore begins to germinate. Long filamentous hyphae enter the ovary and remainder of the pistil, resulting in envelopment of the organs by mycelia within a week. The twisted strands of hyphae give rise to asexual spores called conidia. The hyphae further excrete a sweet mucilaginous substance — honeydew — which envelops numerous conidia. The honeydew attracts insects and also spreads the infection by dripping on to spikelets of grain below. Meanwhile, the hyphae replace the entire kernel and the density increases. Maturation of the new sclerotium is encouraged by warm weather.

Satisfying present-day commercial demand for ergot required some ingenuity since the rye plant is only susceptible to infection during the few days of flowering. Stroking the flowers at the right stage induces instant opening and machines which did this and simultaneously sprayed them with cultures were developed. More efficient are the recent automated techniques for direct inoculation of the rye with spore suspensions. Limiting factors are weather dependence and a single crop per year.

Significant progress on the laboratory culture of Claviceps is relatively recent and several methods have been patented. Submerged cultures give satisfactory yields of clavine alkaloids and simple lysergic acid derivatives but not as yet of the peptide alkaloids. The first step of commercial significance was the discovery that the fungus strain *Claviceps paspali* (Stevens and Hall) is capable of producing considerable quantities of lysergic acid derivatives

References p. 94

especially lysergic acid amide by means of saprophytic submerged culture. Lysergic acid, which is used as starting material for the production of pharmaceuticals, is not obtained by this process on an industrial scale. Another strain of *Claviceps paspali* obtained from the ergot of *Paspalum dilatatum* of Portuguese origin produces free isomeric lysergic acids in good yield by the submerged culture technique.

A detailed survey of the production of ergot alkaloids has been made by Gröger [7].

Analysis for the toxic metabolites

Numerous methods have been described for the qualitative and quantitative estimation of these compounds.

Characteristic colour reactions are given by ergot alkaloids with sulphuric acid, Keller reagent and van Urk's reagent. A representative formula for the Keller reagent would be 0.1% glyoxylic acid and 0.5% ferric chloride to glacial acetic acid. The test involves the observation of the colour that develops when the reagent is mixed with the alkaloid and concentrated sulphuric acid (see Table II).

One of the simplest and most sensitive tests for the indole nucleus of the ergolines is the van Urk test. Originally van Urk used a 1% solution of paradimethylaminobenzaldehyde in alcohol. Allport and Cocking modified the reagent so that the usual solution is now made by dissolving 125 mg paradimethylaminobenzaldehyde in a 65% solution of nitrate-free concentrated sulphuric acid to which 0.1 ml of a 5% aqueous ferric chloride solution is added [35]. The reagent must be freshly prepared at least once weekly.

Most quantitative methods are based on the modified van Urk reaction. Ergolines are also quantitatively estimated by titration, ultraviolet absorbance and spectrophotofluorometric methods [24,36,37].

The estimation of the total alkaloid content is insufficient for pharmaceutical preparations because these lysergic acid derivatives easily isomerise to the isolysergic acid derivatives in which the pharmaceutical effect is practically lost. Ideally all components of the alkaloid mixture should be separated and separately estimated. Chromatographic procedures are the best for separation of the individual alkaloids.

Hofmann has used aluminium oxide columns to separate certain of the simple peptide derivatives of lysergic acid [24]. It was by use of this method that the ergotoxine complex was separated.

Paper chromatographic methods have been extensively and successfully used. Ergometrine and ergometrinine were separated by Forster on unbuffered paper [38]. By the use of impregnated papers the pharmaceutically important ergotamine and ergotoxine group and the corresponding dihydroderivatives were successfully separated [39]. The clavine group of alkaloids has also been separated by paper chromatography [40—42].

Thin-layer chromatographic procedures for the resolution and quantitative determination of ergot alkaloids have also been employed. This would appear to be the method of choice and is receiving most attention in the current literature. The literature has been reviewed by Agurell [43] and Santavý [44]. Pioneering work in this field was done by Rachelmeyer *et al.* [45—47]. The thin-layer chromatographic behaviour of lysergide and 14 related ergot alkaloids has been investigated by Fowler *et al.* [48].

Methods for the quantitative measurements of single alkaloids after chromatographic separation have been repeatedly described. Densitometry, spectrophotofluoremetry and colorimetric methods have been successfully applied [47—53].

Natural occurrence of metabolites

For a long time the ergotine derivatives have been regarded as specific for the genus Claviceps. Recently ergot alkaloids were isolated from fungi outside the genus Claviceps (*Aspergillus fumigatus, A. clavatus, A. nidulans, Rhizopus nigricans, Penicillium chermesinum, P. concavo-rugulosum*) and from higher plants (*Rivea corymbosa, Ipomoea violacea, I. argyrophylla* Vatke, *I. hildebrandtii* Vatke, *I. tricolor* Cav.).

The highly organospecific ergot fungal infection can only develop in the female sex organs of the grasses [54]. The biology of the ergot fungi has been summarized by Gröger [7].

Taxonomically the genus Claviceps presents many problems. Critical reviews have been given by Langdon [55] and Skalický and Starý [56].

Lists of the host plants to various Claviceps species have been published repeatedly [57]. One of the pioneers of ergot research classified the ergot fungi into 16 races [58].

According to Gröger [59] the genus Claviceps may be divided into 3 groups according to their alkaloid-producing ability. Ergot poisoning in animals is usually due to infestation of the growing grasses: perennial rye grass (*Lolium perenne*), cocksfoot (*Dactylis glomerata*) timothy (*Phleum pratense*), crested dog's tail (*Cynosurus cristatus*), oat grass (*Avena pubescens*) and Yorkshire fog (*Holcus lanatus*) are among British grasses susceptible to attack [60]; while in America wheat grasses (*Agropyron* spp.) redtop (*Agrostis alba*), smooth bromegrass (*Bromus inermis*), reed grasses (*Calamagrostis* spp.), wild rye (*Elymus* spp.) red canary grass (*Phalaris arundinacea*), and blue grasses (*Poa* spp.) have been implicated [61].

The 1962 list by Lynn R. Brady, the 1957 publication by Vincenzo Grasso and the 1947 list by Toyohiko Kawatani record all host plants known [57,62]. Apart from 4 species in the family Juncacea and 17 in the family Cyperaceae the almost thousand other host plants for Claviceps fungi belong to the family Gramineae — the grass family. This family (Gramineae) embraces about 620 genera and their thousands of species distributed throughout the world supply us with our cereal and forage crops — Festuceae for fodder grasses, Hordeeae for wheat, barley and rye, Avenae for oats, Oryzae for rice, Paniceae for millet and Maydeae for Indian corn. Ergot is a parasite on all of these [6].

Both corn and rice are reported to be parasitized by ergot. *Claviceps gigantea*, the species infesting corn was described in 1964 [63]. The sclerotium is longer than that of other ergots being up to 8 cm long and 5 cm thick. Chemical analysis by Mas showed it to contain two substances resembling ergot alkaloids [14]. Rice is attacked by fungi which develop sclerotia but whether these sclerotia contain alkaloids is uncertain [65]. Claviceps attacking rice is a distinct species — *Claviceps zizaniae* (Fyles) Comb. Nov. [66].

Nowadays ergot is harvested by the ton. Mechanical inoculation has made possible the controlled production of large quantities of crude ergot in all parts of the world.

Saprophytic cultivation in glass tanks sounds deceptively easy; nothing is needed except a fungus and a culture medium. Wang in 1966 made an excellent review of the subject in her thesis on the saprophytic production of ergot alkaloids [67].

Biological activity in experimental systems

The French physician Tullier dosed various farm animals with ergot in 1630 and reported that they all died. For the next three centuries unbelievable difficulties with biological work were encountered and much controversy raged. This was partly due to the complex nature of ergot and its many extracts, but mainly due to the fact that the science of animal experimentation did not exist. Principles of experimental design, selection of appropriate animals, dosage and environment had to be developed. Ergot, by initially causing chaos, contributed greatly to the principles of pharmacological testing and standardization since it was a desperately needed pharmaceutical. The systematic chemical hunt for active principles which started well into the 19th century provided a great impetus.

Peripheral vasoconstriction

For centuries ergot had been used to stop haemorrhage at childbirth. When it became clear that gangrenous ergotism followed circulatory disturbances, experimental studies of the effects on the circulation commenced.

Frogs were injected subcutaneously and the effects on the circulation observed in the web, tongue and mesentery. Some used rabbits and observed the ear, retina or mesentery. Cyanosis of the rooster's comb, used qualitatively by early European workers, was adopted for quantitative assay by American manufacturers at the turn of the century. It soon became obvious that transient vasoconstriction was a particularly non-specific parameter. A century ago German workers at Strasbourg began to notice other effects and measured them objectively, even using controls. They studied the calibre of blood vessels, heart and muscular activity, salivation, respiration, reflexes and pupillary changes.

Rat tails readily undergo necrosis following the administration of various ergot alkaloids. Anderson and Wells [68] induced gangrene by the subcutaneous or intraperitoneal injection of ergotamine tartrate and found no difference in the response by either route. The sex of the rat had no influence on the response. When single injections ranging between 3.3 and 50mg/kg were given it was observed that the incidence of gangrene with the smallest dose was very low, while with the largest dose a high mortality occured. In the process of developing gangrene the appearance of the rats' tails passed progressively through several poorly defined stages. The earliest indication of impending necrosis was the appearance of a demarcating region of redness and tenderness, distal to which there was an area of extreme pallor, with dark discolouration of the tip of the tail. This change was designated arbitrarily as the onset of gangrene. Half the animals which developed gangrene began to do so within 3 days and 100% within 7 days of the injection.

Epinephrine reversal

In England in 1905, Dale noticed that epinephrine injected into cats after they had been given ergot produced a drop in blood pressure, which was exactly the reverse of what should have happened, and one of the most specific actions of ergot known today. Simultaneously Sollman and Brown in America observed the same reaction. Large doses of ergot extracts given intravenously to dogs obliterated the pressor effects of epinephrine. When epinephrine was injected into the spleen of an ergotized animal it became dilated.

Many attempts were made to perfect the simple blood pressure assay for ergot, with consistently unreliable results. Swiss workers then refined the blood pressure reversal test and it is widely used today.

Epinephrine will normally inhibit peristalsis of the small intestine. Spanish workers demonstrated that when guinea pig intestinal strips had been exposed to ergot and epinephrine was added, instead of being inhibited the contractions became stronger than ever. Thus ergot not only reverses the stimulating action of epinephrine, but also its inhibitory actions.

Uterine tests

Longitudinal strips of muscular structures from the uterus were used in a biological test advocated by Kehrer in Germany in 1907. Strips obtained from the uterus of cats were placed in Ringer's solution and small amounts of ergot were added. The amount which caused an increase in the contractions of the strip was noted. The response of material taken from cats in various stages of reproduction was excessively variable and the lack of a standard preparation at the time limited the value of the tests.

The methods utilizing the isolated cat or guinea pig uterus were better than most existing assays but unless pure alkaloids were used the amines and other extraneous constituents of ergot preparations interfered with the test. It was still not possible to correlate test findings with therapeutic effects.

Clarke and Broom abolished this unsatisfactory state of affairs when they reported their classical method in 1923. They used the isolated uterus of pregnant or non-pregnant rabbits. Longitudinal strips of about 1 cm were cut while the uterus was immersed in Ringer's solution. Strips were placed in separate baths and epinephrine solutions of 1:1 000 000 concentration added. After observing the amount of contraction the epinephrine solution was washed out. 5 minutes later the ergot preparations were added and left to stand for 5 more minutes and then the bath fluid was changed. Then epinephrine was again added and the uterine response — if any — recorded. The end point in this biological titration was that amount of ergot preparation which abolished the response to epinephrine in one dose. By using a known strength of ergotamine in the duplicate bath a quantitative determination could be made. The test was rapid, convenient, inexpensive and exceedingly sensitive, but not very specific.

Thus a mere 45 years ago the best qualitative assay method was epinephrine reversal of blood pressure in the cat. For quantitation, the method used was the Clark and Broom epinephrine reversal in the isolated rabbit uterus, using ergotamine, the only available pure alkaloid, as a standard. With these methods it was finally established that only alkaloidal fractions were pharmacologically active. Soon more alkaloids were isolated. The question of relative physiological actions arose, and again the bio-assay field was plunged into chaos.

In 1938 Rothlin reported his now standard method of assaying the oxytocic components of ergot. *In situ* contractions of the uterus and vagina as one organ were recorded. Fixed doses of a standard were injected alternately with the unknown, which was varied until the two responses were similar. Assays using the isolated uterus were improved. The state of the organ was standardized by using uteri from immature rabbits which had been pretreated with oestrogens. The latent time for a response, expressed as the logarithm, was used rather than attempting biological titration as the measure of potency and this improved the precision of assay tremendously.

Seminal vesicle standardization

The Swiss workers of Basle devised an almost ideal method in 1945, which was a modification of the Clark and Broom method of 1923. Instead of measuring the adrenergic blocking activity in the isolated rabbit uterus, seminal vesicles from guinea pigs were substituted. The test organs were small, sensitive and obtained in pairs of similar size. They displayed virtually no spontaneous activity, always responded to epinephrine by contraction and kept their sensitivity for prolonged periods.

As late as 1947 large collaborative studies were organised in order to identify ideal assay methods of ergot powder and fluid extracts of ergot. They failed, because ergot has a variety of physiological actions, some of which they were not aware of at the time. No single biological assay method can measure them all, especially in samples whose chemical composition was variable.

Mode of action of ergot constituents

Prior to the isolation of ergotamine in 1918, many alkaloids and acids had been isolated but their biological activity was often disappointing. Many workers focussed their attention on other groups of compounds, particularly the basic amines. Fractions containing variable mixtures of amino acids were equally disappointing.

Burger and Dale isolated isoamylamine and tyramine in 1909 and a short time later histamine and acetylcholine. All these amines were active and some were even supplied commercially for obstetrical use. Meanwhile, the ergotoxine complex — a variable combination of ergocornine, ergocryptine and ergocristine — was being investigated. Fortunately, all three alkaloids had similar pharmacological properties and most of the effects of ergot were unravelled. All the work had to be repeated with ergotamine. The results were similar but some minor effects were important clinically. For instance, ergotamine had a weaker suppressive action on the respiratory centre.

The description of the action of the many compounds isolated in recent times has become an exceedingly involved field, with numerous tests having to be performed on each new compound. The task was simplified by Cerletti in 1958 who proposed an activity spectra scheme whereby any known alkaloid could be classed into one of 6 groups. These criteria are essentially the ability to cause vasoconstriction, uterine contraction, adrenergic blockade and serotonin antagonism. Central effects were classed into two groups; firstly, vomiting, bradycardia, inhibition of vasomotor centre and baroceptive reflexes; secondly, excitatory syndrome, mydriasis, hyperglycaemia and hyperthermia. These 6 categories can be further broken down into 3 action groups.

Peripheral effects

These effects of ergot alkaloids, due to smooth-muscle stimulation, include vasoconstriction and oxytocic effects. All smooth muscle, vascular or non-vascular, is contracted. The action is a direct one since there is no relation to innervation nor to any of the chemical mediators, but the actual mechanism is not known. The smooth-muscle excitant action is pronounced in the natural alkaloids but greatly reduced in the hydrogenated derivatives and some hydrogenated alkaloids have actions which are the reverse of natural alkaloids. They may produce vasodilation, lower the blood pressure and even inhibit the uterotonic effects of natural alkaloids.

The constriction of intracranial arteries produced by ergot is useful in the treatment of migraine, but the decreased flow through peripheral, mesenteric or coronary arteries may cause local ischaemia. Ergotamine tartrate is the most effective ergot alkaloid in the treatment of migraine and, if given during prodrome, the attack may be completely aborted. Caffeine is sometimes given as well because it is a vasoconstrictor of intracranial vessels and enhances absorption of ergotamine from the gastrointestinal tract.

References p. 94

As a migraine prophylactic methysergide (a synthetic alkaloid with a potent antiserotonin action) has measurable value in selected patients. As many as 20% of patients cannot tolerate methysergide and its chronic use has been associated with retroperitoneal fibrosis, a previously extremely rare syndrome. This reaction is assumed to be due to the chronic arterial constrictive action of methysergide.

Only at term is uterine muscle more sensitive to ergot than is other smooth muscle. In the absence of pregnancy or early in pregnancy, dangerous amounts of ergot are required to demonstrate the uterine stimulating effect and even then the effect is stronger on the cervix. Ergot cannot therefore be used in any context as an abortifacient. During the third trimester the sensitivity of the uterus gradually increases and ergometrine can be used to induce labour and contract the uterus postpartum. For the induction and stimulation of labour it is much easier to adjust the dosage of oxytocin than that of an ergot alkaloid. For the prevention of postpartum bleeding ergometrine or methylergometrine is most commonly used.

Hypertension is the important side-effect of ergometrine. Oxytocin does not have this effect and methylergonovine is supposed to cause less elevation of blood pressure.

Neurohormonal actions

The neurohormonal effects of ergot are those of serotonin antagonism and adrenergic blockage. They are produced by interference with amines at the receptor site of the myoneural junctions. The antiserotonin effect is exploited clinically by methysergide prophylaxis for migraine and other vascular headaches, in the treatment of the carcinoid syndrome and in certain rheumatic affections. The hydrogenated ergot alkaloids, particularly of the ergotoxine group, cause vasodilation and are used clinically to treat high blood pressure and cerebral circulatory disorders. They also improve tissue metabolism by normalizing and stabilizing the phosphatases.

Central effects

The central effects of the ergot alkaloids are complicated. Stimulation or inhibition of the medulla oblongata produces bulbomedullary effects which include bradycardia, vomiting and inhibition of the vasomotor centre. LSD-25, D-lysergic acid diethylamide and lysergide are synonyms for the compound with the most potent central action. The drug affects the midbrain to produce the symptoms of hyperthermia, hyperglycemia, mydriasis, piloerection, tachypnoea and hyperreflexia.

Bové has named LSD-25 the "drug of the century" [6]. Since Hoffmann's discovery of the hallucinogenic properties of this synthetic compound in 1943 more than 2000 scientific papers and numerous reports in other journals have appeared. Excellent reviews on the status of lysergide in medicine and psychotherapy have been made by Hoffer and Osmond [4].

The action of LSD on species other than man has also been investigated. One extraordinary study showed that after administration of the drug to a spider it became less susceptible to external distractions and spun a web that was more meticulously regular in pattern than is usually the case.

Shelesnyak *et al.* showed in 1963 that a single dose of ergocornine during the post-ovulatory phase of the menstrual cycle interfered with the production of progesterone and led to the prevention of decidualization and therefore nidation — the nesting of the fertilized egg in the wall of the uterus [69].

Other articles in 1965 and 1966 showed that there is a very clear relationship between structure of the alkaloids and their ability to affect progesterone metabolism.

The clavine alkaloids have not yet been used therapeutically.

Poisoning following therapeutic use

Considering the extremely widespread therapeutic use of ergot alkaloids, particularly in obstetrics and for migraine, reports of serious toxicity are not excessive. Its manifestation may, however, be disastrous and interesting information has been obtained from such episodes. Only rare, isolated instances of natural poisoning by ergot have been reported during the last few decades; the widely reported outbreak at Port St. Esprit in 1951 was later attributed to an organic phosphorus preparation used to spray the grain.

The tolerance to ergot alkaloids undoubtedly varies enormously between patients. Toxic symptoms are not always a consequence of errors of dosage, but have been ascribed to unusual sensitivity. Less than 4 mg ergotamine tartrate orally, has been reported to cause gangrene of both legs [70]. The extreme pain of migraine has provided an incentive to take massive doses, but more often poisoning is a consequence of excessive use of small doses over many years. Provocative tests with such patients using doses as small as 1 mg have elicited serious symptoms [71]. No scientific proof of increasing sensitivity exists, but nevertheless, such patients are reminiscent of reports by early chroniclers of a progressive increase in sensitivity to ergot. Another understandably frequent cause of poisoning is the practice in delivery rooms of administering ergotamine and vitamin K to the mother and newborn respectively. If the syringes are kept on the same tray, a simple mistake results in the infant receiving the ergotamine at a dosage 20 times that intended for the mother.

Respiratory depression and other neurological symptoms may develop rapidly in newborn infants receiving accidental doses [72]. Erythema, ecchymoses on the extremeties, oedema and pulmonary manifestations have also been reported [73].

Most of the classical symptoms, including sensations of intense burning alternating with those of icy coldness, have been confirmed in recent severe therapeutic accidents in adults. In patients using ergot alkaloids regularly for migraine, insidious symptoms, which may not be recognized for many years, include pains in the legs and arms, cold hands and feet and nervous disturbances [71]. Acute intermittent attacks of these symptoms have been described [74]. Continued use of ergotamine — sometimes daily for many months — may result in the acute onset of arterial insufficiency in the limbs [75]. Arteriography may reveal local or general narrowing of the aorta, renal, iliac and femoral arteries. The arterial walls are smooth and there is no evidence of atheroma or occlusion. Plethysmography has revealed reduced flow and increased peripheral resistance in the calf and ankle. Biopsies of the skin, muscle and vasculature may be normal and the conduction velocity of nerves remains unchanged.

Acute poisoning with gangrene has been reported after the patients had taken 5 and 72 mg ergotamine and 9 mg ergometrine [76—78]. Treatment has included the use of vasodilators, anticoagulants and infusion of low molecular weight dextran. The use of sympathetic block in ergot toxicity is controversial since the main vasoconstrictor effect seems to be due to a direct action on the receptors of the blood vessels proper. However, sympathetic denervation, in the form of either paravertebral block, sympathectomy or spinal anaesthesia appears to give impressive results. In an unusual case where the upper limbs were the most affected, there seems little doubt that brachial plexus blocks were of considerable value [79].

Ergotism and complications occur readily when used in the presence of contraindications. These are: (1) peripheral vascular disease; (2) hypertension; (3) ischaemic heart disease; (4) liver diseases; (5) severe infections, especially puerperal; (6) renal disease; (7) old age; (8) anaemia; (9) thyrotoxicosis. Experimentally, thyrotoxicosis increases the incidence of ergotamine-induced gangrene of the tail of the rat. There is evidence that thyroxine acts by prolonging the vasoconstrictor action of the alkaloid [80].

References

1 G. Barger, Ergot and Ergotism, Gurney and Jackson, London, 1931, p. 1.
2 J.E. Saxton, The indole alkaloids, in R.H.F. Manske (Ed.), The Alkaloids, Vol. VII, Academic Press, New York, 1960, pp. 4—199.
3 A. Stoll and A. Hofmann, The ergot alkaloids, in R.H.F. Manske (Ed.), The Alkaloids, Vol. VIII, Academic Press, New York, 1965, pp. 726—783.
4 A. Hoffer and H. Osmond, The Hallucinogens, Academic Press, New York, 1967, p. 83.
5 A. Stoll and A. Hofmann, The chemistry of the ergot alkaloids, in S.W. Pelletier (Ed.), Chemistry of the Alkaloids, Van Nostrand Reinhold, New York, 1970, pp. 267—300.
6 F.J. Bové, The Story of Ergot, Karger, Basel, 1970, p. 3.
7 D. Gröger, Ergot, in S. Kadis, A. Ciegler and S.J. Ajl, (Eds.), Microbial Toxins, Vol. VIII, Academic Press, New York, 1972, pp. 321—373.
8 R.D. Barr, A.D. Roy and J.R.M. Miller, Brit. Med. J., 4 (1972) 273.
9 D. Bhana and H. Baddeley, East. Afr. Med. J., 47 (1970) 506.
10 A. Stoll, Swiss Patent 79 879 (1918).
11 W.A. Jacbos and R.G. Gould, J. Biol. Chem., 120 (1937) 141.
12 E.C. Kornfeld, E.J. Fornefeld, G.B. Kline, G.B. Mann, M.J. Jones and R.B. Woodward, J. Am. Chem. Soc., 76 (1954) 5256.

13 H. Kobel, E. Schreier and J. Rutschmann, Helv. Chim. Acta, 47 (1964) 1052.
14 A. Hofmann and H. Tscherter, Experientia, 16 (1960) 414.
15 W.A. Jacobs and L.C. Craig, Science, 82 (1935) 16.
16 A. Stoll and A. Hofmann, Helv. Chim. Acta, 26 (1943) 944.
17 A. Stoll, Helv. Chim. Acta, 28 (1945) 1283.
18 S. Smith and G.M. Timmis, J. Chem. Soc., (1937) 396.
19 A. Stoll and A. Hofmann, Helv. Chim. Acta, 26 (1943) 1570.
20 W. Schlientz, R. Brunner, A. Rüegger, B. Berde, E. Stürmer and A. Hofmann, Experientia, 23 (1967) 991.
21 W. Schlientz, R. Brunner, P.A. Stadler, A.J. Frey, H. Ott and A. Hofmann, Helv. Chim. Acta, 47 (1964) 1921.
22 P.A. Stadler, S. Guttmann, H. Hauth, R.L. Huguenin, E.D. Sandrin, G. Wersin and A. Hofmann, Helv. Chim. Acta, 52 (1969) 1549.
23 W. Schlientz, R. Brunner, F. Thudium and A. Hofmann, Experientia, 17 (1961) 108.
24 A. Hofmann, Die Mutterkornalkaloide, Enke, Stuttgart, 1964, p. 1.
25 J.E. Robbers and H.G. Floss, Arch. Biochem. Biophys., 126 (1969) 967.
26 B. Franck and I. Zimmer, Chem. Ber., 98 (1965) 1514.
27 B. Franck, G. Baumann and U. Ohnsorge, Tetrahedron Letters, (1965) 2031.
28 A. Freeborn, Pharm. J., 88 (1912) 586.
29 J.W. Apsimon, J.A. Corran, N.G. Greasey, K.Y. Sim and W.B. Whalley, Proc. Chem. Soc., (1963) 209.
30 A.T. McPhail, G.A. Sim, J.D.M. Asher, J.M. Robertson and J.V. Silverton, Proc. Chem. Soc., (1963) 210.
31 B. Franck and F. Hüper, Angew. Chem. Intern. Ed., 5 (1966) 728.
32 B. Franck, F. Hüper, D. Gröger and D. Erge, Chem. Ber., 101 (1968) 1954.
33 D. Gröger, D. Erge, B. Franck, U. Ohnsorge, H. Flasch and F. Hüper, Chem. Ber., 101 (1968) 1970.
34 W.B. Turner, Fungal Metabolites, Academic Press, New York, 1971, p. 167.
35 N.L. Allport and T. Cocking, Quart. J. Pharm. Pharmacol., 5 (1932) 341.
36 I. Gyenes and J. Bayer, Pharmazie, 16 (1961) 211.
37 G.E. Foster, J. Pharm. Pharmacol., 7 (1955) 1.
38 G.E. Foster, J. McDonald and T.S.G. Jones, J. Pharm. Pharmacol., 1 (1949) 802.
39 A. Stoll and A. Rüegger, Helv. Chim. Acta, 37 (1954) 1725.
40 R. Voigt, Pharmazie, 14 (1959) 607.
41 S. Yamatodani, Ann. Rept. Takeda Res. Lab., 19 (1960) 1.
42 S. Agurell and E. Ramstad, Lloydia, 25 (1962) 67.
43 S. Agurell and E. Ramstad, Acta Pharm. Suec., 2 (1965) 231.
44 F. Santavý, in E. Stahl (Ed.), Dunnschichtchromatographie, Ein Laboratoriumshandbuch, Springer, Berlin, 1967, pp. 405—449.
45 H. Rochelmeyer, Pharm. Ztg. Ver. Apotheker-Ztg., 103 (1958) 1269.
46 K. Teichert, E. Mutschler and H. Rochelmeyer, Deut. Apotheker-Ztg., 100 (1960) 283 and 477.
47 M. Klavehn, H. Rochelmeyer and J. Seyfried, Deut. Apotheker-Ztg., 101 (1961) 75 and 477.
48 R. Fowler, P.J. Gomm and D.A. Patterson, J. Chromatog., 72 (1972) 351.
49 K. Genest, J. Chromatog., 19 (1965) 531.
50 K. Genest and C.G. Farmilo, J. Pharm. Pharmacol., 16 (1964) 250.
51 L.J. McLaughlin, J.E. Goyan and A.G. Paul, J. Pharm. Sci., 53 (1964) 306.
52 A. Hofmann, Pharm. Weekblad, 100 (1965) 1261.
53 K. Roder, E. Mutschler and H. Rochelmeyer, Pharm. Acta Helv., 42 (1967) 407.
54 E. Gäumann, Pflanzliche Infektionslehre, Birkhäuser, Basel, 1951, p. 1.
55 R.F.N. Langdon, Univ. Queensland Papers, Dept., Botany, 3 (1954) 61.
56 V. Skalický and F. Starý, Preslia, 34 (1962) 229.
57 L.R. Brady, Lloydia, 25 (1962) 1.
58 M. Abe and S. Yamatodani, Progr. Ind. Microbiol., 5 (1964) 205.

59 D. Gröger, in M. Lange-de la Camp (Ed.), Das Art- und Rassenproblem bei Pilzen in taxonomischer, morphologischer, cytologischer, psychologischer, biochemischer und genetischer Sicht, Fischer, Jena, 1968, pp. 169—179.
60 C.M. Edwards, Vet. Rec., 65 (1953) 158.
61 J.M. Kingsbury, Poisonous Plants of the United States and Canada, Prentice-Hall, New Jersey, 1964.
62 V. Grasso, Boll. Staz. Patol. Veg. (3), 15 (1957) 317.
63 S.F. Fuentes, M.L. de la Isla, A.J. Ullstrup and A.E. Rodriguez, Phytopathology, 54 (1964) 379.
64 E.E. Mas, Bol. Soc. Quim. Peru, 4 (1938) 3.
65 G.T. Sinner and L.C. Schramm, J. Pharm. Sci., 57 (1968) 889.
66 M.E. Pantidou, Can. J. Botany, 37 (1959) 1233.
67 C.J. Wang, Diss., University of Michigan, Ann Arbor, 1966.
68 L.H. Anderson and J.A. Wells, Proc. Soc. Exptl. Biol. Med., 67 (1948) 53.
69 M.C. Shelesnyak, B. Lunenfeld and B. Honig, Life Sci., 1 (1963) 73.
70 W. Bross, T. Cisek, T. Czerada and S. Kaziminski, Lancet, 1 (1963) 85.
71 I. Enge and E. Sivertssen, Am. Heart J., 70 (1965) 665.
72 G.G. Brereton-Stiles, W.S. Winship, N.M. Goodwin and R.F. Roos, S. Afr. Med. J., 46 (1972) 2052.
73 W.M. Edwards, Clin. Pediat., 10 (1971) 257.
74 G. Glazer, K.A. Myers and E.R. Davies, Postgrad. Med. J., 42 (1966) 562.
75 M.S. Fedotin and C. Hartman, New Engl. J. Med., 283 (1970) 518.
76 M. London, F. Magora, S. Rogel and H. Romanoff, Angiology, 21 (1970) 565.
77 J.J. Cranley, R.J. Krause, E.S. Strasser and C.D. Hafner, New Engl. J. Med., 269 (1963) 727.
78 R.E. Devitt, E. Chater and D.S. Colbert, J. Ir. Med. Ass., 63 (1970) 441.
79 P. Hudgson and J.A.L. Hart, Med. J. Aust., 2 (1964) 589.
80 J.A. Wells and L.H. Anderson, Proc. Soc. Exptl. Biol. Med., 74 (1950) 374.
81 F. Arcamone, C. Bonino, E.B. Chain, A. Ferretti, P. Pannella, A. Tonolo and L. Vero, Nature, 187 (1960) 238.
82 A. Stoll and E. Burckhardt, Z. Physiol. Chem., 250 (1937) 1.
83 H.W. Dudley and C. Moir, Brit. Med. J., 1 (1935) 520.
84 M.S. Kharasch and R.R. Legault, Science, 81 (1935) 388 and 614.
85 A. Stoll and E. Burckhardt, Compt. Rend., 200 (1935) 1680.
86 M.R. Thompson, Science, 81 (1935) 636.
87 M. Abe, T. Yamano, S. Yamatodani, Y. Kozu, M. Kusumoto, H. Komatsu and S. Yamada, Bull. Agr. Chem. Soc. Japan, 23 (1959) 246.
88 W. Schlientz, R. Brunner and A. Hofmann, Experientia, 19 (1963) 397.
89 M. Abe, Ann. Rept. Takeda Res. Lab., 10 (1951) 73.
90 M. Abe, S. Yamatodani, T. Ymamano and M. Kusumoto, Bull. Agr. Chem. Soc. Japan, 20 (1956) 59.
91 A. Hofmann, R. Brunner, H. Kobel and A. Brack, Helv. Chim. Acta, 40 (1957) 1358.
92 D. Stauffacher and H. Tscherter, Helv. Chim. Acta, 47 (1964) 2186.
93 D. Stauffacher, P. Niklaus, H. Tscherter, H.P. Werner and A. Hofmann, Tetrahedron, 25 (1970) 5879.
94 M. Abe, T. Yamano, Y. Kozu and M. Kusumoto, J. Agr. Chem. Soc. Japan, 25 (1952) 458.
95 H.G. Floss, H. Günther, U. Mothes and I. Becker, Z. Naturforsch., 22b (1967) 399.
96 M. Abe and S. Yamatodani, J. Agr. Chem. Soc. Japan, 28 (1954) 501.
97 J.F. Spilsbury and S. Wilkinson, J. Chem. Soc., (1961) 2085.
98 S. Agurell, Acta Pharm. Suec., 3 (1966) 7.
99 M. Abe, S. Yamatodani, T. Yamano and M. Kusumoto, Agr. Biol. Chem. (Toyko), 25 (1961) 594.
100 M. Abe and S. Yamatodani, Bull. Agr. Chem. Soc. Japan, 19 (1955) 161.
101 E. Ramstad, W.N. Chan Lin, R. Shough, K.J. Goldner, R.P. Parikh and E.H. Taylor, Lloydia, 30 (1967) 441.

Chapter 4

RHIZOCTONIA LEGUMINICOLA — SLAFRAMINE

STEVEN D. AUST

Department of Biochemistry, Michigan State University, East Lansing, Mich. (U.S.A.)

Numerous reports of excessive salivation in cattle consuming certain forages have led to the isolation and characterization of an interesting parasympathomimetic alkaloid called slaframine. Slaframine is synthesized by *Rhizoctonia leguminicola* growing primarily on red clover. The fungus will also produce the toxin while growing on extracts of red clover. The compound is an acetate ester of a unique bicyclic amine synthesized in part from lysine. Slaframine itself has no physiological activity, however, it can be transformed biologically or photochemically to a quaternary amine that is both chemically and physiologically very similar to acetylcholine.

Occurrence

Throughout the Midwest states during the 1950s numerous samples of red clover hay were submitted to Experimental Stations for testing because the hay had caused slobbering and symptoms of toxicity in cattle and other livestock. The problem was first documented in Missouri during the period 1949 to 1958 by O'Dell *et al.* [1]. The cattle would consume from one to three feedings of such forage (termed "slobber forage"), salivate profusely, and refuse further feed. This represented a loss to the farmer including loss of production and the necessity to purchase replacement forage, resulting eventually in the elimination of red clover as a forage crop in the Midwest. The problem occurs almost exclusively on second cutting red clover or in a new seeding of red clover after the nurse crop, or catch crop, has been harvested. Widespread outbreaks of the problem seem to occur under certain climate conditions. Cool moist weather almost always seems to precede the occurrence of slobber forages.

Rhizoctonia leguminicola has also been identified as the causative agent in a red clover disease called Blackpatch Disease [2]. This disease was first reported in Kentucky in 1933 but has been reported throughout the Midwest and Southeast. The dark brown or black, separate filaments give the mycelia its black appearance. The occurrence of the disease, like the occurrence of slobber forages, is favoured by high humidity and mild temperature ($25-29°$). While other legumes can be infected, the natural disease seems restricted to red clover. All attempts to control the forages or find resistant individual plants have been unsuccessful. The fungus has never been observed to produce spores, characteristic of the genus Rhizoctonia, but it can form prominent sclerotia under proper culture conditions.

Isolation of slaframine

Smalley et al. [3] and Crump et al. [4] examined slobber forages in Wisconsin and observed that while the hay was not visibly mouldy, upon microscopic examination all samples proved to be infested with the heavy mycelium of a dark-brown coloured fungus. They succeeded in isolating the fungus in pure culture, identified it as *Rhizoctonia leguminicola*, and connected the fungus with Blackpatch disease of red clover. Smalley et al. [3] grew the fungus on extracts of non-toxic red clover and then force fed the mycelium to guinea pigs which began to salivate profusely after about 60 min. Similar results were obtained when the mycelium was introduced directly into the rumen of fistulated dairy cows. They also found that the culture filtrate contained no toxic material.

With the knowledge that the toxic principle was produced by *Rhizoctonia leguminicola* and the fact that the mycelia grown in pure culture contained the toxic principle, groups at the University of Wisconsin [5] and the University of Illinois [6] endeavoured to isolate the metabolite from the mycelium. A bioassay based on the salivary response of guinea pigs was developed. The isolation procedure developed by Aust [6] started from an ethanol extract of mycelium grown on a 20% cold water infusion of second cutting red clover hay. The mycelium was grown in stationary cultures (Roux bottles) for 3 weeks at 25°. At temperatures higher than 25° growth and the production of slaframine are drastically decreased. All attempts to produce slaframine by growing mycelia on refined culture media were unsuccessful and the production of slaframine by the fungus on extracts of forages other than red clover was minimal.

The procedure finally perfected for the isolation of slaframine is shown in Fig. 1. The mycelium was pressed free of culture media and finely ground in ethanol in a large Waring blender. The mycelium was then filtered into large Soxhlet thimbles and extracted with the ethanolic filtrate for 24 h in Soxhlet extractors. The ethanol was removed from the extract under reduced pressure and the resulting aqueous solution treated with lead diacetate to precipitate a majority of the contaminants. After centrifugation the solution was extracted several times with an equal volume of chloroform. The pH of the solution was then raised to 10, centrifuged and extracted three times with an equal volume of chloroform. Only at pH above 9.0 will slaframine partition into the chloroform phase and therefore this procedure can be repeated as necessary to further purify the toxin. Chloroform-soluble impurities can be extracted into chloroform at acidic pH and slaframine extracted into chloroform at pH 10. The final extract can be used to prepare salts or derivatives of slaframine. For chemical studies the final extract may be acetylated with acetic anhydride and purified by sublimation. For

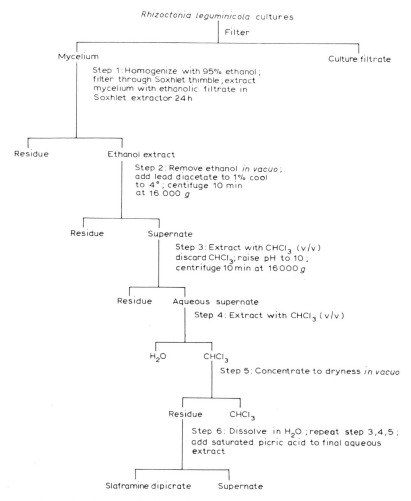

Fig. 1. Slaframine purification scheme.

physiological studies the final extract may be dissolved in dry diethyl ether and the dicitrate salt prepared by precipitation with a saturated solution of citric acid in dry diethyl ether. Crystalline dipicrate is formed by precipitating the base from an acidic solution (HCl) with a saturated solution of picric acid. The dipicrate can then be recrystallized to purity in 20% ethanol. The picrate is fairly stable in the dry state; however, in acidic aqueous solution slaframine undergoes fairly rapid hydrolysis. The free base of slaframine is a clear oil; however, it undergoes oxidation in air at a rapid rate. The alkaloidal nature of the toxin and its physiological action led to the name slaframine (from the Old Norse *slafra*, to slaver).

References p. 109

Molecular structure of slaframine

Slaframine was found to have a molecular weight of 198 as determined by the molecular ion at m/e 198 in the mass spectrum of the free base [7]. This agrees with the elemental analysis of the dipicrate $C_{10}H_{20}N_2O_3 \cdot 2C_6H_3N_3O_7$, m.p. 183—184. Treatment with weak base results in the loss of physiological activity and the release of acetic acid. Van Slyke amino nitrogen analysis indicated that one of the nitrogen atoms was a primary amine and slaframine is ninhydrin-positive. Treatment of slaframine with acetic anhydride results in a diacetylated compound $C_{12}H_{20}N_2O_3$, m.p. 143—146, called *N*-acetylslaframine. As slaframine forms a dipicrate derivative, the second nitrogen must be basic. The presence of a tertiary nitrogen was confirmed for slaframine is positive to Dragendorff's reagent and to citric acid acetic anhydride. As slaframine possesses only one methyl group, located on the acetoxy moiety, the second nitrogen must be a bridgehead nitrogen joining two fused rings [7].

Application of high resolution mass spectrometry, in conjunction with deuterium oxide exchange, to slaframine and a cyanogen bromide cleavage product of slaframine, resulted in the structure and fragmentation pattern shown in Fig. 2 [7,8]. The m/e 142 fragment did not contain deuterium when the exchangeable hydrogens of slaframine were exchanged with deute-

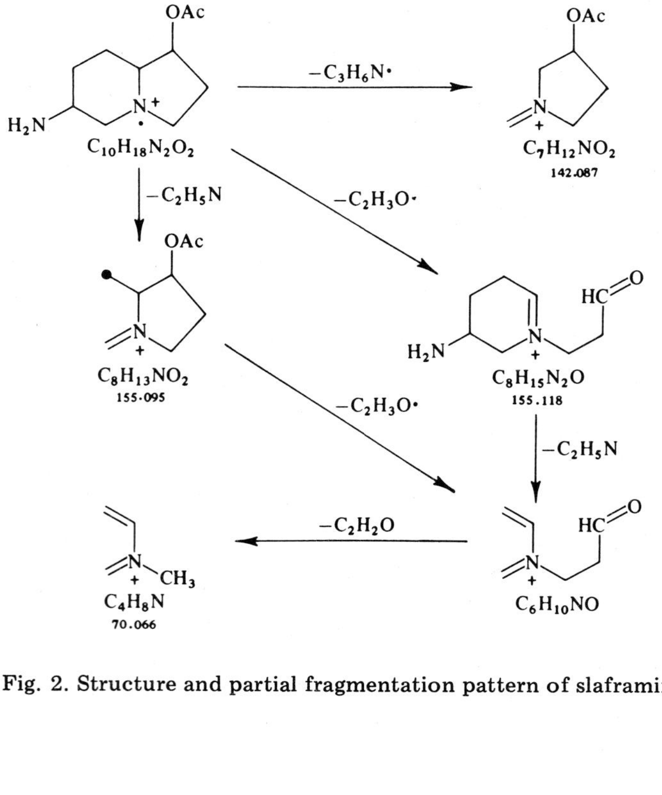

Fig. 2. Structure and partial fragmentation pattern of slaframine.

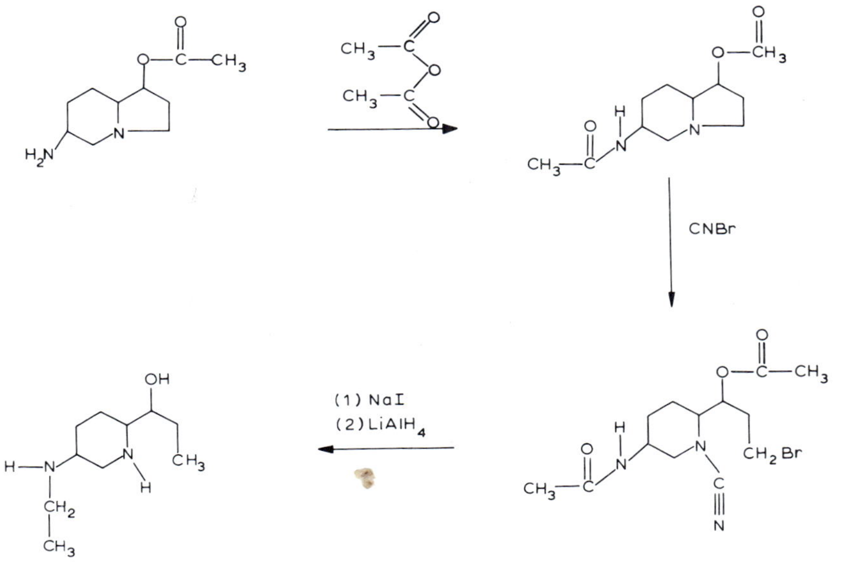

Fig. 3. Cyanogen bromide cleavage of *N*-acetylslaframine.

rium. Thus the fragment must arise from the molecular ion by the loss of C_3H_6N, the primary nitrogen plus a three-carbon fragment. Two m/e 155 fragments were resolved by high resolution mass spectrometry and only one was found to contain exchangeable hydrogens. The other results from the loss of acetyl from the molecular ion. The pyrrolidine ring of *N*-acetyl-slaframine was cleaved with cyanogen bromide and the cleavage product reduced with lithium aluminium hydride after bromide replacement by iodide (Fig. 3). Nuclear magnetic resonance of this product showed a $C\text{-}CH_2\text{-}CH_3$ group which was shown to be part of $CHOH\text{-}CH_2\text{-}CH_3$ by loss of C_3H_7O in the mass spectrum of the reduced cyanogen bromide cleavage product. Since the terminal group was ethyl the acetoxy group of slaframine must be placed on C-1.

The absolute configuration of slaframine (Fig. 4) was deduced in the following manner. The absolute configuration of the acetoxy group was found by reaction of internally racemic 2-phenylbutyric anhydride with *N*-acetyl-*O*-deacetyl-slaframine [9]. The residual 2-phenylbutyric acid was found to be levorotatory, which empirically results in the assignment of S configuration. Since the C-1 proton and the bridgehead proton are *cis* the absolute configuration at C-8a was deduced to be S. The primary amino group was placed by nuclear magnetic resonance spin decoupling studies. The H-5 axial proton was found to be coupled only to the H-5 equatorial proton and one H-6 proton. If two protons occupied the C-6 position two couplings would be expected. Therefore, the amino group must occupy the remaining position on C-6. The band width of the H-6 proton is typical of an

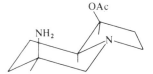

Fig. 4. Stereochemistry of slaframine.

equatorial proton, thus the amino group is axially directed and of S configuration. This information results in the structural assignment of slaframine
as (1S, 6S, 8aS)-1-acetoxy-6-aminooctahydroindolizine (Fig. 4). Final confirmation of this structure has been obtained by synthesis of N-acetyl-
slaframine and comparison with the natural product [8].

Analysis for slaframine

The most obvious physiological manifestation of slaframine is profuse
salivation (although all cholinergic exocrine glands are stimulated). Therefore
a sensitive bioassay can be developed with almost any laboratory animal by
simply injecting animals with various doses of the test material and observing
for salivation after a period of time [6]. The guinea pig is the usual animal of
choice for a bioassay for it can be used to assay toxic feeds or solutions
containing slaframine. The feed is usually finely ground in a hammer mill
and mixed with an equal weight of a standard guinea pig ration. Guinea pigs
are then fed the mixture for at least one week and observed daily for excessive salivation. As an index of the degree of salivation, the response is graded
1+ for slight, 2+ for intermediate, and 3+ for the most severe degree.
Animals which graded 3+ drool so much that the saliva keeps the fur on the
throat constantly wet.
    Solutions containing slaframine can be tested by the guinea pig bioassay
providing they contain no other toxic principles. The solution is injected
intraperitoneally into animals of comparable age, sex and weight. The
animals are then observed periodically after injection for the degree of salivation, rated 0 to 3+, as described above. Salivation usually occurs in 10 to 15
min and if it does not occur in about 30 min it usually will not appear and
the test can be regarded as negative, although low doses may give a non-
detected response.
    Samples for bioassay can be concentrated or deionized very simply
owing to the fact that slaframine can be extracted into chloroform at basic
pH. A solution to be tested can first be extracted with chloroform at neutral
pH and then at pH 10. The pH 10 extract can then be dried under vacuum
and dissolved in an appropriate amount of saline and the pH adjusted to
neutrality. Extraction is practically quantitative and the slaframine remains
active providing the pH does not exceed 10, at which time hydrolysis of
slaframine occurs.
    The species used and other factors such as age, prior treatment, etc. are
also important. This stems from the fact that slaframine must be activated

by liver microsomal enzymes (see Section VII) which show variation between species, with age and prior drug treatment.

Chemical analysis for slaframine can be best achieved by gas chromatography (OV-17 or SE-30); however, qualitative determinations can be made by paper or thin-layer chromatography. Slaframine can be detected on paper and thin-layer plates by ninhydrin or, more specifically, Dragendoff's reagent.

Biosynthesis of slaframine

Biosynthetic studies were actually initiated prior to complete characterization of slaframine in hopes of obtaining information useful in assigning chemical structure. The systematic addition of labeled amino acids to growing cultures of *Rhizoctonia leguminicola* and subsequent measurement of radioactivity in the isolated alkaloid revealed that lysine and serine were involved in the biosynthesis of slaframine [10]. Both C-1 and C-6 of lysine were incorporated but only carbons 2 and 3 of serine. The involvement of lysine seemed logical when the piperidine ring in slaframine was elucidated. This finding suggested the involvement of pipecolic acid in slaframine biogenesis and indeed pipecolic acid was most effective in diluting out the incorporation of labeled lysine into slaframine and is itself directly incorporated into slaframine [11]. α-Aminoadipic acid could also be shown to be involved by the fact that it was incorporated into slaframine. Aminoadipic acid was also found to be involved in the biosynthesis of lysine by this organism [11].

Physiological mode of action of slaframine

The biological activity of slaframine appears to be limited to the cholinergic stimulation of exocrine glands, the most obvious being salivation [12]. No effect of slaframine can be demonstrated on ganglionic transmission, peripheral blood flow or neuromuscular transmission [12]. Also, at reasonable doses no change in heart rate, respiration or blood pressure can be observed. The salivary response to slaframine can be blocked by prior administration of atropine, indicating the involvement of the cholinergic system. However, it is important that the atropine is administered prior to slaframine, for once salivation has started it cannot be reversed by atropine. This may have some implications concerning the affinity of the two compounds for the cholinergic receptor. The compound must be binding directly to the receptor for if it were increasing acetylcholine concentration, by inhibiting cholinesterase, for example, atropine should readily reverse the effects of slaframine. Quantitative salivation studies, in which the salivary ducts of anaesthetized cats were cannulated so that salivary flow due to slaframine could be measured by means of a photoelectric drop counter, indicated that a single dose of slaframine (0.3 mg/kg) was capable of stimulating salivation for several hours (Fig. 5) after which time the animals were

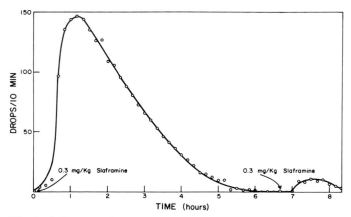

Fig. 5. Salivary activity (submaxillary) of an anaesthetized cat after the administration of slaframine.

almost unable to respond to a second dose of slaframine. Quantitative pancreatic studies were conducted by cannulating the pancreatic duct of goats and calves [13], After a brief delay, presumably required for the bioactivation of slaframine, there is a substantial and prolonged increase in the rate of secretion of pancreatic fluid (Fig. 6).

Analysis of the pancreatic fluid obtained *via* a permanent pancreatic cannula in calves given slaframine yielded some interesting results concerning the selectivity of stimulation by slaframine [13]. As the rate of secretion increased, the concentration of the various components in the fluid changed. The concentration of N-acetylneuraminic acid (NANA) decreased while the specific activity of the digestive enzymes increased. Since the concentration of total protein did not significantly change these results suggested that there was a selective secretion of digestive enzymes. Since the stimulated secretion rate persists for such a long period of time there must be an increased rate of synthesis of digestive enzymes. This possibility was strengthened by the fact that there was an increased incorporation of $[^{14}C]$-leucine into pancreatic proteins by an animal treated with slaframine (Fig. 7).

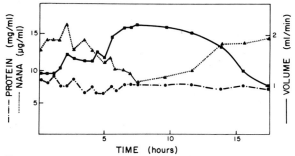

Fig. 6. Changes, with time, in the volume, protein and NANA content of pancreatic fluid secreted by a calf stimulated with slaframine (0.2 mg/kg).

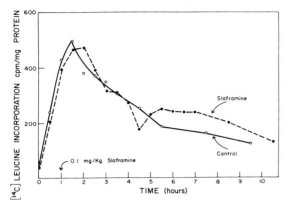

Fig. 7. The effect of slaframine (0.1 mg/kg) on the incorporation of [^{14}C]leucine into pancreatic protein by the calf.

All attempts to demonstrate an effect of slaframine in cholinergic *in vitro* test systems have met with failure. This fact, coupled with the observation that slaframine has no immediate effect *in vivo* suggested that the compound required bioactivation. Indeed, it could be shown that the compound was most active when administered *via* the portal vein and was not active at all when the vena cava was ligated (Table I) [12]. This suggested that activation was accomplished by the liver.

Once the liver was implicated, the involvement of the mixed-function oxidase responsible for most biological drug transformations was investigated by pretreating the animals with specific inducers and inhibitors of this sys-

TABLE I

EFFECT OF ROUTE OF ADMINISTRATION ON THE DELAY OF SALIVATION AFTER SLAFRAMINE ADMINISTRATION

Method and route of administration	Species	Dose (mg/kg)	Delay [a]
i.v. (tail vein)	mouse	2	24 ± 3.0
i.p.	mouse	2	16 ± 2.0
Portal vein [b]	mouse	2	9 ± 1.0
i.v. [b]	mouse	2	23 ± 3.5
Vena cava liver ligate [b] 3 min [c]	rat	4	16 ± 1.0
Vena cava liver ligated 5 min [c]	rat	4	21.5 ± 2.5
Vena cava liver ligated 7 min [c]	rat	4	29 ± 4.0
Vena cava liver ligated ∞ min [c]	rat	4	∞
Sham operated control 0 min [c]	rat	4	10.5 ± 1.25

[a] Times are minutes ± S.D. that salivation occurred after the administration of slaframine. All tests include at least 4 animals per group.
[b] Animals were anaesthetized with pentobarbital (100 mg/kg) and an abdominal incision was made for slaframine injections or sham operations.
[c] Minutes the vena cava was ligated following the administration of slaframine.

References p. 109

TABLE II

EFFECT OF VARIOUS INDUCERS AND INHIBITORS OF DRUG-METABOLIZING ENZYMES ON THE DELAY[a] OF SALIVATION AFTER SLAFRAMINE ADMINISTRATION

Treatment	Dose (mg/kg)	Delay
Control		15.3 ± 0.8
Slaframine	5 (10 days)	16.8 ± 0.8
Phenobarbital[b]	100	8.8 ± 0.5
DDT[b]	30	8.0 ± 0.5
Chlordane[b]	40	8.0 ± 0.5
SKF-525-A[c]	50	30 ± 1.0
SKF-8742-A[c]	50	27 ± 2.0
DPEA[c]	7.5	20 ± 1.0
Lilly 18947[c]	50	27 ± 1.7

[a] Times are minutes ± S.D. that salivation started after the administration of slaframine (5 mg/kg). All tests included at least 4 mice per group and all injections were made i.p.
[b] The inducers were given daily for 5 days prior to testing with slaframine.
[c] The inhibitors were given 30 min prior to testing.

tem [12]. Table II shows that pretreating animals with compounds known to induce these enzymes (phenobarbital or DDT) greatly enhances the activity of slaframine, measured as the time that salivation occurred after the administration of slaframine. Interestingly, slaframine was unable to induce its own activation. Various compounds known to inhibit the mixed-function oxidase of liver endoplasmic reticulum were able to delay the onset of salivation in response to slaframine. These results suggested that slaframine is activated by the drug metabolism enzymes of liver endoplasmic reticulum (microsomes).

Incubation of slaframine with rat liver microsomes in the presence of NADPH resulted in the production of a product or products capable of stimulating the contraction of the guinea pig ileum (Fig. 8A) [14]. As was the case *in vivo*, atropine was not able to reverse the effect of the product on the guinea pig ileum *in vitro* (Fig. 8B), but was able to prevent the effect if previously administered (Fig. 8C). Incubating the reaction under an atmosphere of 80% carbon monoxide and 20% oxygen did not inhibit the production of the compound capable of contracting the guinea pig ileum, indicating that cytochrome P_{450} may not be involved.

During the activation experiments, it was observed that boiled microsomes were able to activate slaframine to a small and variable extent [15]. Further investigation of this phenomenon revealed that the reaction was dependent upon light and flavins. The most efficient flavin was flavin mononucleotide (FMN). The FMN-catalyzed activation also occurred anaerobically and during the anaerobic reaction the reduction of FMN could be observed [13]. The reduction of FMN could be equated with the production

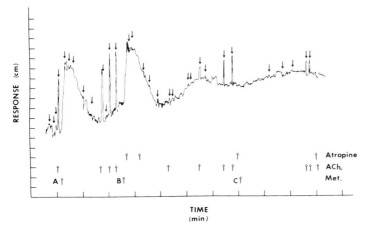

Fig. 8. Effect of slaframine metabolites on the isolated guinea pig ileum. Additions of the metabolites (Met.), acetylcholine (ACh.) and atropine are indicated the the lower arrows. Washes are indicated by the upper arrows.

of the active metabolite (Fig. 9). Further investigation of this phenomenon revealed that maximum activation of slaframine and reduction of FMN by slaframine occurred at the absorption maximum for FMN (455 nm).

During photochemical activation of slaframine (by light-activated FMN) 1 mole of ammonia is produced per mole of slaframine consumed (unpublished results). The deaminated products of slaframine cannot be extracted from the aqueous phase by organic solvents, but they can be isolated by cation exchange chromatography. Upon reduction with sodium borohydride (at which time biological activity is lost) two compounds can be extracted with chloroform. These products have been identified as derivatives of slaframine in which the primary amino group has been replaced by either

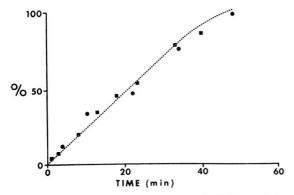

Fig. 9. Reduction of photoactivated FMN by slaframine and the activation of slaframine. Slaframine $(1 \cdot 10^{-3} M)$ was mixed with FMN $(1 \cdot 10^{-4} M)$ under reduced light and assayed for photoreduction of FMN (anaerobically at 450 nm) and activation of slaframine (aerobically with the guinea pig ileum assay). Data are expressed as per cent of final activity. ●, slaframine *in vitro* activity; ■, FMN reduction.

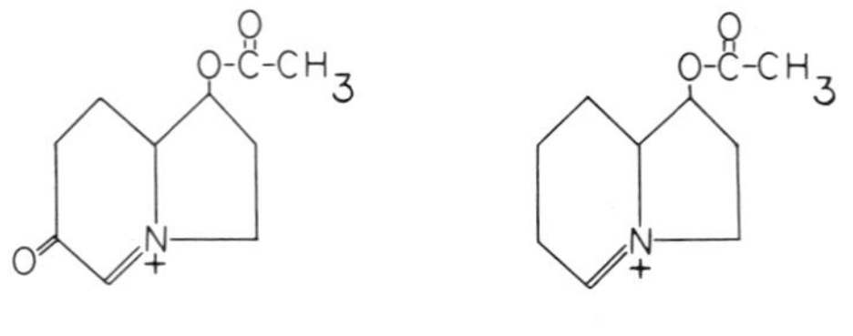

Fig. 10. Proposed deaminated products of slaframine.

a hydrogen, to give deaminoslaframine, or a hydroxyl group, to give hydroxyslaframine. When the photochemical reaction products are reduced with sodium borodeuteride the deaminoslaframine isolated was found to contain one deuterium atom (determined by mass spectrometry) whereas the hydroxyslaframine was found to contain two deuterium atoms. These results would be obtained only if the photochemical reaction products were deaminated forms of slaframine in which the tertiary amine has been oxidized to a quaternary amine. Since the hydroxyslaframine isolated after sodium borohydride reduction contained two deuteriums, the hydroxyl group must have been a ketone. The second deuterium would be on carbon-6 and the hydroxyl deuterium would have exchanged with the solvent. The two cations would therefore have the structures shown in Fig. 10. Preliminary studies have shown that the ketone (I) is the biologically active product [16]. The cations were separated by cation exchange chromatography and assayed for biological activity using the guinea pig ileum assay. Only the ketone was found to have biological activity. Evidence has also been obtained to suggest that rat liver microsomes are capable of producing the ketone, suggesting that the rat liver can activate slaframine.

Summary

The agent in certain legume forages that causes excessive salivation in cattle consuming these forages has been identified as 1-acetoxy-6-aminooctahydroindolizine and named slaframine. The compound is synthesized from lysine by the fungus *Rhizoctonia leguminicola* and can be produced in pure cultures of the fungus growing on extracts of red clover.

Slaframine is activated by enzymes located in the endoplasmic reticulum of the liver which can be induced by pretreating the animals with a variety of compounds, including phenobarbital. The active metabolite can also be generated photochemically with FMN. During anaerobic activation the FMN can be shown to be reduced and free ammonia is produced. The active metabolite cannot be extracted into organic solvents; however, following reduction with sodium borohydride (at which time activity is lost) two products can be extracted into chloroform and have been identified as deaminoslaframine and hydroxyslaframine. Therefore the deaminated products of slaframine must be quaternary amines in which the primary amino group has been either replaced by a keto group (I) or removed (II). Preliminary results suggest that the ketone has biological activity and can be formed from slaframine by rat liver.

References

1   B.L. O'Dell, W.O. Regan, and T.J. Beach, Missouri Univ. Agric. Exptl. Stat., Res. Bull., 702 (1959) 12.
2   F.J. Gough and E.S. Elliott, West Va. Univ. Agric. Exptl. Stat., Bull., 387T (1956).
3   E.B. Smalley, R.E. Nichols, M.H. Crump and J.N. Henning, Phytopathology, 52 (1962) 753.
4   M.H. Crump, E.B. Smalley, J.N. Henning and R.E. Nichols, J. Am. Vet. Med. Assoc., 143 (1963) 996.
5   D.P. Rainey, E.B. Smalley, M.H. Crump and F.M. Strong, Nature, 205 (1965) 203.
6   S.D. Aust and H.P. Broquist, Nature, 205 (1965) 204.
7   S.D. Aust, H.P. Broquist and K.L. Rinehart Jr., J. Am. Chem. Soc., 88 (1966) 2879.
8   R.A. Gardiner, K.L. Rinehart Jr., J.J. Snyder and H.P. Broquist, J. Am. Chem.Soc., 90 (1968) 5639.
9   A. Horeau, Tetrahedron Letters, 506 (1961).
10 S.D. Aust, Doctoral Thesis, University of Illinois, Urbana, Ill., 1965.
11 F.P. Guengerich, Federation Proc., 30 (1971) 1067.
12 S.D. Aust, Biochem. Pharmacol., 18 (1969) 929.
13 S.D. Aust, Biochem. Pharmacol., 19 (1970) 427.
14 T.E. Spike and S.D. Aust, Biochem. Pharmacol., 20 (1971) 721.
15 T.E. Spike, M.S. Thesis, Michigan State University, East Lansing, Mich., 1969.
16 F. Peter Guengerich, Ph.D. Thesis, Vanderbilt University, Nashville, Tenn., 1973.

Chapter 5

PHOMOPSIS LEPTOSTROMIFORMIS

W.F.O. MARASAS

Plant Protection Research Institute, Pretoria (Republic of South Africa)

The fungus *Phomopsis leptostromiformis* (Kühn) Bubák ex Lind was recently shown to be the cause of lupinosis in sheep [1—3] and mice [4]. Lupinosis is a hepatotoxic condition characterized by severe liver damage and icterus, and should be clearly distinguished from lupin poisoning which is a nervous disorder caused by alkaloids present in bitter lupins [5—8]. A comprehensive review of lupinosis was published by Gardiner in 1967 [8].

The first intensive studies on lupinosis were carried out in Germany during the latter half of the previous century when this disease became a serious threat to sheep farming in that country [5,9,10]. These early German workers carefully distinguished between lupinosis and alkaloid poisoning, and from the very beginning suggested that toxic fungi growing on the lupin plants may have been the cause of lupinosis [5]. In feeding trials with sheep, Liebscher [9] demonstrated that lupin alkaloids could cause nervous disorders, but not icterus and liver damage. He found that the hepatotoxic factor in certain lots of lupin hay was destroyed by steaming and could be extracted with alkali. This principle was named "ictrogen" [9]. Subsequently, several investigators confirmed that the lupin alkaloids do not cause the hepatic damage and icterus characteristic of lupinosis [7,11—14]. Nevertheless, the confusion between alkaloid poisoning and lupinosis persisted for a long time, and even in 1961 it was stated that: "To the present author, it seems clear that alkaloids are the cause of lupinosis" [15]. The theory that the lupinosis toxin is in some way related to the lupin alkaloids, led to the erroneous supposition that alkaloid-poor or "sweet" lupins do not cause lupinosis [15]. It is now well established that bitter as well as sweet lupins are capable of causing lupinosis [1,2,3,8,11,16] and that the lupin alkaloids are in no way involved in the aetiology of the disease [1,2,4].

The second major phase in lupinosis research was initiated in Australia when serious and widespread outbreaks started to occur around 1950 [7,8,12]. At the beginning of these investigations, dietary deficiencies and copper excess were investigated as possible aetiological factors. The possibility that lupinosis may be a manifestation of dietary deficiency was investigated because a number of factors had contributed to the grazing of sheep almost exclusively on lupins during the dry summer and autumn months [12]. Based on pathological similarities, deficiencies of choline and methionine in the lupin diet were considered as possible causes of the liver damage in lupinosis [12]. The possible importance of other nutritional factors such as vitamin E deficiency were also investigated [8]. The same line of investigation was followed in South Africa at that time, and methionine deficiency was incriminated in the aetiology of lupinosis [11,17]. In the second place, the possible rôle of copper excess was investigated in Australia [18—20]. This was done because of the fact that the first serious outbreaks of lupinosis occurred after the establishment of the practice of using copper fertilizers on the copper-deficient sand plain soils in the lupin-growing areas of Western Australia [8,19,20]. Moreover, sheep affected by lupinosis were found to accumulate excessive amounts of copper in their livers [18,20]. Gardiner [20] concluded that the pathogenesis of lupinosis is basically concerned with toxic factors in the lupin feeding, but that liver copper storage may add an important complicating feature.

From 1960 onward, the possibility that fungi growing on the lupin plants may render them toxic, which was originally suggested by the early German workers, was intensively investigated in Western Australia [21—23]. The same approach was taken in South Africa following a field outbreak of lupinosis in sheep during October, 1969 [1—3]. These investigations culminated in the independent discovery in South Africa [1—3] and Australia [4] that the fungus *Phomopsis leptostromiformis* is the cause of lupinosis.

Lupinosis — the syndrome

(a) Geographical distribution

(i) Europe

Yellow-flowered, bitter lupins (*Lupinus luteus* L.) were sown in Germany for the first time in 1841 and the first outbreak of lupinosis in sheep grazing these plants occurred in 1862 [15]. These outbreaks increased in severity and by 1880 lupinosis had become a major sheep disease in Germany with the result that the practice of feeding lupins to sheep declined [5,10,15]. Following the development of alkaloid-poor or sweet varieties of *L. luteus*, *L. albus* L. and *L. angustifolius* L., the cultivation of these sweet varieties has increased rapidly since 1934 in Germany, Poland and Hungary, and only sweet lupins are fed to stock [15]. According to Hackbarth [15], lupinosis ceased to be a problem in Europe after the use of sweet lupins became general. Sporadic outbreaks of lupinosis have, however, occurred in Europe since 1934, *e.g.* in dairy cattle in Germany [24] and in cattle, pigs and horses in Poland [8].

(ii) North America

The only report indicating that lupinosis may occur in North America is the brief description of suspected equine lupinosis in Montana by Marsh *et al.* [6]. During the period 1940 to 1960, yellow lupins were extensively grazed by cattle in the south-eastern United States without any health problem [25].

(iii) New Zealand
Three outbreaks of lupinosis in sheep grazing *L. angustifolius* were reported in New Zealand during the summer of 1943 [26].

(iv) Australia
Although bitter blue lupins (*L. varius* L. and *L. angustifolius*) have since about 1920 been used extensively in Western Australia as a sheep fodder, the first authenticated outbreak of lupinosis occurred in the autumn of 1948 [12]. Lupinosis has become an increasingly serious problem since 1950, and widespread and severe outbreaks have occurred in sheep [7,8,12,13,22,27, 28] and to a lesser extent in cattle [29] and horses [30].

(v) South Africa
Lupins (*L. luteus* and *L. angustifolius*) were introduced into the winter-rainfall Swartland area of the western Cape Province shortly after World War II, and their cultivation as a green manure crop in rotation with wheat and as a forage crop for sheep increased rapidly in the 1950s [31—33]. The first recorded outbreak of lupinosis occurred in April, 1954 in sheep grazing dry lupin lands [11,17]. Outbreaks occurred regularly between 1954 and 1965 [8], and a severe outbreak was recorded in October 1969 in sheep grazing sweet white lupins (*L. albus*) [1,2]. In recent years there has been a great decrease in the area planted to lupins in the western Cape [33], but sweet blue *L. angustifolius* is being planted to an increasing extent by dairy farmers in the Alexandria-Bathurst areas of the eastern Cape Province [34].

(b) Clinical signs and pathological changes

(i) Sheep
Lupinosis is known primarily as a sheep disease, probably not because sheep are more susceptible than other animal species, but because lupins have been used most extensively as a forage crop for sheep. The disease has been experimentally induced in sheep by several investigators [1,2,8,10,13, 20,23,35,36].
The first clinical sign of acute lupinosis is inappetence which may amount to complete anorexia, followed by icterus and death within 2 days to 2 weeks after the ingestion of toxic lupins [1,2,6—8,10—12,15,26,35,36]. In less acute cases the anorexia is only partial and is associated with dullness, depression, loss of weight and icterus which may progressively become less pronounced and eventually disappear while deaths may continue for 2 months after removal from toxic lupin paddocks [8,35,36]. In some instances, evidence of photosensitization involving lesions of the ears, muzzle and other parts of the body not protected by pigment or wool, have been reported [7,12,26,37]. Other workers have stated that photosensitivity is seldom observed in sheep affected by lupinosis [8,16,35,38].
Chemical pathological changes include increases in serum glutamic oxalacetic transaminase, lactic dehydrogenase, glutamic dehydrogenase, alkaline

phosphatase and bilirubin [1,2,8,13,36]; pronounced elevations in serum copper, total iron-combining capacity, liver iron content and urinary copper excretion [13,19,20]; decline in liver cobalt values, elevation of serum vitamin B_{12} followed by decreases to below the initial level [19]; elevation of blood sugar levels immediately before death, and urine at necropsy positive for ketone bodies [1,2].

In acute cases there is pronounced generalized icterus in association with an enlarged, friable, bright yellow, fatty liver [1,2,7,12,35]. Subacute and chronic cases are characterized by atrophy, fatty degeneration and cirrhosis of the liver which is often shaped like a boxing-glove [1,2,7,11,12,26, 35]. Other necropsy findings include enlargement of the gall bladder, splenomegaly, nephrosis, impaction of the large intestine, ascites, hydrothorax and hydropericardium [1,2,7,10,12,26,35].

The liver is the primary organ affected and characteristic histopathological lesions include fatty metamorphosis and formation of yellowish-brown granules composed of lipofuscins and haemosiderin in hydropic centrilobular liver cells in acute stages; accumulation of granular degenerated hepatocytes in centrilobular areas with prominent stellate fibrosis in subacute stages; progressive fibrosis, bile duct cell and Kupffer cell proliferation, and cirrhosis in chronic stages [8,35,36]. The acute stage of lupinosis is also characterized by a fundamental disturbance in nuclear division of hepatocytes involving an interference with late metaphase followed by fragmentation of chromatin [8,35,36]. Abnormal mitotic figures are also known to occur in hepatocytes of mice injected with extracts of toxic lupins [4,39].

Characteristic hepatic lesions in sheep dosed with pure cultures of *P. leptostromiformis* in the acute phase include massive fat metamorphosis (Figs. 4,6,8); formation of eosinophilic globules in the cytoplasm of hepatocytes (Figs. 4,7); karyorrhexis (Figs. 8,9,10); and vesiculation of nuclei (Figs. 11,12) [1,2]. In subacute to chronic cases, the most pronounced lesions are accumulation of pigment in the hepatocytes (Figs. 2,5); megalocytosis; proliferation of bile ducts and bile duct epithelial cells (Fig. 3); and centrilobular fibrosis (Fig. 1) [1,2].

The most consistent change in other organs is haemosiderosis of the spleen, kidneys and lymph nodes [1,2,35].

The hepatic changes in sheep affected by lupinosis have been compared to those caused by diets deficient in methionine [11], choline [7,12] and

Plate 1. Hepatic lesions of lupinosis in sheep.
Fig. 1. Fibrotic tracts. × 30. T. & E.; Fig. 2. Pigment accumulation. Variation in size of nuclei. × 200. H. & E.; Fig. 3. Proliferation of bile ducts and bile duct epithelial cells. × 200. H. & E.; Fig. 4. Eosinophilic globules and fatty metamorphosis. × 350. H & E.; Fig. 5. Spongy pigment-laden cells. × 200. Sudan Black; Fig. 6. Degenerated and fat-laden hepatocytes. × 1200. H. & E.; Fig. 7. Eosinophilic globules. × 1200. H. & E.; Fig. 8. Fatty degeneration, mitotic figure and karyorrhexis. × 1200. H. & E.; Fig. 9. Karyorrhexis. × 1200. H. & E.; Fig. 10. Karyorrhexis. × 1200. H. & E.; Fig. 11. Vesiculated nucleus and variation in nuclear size. × 1200. H. & E.; Fig. 12. Vesiculation of nucleus. × 1200. H. & E. (Reproduced by permission of the South African Veterinary Medical Association, J.S. Afr. Vet. Med. Ass., 41 (1970) 242).

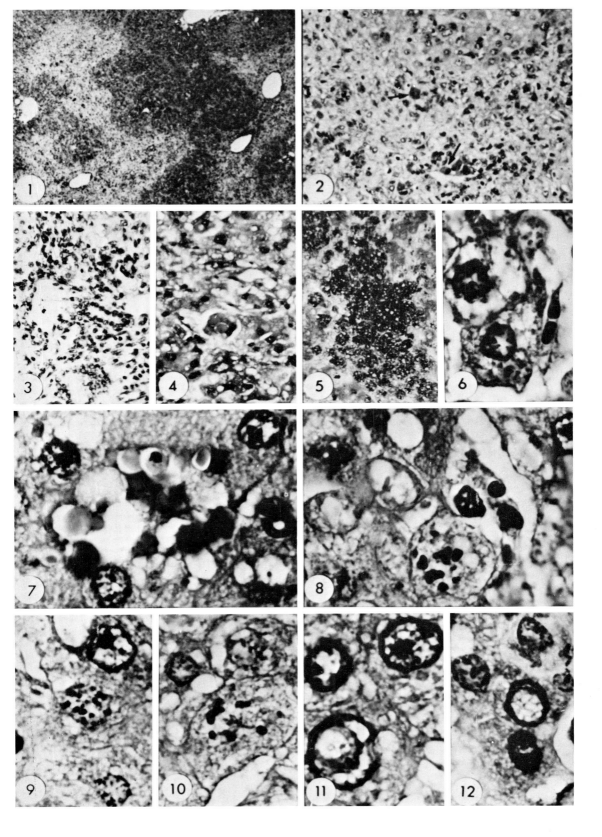

protein [35]. The similarities with aflatoxicosis, seneciosis and enzootic icterus have also been pointed out [1,2].

(ii) Other animals

Field outbreaks of lupinosis also occur in cattle [7,8,24,29], horses [6—8,30] and pigs [8]. In addition to sheep, the disease has been induced experimentally in the goat [10], horse [10], dog [10], rabbit [13,40], and mouse [4,39].

(1) Cattle. Relatively few field outbreaks of lupinosis in cattle have been reported in Germany [24], Poland [8], Australia [7,29] and South Africa [8,16,37].

Clinical signs in acute cases are inappetence, lacrimation, salivation, lethargy and intense icterus of conjunctival and mucous membranes [29]. Signs of photosensitization have been observed in cattle having access to green feed [7,29,37]. The chronic stage is characterised by inappetence, icterus, loss of condition, reduction in milk yield and few or no deaths [16,24,37].

In acute cases the liver is greatly enlarged, bright yellow to rich orange-saffron in colour, extremely fatty and friable; the gall bladder is greatly distended and the kidneys bile-stained and fatty [29]. Icterus is a characteristic gross pathological feature and is sometimes very pronounced with terminal serum bilirubin levels of 10 to 25 mg/100 ml [29]. In the chronic stage, the liver is smaller than normal, dull greyish-brown, coarsely nodular and streaked with connective tissue [29].

The predominant histopathological lesion in acute bovine lupinosis is extremely severe fatty degeneration of the liver [29]. Other hepatic lesions include bile ductule cell proliferation, increase in small bile ductules accompanied by portal fibrosis, and abnormal mitotic figures [29]. Extensive liver damage characterized by connective tissue containing proliferating bile ductule cells and bile ductules and regenerative nodulation of hepatic tissue, is found in chronic cases [29].

(2) Horses. Only a few cases of equine lupinosis have been described in the literature. Lupinosis was experimentally produced in a horse by Roloff [10] and a field outbreak of suspected equine lupinosis has been recorded in North America [6]. According to Gardiner [8], occasional outbreaks in horses have occurred in Poland. Isolated outbreaks have been reported in Australia [7] and a detailed description of five cases was given by Gardiner and Seddon [30]. According to Schneider [16], mortalities have occurred on a large scale in the western Cape Province of South Africa and he considers horses to be even more susceptible to lupinosis than sheep. He also mentions that horses frequently die of lupinosis when lupin hay is used as bedding in their stables.

Clinical signs include inappetence, depression, marked icterus of the visible mucous membranes, haemorrhages in the conjunctiva and inability to rise [10,16,30]. Death occurs within a few days following the appearance of the first clinical signs [16,30]. Signs of nervous disorder are often seen before the animals die [16].

The most pronounced gross pathological change is severe generalized icterus [30]. The liver is enlarged, light yellow or ochre with a firm larda-ceous consistency and the urinary bladder enlarged and filled with red or reddish-brown urine [30].

Characteristic histopathological lesions in the liver are fatty degenera-tion, bile ductule proliferation, generalized fibrosis and scattered degenerat-ing liver cells containing pigmented granular masses [30]. The kidneys are also severely damaged and contain haemoglobin casts and cell debris in the tubular lumens [30].

(3) *Pigs*. According to Gardiner [8], the only recorded outbreak of porcine lupinosis occurred in Poland and was characterised by inappetence, dullness, temperature rises, icterus, gastro-intestinal tympany and degenera-tive changes in the liver. Hackbarth [15] also mentions cases of poisoning of pigs in Poland which he ascribes to contamination of sweet lupins with bitter varieties. Lupinosis is unknown in pigs in Western Australia [8], and no field outbreaks in South Africa have come to our attention.

(4) *Goats*. Lupinosis was experimentally produced in a goat by Roloff [10]. He dosed the goat with an alkali extract of toxic lupins. Clinical signs were inappetence, depression and icterus. The goat died 8 days after dosing.

(5) *Dogs*. Lupinosis was experimentally produced in dogs by dosing them with ground toxic lupins or extracts of lupin material by two groups of early German research workers [10,41]. Clinical signs included inappetence, pronounced icterus, depression, loss of weight and death [10]. Necropsy findings were generalized icterus, haemorrhages in the gastro-intestinal tract, enlargement of the spleen and fatty infiltration of the heart [10].

(6) *Rabbits*. In feeding trials with small animals, Roloff [10] was un-able to produce typical lupinosis in either guinea pigs or rabbits.

Chronic lupinosis in rabbits was first described by Dobberstein and Walkiewicz [40] in rabbits fed bitter *L. luteus* for periods up to 11 months. No icterus or other clinical signs of disease were noticed, but the livers of some animals were small, shrunken and had a granular surface. Histopatho-logical lesions included a decrease in glycogen content and atrophy of hepa-tocytes, accumulation of brown, granular, lipofuscin-like pigment in the cytoplasm, megalocytosis, formation of polykaryocytes and abnormal mi-totic figures [40]. After prolonged periods of feeding, the architecture of the liver became greatly disturbed by progressive fibrosis. These hepatic changes were compared with those of phosphorus poisoning [40].

Gardiner [13] also reported characteristic changes of lupinosis in rab-bits, but not guinea pigs and rats, fed diets containing toxic lupins for periods up to 31 days. Hepatic lesions included hydropic degeneration and atrophy of centrilobular hepatic cells with accumulation of yellowish-brown, lipofuscin-like granules, karyorrhexis, multinucleation of liver cells and abnormal mitosis [8,13].

(7) *Mice.* A mouse toxicity test has been developed by the Australian workers for assaying extracts of toxic lupins [4,39]. The mice used in these tests display virtually no clinical signs, except that deaths occasionally occur and these moribund mice are always markedly icteric [4]. The most prominent histopathological lesions in the liver are nuclear and cellular hypertrophy, mitotic stimulation, abnormal mitosis, multiple nuclear inclusions, formation of polykaryocytes and bile duct proliferation [4,39]. It appears from these studies that the lupinosis toxin has a mitosis stimulating effect and that the cells that have not commenced mitosis develop nuclear aberrations [4,39]. These pathological changes in mice have been compared with the nuclear effects of colchicine [39], thioacetamide [4] and plants containing pyrrolizidine alkaloids [4,39].

(c) *Weather conditions associated with outbreaks of lupinosis*

Kühn [5] noted that severe outbreaks of lupinosis followed the ingestion of lupins which had been exposed to rain after harvesting. Since that time, an intimate association between periods of rain and high humidity and the onset of field outbreaks of lupinosis has become apparent [1,2,8,10,20, 22,23,28,30]. In Western Australia, the summer months are usually dry and outbreaks of lupinosis typically occur a week or two after cyclonic disturbances accompanied by rainfall of 50 points or more over a period of 1 to 3 days of high humidity [20,22,30]. The exact amount of rain and/or the duration of the period of high humidity necessary for the precipitation of lupinosis outbreaks are not well enough defined at present to forecast field outbreaks with certainty [22].

The observation that field outbreaks of lupinosis were associated with summer rainfall, was one of the factors that led the early investigators in Germany to consider the possible rôle of fungi in the aetiology of the disease [5,10].

(d) *Fungi in the aetiology of lupinosis*

The possible rôle of fungi in the aetiology of lupinosis was considered in Germany as early as 1876 when O. Brefeld examined toxic lupin plants for parasitic fungi and found only a rust [10]. In 1879, Zürn [42] advanced the theory that fungi were directly responsible for the disease by developing in the animal following the ingestion of fungal spores on lupin plants.

In 1880, Julius Kühn [5] first suggested that toxic metabolites of fungi growing on lupin plants may be the cause of lupinosis. He rejected Zürn's [42] theory because the toxin could be extracted from the lupins by a process that killed the spores. In an ingenious attempt to determine the causal fungus, Kühn compared the mycoflora of toxic and non-toxic lupins. He found 10 species of fungi on toxic lupins and 6 on non-toxic lupins. A fungus which he described as a new species, *Cryptosporium leptostromiforme* Kühn, was found fairly frequently (ziemlich häufig) on toxic, but only moderately frequently (nur mässig häufig) on non-toxic lupins [5]. Kühn apparently did not test individual fungi for toxicity. Roloff [10] reported that fungi (unspecified) isolated from toxic lupins were cultured artificially and fed to dogs with negative results. Thus the fungal aetiology of lupinosis was not proved by the early German research workers, although Julius Kühn came remarkably close to solving the problem in 1880 [5].

After 1880 the practice of feeding lupins to sheep declined in Germany, because of the high incidence of lupinosis [8,15] and investigations on the fungal aetiology of the disease were discontinued. In 1916, Marsh *et al*, [6] stated that the theory that lupinosis is caused by a hepatotoxin (ictrogen) produced as a result of the growth of micro-organisms on the lupin plant, was generally accepted as the most probable one. In 1943, Brash [26] observed "mouldy seeds" in a toxic lupin paddock in New Zealand. Since 1960, serious attention has been given to the possible fungal aetiology of lupinosis in Western Australia [4,7,8,13,20—23,30,36,39,43].

Gardiner and Nottle [21] attempted to induce lupinosis experimentally by culturing fungi on non-toxic lupin material and feeding to sheep. They were unable to reproduce the typical lesions of the disease in these experiments, but a few early hepatic changes suggestive of lupinosis were observed [21]. This work was continued with more than 20 species of fungi isolated from toxic lupins, but only slight liver changes of lupinosis could be elicited [13,20,22].

In 1966, Gardiner [23] succeeded in reproducing typical lupinosis in sheep by feeding them unsterilized, non-toxic lupins inoculated and incubated with a mixed fungal suspension from toxic lupins. He was, however, unable to induce toxicity by inoculating individual fungal species on to the lupins. *Pleospora herbarum* (Pers. ex Fr.) Rab., and *Cytospora* sp. were inoculated individually on to unsterilized or propylene-oxide-sterilized lupins, but the fungi either failed to grow or were overrun by *Rhizopus stolonifer* (Ehrenb. ex Fr.) Lind. Gardiner [23] concluded that "the fungi responsible for toxic formation were not definitely established, although either *Cytospora* or *Pleospora* spp. appeared to be a probable cause".

In 1970, lupinosis was experimentally reproduced with a pure culture of a single fungus for the first time by Van Warmelo *et al.* [1,2]. They isolated the fungus *Phomopsis leptostromiformis* (Kühn) Bubák ex Lind from sweet white lupins (*L. albus*), which had caused a field outbreak of lupinosis. Typical lupinosis was induced in sheep fed pure cultures of this fungus incubated on autoclaved, non-toxic white lupin seeds [1,2]. Gardiner and Petterson [4] independently demonstrated that typical lesions of lupinosis are induced in mice by extracts of pure cultures of *Phomopsis rossiana* (Sacc.) Sacc. et D. Sacc. (= *Cytospora* sp.) incubated on gas-sterilized, non-toxic lupins. (*P. rossiana*, as well as *Cryptosporium leptostromiforme* Kühn, are synonyms of *P. leptostromiformis* [3]).

Thus the search for the causal fungus of lupinosis was concluded exactly 90 years after Julius Kühn [5] first described *C. leptostromiforme* in one of the first publications dealing with lupinosis.

Phomopsis leptostromiformis — the causal fungus

(a) *Taxonomy and morphology*

P. leptostromiformis is a plant pathogenic pycnidial fungus which belongs in the Coelomycetes, Sphaeropsidales, Sphaeropsidaceae [44,45].

Black, stromatic pycnidia are produced on lupin stems, pods and seeds [1—3,5,44—50] (Plate 2). The fungus is easily isolated in culture, grows rapidly, and produces a thick white aerial mycelium in which black, stromatic pycnidia (pycnosclerotia) eventually form in concentric zones [1—3,48,50] (Plate 3). Pycnidiospores are produced on conidiophores in flattened or ellipsoidal fertile locules in the upper part of the stromatic bodies [3,45]. The spores are commonly extruded through the ostioles in cream-coloured to salmon-pink droplets and are unicellular, hyaline, cylindrical with tapered ends, biguttulate, $5—12 \times 1.5—2.5\mu$ [3,44,45,48,50]. Van Jaarsveld [51] recently reported the formation of stylospores on incubated, naturally infected lupin stems and sterilized stems inoculated with stylospores produced on naturally infected stems. According to him, stylospores and pycnidiospores are produced in separate pycnidia on the same stem. Although the stylospores were shown to contain nuclear material, germination of stylospores was not observed [51].

The optimal temperature for vegetative growth ranges from 24 to 28° [3,45,50,51] (Plate 4). In our experience, spore release from the pycnosclerotia occurs only when the cultures are incubated at 20° or lower and exposed to light [1—3] (Plate 3). Van Jaarsveld [51] found that the optimal

 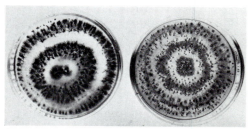

Plate 2. Pods and seeds of *Lupinus albus* infected by *Phomopsis leptostromiformis*.

Plate 3. Cultures of *Phomopsis leptostromiformis* on 1.5% malt extract agar. The sterile culture on the left was incubated at 25°. The sporulating culture on the right was incubated at 18° in diffuse light for 21 days.

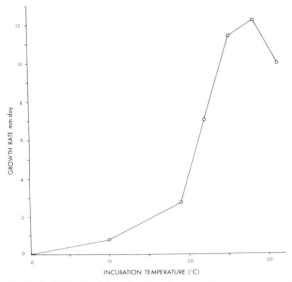

Plate 4. Effect of incubation temperature on vegetative growth (linear extension on 1.5% malt extract agar) of *Phomopsis leptostromiformis*.

temperature for the formation of pycnidia and spore release was $18°$, but sporulation also occurred in illuminated cultures at incubation temperatures as high as $28°$. In the western Cape Province, no sporulating pycnidia were found on lupin plants during the summer, but they were abundant on lupin litter collected in winter [3]. In culture, spore discharge is also enhanced by ultraviolet light radiation [50], by media containing 1 mg/ml of $NH_4 NO_3$ [50] and by a C : N ratio of 16 : 1 [51]. The optimal pH for vegetative growth is between 5.0 and 5.5 [50].

Although it has been suggested that the ascomycetous state of *P. leptostromiformis* may be *Diaporthe lupini* Harkness [44,45], which was originally described in California [52], attempts by several investigators to induce the formation of a perfect state have been unsuccessful [3,45—47]. Strukchinskas [50] reported that ascocarps of *D. lupini* were produced along with pycnidia of *P. leptostromiformis* on overwintered lupin stems incubated in a moist chamber for a few days. No experimental proof of the genetic connection between the two fungi has, however, been presented.

(b) Synonymy

The fungus found parasitizing *L. albus* in South Africa and capable of inducing lupinosis when dosed to sheep was compared with various European collections of fungi parasitic on lupins [3]. On the basis of these morphological studies, Van Warmelo and Marasas [3] concluded that the correct name of the causal fungus of lupinosis is *Phomopsis leptostromiformis* (Kühn) Bubák ex Lind [53], and that the following names are synonyms: *Cryptosporium leptostromiforme* Kühn [5,54], *Phomopsis lepto-stromoides* (Kühn) Bubák apud Kabát et Bubák [55], *Phoma rossiana* Sacc. apud Maire et Sacc. [56] and *Phomopsis rossiana* (Sacc.) Sacc. et D. Sacc. [57].

No material of the fungus identified as *P. leptostromiformis* in the U.S.A. [48,49] could be examined by Van Warmelo and Marasas [3], since viable cultures were no longer available [25]. Consequently it could not be established whether the North American fungus is conspecific with the South African and European material of *P. leptostromiformis* [3]. Judging by the description of the North American fungus [48] and by the fact that *D. lupini*, which is considered to be the perfect state of *P. leptostromiformis* [50], was first described in California [52], it seems very likely that *P. leptostromiformis* does occur in North America.

Cultures of the fungus identified as *P. rossiana* in Western Australia and shown to cause lupinosis in mice [4] were not examined by Van Warmelo and Marasas [3]. The type material of *P. rossiana* was, however, found to be identical with *P. leptostromiformis* [3].

(c) Geographical distribution

P. leptostromiformis is known to occur on *Lupinus* spp. in Germany [5,46,47,58—61]; Italy [62]; France [63]; Holland [64]; England [44]; Portugal [65]; Denmark [44,53]; U.S.A. (Georgia and Florida) [48,49]; Australia [4]; and South Africa [1—3].

It is interesting to compare the known geographical distribution of *P. leptostromiformis* with that of lupinosis. The only country where outbreaks of lupinosis have been reported [26] but where *P. leptostromiformis* is not known to occur, is New Zealand. On the other hand, the fungus occurs in a number of countries where outbreaks of lupinosis have not to my knowledge been reported, *e.g.* the Soviet Union where *P. leptostromiformis* has been known to occur since 1912 and has caused severe epiphytotics in lupin fields [50]. The situation in North America is also puzzling, because outbreaks of lupinosis in grazing animals have not been reported in Georgia and Florida where the fungus has been known to cause a disease of *L. luteus* since 1954 [48]. In 1960 it was reported that the lupin stem blight disease was not serious in these States and that less than 1% of the plants in a field were usually infected [48]. The disease increased in prevalence during the early 1960s until it became one of the most severe fungal diseases of yellow lupin in the south-eastern U.S. [25]. By the time that this disease became severe, the Bean Yellow Mosaic Virus had, however, limited yellow lupin cultivation to the extent that it was no longer being grown as a pasture plant [25].

(d) Pathogenicity

P. leptostromiformis causes a disease of lupins known as "stem blight" [48] or "stem spot" [45] which is characterized by brownish, sunken, linear stem lesions on which black stromata develop, circular lesions on the pods, and brownish discolouration of infected seeds [1–3,5,45–48,50,51,58, 60,61,64] (Plate 2). The disease is most severe on dry, sandy soils during hot weather [45].

A definite parasitic and saprophytic stage occurs in the life cycle of the causal fungus [45–47,51]. The parasitic stage infects living lupin plants in summer, and during the winter and spring the fungus develops as a saprophyte on dead lupin plants [45–47]. In the winter-rainfall areas of South Africa, the fungus parasitizes living lupin plants during the growing season in winter, and grows saprophytically on mature and dead plants during the spring and summer [51]. The fungus survives in infected lupin debris for at least 3 years and the pycnidiospores, and possibly ascospores, produced in spring (or autumn) are the source of primary inoculum [45–47,50,51]. P. leptostromiformis is also seed-borne and remains viable in infected seeds of L. luteus for 2 years or longer [48–50].

In nature, P. leptostromiformis is best known as a pathogen of L. luteus [45–48]. Natural infections of L. angustifolius var. leucospermus [46,47] and L. albus [1–3] have also been reported. The disease has been reproduced by artificial inoculation in L. luteus [45,48,51], L. albus [45,48,51] and L. angustifolius [51]. Ostazeski and Wells [48] reported that L. albus and L. luteus are susceptible and L. angustifolius is resistant to infection by P. leptostromiformis. In field experiments in Poland, it was found that the fungus infects living plants of L. luteus, but not of L. albus or L. angustifolius [45]. The fungus does, however, develop as a saprophyte on the fully mature plants of L. albus and L. angustifolius at the end of the growing season [45]. Recent investigations in South Africa have established that commercial cultivars of L. albus are very susceptible and cultivars of L. luteus and L. angustifolius resistant to infection by P. leptostromiformis during the growing season [51]. After maturity, however, saprophytic colonization by the fungus occurs to an equal extent on plants of all three species [51]. The highest level of resistance to infection was found in living plants of L. mutabilis [51]. Although P. leptostromiformis is capable of colonizing dead plants of L. mutabilis, sporulation on litter of this species is severely restricted [51].

The occurrence of P. leptostromiformis as a parasite primarily on L. luteus [45–48] and L. albus [1–3,51] and as a rapidly growing saprophyte on these Lupinus spp. as well as L. angustifolius [1–3,45,51], has an important bearing on the occurrence of outbreaks of lupinosis. In Germany, severe epiphytotics of Phomopsis stem blight occurred in the very susceptible L. luteus [46,47,58,60,61] simultaneously with severe field outbreaks of lupinosis in sheep grazing these lands [5,10]. The same situation apparently applies in Poland where up to 100% of the plants in L. luteus lands may be infected [45]. It is not clear why outbreaks of lupinosis have not occurred in L. luteus lands infected by P. leptostromiformis in Georgia and Florida [25]. Apparently Phomopsis blight was only a minor disease of L. luteus in the south-eastern U.S. until 1960 [48] and shortly after it became an important disease, the cultivation of yellow lupin as a forage crop was discontinued because of the high incidence of Bean Yellow Mosaic Virus [25]. The fact that lupinosis ceased to be a problem in Germany after the introduction of sweet lupins in the 1930s [15] cannot be explained on the basis of increased resistance in sweet yellow lupins to *P. leptostromiformis*, since there is no difference in the susceptibility of sweet and bitter *L. luteus* [45].

In Australia and South Africa, where the susceptible *L. luteus* and *L. albus* are cultivated to a limited extent only, the resistance of living plants of *L. angustifolius* may have played a rôle in delaying the build-up of inoculum of *P. leptostromiformis*. This in turn may be one of the reasons why lupins were cultivated in these countries for several years before the first outbreaks of lupinosis occurred [7,8,11,12,17]. The suggestion that outbreaks of lupinosis may be prevented by planting resistant varieties of *L. angustifolius* [2,33,34] is, however, not supported by the fact that this resistance breaks down at maturity when the fungus develops rapidly as a saprophyte on the dead *L. angustifolius* plants [45,51]. It is at this very stage that outbreaks of lupinosis usually occur in animals grazing the dry lupin stubble [1,2,7,8,12]. It stands to reason, however, that the build-up of inoculum will be much more rapid on susceptible plants such as *L. albus* that are infected during the growing season, than on *L. angustifolius* plants that are colonized only at full maturity [51].

Little information is available on the control of the lupin disease caused by *P. leptostromiformis*. The importance of incorporating resistant species such as *L. mutabilis* in lupin breeding programs is self-evident. Crop rotation was suggested as early as 1893 [46,47], but the efficacy of this method is doubtful because the fungus is capable of surviving in infected debris for several years [45,50]. The maintenance of high vitality in the plants by means of an adequate water supply was considered the most important factor in resistance to infection by Kochman [45]. Hot air treatments failed to eradicate the fungus from infected seed [49], but seed treatment with 50% tetramethyl thiuram disulphide greatly improved the crop [50]. This chemical, as well as mercuran, ethyl mercuric chloride, tetracycline and nystatin, has also been shown to inhibit mycelial growth *in vitro* [50]. Attention should be given to the spraying of lands of mature, dry lupins with fungicides to control the saprophytic development of *P. leptostromiformis*, and in so doing, to prevent outbreaks of lupinosis in animals grazing the dry lupin stubble. In this regard, the exact climatic conditions under which rapid saprophytic development of the fungus on lupin stubble takes place, should also be clearly defined so that outbreaks of lupinosis can be predicted and thus avoided.

(e) *Toxicity*

Toxicity of *P. leptostromiformis* to animals was first demonstrated in 1970 when it was shown that pure cultures of this fungus incubated on autoclaved *L. albus* seeds cause typical lupinosis in sheep [1,2]. Gardiner and Petterson [4] independently established that this fungus (as *P. rossiana*) causes typical histopathological changes of lupinosis when extracts of pure cultures incubated on gas-sterilized *L. varius* stalks and leaves in micro-chambers are injected intraperitoneally in mice. These authors [4] suggested that only the genus Lupinus may contain the basic substrates necessary for toxin production by *P. leptostromiformis*. We have, however, found that the lupinosis toxin is also produced by *P. leptostromiformis* on autoclaved kernels of yellow maize (*Zea mays* L.) [66].

The chemical nature of the lupinosis toxin produced by *P. leptostromiformis* is still unknown. The early German workers demonstrated that the toxin remains stable in stored dry lupins for at least 18 months, that it is not destroyed by dry heat but that steaming under pressure breaks down the toxin which is soluble in alkali but not in glycerine, alcohol or ether [5,8,9, 10,41]. On the basis of certain similarities between lupinosis and phosphorus poisoning, Roloff [10] suggested that the toxin may contain phosphorus.

Very little new information on the chemistry of the toxin has become available. It has been established that the toxic principle is not aflatoxin [1,43]. The toxin is extractable from toxic lupins with water at a slightly alkaline pH (7.0—7.2) and can be extracted from the aqueous phase with *N*-butanol but not with chloroform, ether or aromatic solvents [43]. Methanol proved to be superior to either ethanol or water in extracting the toxin from pure cultures of *P. leptostromiformis* incubated on lupin material [4]. On the basis of solubility characteristics, the workers in Western Australia have concluded that the toxin is not an alkaloid and that it is phenolic and/or acidic in nature [43].

Acknowledgement

The writing of this chapter would not have been possible without the assistance and advice of my colleagues at the Plant Protection Research Institute, Onderstepoort Veterinary Research Institute, the Rand Afrikaans University and the University of Stellenbosch. Their co-operation is much appreciated.

References

1   K.T. van Warmelo, W.F.O. Marasas, T.F. Adelaar, T.S. Kellerman, I.B.J. van Rensburg and J.A. Minne, J. S. Afr. Vet. Med. Ass., 41 (1970) 235.
2   K.T. van Warmelo, W.F.O. Marasas, T.F. Adelaar, T.S. Kellerman, I.B.J. van Rensburg and J.A. Minne, Experimental evidence that lupinosis of sheep is a mycotoxicosis caused by the fungus *Phomopsis leptostromiformis* (Kühn) Bubák, in I.F.H. Purchase (Ed.), Mycotoxins in Human Health, Macmillan, London, 1971, pp.185—193.
3   K.T. van Warmelo and W.F.O. Marasas, Mycologia, 64 (1972) 316.
4   M.R. Gardiner and D.S. Petterson, J. Comp. Pathol. Therap., 82 (1972) 5.
5   J. Kühn, Ber. Landw. Inst. Univ. Halle, 2 (1880) 115.
6   C.D. Marsh, A.B. Clawson and H. Marsh, U.S.D.A. Bull., 405 (1916) 1.

7 H.W. Bennetts, J. Agric. West. Aust., 1 (1960) 47.

8 M.R. Gardiner, Advan. Vet. Sci. 11 (1967) 85.

9 G. Liebscher, Ber. Landw. Inst. Univ. Halle, 2 (1880) 53.

10 F. Roloff, Arch. Wiss. Prakt. Tierheilk., 9 (1883) 1.

11 J.W. Groenewald, J.D. Smit and T.F. Adelaar, J. S. Afr. Vet. Med. Ass., 25 (1954) 29.

12 H.W. Bennetts, Aust. Vet. J., 33 (1957) 277.

13 M.R. Gardiner, Aust. Vet. J., 37 (1961) 135.

14 J.E. Peterson, Aust. J. Exptl. Biol. Med. Sci., 41 (1963) 123.

15 J. Hackbarth, J. Aust. Inst. Agric. Sci., 27 (1961) 61.

16 D.J. Schneider, Veterinary Practitioner, Wellington, Cape Province, South Africa, 1972 (Personal communication).

17 C.H. Flight, Farming S. Afr., 32 (1956) 37.

18 A.B. Beck and H.W. Bennetts, J. Roy. Soc. West. Aust., 46 (1963) 5.

19 M.R. Gardiner, J. Comp. Pathol. Therap., 75 (1965) 397.

20 M.R. Gardiner, J. Comp. Pathol. Therap., 76 (1966) 107.

21 M.R. Gardiner and M.C. Nottle, J. Agric. West. Aust., 1 (1960) 1021.

22 M.R. Gardiner, J. Agric. West. Aust., 5 (1964) 890.

23 M.R. Gardiner, Brit. Vet. J., 122 (1966) 508.

24 W. Meyer-Bahlburg, Mitt. Deut. Landw. Ges., 72 (1949) 3.

25 H.D. Wells, Georgia Coastal Plain Experiment Station, Tifton, U.S.A., 1971 (Personal communication).

26 A.G. Brash, N.Z. J. Agric., 67 (1943) 83.

27 H.G. Neil, W.J. Toms and C.M. Ralph, J. Agric. West. Aust., 1 (1960) 565.

28 M.R. Gardiner and A.W. Williams, J. Agric. West. Aust., 1 (1960) 1025.

29 M.R. Gardiner, J. Comp. Pathol. Therap., 77 (1967) 63.

30 M.R. Gardiner and H.D. Seddon, Aust. Vet. J., 42 (1966) 242.

31 P.D. Henning, Farming S. Afr., 24 (1949) 227.

32 L.G. Swart and C.R. Liebenberg, Farming S. Afr., 29 (1954) 227.

33 W.S. Grobbelaar, Farming S. Afr., 47 (1971) 59.

34 H.J. Straatman, E.B. Birch and R.J. Marshall, Farming S. Afr., 47 (1972) 37.

35 M.R. Gardiner, Pathol. Vet., 2 (1965) 417.

36 M.R. Gardiner and W.H. Parr, J. Comp. Pathol Therap., 77 (1967) 51.

37 R.F.O. Visser, Veterinary Practitioner, Piketberg, Cape Province, South Africa, 1972 (Personal communication).

38 T.S. Kellerman, P.A. Basson, T.W. Naude, I.B.J. van Rensburg and W.G. Welman, Onderstepoort J. Vet. Res., 40 (1973) 115.

39 M.R. Gardiner, J. Pathol. Bacteriol., 94 (1967) 452.

40 J. Dobberstein and W. Walkiewicz, Arch. Pathol. Anat., 291 (1933) 695.

41 C. Arnold and G. Schneidemuhl, Jahresber. Königl. Tierarzneischule, Hannover, 15 (1883) 108 (cited by M.R. Gardiner, Advan. Vet. Sci., 11 (1967) 134).

42 F.A. Zürn, Vort. Thierärzte, 7 (1879) 251 (cited by M.R. Gardiner, Advan. Vet. Sci., 11 (1967) 138).

43 D.S. Petterson and W.H. Parr, Res. Vet. Sci., 11 (1970) 282.

44 W.B. Grove, British Stem- and Leaf-fungi, Vol. 1, University Press, Cambridge, 1935, p. 488.

45 J. Kochman, Acta Agrobot., 6 (1957) 117.

46 M. Fischer, Das *Kryptosporium leptostromiforme* J. Kühn. Ein Kernpilz, der eine ernste Gefahr für den Lupinenbau bedeutet, Landwirtschaftl. Thierzucht, Bunzlau, 1893 (cited by M. Fischer, Bot. Zbl., 54 (1893) 289).

47 M. Fischer, Bot. Zbl., 54 (1893) 289.

48 S.A. Ostazeski and H.D. Wells, Plant Dis. Reptr., 44 (1960) 66.

49 S.A. Ostazeski and P. Decker, A. Rept. Exptl. Stat. Florida. 1959—1960, (1961) 364.

50 M. Struckchinskas, Mikol. i. Fitopatol., 5 (1971) 443.

51 A.B. van Jaarsveld, Lupienskroeisiekte in Wes-Kaapland: weerstand en epidemiologie, M.Sc. (Agric.) Thesis, University of Stellenbosch, Stellenbosch, 1973.

52 P.A. Saccardo, Sylloge Fung., 9 (1891) 722.
53 J. Lind, Danish Fungi as represented in the Herbarium of E. Rostrup, Copenhagen, 1913, p. 648.
54 P.A. Saccardo and A. Trotter, Sylloge Fung., 22 (1913) 1233.
55 J.F. Kabát and F. Bubák, Ann Mycol., 10 (1912) 240.
56 R. Maire and P.A. Saccardo, Ann. Mycol., 1 (1903) 220.
57 P.A. Saccardo and D. Saccardo, Sylloge Fung., 18 (1906) 265.
58 K.F. von Tubeuf, in W.G. Smith (Ed.), Diseases of Plants (English ed.), Longmans Green, London, 1897, p. 598.
59 O. Appel, Beitr. Pflzucht, 1912 (1912) 31.
60 H. Pape, IIIte Landwirtschaftl. Ztg., 47 (1927) 316.
61 H. Pape, Melanconiales, in P. Sorauer (Ed.), Handbuch der Pflanzenkrankheiten, 5th ed. Vol. 3, Parey, Berlin, 1932, p. 494—577.
62 T. Ferraris, Trattato di Patologia e Terapia Vegetale, Hoepli, Milan, 1926, p. 1004 (cited by J. Kochman, Acta Agrobot., 6(1957) 117).
63 E. Viennot-Bourgin, Les Champignons Parasites des Plantes Cultivées, Masson, Paris, 1949, p. 1389 (cited by J. Kochman, Acta Agrobot., 6 (1957) 117).
64 J. Ritzema Bos, Ziekten en Beschadigingen der Landbouwgewassen, Deel II, Tweede Stuk, Wolters, Groningen, 1915, p. 189—190.
65 M.T. Lucas and E. de S. Da Camara, Agron. Lusit., 16 (1954) 81.
66 W.F.O. Marasas, T.S. Kellerman, L.A.P. Anderson and I.B.J. van Rensburg, 1971 (Unpublished data).

Chapter 6

OESTROGENIC MYCOTOXINS SYNTHESIZED BY FUSARIUM

C.J. MIROCHA and CLYDE M. CHRISTENSEN

Department of Plant Pathology, University of Minnesota, Minneapolis, Minn. (U.S.A.)

More than 40 years ago McNutt *et al.* [1] reported that sows which had consumed mouldy feed developed enlarged vulvas and mammary glands, plus at times prolapsed vaginas or rectums. They did not establish experimentally a cause-and-effect relationship between the mouldy feed and the symptoms of oestrogenism in the swine. McErlean [2], in 1952, reported similar symptoms in swine in Ireland, associated with the consumption of barley infected with Fusarium; the barley was invaded by other fungi as well. In years of moist weather during the time that seeds are maturing, barley, oats, and wheat kernels may be relatively heavily invaded by various species of Fusarium; if the invasion is severe enough, it may cause a reddish discolouration of a portion of the kernels, non-development of the kernels and shrivelling, or decay of the partly developed kernel; the disease is known as "scab" or "blight". Kernels of wheat decayed by the fungus just before harvest may have a chalky texture and bleached appearance; in grain inspection parlance these are designated "tombstone damage" kernels. Both *Fusarium roseum* and *F. moniliforme* may invade kernels of corn on the cob, or in the ear, just before maturity of the grain, or just after maturity but before the corn is harvested. Such infection, by *F. roseum* (= *Gibberella zeae*) may result in production of "refusal factor", so named because swine refuse to eat such corn, or rations compounded of it, but will not result in development of biologically significant amounts of zearalenone. Evidence for this will be cited later. The oestrogenic syndrome in swine has been found in many countries and a chronological summary of first reports is listed in Table I.

It has been known for several decades that grain invaded by some species of Fusarium, under some conditions might be toxic. Christensen and Kernkamp [3] in the 1930s reported that barley blighted by *F. saubinetti* was toxic to swine, but they did not mention oestrogenism. In corn, Fusarium produced the oestrogenic compound almost only when the fungus grows on corn stored on the cob in cribs, and so it is not likely that Christensen and Kernkamp, dealing with barley blighted in the field before harvest, would have been concerned with oestrogenism. They stated, "... more than a dozen species of Fusarium ... cause blight of barley grains" and also that "*F. graminearum* Schwabe was by far the most prevalent cause of

TABLE I

THE OCCURRENCE OF THE OESTROGENIC SYNDROME IN SWINE IN VARIOUS COUNTRIES

Year	Country	Author	Diagnosis	Feedstuff
1928	U.S.A.	McNutt [1]	Porcine vulvovaginitis	Maize
1937	Australia	Pullar [58]	Porcine vulvovaginitis	Maize
1952	Ireland	McErlean [2]	Porcine vulvovaginitis	Barley
1962	France	Lelievre [59]	Porcine vulvovaginitis	Maize
1962	Italy	Paita [60]	Porcine vulvovaginitis	Maize
1963	Yugoslavia	Stamatovic [61]	Porcine vulvovaginitis	Maize and barley
1967	Romania	Bugeac [62]	Porcine vulvovaginitis	Maize
1968	Hungary	Danko [30]	Porcine vulvovaginitis	Maize
1968	Denmark	Eriksen [63]	Porcine vulvovaginitis	Barley
1969	Hungary	Debreczeni [31]	Vulvovaginitis, refusal, emesis, fever, diarrhoea	Maize
1971	Canada	Bristol [29]	Vulvovaginitis	Maize

barley blight in Minnesota." What they designated as *F. graminearum* we would call *F. roseum*, and some others would call *F. roseum* forma species "graminearum." Joffe [4] in 1965 reported that outbreaks of alimentary toxic aleukia (ATA) in man in Russia in the 1940s were caused by consumption of millet that had overwintered in the field and that had been invaded by *F. poae* and *F. sporotrichioides*. We would include both of these species in *F. tricinctum*.

Possibly a few comments are in order concerning the taxonomy and nomenclature of species of Fusarium. Species of this genus are common and widespread in nature; they cause a number of diseases such as wilts, blights, and rots of economically important plants, and for this reason they have been studied intensively by plant pathologists throughout the world, for more than a century. By the early 1930s probably more than 1000 species of Fusarium had been described, most of them identifiable, if at all, only to those who had described them. Wollenweber and Reinking [5], in Germany, tried to reduce this chaos to some order, and in their monograph they described 65 species and 78 named varieties. Snyder and Hansen [6,7] shortly thereafter reduced the number of species of Fusarium to 9, all of them identifiable by morphological characteristics as observed in culture on a standard agar medium. Toussoun and Nelson [8] in 1968 published a pictorial guide to the identification of these species. Gordon [9], in Canada, developed a system of classification of Fusarium in which he followed Snyder and Hansen in part, Wollenweber and Reinking in part, and, in part, his own system. Still other systems of classifying Fusarium species have been devised by workers in Italy, France, and the U.S.S.R., each system different from its predecessors. The latest system is that of Booth [10], which appeared in 1971; it includes about 50 species. Obviously, no system so far devised by a specialist in the taxonomy of Fusarium has satisfied another specialist in this field. We have followed the system of Snyder and Hansen because it has the virtue of being understandable to and usable by the non-specialist in Fusarium taxonomy.

Chemistry and physiological activity

The chemical isolation and structural formula of zearalenone (I) (Table II) were reported in patents issued to Andrews and Stob [11,12]. The patents claimed that zearalenone (referred to as the fermentation oestrogenic substance, FES) had both anabolic and oestrogenic activity, and would promote weight gain in animals such as lambs, cattle, and swine. The isolate of *Fusarium roseum* "graminearum" = *Gibberella zeae* (Gordon) used was deposited at the fungus culture collection at the NRRD laboratory in Peoria, Ill., U.S.A., and designated NRRL 2380. Hodge *et al.* [13] described derivatives of (II) as active, where carbons $1',2'$ are $-CH_2CH_2-$ or $-CH=CH-$ and R_1 and R_2 are hydrogen or a substituted or unsubstituted alkyl; R_3 is either O, H_2, or OH.

When the ketone in the $6'$ position is reduced to a hydroxyl group and the double bond at carbons $1',2'$ is reduced (III), two diastereoisomers result because of the introduction of a new asymmetric carbon atom in the $6'$ position. One of the isomers has a melting point of $178-180°$ and is about four times more active than the parent molecule (zearalenone) and the other melts at $146-148°$ and is only slightly more active [13,14] (see U.S. Patent 3,239,345). The isomer with the higher melting point is being marketed as a growth promoting agent in beef cattle (RALGRO) and is applied as an implant behind the ear of the animal. This same derivative is also being tested by Sandoz as a chemotherapeutic agent for the alleviation of post-menopause discomfort. Partial reduction of the ketone with sodium borohydride without reduction of the vinyl group at carbons $1',2'$ yields (IV) composed of two diastereoisomers (mixture melts at $90-102°$) and is less active than the analogous dihydro derivatives [13,14] (see U.S. Patent 3,239,348). Total reduction of the ketone does not increase the activity [13,14] (see U.S. Patent 3,239,354). The derivative dideoxyzearalane (V) loses all oestrogenic activity. On the other hand (VI) deoxytetrahydro-zearalane [14], is just as active as zearalenone (I) and apparently the phenolic hydroxyl groups are important for oestrogenic activity whereas the $6'$ ketone is not. The oestrogenic activity of dimethylzearalene (VII) and *p*-methylzearalene (VII), was reduced to about 6% of the parent compound (I), while *O*-methylzearalene was about 20% as active as (I) [14].

Johnston *et al.* [15] determined the effect of phenolic OH addition or removal on the biological activity of the parent compound. The dideoxyzear-alanone derivative (X), had less than 0.1 of the activity of zearalanone (XIV). The 2-deoxyzearalanone derivative (XI), retained about two-thirds of the activity of zearalanone (XIV) while the 4-deoxyzearalanone (XII) derivative had less than 0.1 the activity of zearalanone. The derivative 5-hydroxy-zearalenone (XIII) had less than one-third the oestrogenic activity of (XIV).

References p. 147

TABLE II

VARIOUS DERIVATIVES OF ZEARALENONE (F-2)

No.	R_1	R_2	R_3	R_4	$C_{1'}$ and $C_{2'}$	Name
I	OH	OH	O	H	CH=CH	Zearalenone (F-2)
II	OH or alkyl	OH or alkyl	O, H_2, OH	H	CH=CH or CH_2—CH_2	
III	OH	OH	OH	H	CH_2—CH_2	Dihydrozearalanol
IV	OH	OH	OH	H	CH=CH	Zearalenol
V	H	H	H_2	H	CH_2—CH_2	Dideoxyzearalane
VI	OH	OH	H_2	H	CH_2—CH_2	Deoxytetrahydrozearalane
VII	O—CH_3	O—CH_3	H_2	H	CH=CH	Dimethylzearalene
VIII	O—CH_3	OH	H_2	H	CH=CH	O-Methylzearalene
IX	OH	OH	H_2	H	CH=CH	p-Methylzearalene
X	H	H	O	H	CH_2—CH_2	Dideoxyzearalanone
XI	H	OH	O	H	CH_2—CH_2	2-Deoxyzearalanone
XII	OH	H	O	H	CH_2—CH_2	4-Deoxyzearalanone
XIII	H	H	O	OH	CH_2—CH_2	5-Hydroxy-2,4-deoxyzearalanone
XIV	OH	OH	O	H	CH_2—CH_2	Dihydrozearalanone
XV	OH	OH	H_2	H	CH_2—CH_2	Zearalane
XVI	O—CH_3	O—CH_3	O	H	CH=CH	Dimethoxyzearalenone

Johnston *et al.* [15] interpreted the greater activity of (XI) which has an OH group in position 2 to be the result of a conformational restriction which H bonding of the lactone imposes on the molecule.

Derivative (VI), tetrahydrozearalane, is reported to prevent pregnancy by inhibiting spermatogenesis in male mammals other than *Homo sapiens* [16].

Hurd and Shah [17] studied the absolute configuration of the methyl group attached to C_{10}, which in the naturally occurring zearalenone (I) has an S (sinister) configuration. They compared the uterotropic activities of zearalanone (XIV) with the synthetic R (rectus) S zearalanone. The S zearalanone was slightly more active (1.1×) than S zearalenone (I) standard while the RS zearalenone had 0.9 of the activity of the standard.

C. Allen Peters [18] compared the uterotropic activity of the *cis* and *trans* configuration of the 1',2' double bond. The *trans* configuration is easily converted to the *cis* by irradiation with ultraviolet from a mercury lamp. In zearalenone (I), the *cis* isomer has about the same activity as the *trans* but in the zearalenols (IV), both *cis* isomers (that of the high and low melting point isomers) had significantly more uterotropic activity than their respective *trans* isomers.

A synthesis of the novel 7'-hydroxymethyl, 7'-formyl, 7'-carboxy and 7'-carbomethoxymethyl derivatives of zearalane (XV) was reported by Jensen *et al.* [19]. Although no detailed uterotropic data were reported, the authors stated that these derivatives retained their oestrogenic and anabolic activity. It was interesting to note that the functional group at the 6' position could be moved on the undecenyl ring and still retain activity and perhaps enhance it.

In general, it appears that optimum uterotropic activity is obtained when the 6' carbonyl is reduced to an OH group and the double bond between carbons 1' and 2' is reduced. Reduction of the olefin also enhances activity in the zearalenone (I) series. Further, hydroxyl groups or some equivalent functional group is necessary at least in position 2 of the benzene ring.

Naturally occurring derivatives of zearalenone

Mirocha *et al.* [56,64] reported the natural occurrence of at least 7 derivatives of zearalenone from *Fusarium roseum* growing in culture on corn. The metabolites were partially characterized by their ultraviolet absorption spectrum and separation by thin-layer and gas-liquid chromatography. The most abundant of these derivatives were designated as F-5-3 and F-5-4, both of which have a mass of 334 and an emperical formula of $C_{18}H_{23}O_6$. The structure of the latter was incorrectly reported as 3'hydroxyzearalenone (alpha and beta isomers) by Mirocha *et al.* [56]. Subsequent studies using mass spectrometry (Jackson *et al.* [57]) revealed that the OH group of both derivatives is in the 8' position rather than the 3'. Another study by Bolliger and Tamm [65], unknown to us at the time, described these same metabolites as 8'-hydroxyzearalenone and 8'-epi-hydroxyzearalenone. In addition, Bolliger and Tamm [64] described 5-formylzearalenone and 7'dehydrozearalenone in cultures of *Gibberella zeae*. Steele [66] reported the occurrence of the 6',8'-dihydroxyl derivative of zearalenone produced by *F. roseum* growing on a solid medium of shredded wheat.

Biological activity

Zearalenone (F-2) and its uterotropically active derivatives can be quite properly classified as oestrogens in the sense that they produce oestrus *i.e.* cornification of the vagina of adult mice, as in natural oestrus. Although not steroids, but rather phenols and of mycological origin, they produce a true oestrus as well as the more general responses such as anabolism associated with oestrogens.

(1) Swine

Swine are perhaps the most sensitive of animals to activity of zearalenone, and it is among these animals that oestrogenism was first discovered. Under experimental conditions, 6-week-old gilts were fed zearalenone at a rate of 5 mg/day for 5 days. Within 5 to 7 days, enlarged vulvas, mammae and nipples, and prolapse of the vagina were noted. The uteri of these gilts were greatly enlarged and the ovaries were atrophied. The animals returned to normal after being taken off the ration.

In another test (Table III), using 6-week-old prepubertal gilts, as little as 1 mg pure crystalline zearalenone per day for 8 days was sufficient to incite pronounced tumefaction of the vulva. Those fed zearalenone *per os* in gelatin capsules at a rate of 5 mg/day developed tumefaction in 4 days. In some animals, a marked atrophy of the ovaries developed. There was a distinct oedema and cellular proliferation in all layers of the uterus [20].

Microscopic changes in these animals described in Table II were reported by Kurtz *et al.* [21]. An interstitial oedema and ductular proliferation of the mammary gland was observed in all treated gilts (Fusarium-invaded corn and zearalenone) as well as a pronounced oedema and cellular proliferation in all layers of the uterus. The thickness of the myometrium was greatly increased. The gilts with zearalenone had significant epithelial changes in the cervix characterized by metaplasia of the mucosal epithelium to a stratified squamous cellular layer sometimes 15 cells thick and irregular in distribution. Squamous metaplasia, similar to that found in the cervix but more severe, was found in the vaginal sections.

According to Nelson *et al.* [22] 4 Yorkshire gilts were bred and fed the following rations: the control gilt received normal sow rations; the others received feedstuff containing respectively 25, 50 and 100% corn invaded by an isolate of Fusarium known to produce an oestrogenic response in immature gilts. The corn substrate colonized by Fusarium was kept at 23—28° for 2 weeks and 12° for 4 weeks. The gilt receiving a ration containing 50% corn

TABLE III

EFFECT OF PURE ZEARALENONE (F-2) AND FUSARIUM-INVADED CORN ON THE UTERUS OF GILTS [a]

Treatment	Weight of gilt [b] (kg)	Weight of uterine horn (g)	
		Total	Per kg body weight
Control, sacrificed after 5 days	34.1	27.8	0.81
Control, sacrificed after 14 days	75.5	28.0	0.37
Oestradiol, 0.5 mg i.m. daily for 4 days	60.5	173.1	2.86
Oestradiol, 0.5 mg i.m. daily for 14 days	81.4	241.6	3.00
F-2, 1 mg daily for 8 days	49.5	47.0	0.95
F-2, 5 mg daily for 5 days	57.7	85.4	1.48
F-2, 10 mg daily for 5 days	53.6	174.6	3.26
F-2, 25 mg daily for 4 days	76.3	271.6	3.56
F-2, 50 mg daily for 4 days	78.6	251.9	3.20
Fusarium-invaded corn, 80 g daily for 6 days	59.5	136.9	2.30
Fusarium-invaded corn, 80 g daily for 14 days	75.5	206.0	2.70

[a] After administration *per os.*
[b] The proprietary material oestradiol (ECP, Upjohn Co.) was administered by i.m. injection; F-2 by gelatin capsule; Fusarium-invaded feed was mixed with normal ration. Prepuberal pigs about 6 weeks of age.

invaded by Fusarium developed an enlarged vulva after 4 days and aborted 21 days later. The control gilt weaned a litter of 10 pigs, and the 3 gilts fed different amounts of corn invaded by Fusarium weaned a total of 11, or an average of 3.7 per gilt.

In another test, Yorkshire gilts were bred and started on a diet, immediately after conception, containing pure crystalline zearalenone at rates of 6 mg (1 gilt), 60 mg (2 gilts) and 150 mg (2 gilts) per day for 15 days. The average number of pigs born in the 6 mg and 60 mg per day treatment was 8 and the 150 mg treatment group averaged 6. The number of pigs in the control treatment (4 gilts) averaged 9.25 [23].

In another similar experiment with open Yorkshire gilts, 2 animals were fed 60 mg zearalenone per day, 2 at 150 mg per day and 3 control animals with no zearalenone in their otherwise comparable diet. The gilts were kept on this diet for 27 days and then sacrificed. The results of this test are shown in Table IV. The gilts on the diet of zearalenone were consistently found to have many follicles on both ovaries but no corpora lutea in contradistinction to the control animals which had a normal distribution of corpora lutea [23].

These results suggest that the reproductive performance of gilts may be influenced by low levels of an exogenous oestrogen. In support of this hypothesis, Young and Teague [24] reported that a lower ovulation rate, fewer live embryos and lower percent of corpora lutea, represented as live embryos, were associated with the addition of low levels of stilboestrol to a legume-free ration. Foote et al. [25] showed that oestradiol suppressed the development and maintenance of follicles when injected subcutaneously. Wilson et al. [26] reported that two of three strains of Fusarium roseum produced weaker pigs and more mummified pigs at birth when fed to gestating gilts.

Christensen et al. [27] studied the effect of rations containing corn invaded by Fusarium roseum on both male and female swine. The feed they used consisted of 50% premix (containing minerals, vitamins and protein), 30—40% of sound corn, and 0.7—20% of corn invaded by F. roseum. Several different batches of corn invaded by F. roseum were used during the course of the trial, and each batch was added to the feed in an amount that

TABLE IV

EFFECT OF ZEARALENONE ON THE ORGANS OF SEXUALLY MATURE GILTS AFTER BEING PLACED ON A DIET CONTAINING ZEARALENONE FOR 27 DAYS

	Control gilts	Zearalenone-treated gilts	
		60 mg per day	150 mg per day
Number of gilts	3	2	2
Average wt. at start of experiment (lb)	250.7	233.5	244.5
Average wt. at end of experiment (lb)	269.3	273	276.5
Average wt. gain (lb)	18.7	39.5	32.0
Average wt. thyroid (g)	9.3	11.8	11.1
Average total wt. of 2 adrenal glands (g)	6.0	4.2	5.7
Average wt. pituitary (g)	0.543	0.487	0.556
Average wt. uterine horn (g)	91.8	275.4	128.7
Average wt. ovary (g)	4.0	5.3	4.8

provided 500–600 parts of zearalenone per million parts of feed. The control animals received a commercial hog ration. For the first 64 days of the trial, the pigs on experiment received a ration containing corn invaded by *F. roseum*, then for the next 60 days they were given a ration of commercial pigfeed.

The pigs were from 10 litters that were farrowed within a few days of each other and which were raised under our care. They were about 6 weeks old at the beginning of the experiment. The control group consisted of 7 males and 15 females, and the other group consisted of 9 males and 14 females.

All of the gilts on the ration containing corn invaded by *F. roseum* developed swollen vulvas and large nipples within 3 to 4 weeks, and 3 of them developed prolapsed rectums after 25, 38 and 47 days, respectively. After the feeding of the ration containing corn invaded by Fusarium was discontinued, the outward sign of oestrogenism regressed. After 124 days the male pigs on the ration containing corn invaded by *F. roseum* averaged 102 lb (*ca.* 46.3 kg) in weight and the females averaged 86.3 lb (*ca.* 39.2 kg); male pigs on the control ration averaged 163 lb (*ca.* 74 kg) and females averaged 157 lb (*ca.* 71.3 kg). In relation to the weight of the animal, the weight of the uterine horn of the gilts receiving the ration with corn invaded by *F. roseum* was nearly double that of the control and the weight of the testes of the males on the zearalenone ration was 30% less than the weight of testes of those on the control ration. The atrophy of the testes shown in this experiment confirms the observations of the authors noted in other experiments and field cases. No significant differences were noted between the pigs in the two treatments in weight of pituitary, thyroid, adrenals or brain (See Table V).

As shown in Table IV and other experiments pure crystalline zearalenone does have some anabolic properties resulting in weight gain of experimental animals. This property is well documented and prompted Commercial Solvents Corporation of U.S.A. to market a derivative of the parent compound (dihydrozearalanol, Table II) as an anabolic agent to increase beef production. On the other hand, a diet of Fusarium-invaded corn, when added to an otherwise balanced diet, will cause the oestrogenic syndrome but in addition will also result in weight loss. It appears that *Fusarium roseum* produces other toxin(s) responsible for the weight loss and perhaps for the abortion noted in field cases and experiments. We have never been able to induce abortion in swine with pure zearalenone but only when Fusarium-invaded corn was added to an otherwise nutritionally adequate diet. However, zearalenone can definitely contribute to infertility in swine by its effect on the ovaries, *i.e.* apparent suppression of the development of corpora lutea.

For the proper understanding of the effects of mycotoxins on swine production in the upper midwest of the U.S.A., it is necessary to divide the problem into 3 distinct facets: (*1*) outward signs of hyperoestrogenism, (*2*) infertility characterized by a lack of production, and (*3*) abortion. The

TABLE V

COMPARATIVE WEIGHTS OF WHOLE ANIMALS AND OF VARIOUS ORGANS OF MALE AND FEMALE PIGS GIVEN A RATION CONTAINING CORN INVADED BY *FUSARIUM ROSEUM* AND THOSE GIVEN A RATION CONTAINING SOUND CORN

	Male		Female	
	Sound corn	Fusarium-invaded corn	Sound corn	Fusarium-invaded corn
Number of pigs	3	7	2	5
Duration of test (days)	126	124	126	124
Average wt. of:				
Whole pigs (lb)	163	102	157	86.3
Pituitary (g)	0.2	0.2	0.3	0.4
Thyroid (g)	5.2	4.5	4.2	3.2
2 Adrenals (g)	3.0	3.8	3.9	3.7
Brain (g)	87.3	79.7	93.2	82.2
2 Testes (g)	283.4	132.3		
2 Ovaries			4.6	2.2
1 Uterine horn (g)			11.6	10.8
Avg. length uterine horn (cm)			32.0	17.6

hyperoestrogenism can to a great extent be attributed to zearalenone and the infertility at least in part. Infertility can be caused by numerous factors, genetic and psychological as well as nutritional. Although zearalenone can contribute to infertility, it is possible that other oestrogens such as diethyl-stilboestrol (DES) as reported by Mirocha *et al.* [28] may be present in the feedstuff and contribute to the problem. We have found both DES and zearalenone occurring together in a commercially prepared, pelleted pig ration [28]. Further, as reported and discussed by Bristol and Djurickovic [29], infertility causes more serious losses where breeding females are affected as it can disrupt a breeding program. They reported that in one instance there were no conceptions in a herd (size not known) due to prolonged oestrus in the females and lack of libido in the males.

The idiopathic abortion portion of the syndrome, at least as related to mycotoxins, is due to another as yet undefined toxin produced by *F. roseum*. The isolate of *F. roseum* used in most of our tests (designated Mapleton No. 10), evidently produces a complex of toxins, and although oestrogenic, on some occasions it has been lethal to white rats in feeding tests.

Nelson *et al.* [22] reported that over a period of 4 years, 72 accessions of aborted swine foetuses were submitted to the University of Minnesota Veterinary Diagnostic Laboratories. His summary shows that 77.5% of the porcine abortions could not be attributed to any known infectious diseases when examined by routine diagnostic techniques. Although zearalenone probably did not cause any of these abortions, we suspect that some other toxin produced by both oestrogen-producing and non oestrogen-producing strains of *F. roseum* may account for a portion of the cases. It is important for Veterinary Diagnostic Laboratories to further define the complex porcine abortion problem.

In the study of case histories and other reports of hyperoestrogenism in swine, vulvovaginitis is only one of the signs reported. For example, Danko and Aldasy [30] in Hungary reported vulvovaginitis resulting from feeding mouldy corn to swine. Continuous feeding of the mouldy foodstuff caused loss of appetite, diarrhoea and liver damage. They further reported that the crude toxic material when fed to young female rats, caused swelling of the uterus and in some cases death. Debreczeni and Rejto [31], also in Hungary, described the occurrence of vulvovaginitis in swine fed mouldy corn along with signs of fever, faintness, nausea, vomiting, diarrhoea and considerable loss of appetite. Voluntir *et al.* [32] reported that corn infected with *F. sporotrichioides* will lead to vulvovaginitis in a short time, and that prolonged feeding leads to an increase in the number of infertile inseminations, decrease in litter size, and an increase in the mortality rate during the first 10 days subsequent to parturition. They correlated mortality with the increase in agalactia in affected sows. As an example, Voluntir *et al.* [32] stated that during April of 1969, when no toxicosis due to Fusarium was prevalent, out of 104 pregnant sows, 101 went successfully to term and only 3 (3%) sows aborted. The total number of piglets born was 792 with an average litter size of 8. The number of piglets that died during the first 10 days after birth was 145 or 18.3%. In September of 1969, when toxicosis from Fusarium was present in this herd, there were a total of 135 pregnant sows and 112 went successfully to term and 23 (17%) aborted. The number of piglets born was 520 with an average litter size of 5. The number of piglets that died during the first 10 days after birth was 190 or 36.5%. Part of the infertility reported by Voluntir *et al.* [32] were attributed to oocyte necrobiosis followed by a drop in the frequency of follicle maturation. Ozegovic [33] also reported complications other than oestrogenic associated with mouldy corn infected primarily with *F. graminearum* in Yugoslavia. He reports refusal of food, loss of weight, abortions, diminished fertility and endometritis.

(2) *Poultry*

The activity of zearalenone on birds is not so pronounced as on swine, but it still can be considerable. Meronuck *et al.* [34] tested the activity of rations containing corn invaded by *F. roseum*. Turkey poults [35] fed 10% corn heaviliy invaded by *F. roseum* for 76 days, ate only slightly less than those fed a ration containing sound corn, and body weight gain and feed efficiency of the two groups were nearly identical. Some of the turkeys on the 10% Fusarium-corn diet developed swollen vents but otherwise the birds seemed normal and did not have any detectable internal lesions when necropsied. Of the turkeys kept on a 40% Fusarium-corn ration for 46 days, feed consumption as well as the final body weight was greatly reduced. Of the 8 turkeys fed this ration, 3 developed prolapsed cloacae. An enlargement of the bursae of Fabricius was also noted [32].

In another experiment, pure zearalenone (300 ppm) was added to the otherwise balanced ration to authenticate the hyperoestrogenic response seen in birds fed 10 and 40% Fusarium-corn. 10- to 12-day-old turkey poults were used and swelling of vents was visible on these young birds within 4 days. Cystic development of the right genital tract, eversion of the cloaca and enlargement of the cloacal bursa and oviduct were noted in some of the treated birds (unpublished results).

Speers *et al.* [36] fed corn colonized by *F. roseum* and pure zearalenone to young chicks and laying hens. In chicks, both final bird weight and average gain of the chicks showed a quadratic response to increasing levels of zearalenone. Growth appeared stimulated by the ration containing 300 and 800 ppm zearalenone, and by the ration containing 10% Fusarium-corn, whereas chicks fed 20% Fusarium-corn (1600 ppm zearalenone) had weight gains similar to the control group. Comb weight and ovary length exhibited a similar quadratic response to zearalenone level in the feed. Although no pathological or physiological changes other than body weight were apparent during the experiment, when the birds were necropsied at the termination of the experiment, a high incidence of cystic development on the genital tract was observed. These cysts appeared as thin-walled, fluid-filled sacs attached to or part of the immature oviduct. The bursa of Fabricius increased in weight at 300 ppm zearalenone while the Fusarium-corn-amended ration caused a reduction in weight. The increase in bursal weight appeared to be related to the presence of cyst development. Feeding a Fusarium-corn-amended ration (10 and 20%) and zearalenone (250 and 500 ppm) to mature laying hens had little influence on the various performance, reproductive or physiological, criteria in the study. Egg shell quality, however, declined when the rations contained Fusarium-contaminated corn.

Sherwood and Peberdy [35] tested the effect of zearalenone on male chicks. They found that incorporation of 40 ppm zearalenone into the feed of male chicks for a period of 10 days had little effect on weight gain or efficiency of food conversion, but did cause a significant increase in the weight of testes and comb. Sherwood and Peberdy [35] point out that zearalenone exhibits oestrogenic properties that the increase in weight of these organs would not be expected to be a direct response because according to Sturkie [37], their development is controlled by the androgen hormones.

Speers *et al.* [38] fed laying hens a ration containing various amounts of corn invaded by *F. roseum*, *F. tricinctum* or *F. roseum* (isolate oxyrose). Single Comb White Leghorn (SCWL), Rhode Island Red (RIR) and New Hampsire (NH) were used in one test and SCWL and RIR hens in the second; the tests lasted for 42 and 49 days, respectively. In experiment one, *F. roseum* (10% of diet, providing 500 ppm zearalenone) had no detrimental effect on performance or egg quality. Corn invaded by *F. tricinctum* and *F. roseum* (isolate oxyrose) at 5% of the diet, caused a marked decrease in feed intake and subsequent body weight loss for all three breeds. Egg production of all birds fed *F. roseum* (isolate oxyrose) ceased in one week as did production of NH and RIR hens fed *F. tricinctum* while production of SCWL continued to be normal. In the second experiment, the incorporation of the *F. roseum*-(oxyrose) and *F. tricinctum*-invaded corn was reduced to 1 and 3% of the diet. This diet again reduced feed intake and egg production dropped in the RIR but not SCWL.

The isolates of *F. tricinctum*, *F. roseum* and *F. roseum* (oxyrose) are all excellent producers of zearalenone and all have been isolated from feeds suspected of harbouring mycotoxins. As is obvious, the traces of *F. roseum* vary in their toxicity although they may produce similar amounts of zearalenone. The data of Speers *et al.* [36,38] and Meronuck *et al.* [34] suggest that hyperoestrogenism in poultry may be confounded by signs of toxicity, especially reduced weight gain and reduced egg production, the latter two being more important. The three isolates of Fusarium used in their experiments (*F. roseum*, *F. roseum* (oxyrose), *F. tricinctum*) all produce copious amounts of zearalenone but in addition they also produce different trichothecene derivatives which we believe account for their toxicity. As an example, the *F. tricinctum* isolate used in the experiments of Speers *et al.* [36,38] is the same isolate that Christensen *et al.* [39] reported as highly toxic to turkey poults, causing pronounced bilateral necrotic lesions at the angles of the mouth, and finally death. The trichothecene called T-2 toxin had been identified as the toxin responsible for the mouth lesions and death. *F. roseum*, isolate oxyrose, produces copious amounts of the toxic scirpene, monoacetoxyscerpenol.

(3) Dairy cattle

There is no conclusive evidence that zearalenone causes the oestrogenic syndrome in dairy cattle, although vulvovaginitis is often reported by veterinarians in dairy herds. There is, however, circumstantial evidence that zearalenone may be involved as reported by Mirocha *et al.* [40] and Roine *et al.* [41]. The former report involved a bad lot of hay which when fed to dairy cattle, resulted in decreased fertility in these animals. Their normal artificial insemination index (AI) was 1.2 but when the 150 cattle were given the bad lot of hay, there was a rise of the AI to 4. After this observation, the animals were quickly taken off the hay and their AI index gradually returned to normal. The hay was badly contaminated with fungi at least in some phase of the operation because ergosterol (good indicator of colonization by Fusarium) was easily detected. In addition, 14 ppm of zearalenone was found in the hay sample. The case report by Roine *et al.* [41] involved a herd of 21 dairy cows showing fertility disturbances and a clinical diagnosis of vaginitis and prolonged heat. There were no clear signs of endometritis; the secretion appeared to originate from the vagina and uterine and vaginal douches seemed to have no effect. Luteinizing hormone (LH) injections given at the assumed start of heat did not affect the situation. The fodder was suspected of harbouring oestrogens and 25 ppm of zearalenone was found in the original feed sample [42]. In addition, *F. graminearum*, *F. tricinctum* and *F. culmorum* and *F. poae* were isolated from the feed sample.

(4) Regulation of sexual reproduction in Gibberella zeae (Fusarium roseum "Graminearum")

The sexual stage of the life cycle of *Gibberella zeae* begins with the union of two nuclei to form the diploid cell. Cells adjacent to the fused nuclei are in the meantime stimulated to divide and morphogenesis is initiated to form the ascus contained within the perithecium. Prior to maturation of the asci, the diploid nucleus undergoes one meiotic division followed rapidly by one mitotic division giving rise to 8 haploid ascospores within the ascus. This complex phenomenon constitutes the sexual stage within the genus Fusarium.

In view of the fact that zearalenone affects the sexual cycle of swine, turkeys, chicks, rats, mice, rabbits, and guinea pigs, perhaps it is not surprising that it should also act as a sex regulating hormone in *F. roseum*, the fungus that produces it. This is the case. Wolf and Mirocha [43] reported that when amounts of zearalenone ranging from 0.1 ng to 10.0 ng are applied to a 1.0 cm diameter disc of Coons' synthetic medium in agar, perithecial formation is enhanced by as much as 100%. Amounts in excess of 10.0 μg inhibit perithecial formation. Zearalenone has a periodicity of action relative to both enhancement and inhibition of perithecial formation. The ability to enhance is greatest up to 4 days of culture age when amounts optimum for enhancement are used. Optimum inhibition of perithecial formation with inhibitory amounts (10.0 μg and 100.0 μg) occur when the culture age is 3 to 4 days old. It appears that *Fusarium roseum* is sensitive to extremely small amounts of this hormone (0.1—10.0 ng) which is in contrast to its activity in mice where it is approximately 0.016 times as active as oestrone.

A limited study on the structure—activity relationships reveals that the hydroxyl or ketone function in the 6' position of the undecenyl ring is necessary for activity. Further, unsaturation at the 1',2' position in the undecenyl ring is associated with inhibitory activity at the higher concentrations (10.0—100.0 μg). The sex-regulating activity of the following derivatives of zearalenone was tested in Fusarium: dideoxyzeralane (V), zearalane (XV), dimethoxyzearalenone (XVI), zearalanone (XIV), zearalanol (III) and zearalenol (IV). The relationship of sex-regulating activity in *Fusarium roseum* closely parallels the sex-regulating activity in female mice and rats. It was surprising to find this relationship in such two dissimilar biological entities.

Occurrence of zearalenone in feedstuffs

Nearly all of the suspect cases of mycotoxicosis investigated at the University of Minnesota are submitted to the Diagnostic Laboratory of the College of Veterinary Medicine where they are thoroughly examined for known animal diseases. If no known causative agent can be assigned, samples of the feedstuff are then submitted for mycological and chemical analyses. Of such samples submitted to our laboratory during the three years, 1968 through 1970, 65 were analyzed for the presence of zearalenone. The complaints by farmers and practicing veterinarians who submitted the samples were: abortions in swine and dairy cattle, sows returning to heat, vaginal and rectal prolapse in swine, infertility in swine and dairy cattle, prolapsed uteri in cows, vulvovaginitis in swine, poor lactation in sows, stillbirth in sows, early farrowing in sows, foetal resorption in swine, abnormally large udder development in cattle and failure of sows to come into heat. The results of analyses for zearalenone are shown in Table VI. 45% of the samples submitted were positive for zearalenone, ranging in amounts from 0.1 to 2909 ppm based on the dry weight of the feed, a surprisingly high percentage.

Normally we consider concentrations between 1 and 5 ppm zearalenone as physiologically significant based on responses obtained in test animals. As an example, an 8- to 10-week-old (35—65 lb) gilt will normally consume 2 lb feed per day. If zearalenone is present at 1—5 ppm, the animal will consume approx. 0.9 to 4.5 mg of actual zearalenone per day. One mg of pure zearalenone administered to a 60—70 lb gilt (10—12 weeks old) per day will cause signs of vulvovaginitis on the 4th day. A mature gilt ready for breeding will weigh approx. 250 lb and consume about 9 lb feed per day. On a ration containing 1 ppm zearalenone, the gilt will consume 4 mg and on a 5 ppm diet, 20 mg zearalenone per day. The distribution of zearalenone in the feedstuff is not normally uniform, and the animal can ingest a large amount from a severely infected portion of kernels while other portions may be relatively free. There is also the possibility that other known or unknown oestrogens might be present in the feed mixture.

Microcha *et al.* [28] reported that approx. 50% of the feed samples associated with oestrogenism in animals and submitted to the University of Minnesota for analysis, did not contain detectable amounts of zearalenone, although when fed to weanling, white female rats, they caused an increase in the weight of the uterus. Five of these samples were selected for more intensive investigation, and all five were found to contain DES in concentrations of 240, 2400, 28, 90 and 1—10 ppb. All five samples, when incorporated into a nutritionally balanced rat diet and fed to 21-day-old female white rats, caused a significant increase in the uterine weight (ranged between 134 to 422 mg). As little as 0.8 to 1.0 μg of authentic DES (total

TABLE VI

SAMPLES OF FEEDSTUFF SUSPECTED OF INVOLVEMENT IN HYPERESTRO-GENISM, SUBMITTED TO THE UNIVERSITY OF MINNESOTA FOR ANALYSIS OF ZEARALENONE

Year	Number examined	Number positive	Range of zearalenone concentration (ppm)
1968	24	9	0.1—2909
1969	23	9	0.5— 2.0
1970	18	10	2.0— 306

dose) increased the uterine weight of our test rats to 150 mg, as compared to 32 mg for control rats.

Biosynthesis of zearalenone

The various isolates of *Fusarium roseum* "graminearum" isolated in our laboratory over the years can be roughly divided into 3 groups, based on zearalenone production [44,45] — the strong producers (2000 ppm), moderate producers (1000 ppm) and the weak producers to none at all (trace). Some of the strong producers have produced in our laboratory concentrations of as high as 38 000 to 40 000 ppm using a solid medium and optimum temperature and duration of culture.

Zearalenone is routinely produced by seeding autoclaved moist rice or corn in quart bottles with a soil suspension of spores of *F. roseum*. Prior to seeding and before autoclaving, the corn is adjusted to a moisture content of 45% and the polished rice to 60% for best results [44]. The culture is incubated at 24—27° for 1 or 2 weeks to develop sufficient mass of fungus mycelium and then incubated at a temperature of 12—14° for 4 to 6 weeks. Glucose may be added (1—6%) to the rice or corn substrate for optimum production [46].

Eugenio *et al.* [44] describe production of zearalenone on maize, rice, wheat, barley and oats. Sherwood and Peberdy [47] studied the effects of moisture content, temperature and length of incubation period in stored grains. It appeared that once grain is sufficiently invaded by the fungus, most rapid production of zearalenone takes place at 12°, confirming the finding of Eugenio *et al.* [44] and Mirocha *et al.* [46].

Caldwell and Tuite [48] concluded that, in nature, zearalenone is produced in physiologically significant amounts (5—10 ppm) after harvest, when corn is stored on the cob in cribs, not in ears on plants in the field. *F. roseum* does invade ears of maize in the field and causes "pink ear rot" that may be involved in the "refusal factor" but such ears contain little or no zearalenone.

The enzymes responsible for biosynthesis of zearalenone are either induced or activated at low temperature (12—14°), a temperature which is not the optimum for growth of the organism, but rather one of stress [44,46,49]. An interesting parallel exists between the low temperature requirement for the production of zearalenone under experimental conditions, and the relationship between the incidence of the oestrogenic syndrome in swine and corn stored at low temperature. In the corn belt of the U.S., much corn is normally harvested dry and stored on the cob in open cribs where it is exposed to weathering. The frequent autumn rains increase the moisture content in certain ears to an extent (22—30%) that *F. graminearum*, as well as other fungi, can colonize the tissue. With decreasing autumn temperatures, the conditions are optimal for biosynthesis of zearalenone. The temperatures may be fairly high by day and low at night, or a prolonged warm period is followed by a cold one, thereby making conditions still more favorable for synthesis. Up to 2000 ppm of zearalenone have been found in corn brought in from the farm for analysis, and when such corn is ground and blended into a feed mixture, the process used in preparation may kill the fungus involved, but the products of its metabolism remain.

Caldwell and Tuite [50] reported the following species and subspecies as producing zearalenone in culture: *F. roseum*, *F. tricinctum*, *F. roseum* "Culmorum", *F. roseum* "Equiseti", *F. roseum* "Gibbosum" and *F. roseum* "Graminearum". Mirocha *et al.* [51] reported that zearalenone is also produced by *F. moniliforme*. Fiussello *et al.* [52] reported that *F. moniliforme*, strain 57, gave a positive uterotrophic response when tested in white mice. Sherwood and Peberdy [53] confirmed the production of zearalenone by an isolate of *F. moniliforme* [53].

To date, our laboratory has found *F. roseum*, *F. tricinctum*, *F. oxysporum*, *F. sporotrichiodes* [54] and *F. moniliforme* as producers of zearalenone. Of the five, *F. moniliforme* produces only trace amounts (1—19 ppm) whereas certain isolates of the others produce copious amounts.

The metabolic pathway of biosynthesis of zearalenone was studied by Steele *et al.* [55]. They found that $[1\text{-}^{14}C]$acetate and $[2\text{-}^{14}C]$diethylmalonate were rapidly (within 1 min) incorporated into zearalenone. $^{14}CO_2$ was also incorporated into zearalenone but at a slower rate. *Fusarium roseum* fixes atmospheric CO_2. DL-$[2\text{-}^{14}C]$mevalonic lactone, $[1\text{-}^{14}C]$-senecioate and $[G\text{-}^{14}C]$shikimate were not incorporated. DL$[2\text{-}^{14}C]$-mevalonic lactone was incorporated into ergosterol within 15 min, indicating that its inability to label zearalenone was not due to impermeability of the mycelium. Diluting $[1\text{-}^{14}C]$acetate with unlabelled diethylmalonate reduces the amount of acetate incorporation. We concluded that zearalenone is synthesized by the acetate-malonyl-CoA pathway of biosynthesis which in effect amounts to a "head-to-tail" condensation. If this is true, then when $[1\text{-}^{14}C]$acetate is used as substrate, then starting with carbon 2, every other carbon of zearolenone should theoretically contain ^{14}C. Steele *et al.* [55] chemically degraded such ^{14}C-labelled zearalenone and found that relative molar activity of the degradation fragments (CO_2, oxalic acid, succinic acid, glutaric acid and the aromatic ring) agreed with the postulate of synthesis *via* the malonyl-CoA-acetate pathway.

Analytical methods

Extraction and clean-up

Grind a 25-g sample of dried maize or barley in a Stein Mill, moisten with water (15% w/v) and then extract in a Soxhlet apparatus with 250 ml ethyl acetate for 7 h at one cycle/20 min. Concentrate the extract to near dryness in a flash evaporator and re-dissolve in 25 ml chloroform. Transfer chloroform into a 250 ml separatory funnel (SF). Add 10 ml of 1 *N* NaOH and roll the SF gently. Do not shake vigorously so as not to cause an emulsion. Drain the chloroform layer into a second 250 ml SF. Add 10 ml

1 N NaOH again to this chloroform, roll gently, equilibrate, and then discard the chloroform. Combine the 2 NaOH layers in the first SF and to this add 5 ml chloroform, roll gently and discard chloroform layer (wash NaOH layer twice in this manner). Next, acidify the NaOH layer with 2 N H_3PO_4 to pH 7.5 and add 15 ml of chloroform to the aqueous solution, shake vigorously and drain the chloroform into the second SF. Repeat this washing process and combine all the chloroform in the second SF and then dry the chloroform by pouring through a small column (10 g) of anhydrous Na_2SO_4. Concentrate the chloroform to near dryness on a steam bath and transfer quantitatively with acetone to an 0.5-dram vial. Concentrate the acetone solution in the vial to near dryness with nitrogen. See ref. 67 for greater detail.

Quantitation

(A) Thin-layer chromatography (TLC)

Dissolve the sample in 0.2 ml acetone and apply on a 20 × 20 cm, 0.25 mm thickness, Silica gel G TLC plate, previously activated at 110° for 1 h. Apply one 5 μl, two 10 μl and one 20 μl spots of the sample at 1-cm intervals on the plate. Apply 1 μl (1 μg/μl) of authentic zearalenone to the second 10 μl spot (becomes the internal standard) and apply the standard (1 μl) alone, next to the 20 μl spot of sample. Develop the TLC plate in a Desaga-Brinkmann (or equivalent) TLC developing tank lined with 2 mm thickness Whatman chromatography paper and add 100 ml of chloroform—ethyl alcohol (97:3 v/v) immediately before development. Develop the plate so that the solvent front travels about 18 cm and then air-dry the plate. Irradiate with a long wavelength (356 nm) UV lamp; zearalenone fluoresces blue-green. When irradiated with short wavelength (254 nm) UV lamp, zearalenone fluoresces more intensely. This dual intensity fluorescence is used as a confirmatory test.

The extraction, and separation procedure as described will separate not only zearalenone but also two naturally occurring derivatives called F-5-3 and F-5-4 [56]. The R_F values in the solvent system as described above are: 0.5, 0.32 and 0.24, respectively. All three components exhibit the dual intensity fluorescence phenomenon and can be easily picked out from among the other fluorescing components. Compounds F-5-3 and F-5-4 have been identified as stereoisomers of 8' hydroxyzearalenone [57].

(B) Gas-liquid chromatography (GLC)

Transfer 200 μl of the sample solution to an 0.5-dram vial or to a Pierce Reactivial® and concentrate to near dryness under a stream of nitrogen. Add 20 μl of one of the following reagents: (*1*) MSTFA® (*N*-methyl-*N*-trimethylsilyl-trifluoroacetamide) (Pierce Chem. Co.); (*2*) Tri-Sil 'BT"® (Bis-trimethylsilyl-acetamide and trimethylchlorosilane) (Pierce Chem. Co.); (*3*) Meth Elute (Trimethyl anilinium hydroxide, 0.2 *M* in methanol (Pierce Chem. Co.)

Reagent 1 and 2 react with the hydroxyl groups on the aromatic ring and form trimethylsilyl (TMS) ethers. Reagent 3 forms the methoxy derivative. Inject 1 μl of this material into a gas chromatograph under the following conditions: *Column:* 3% OV-1 on 100/120 Gas Chrom Q, 3 ft by 1/8" SS; *Carrier gas:* N_2, 20 ml/min; *Temperature program:* 150—260° at 8°/min; *Detector:* H_2 flame ionization detector.

When making the TMS ether derivative of zearalenone to be used as a standard, use between 50 and 250 ng/μl in concentration. The limit of sensitivity on the column is about 5 ng.

Acknowledgment

This research was supported in part by USPHS Research Grant No. 2RO1-FD-00035-06.

References

1 S.H. McNutt, P. Purevin and C. Murray, J. Am. Vet. Med. Assoc., 73 (1929) 484.
2 B.A. McErlean, Vet. Rec., 64 (1952) 539.
3 J.J. Christensen and H.C.H. Kernkamp, Minn. Agric. Exptl. Stat. Tech. Bull., 113 (1936).
4 A.Z. Joffe, In Mycotoxins in Foodstuffs, ed. G.N. Wogan, M.I.T. Press (1965) 77.
5 H.W. Wollenweber and O.A. Reinking, Die Fusarien — ihre Beschreibung, Schadwirkung and Bekämpfung, Verlagsbuchhandlung Paul Parey, Berlin, 1935.
6 W.C. Snyder and H.N. Hansen, Am. J. Bot., 27 (1954) 64.
7 W.C. Snyder and H.N. Hansen, Am. J. Bot., 32 (1945) 657.
8 T.A. Toussoun and P.E. Nelson, A Pictorial Guide to the Identification of Fusarium Species, Pennsylvania State University Press, Philidelphia, 1968.
9 W.L. Gordon, Can. J. Res. C., 22 (1944) 282.
10 C. Booth, The Genus Fusarium, Commonwealth Mycological Inst., Kew, Surrey, Great Britain, 1971.
11 F.N. Andrews and M. Stob, Belgian Patent 611, 630 (1961).
12 F.N. Andrews and M. Stob, U.S. Patent 3,196,019 (1965).
13 E.B. Hodge, P.H. Hidy and H.L. Wehrmeister, U.S. Patents 3,239,341; 3,239,342; 3,239,345, through 3,239,349; 3,239,351 through 3,239,357 (1966).
14 Britsh Patent 1,152,678 (1969).
15 David B.R. Johnston, Carol A. Sawicki, T.B. Windholz and A.A. Patchett, J. Med. Chem., 13 (1970) 941.
16 British Patents 1,152,674 through 1,152,677 (1969).
17 Richard N. Hurd and D.H. Shah, J. Am. Chem. Soc., 38 (1973) 390.
18 C. Allan Peters, J. Med. Chem., 15 (1972) 867.
19 Norman P. Jensen and T.B. Windholz, U.S. Patent 3,621,036 (1971).
20 G.H. Nelson, D.M. Barnes, C.M. Christensen and C.J. Mirocha, Symp. Proc. 70-0 Extension Service, Univ. of Nebraska, 1970, p. 90.
21 H.J. Kurtz, M.E. Nairn, G.H. Nelson, C.M. Christensen and C.J. Mirocha, Am. J. Vet. Res., 30 (1969) 551.
22 G.H. Nelson, C.M. Christensen and C.J. Mirocha, Proc. 70th Annual Meeting U.S. Livestock Sanitary Assoc., 1966, p. 614.
23 R. Zemjanis and C.J. Mirocha, (1973) (Unpublished data).
24 E.P. Young and H.S. Teague, J. Anim. Sci., 17 (1958) 1224.
25 W.C. Foote, D.P. Waldorf, H.L. Self and L.E. Casida, J. Anim. Sci., 17 (1958) 534.
26 R.F. Wilson, V.D. Sharma, L.E. Williams, D.P. Sharda and H.S. Teague, J. Anim. Sci., 26 (1967) 1479.

27 C.M. Christensen, C.J. Mirocha, G.H. Nelson and J.F. Quast, Appl. Microbiol., 23 (1972) 202.

28 C.J. Mirocha, C.M. Christensen, George Davis and G.H. Nelson, J. Agr. Food Chem., 21 (1973) 135.

29 F.M. Bristol and S. Djurickovic, Can. Vet. J., 12 (1971) 132.

30 Gy. Danko and P. Aldasy, Magyar Allatorvosok Lapia, 24 (1969) 517.

31 I. Debreczeni and Rejto, Magyar Allatorvosok Lapia, 24 (1969) 520.

32 V.I. Voluntir, I. Popescu, R. Jivanescu, Moga Minzat, Maria Purcel Vlah, Sanda Constant Inescu and M. Filip, Rev. Zooteh. Med. Vet., 21 (1971) 68.

33 L. Ozegovic, Veterinaria, 19 (1970) 525.

34 R.A. Meronuck, K.H. Garren, C.M. Christensen, G.H. Nelson and Fern Bates, Am. J. Vet. Res., 31 (1970) 551.

35 R.F. Sherwood and J.F. Peberdy, British Poultry Science, 14 (1973) 127.

36 G.M. Speers, R.A. Meronuck, D.M. Barnes and C.J. Mirocha, Poultry Science, 50 (1971) 627.

37 P.D. Sturkie, Avian Physiology, Comstock, Ithaca, N.Y., 1965.

38 G.M. Speers, C.J. Mirocha and C.M. Christensen, Poultry Science, 51 (1972) 1872.

39 C.M. Christensen, R.A. Meronuck, G.H. Nelson and J.C. Behrens, Appl. Microbiol., 23 (1972) 177.

40 C.J. Mirocha, J. Harrison, Agnes A. Nichols and Mary McClintock, Appl. Microbiol., 16 (1968) 797.

41 K. Roine, E.L. Korpinen and K. Kallela, Nord. Vet-Med., 23 (1971) 628.

42 E.L. Korpinen, Proc. IUPAC-Sponsored Symposium on Control of Mycotoxins, Göteborg, Sweden, 1972 p. 21.

43 J.C. Wolf and C.J. Mirocha, Can J. Microbiol., 19 (1973) 725.

44 P. Eugenio Cesaria, C.M. Christensen and C.J. Mirocha, Phytopathology, 60 (1970) 1055.

45 P. Eugenio Cesaria, Ph.D. Thesis, Dept. Plant Pathol. Univ. of Minnesota, 1968.

46 C.J. Mirocha, C.M. Christensen and G.H. Nelson, Biotechnol. Bioengr., 10 (1968) 469.

47 R.F. Sherwood and J.F. Peberdy, J. Stored Prod. Res., 8 (1972) 71.

48 R.W. Caldwell and J. Tuite, Phytopathology, 60 (1970) 1696.

49 C.J. Mirocha, C.M. Christensen and G.H. Nelson, Appl. Microbiol., 15 (1967) 497.

50 R.W. Caldwell and J. Tuite, Appl. Microbiol., 20 (1970) 31.

51 C.J. Mirocha, C.M. Christensen and G.H. Nelson, Appl. Microbiol., 17 (1969) 482.

52 N. Fiussello, J.C. Scurti and G. Cantini, Allionia, 16, (1970) 43.

53 R.F. Sherwood and J.F. Peberdy, Feed and Farm Supplies, 69 (1972) 9.

54 *Fusarium sporotrichioides* is the nomenclature used by U.S.S.R. scientists from which this culture was obtained. This is one of the original isolates reported by Russian scientists as causing alimentary toxic aleukia. This species is thought to be synonymous with *F. tricinctum*.

55 J.A. Steele, J.R. Lieberman and C.J. Mirocha, Can. J. Microbiol., 20 (1974) 531.

56 C.J. Mirocha, C.M. Christensen and G.H. Nelson, in S. Kadis, A. Ciegler and S.J. Ajl (Eds.), Microbial Toxins, Vol. VII, Chapter 4, Academic Press, New York, 1971.

57 R.S. Jackson, S. Fenton, C.J. Mirocha and G. Davis, J. Agric. Fd Chem., (In press).

58 E.M. Pullar and W.M. Lerew, Aust. Vet. J., 13 (1937) 28.

59 J. Lelievre, J. Bremond and J. Rebour, Bull. Soc. Vet. Pratique de France, 46 (1968) 18.

60 C. Paita, Vet. Ital., 4 (1965) 195.

61 S. Stamatovic, Z. Ljesevic and S. Durickovic, Vet. Glasnik, 17 (1963) 507.

62 T. Bugeac and C. Berbinschi, Rev. Zooteh. Med. Vet., 17 (1967) 56.

63 E. Eriksen, Nord. Vet. Med., 20 (1968) 396.

64 C.J. Mirocha, C.M. Christensen and G.H. Nelson, Cancer Res., 28 (1968) 2319.

65 G. Bolliger and Ch. Tamm, Helv. Chim. Acta, 55 (1972) 3030.

66 J.A. Steele, Biogenesis and Metabolism of Zearalenone by *Fusarium roseum*, Ph.D. Thesis, Dept. Plant Pathol., Univ. of Minnesota, St. Paul, Minn. (U.S.A.) 1974.

Chapter 7

PENICILLIUM CYCLOPIUM

I.F.H. PURCHASE

Central Toxicology Laboratories, Imperial Chemical Industries Ltd., Alderley Park, Macclesfield, Cheshire (Great Britain)

The fungus *Penicillium cyclopium* has been isolated from a variety of foodstuffs in many parts of the world. With the upsurge of interest in mycotoxicoses in the last decade, reports have appeared of the isolation of toxigenic strains. Scott [1] reported the isolation of 228 mould strains (representing 59 species) from legumes and cereals and tested them for toxicity in 1-day-old Pekin ducklings. 46 of these strains were toxic, but of the 5 strains of *P. cyclopium* tested, none proved to be toxic. One strain of *P. urticae*, later shown to be *P. cyclopium* [2], isolated from groundnuts, was toxic to rats, mice and ducklings. Cyclopiazonic acid was isolated from this strain. Later work on this strain showed that it was toxic to mice (100% mortality in 14 days) producing lesions very similar to those of pure cyclopiazonic acid [3].

Wilson *et al.* [4] reported the isolation of 2 strains of *P. cyclopium* from compounded feedstuffs which caused mortality and 1 from groundnuts. All 3 strains produced a neurotoxin designated penitrem A. Later work cast some doubt on the identity of these strains which were then classified as *P. crustosum* [5]. The same toxin was, however, recovered from strains of *P. cyclopium* [5].

Penicillium cyclopium was isolated from wheat grain and katsuibushi in Japan [6]. Assay of cultures of *P. cyclopium* on various semi-synthetic media did not reveal any production of cyclopiazonic acid. Extracts of mycelium and filtrate of all 3 strains examined were toxic producing a cytotoxic effect on the bone marrow and liver damage in mice [7]. Martin *et al.* [8] examined 635 food samples collected in Southern Africa for fungal flora. *Penicillium cyclopium* was regularly present on all foodstuffs examined (maize, groundnuts, sorghum, various legumes, maize meal, groundnut meal) but had a maximum affinity for maize meal and beans and the least affinity for groundnuts. A total of 531 of these isolates was tested for toxicity in day-old ducklings and 6 of the 21 isolates of *P. cyclopium* were found to be toxic.

Thus, *P. cyclopium*, which is a common isolate from foodstuffs, is frequently capable of producing toxic metabolites and 12 out of 32 strains tested by the authors mentioned above were toxic. For this reason alone, *P. cyclopium* is potentially of importance as a food and feed contaminant.

Diseases caused by *P. cyclopium*

In 1968 Wilson *et al.* [4] reported two outbreaks of disease which occurred after the ingestion of mouldy food which was contaminated with *P. cyclopium*. Several prize sheep died after exhibiting symptoms which included anorexia, depression, humping of the back, diarrhoea, slobbering, generalized weakness and convulsions. One surviving animal had residual nervous symptoms, and its owner classed it as a "dummy" because of its lack of response to stimuli. The other outbreak occurred when two horses died after eating pelletted feed which had become wet and mouldy during delivery to the farm. The horses died suddenly and no details of symptoms or of post-mortem findings were available [9].

A third outbreak of disease in England was reported in 1970 [10,11]. 12 calves showed ataxia and muscular tremors in the hind quarters which were followed by convulsions and subsequently death. *Penicillium cyclopium* was isolated from the feedstuff (crushed barley) and cyclopiazonic acid, identified by thin layer chromatograph (TLC) analysis, was isolated.

Cyclopiazonic acid

Holzapfel [12] was the first to report the isolation of a major toxic metabolite of *P. cyclopium*. The strain of *P. cyclopium* used in this study was isolated from groundnuts and was toxic to ducklings, mice and rats [1]. It was grown on sterilized moistened maize meal. The toxin was extracted with chloroform-methanol, purified by solvent partition and chromatography on formamide-impregnated cellulose powder followed by ion-exchange column chromatography. The major toxic compound was cyclopiazonic acid ($C_{20}H_{20}N_2O_3$) with the structural formula shown in Fig. 1.

Subsequently two metabolites chemically related to cyclopiazonic acid were isolated from shake cultures of *P. cyclopium* [13], but neither have been tested for toxicity. Cyclopiazonic acid imine (molecular formula $C_{20}H_{21}N_3O_2$) and bissecodehydrocyclopiazonic acid (molecular formula $C_{20}H_{22}N_2O_3$) have the structural formulae presented in Figs. 2 and 3.

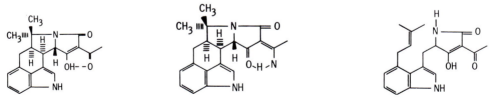

Fig. 1. The structure of cyclopiazonic acid.

Fig. 2. The structure of cyclopiazonic acid imine.

Fig. 3. The structure of bissecodehydrocyclopiazonic acid.

TABLE I

PHYSICAL CHARACTERISTICS OF CYCLOPIAZONIC ACID AND ITS DERIVATIVES

	Mol. wt.	Melting point °C	UV absorption	
			λ max	Log ϵ
Cyclopiazonic acid	336	245—246	225	4.60
			253	4.22
			284	4.31
Cyclopiazonic acid imine	336	277—278	224	4.58
			244	4.03
			275	4.16
			286	4.32
			293	4.35
Bissecodehydrocyclopiazonic acid	338	168—169	225	4.55
			276	4.28
			296	4.07

The physical characteristics of these three compounds are presented in Table I.

Cyclopiazonic acid gives a blue-violet Ehrlich colour reaction and an orange-red $FeCl_3$ colour reaction. Bissecodehydrocyclopiazonic acid gives the same colour reaction with $FeCl_3$ but may be differentiated from cyclopiazonic acid by the absence of a colour reaction with Ehrlich's reagent.

Biosynthesis of cyclopiazonic acid

Holzapfel and Wilkins [14] provided evidence that cyclopiazonic acid is derived from 1 molecule of tryptophan, 1 molecule of mevalonic acid and 2 molecules of acetic acid. The toxic strain (CSIR 1082) of *P. cyclopium* was grown on a synthetic medium with sodium nitrate as the nitrogen source and glucose as the carbon source. In a shake culture the maximum yield of cyclopiazonic acid was obtained when mycelial growth had practically ceased.

In this system 24.7% of the ^{14}C from D L-tryptophan universally ^{14}C-labelled in the benzene ring was incorporated into the cyclopiazonic acid. This efficient incorporation of D L-tryptophan was regarded as evidence that it is a direct precursor of cyclopiazonic acid. Incorporation from [1-^{14}C]-sodium acetate and [2-^{14}C] mevalonic acid was 3.5 and 7.0% respectively. Cyclopiazonic acid, labelled from [2-^{14}C] mevalonic acid was degraded chemically. It was found that the majority (98.4%) of the activity of the labelled cyclopiazonic acid was retained in the amino acid, and its methyl ester, derived by cleavage of ring E of *O*-methyldesacetylcyclopiazonic acid by a retro-Claisen reaction. Oxidation of *O*-methyldesacetylcyclopiazonic acid yielded acetic acid which represented the carbon 10 and *gem*-dimethyl-group and which contained 47.6% (representing 95.2% of the original label because of the presence of 2 methyl-groups) of the activity. Degradation of this acetic acid indicated that the label was present in the *gem*-dimethyl-group and that none was derived from carbon 10. Holzapfel and Wilkins considered that mevalonic acid is incorporated into cyclopiazonic acid, after conversion into γ,γ-dimethylallyl pyrophosphate.

Similarly, cyclopiazonic acid, labelled from [1-^{14}C]sodium acetate, was degraded by chemical means. Cyclopiazonic acid reacted with 0.1 N H_2SO_4 in methanol to give desacetylcyclopiazonic acid (64% of original activity) and acetic acid (35% of original activity). Degradation of the acetic acid indicated that the activity was derived from carbon 17. Hydrolysis of O-methyldesacetylcyclopiazonic acid, derived from desacetylcyclopiazonic acid, yielded the amino acid which was isolated as its methyl ester (48% of original activity) and acetic acid (33% of original activity). This acetic acid derived its label from carbon 8. Oxidation of O-methyl-desacetylcyclo-piazonic acid yielded acetic acid. Degradation of this acetic acid indicated that the label was derived from carbon 10 (14% of original activity) and not the *gem*-dimethyl-group.

From the above evidence the incorporation of acetate and mevalonate into cyclopiazonic acid can be represented by the scheme in Fig. 4.

Holzapfel and Wilkins [14] noted that the concentration of bissecode-hydrocyclopiazonic acid (given the trivial name β-cyclopiazonic acid as opposed to cyclopiazonic acid with the trivial name α-cyclopiazonic acid) in shake cultures increases rapidly in the early stages of fermentation when cyclo-piazonic acid is present in trace amounts. The concentration of β-cyclopia-zonic acid decreases rapidly as soon as cyclopiazonic acid concentration increases. This observation suggested that β-cyclopiazonic acid was a precursor of cyclopiazonic acid.

Radioactively labelled β-cyclopiazonic acid was prepared in a shake culture with a synthetic medium containing [1-^{14}C]acetate and a low zinc ion or ferrous ion concentration (which favoured accumulation of β-cyclopia-zonic acid at 7 days).

The β-cyclopiazonic acid prepared in this way was added to a shake

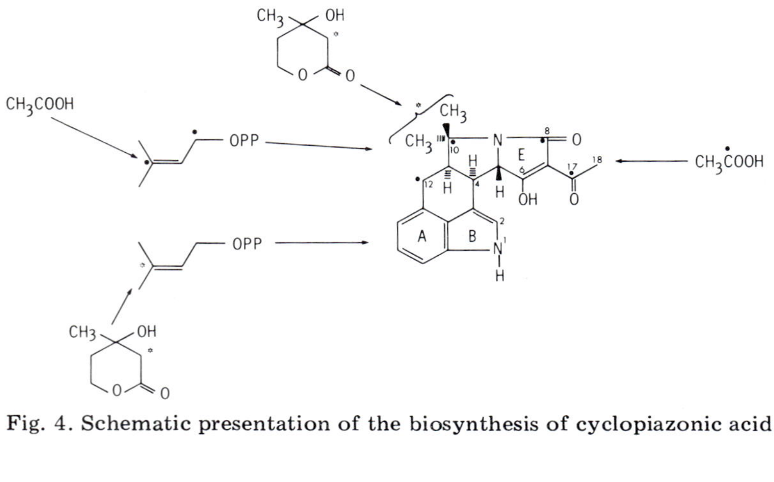

Fig. 4. Schematic presentation of the biosynthesis of cyclopiazonic acid.

culture of *P. cyclopium* on a complete synthetic medium at 6 days. Cyclopiazonic acid is synthesized rapidly from this stage and the cyclopiazonic acid isolated 48 h later contained 67% of the added label. A further 18% of activity was found in the β-cyclopiazonic acid recovered at the end of fermentation. This result provided further evidence that β-cyclopiazonic acid is a direct precursor of cyclopiazonic acid. Confirmation was obtained by Schabort [15] who isolated 5 iso-enzymes from *P. cyclopium* capable of converting β-cyclopiazonic acid into cyclopiazonic acid.

Schabort and co-workers [16—18] proposed a two-stage conversion of β- into α-cyclopiazonic acid. The kinetics of the first step, the dehydrogenation of β-cyclopiazonic acid, were estimated spectrophotometrically using either 2,6-dichlorophenolindophenol or cytochrome C as the electron acceptor. The conversion of β- to α-cyclopiazonic acid was assayed either by the Ehrlich colour reaction (given only by α-cyclopiazonic acid), paper chromatography or ultraviolet (UV) spectrophotometry. Confirmation that β-cyclopiazonic acid was converted to α-cyclopiazonic acid was obtained by mass spectrometry.

Five iso-enzymes of the enzyme responsible for converting β- into α-cyclopiazonic acid (β-cyclopiazonate oxidocyclase) were isolated and purified from the mycelium of *P. cyclopium*. They had an optimum pH of 6.8 and did not require zinc or ferrous ion for their activity. The enzymes were flavoproteins with molecular weights of approx. 50 000.

The biochemical and biophysical characteristics of these five iso-enzymes have been described in detail by Schabort and co-workers [16—18]. Apart from 2,6-dinitrophenolindophenol and cytochrome C, phenazine methosulphate was an effective electron acceptor and oxygen could act as an electron acceptor with phenazine methosulphate as an intermediate electron carrier. Most bivalent metal ions inhibited the reaction as did 2,4-dinitro-1-fluorobenzene and L-1-tosylamide-2-phenylethylchloroethyl ketone. The latter effects indicate that amine groups and histidine imidazole groups play a role in these reactions.

Environmental conditions favouring production of cyclopiazonic acid

No details of the conditions which encouraged the production of toxin in fatal cases of poisoning are available. In the one reported outbreak of poisoning by cyclopiazonic acid crushed barley was implicated as the substrate.

In the initial screening by Scott [1] of *P. cyclopium* (referred to in his article as *P. urticae* strain G391 and later reclassified as *P. cyclopium* and held in the CSIR collection as 1082) for toxicity, the fungus was grown on sterilized maize meal with a moisture content of about 40%. The culture was incubated at 26° for 2 to 3 weeks until abundant mycelial growth and sporulation occurred. Later work by Wilkins and others [13,14,16,] showed that a shake culture (150 rev./min at 25°) using a Czapek-type medium supported growth of *P. cyclopium*. The medium contained: glucose (60 g); $NaNO_3$ (4.2 g); $MgSO_4$ $7H_2O$ (0.5 g); KCl (0.5 g); K_2HPO_4 (1 g);

$Na_2 B_4 O_7 \cdot 1 OH_2 O$ (0.7 mg); $(NH_4)_6 Mo_7 O_{24} \cdot 4 H_2 O$ (0.5 mg); $CuSO_4 \cdot 5 H_2 O$ (0.3 mg); $MnSO_4 \cdot H_2 O$ (0.11 mg); $ZnSO_4 \cdot 7 H_2 O$ (17.6 mg); $FeSO_4 \cdot 7 H_2 O$ (10 mg); deionized water (1 l). In the early stages of growth β-cyclopiazonic acid was present in high concentration. Later (5—7 days) when growth rate, measured as the rate of increase of total mycelial nitrogen, had practically ceased, α-cyclopiazonic acid levels increased rapidly and simultaneously with a decrease in β-cyclopiazonic acid levels. Cyclopiazonic acid imine accumulated during later stages of the fermentation. Higher levels of β-cyclopiazonic acid were obtained by culturing under the same conditions using the synthetic medium described above except that the $ZnSO_4 \cdot 7H_2 O$ was reduced to 0.07 mg/1.

Analytical methods. In the early work on the structure and biosynthesis of cyclopiazonic acid several methods of isolation and identification were described. Thus, $CHCl_3$:MeOH (1+1) was used to extract the toxin from moulded maize meal and chloroform was used in the case of synthetic media [14,18]. Purification of the primary extract was achieved by solvent partition between chloroform and water and 95% MeOH and hexane. A more effective purification stage depends on the fact that cyclopiazonic acid behaves as a monobasic acid. Extraction of a chloroform solution by aqueous $NaHCO_3$, acidification of the bicarbonate extract and re-extraction with chloroform is an effective purification procedure [12]. For preparative work column chromatography on formamide-impregnated cellulose or ion-exchange resin has been described [12].

Quantitation of the cyclopiazonic acid in the purified extract can be achieved in several ways. The colour reaction with Ehrlich's reagent was used by Schabort [18] to estimate the quantity of cyclopiazonic acid. To a methanolic solution of the extract (2 ml), 1 ml 10% 4-dimethylamino-benzaldehyde (Ehrlich's reagent) and 2 ml concentrated HCl were added. The absorbance of this mixture was read after 15 min at 580 nm and the quantity of toxin estimated by comparison with a standard curve.

Ultraviolet spectrophotometry [18] has also been used and the quantity may be estimated from the absorbance at 225, 253 or 284 nm and the log ε values of 4.60, 4.22 and 4.31 for a methanolic solution.

A qualitative method for the detection of cyclopiazonic acid has been described by Steyn [19]. The water-soluble and lipid material was removed from the primary extract by liquid-liquid partition. Cyclopiazonic acid was further purified by the removal of neutral material as discribed above. Chromatoplates were prepared from a slurry of Silica Gel G (Merck) and 0.4 N aqueous oxalic acid (1+2). Glass plates were spread with a 0.25 mm layer of the slurry and air-dried. The purified extract was spotted on to the plate and the plate developed in chloroform:methylisobutylketone (4+1) in a saturated tank. After drying at room temperature, the plate was viewed under long-wave UV light. Cyclopiazonic acid had an R_F value of 0.65 and green fluorescent colour. Its identity could be confirmed by colour reagents. After spraying with concentrated $H_2 SO_4$ and heating to 110° for 10 min, cyclopiazonic acid became red-brown. Spraying with 1% ethanolic solution of ferric chloride gave a red-brown colour and Ehrlich's reagent gave a violet colour. On prolonged standing of the silica gel plates impregnated with oxalic acid, cyclopiazonic acid gave a violet-red colour. Further confirmation may be obtained by omitting oxalic acid from the slurry, when cyclopiazonic acid loses its mobility.

No sensitive quantitative method has been described for the estimation of cyclopiazonic acid.

Biological activity of cyclopiazonic acid

Cyclopiazonic acid is toxic to rats [2] and by implication to mice [1], ducklings [1] and calves [10]. The only published detailed work on its toxicity concerns the rat.

Cyclopiazonic acid, dissolved in 1 N sodium bicarbonate, was dosed intraperitoneally to male Wistar-derived rats (mean weight 108 g) [2]. Rats receiving doses greater than 8 mg/kg died within 2 h. After about 5 min the rats were disinclined to move, became ataxic and resented being handled. 10 to 30 min later the rats became immobile and died within the next 90 min in extensor spasm with cyanotic mucous membranes. Rats receiving 2.5 or 4.5 mg/kg died 1—3 days after dosing and those receiving lower doses recovered and survived to 10 days. The LD_{50} was estimated as 2.3 mg/kg.

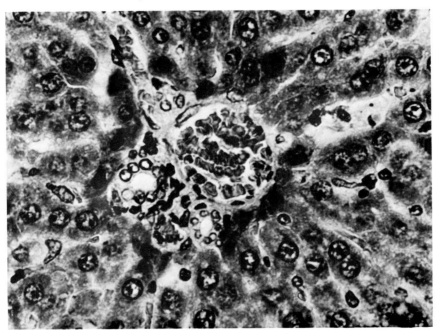

Fig. 5. Portal tract from the liver of a rat killed 10 days after receiving cyclopiazonic acid (67.5 mg/kg *per os*). Single cell necrosis and changes in the arrangement of chromatin in the hepatocyte nuclei are present. Margination of the chromatin in bile duct cell nuclei is also prominent. H and E. × 720. (From ref. 2 by permission from Academic Press, New York).

Oral administration produced a different syndrome. Mortality occurred up to 6 days after dosing and the animals were prostrate for several hours before death. The LD_{50} values were 36(32—41) mg/kg in males and 63(52—67) mg/kg in females. The difference in the symptoms and toxicity of cyclopiazonic acid when given orally or intraperitoneally was ascribed to the acidic nature of the toxin. After intraperitoneal administration, absorption was rapid producing nervous symptoms. Oral administration into the acidic stomach probably rendered the cyclopiazonic acid insoluble with absorption occurring only from the more alkaline small intestine. Ducklings receiving oral doses died quickly with nervous symptoms indicating a rapid absorption from the crop.

Histopathological examination of organs from cyclopiazonic acid-treated rats revealed two basic types of lesions. At high doses single cell necrosis was seen in many organs. At these and lower doses the major effect was on cells lining ducts. Rats dying within 36 h of dosing were all females. There was focal or single cell necrosis of hepatocytes. Cells lining the bile ducts were enlarged with large nuclei containing a thin marginal band of chromatin (Fig. 5). At higher doses (67 and 83 mg/kg) this enlargement of bile duct lining cells, particularly in the smaller bile ducts, appeared to occlude the lumen and necrosis of these cells occurred. Post-mortem autolysis affected the bile duct cells and in some cases they had deteriorated to such an extent that no bile ducts were visible. Tubular cells in the cortico-

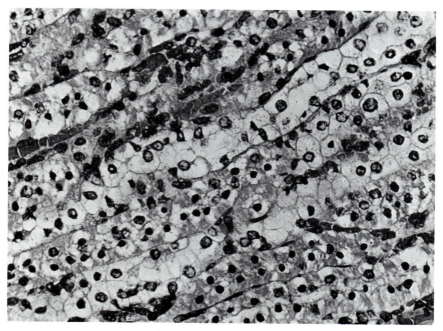

Fig. 6. Degenerative changes in the renal medulla of a rat killed 10 days after receiving cyclopiazonic acid (45 mg/kg *per os*). Two distinct changes are visible — pyknosis and enlargement of the nucleus with margination of the chromatin. H and E. × 450. (From ref. 2 by permission from Academic Press, New York).

medullary junction of the kidneys were necrotic. The tubular epithelial cells in the medulla were either necrotic or enlarged with swollen nuclei containing marginal chromatin (Fig. 6).

The islets of Langerhans had distinct degenerative changes and necrosis particularly of cells in the periphery of the islets. The exocrine portion of the pancreas was also affected with necrosis of acinar cells and enlargement of nuclei and cytoplasm of duct cells (Fig. 7).

The rats dying 4—6 days after dosing were males. The lesions observed were very similar to those described for females. In addition the spleen showed signs of atrophy with a folded capsule. The red and white pulp were less cellular than controls and necrosis of lymphocytes was observed (Fig. 8). Extensive hyaline degeneration of the myocardium and necrosis of myocardial fibres with haemorrhage into the necrotic areas was also seen. The lesions in animals killed at 10 days were less severe. Rats receiving an oral dose of 50 mg/kg were killed at intervals. The liver lesions were the first to appear at 8 h and were most severe at 24 h. The renal tubular lesions were observed between 8 and 192 h. The pancreatic changes were first observed at 16 h and were still present, although mild, at 192 h. There was an absence of granules in the islets of Langerhans, suggesting that the coma preceding death in orally dosed animals was the result of hypoglycaemia. Lesions in other organs were also recovering by 96 and 192 h.

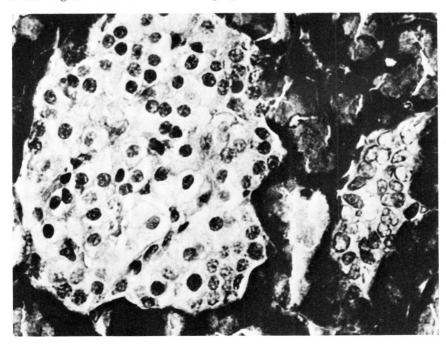

Fig. 7. Section of the pancreas of rat killed 8 days after receiving a dose of cyclopiazonic acid (50 mg/kg *per os*). Cell nuclei on the periphery of the islet show degenerative changes and characteristic nuclear abnormalities are present in the pancreatic ducts. H and E. × 720. (From ref. 2 by permission from Academic Press, New York).

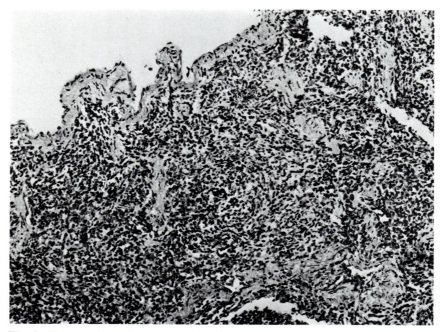

Fig. 8. Degenerative changes in the spleen of a male rat dying 6 days after a dose of cyclopiazonic acid (36.7 mg/kg *per os*). H and E × 180. (From ref. 2 by permission from Academic Press, New York).

The effect of three mycotoxins, one of which was cyclopiazonic acid, on deoxyribonuclease was studied by Schabort and Roberts [20]. Cyclopiazonic acid activated bovine pancreas deoxyribonuclease slightly, the increase in activity being 10% at a concentration of 10 μM. Binding to DNA or DNAase could not be measured for technical reasons.

Penitrem A

Wilson and co-workers [4,5,9] have isolated a toxin from the strain of *P. cyclopium* causing disease in sheep and horses in Tennessee.

The toxin, known as penitrem A, was prepared by extraction from the culture with ethyl ether, purification by column chromatography using 200 mesh Florisil as the stationary phase, and 3% methanol in chloroform or ethyl ether:petroleum ether 1:1. Final purification was by recrystallization from ethyl ether:petroleum ether. It is a white feather-like crystalline substance with the molecular formula $C_{37}H_{44}O_6NCl$ with a molecular weight of 633. The nucleus of this toxin is probably a steroid. Crystals decomposed at 180—200° without melting. Penitrem A is soluble in polar solvents, sparingly soluble in aliphatic hydrocarbons and very slightly soluble in acidified and alkalinized water. It decomposed rapidly in alcohol at 25° when treated with acid changing to a yellow and then green non-toxic compound. Similarly, it was unstable in chloroform exposed to light changing from brown to green to dark blue.

The UV absorption spectrum of this compound had peaks at 233 and 295 nm with ϵ values of 31 500 and 16 200 respectively. Slight variations in the spectra are observed in 0.1 N NaOH and 0.1 N HCl.

Extracts from cultures of *P. cyclopium* frequently contain a compound which has a similar R_F value to penitrem A. It turns pink on spraying with 2% ferric chloride in butanol and later turns brown. It is weakly fluorescent under UV light and has been tentatively identified at $9\alpha,11\beta$-ergosta-5,7,22-trien-3β-ol. The crystals are larger than those of penitrem A and colourless. Penitrem A is more soluble in methanol than the steroid and this may help in purification.

Environmental conditions favouring production of the toxin

Penitrem A is produced on several natural substrates and synthetic media. Wilson [9] describes a liquid medium (3% dehydrated mashed potatoes, 2% skimmed milk solids, 2% sucrose and streptomycin 10 μg/ml) which supports growth and toxin production. *P. cyclopium* is grown on a stationary culture with a large surface area at 25°. Toxin production commences at 7 days and is maximal at 14 days (about 30 mg/l) when the pH is 6.5—7.2. Toxin concentration decreases on longer incubation. No toxin production occurs in shake cultures.

The toxin is present in the mycelial mat only. Mycelial mat weight is not necessarily related to toxin yield although when toxin is formed higher yields are obtained with optimal mycelial growth.

The bromo-toxin may be prepared by substituting KBr for KCl in a synthetic medium and traces of the iodo-toxin and the dehalogenated molecule have been detected in media with KI substituted for KCl and without Cl respectively.

Although *Penicillium cyclopium* is used by Wilson [9] to describe the species producing the tremorigenic toxin, there is some confusion about the exact identification of the strain. It was classified as *P. crustosum* by the Centraal Bureau voor Schimmelcultures [9]. Penitrem A has been isolated from authentic strains of *P. cyclopium* and also apparently from strains of *P. palitans* and *P. viridicatum*.

Analysis. The toxin may be extracted with ethyl ether. The quantity of toxin may be assayed by TLC using silica gel G as the stationary phase and ethyl ether-cyclohexane (3+1) as the developing solvent. After developing the plate, it is sprayed with 2% ferric chloride in butanol when the toxin appears as a green spot at R_F 0.6. The green colour is probably a degradation product and similar colours are produced by spraying with acids. The quantity of toxin may be estimated by UV in the visible region after spraying with acid. Visual comparison with standards provides a less accurate method.

Biological activity. Penitrem A produces neurological and renal effects in experimental animals [4,5,9]. The neurological effects are seen in several

References p. 161

species, with chickens, rabbits and guinea pigs being more resistant than rats and mice to the lethal effects and hamsters being the most resistant species tested. Sheep, cattle and horses are also susceptible, with mortality occurring under field conditions. The neurotoxicity begins within 5 to 30 min of administration. In mice small doses administered orally or intraperitoneally (250 μg/kg in propylene glycol or in aqueous suspension) cause perceptible tremors which persist for several hours. Larger doses produce an increase in irritability, weakness of the limbs, inability to grasp, pinna loss, slight exophthalmos and marked tremors. With doses of 2.5 mg/kg or higher the tremors progressed to clonic or tetanic convulsions and often death. The LD_{50} appears to be greater than 10 mg/kg, since only 1 out of 3 animals died after receiving this dose. Symptoms seen may also include tachypnoea, diarrhoea, diuresis, loss of corneal reflex, mydriases and lacrymation. Survivors' convulsions reverted back to tremors and tremblings which persisted for 24 or 72 h. Repeated small doses produced no tachyphyllaxis. Rats are less susceptible to the early convulsive effects but other neurological signs may be more severe. A dose of 2.5 mg/kg may cause difficulty in locomotion for 2—3 weeks. Coarse tremors, spastic paralysis and circling are also seen.

The toxin is about 10 to 20 times more potent as a tremorgen than acroline or tremorine. It has no anti-pseudocholinesterase activity.

Stern [21] has studied the influence of a number of compounds on the tremors produced by Penitrem A. Of the 46 compounds administered only glycine, mephenesin, oxyaminoacetic acid, and diazepam abolished the tremors. Amino acids had no effect. Anticholinergic compounds (atropine, scopolamine, ethylbenztropine, coramiphen) had no effect at low doses but increased the tremor at higher doses. There was an additive effect with known tremorgens (oxotremorine, physostigmine, armine and acediclin). A large number of pharmacologically active compounds had no effect (catecholamines, α -and β-blockers, histidine, 5-hydroxytryptophan, L-dopa, γ-hydroxy-butyrate, α-methyl-*m*-tyrosine, hexamethonium). Penitrem A reduced brain acetylcholine levels but had no effect when injected directly into the brain in the globus pallidus, nucleus caudatus or nucleus niger. No tremors occurred below the level of spinal transection. Stern suggested that penitrem A generated the tremors at the level of the spinal cord. It inhibited the interneurones inhibiting the α-motor cells of the anterior horn. The α-motor cells have a spontaneous rhythm with a frequency very similar to that of the tremor produced by penitrem A which is normally masked by inhibiting mechanisms. When the inhibition is abolished by penitrem A, spontaneous activity develops.

Using the rat phrenic nerve-diaphragm preparation, Wilson *et al.* [22] studied the effect of penitrem A on the neuromuscular junction. The toxin was given intraperitoneally (2.5 mg/kg) and the animals killed 2 h later. Penitrem A produced an increase in all four parameters estimated (resting potential, end plate potential amplitude, end plate potential duration and miniature end plate potential frequency). When the toxin was applied directly to the preparation there was a marked increase in miniature end plate potential frequency which persisted after normal bathing fluid was returned. Other effects were similar to those produced by intraperitoneal administration. As penitrem A had little effect on diaphragm cholinesterase activity, it probably produced its effect by influencing pre-synaptic transmitter release. The increased frequency of miniature end plate potentials suggests stimulation of release of transmitter packets. The increase in amplitude could be due either to increased transmitter packets or increased sensitivity of the post-synaptic membrane. The fact that 24 h after an intraperitoneal dose of penitrem A no neuromuscular effects were seen in spite of the presence of tremors, confirms Stern's suggestion of a spinal site of action.

Diuresis is produced in both rats and mice by intraperitoneal or oral administration of penitrem A [5,9]. Over a period of 4.5 h after administration of the toxin (2.0 mg/kg) in 10—15 ml saline total urinary electrolytes excreted increased (Na^+ 345 to 1830 micro equivalents, K^+ 150 to 505, Ca^{2+} 7 to 81, Mg^{2+} 30 to 151, Cl^- 370 to 1902, glucose 2 mg to 60 mg, volume 1.95 ml to 18.05 ml). This diuretic effect continued for several hours and was less marked at a lower dose (0.5 mg/kg). Wilson [5,9] suggests that the toxin prevents kidney tubular reabsorption.

References

1 D.B. Scott, Mycopathol. Mycol. Appl., 25 (1965) 213.
2 I.F.H. Purchase, Toxicol. Appl. Pharmacol., 18 (1971) 114.
3 S.J. van Rensburg, I.F.H. Purchase and J.J. van der Watt, Hepatic and renal pathology induced in mice by feeding fungal cultures, in I.F.H. Purchase (Ed.), Mycotoxins in Human Health, Macmillan, London, 1971, pp. 153-161.
4 B.J. Wilson, C.H. Wilson and A.W. Hayes, Nature, 220 (1968) 77.
5 B.J. Wilson, Recently discovered metabolites with unusual toxic manifestations, in I.F.H. Purchase (Ed.), Mycotoxins in Human Health, Macmillan, London, 1971, pp. 223-229.
6 H. Kurata, S. Udagawa, M. Ichinoe, S. Natori and S. Sakaki, Field survey of mycotoxin-producing fungi contaminating human foodstuffs in Japan with epidemiological background, in I.F.H. Purchase (Ed.), Mycotoxins in Human Health, Macmillan, London, 1971, p. 101-106.
, 7 M. Saito, M. Enomoto, M. Umeda, K. Ohtsubo, T. Ishiko, S. Yamamoto and H. Toyokawa, Field survey of mycotoxin-producing fungi contaminating human foodstuffs in Japan with epidemiological background, in I.F.H. Purchase (Ed.), Mycotoxins in Human Health, Macmillan, London, 1971, pp. 159-183.
8 P.M.D. Martin, G.A. Gilman and P. Keen, The incidence of fungi in foodstuffs and their significance based on a survey in the eastern Transvaal and Swaziland, in I.F.H. Purchase (Ed.), Mycotoxins in Human Health, Macmillan, London, 1971, pp. 281-290.
9 B.J. Wilson, Miscellaneous Penicillium toxins, in Microbial Toxins, Vol. 6, Academic Press, New York, 1971, pp. 479-485.
10 J. Harrison, Trop. Sci., 13 (1971) 57.
11 A. Hacking, personal communication (1973).
12 C.W. Holzapfel, Tetrahedron, 24 (1968) 2101.
13 C.W. Holzapfel, R.D. Hutchinson and D.C. Wilkins, Tetrahedron, 26 (1970) 5239.

14 C.W. Holzapfel and D.C. Wilkins, Phytochemistry, 10 (1971) 351.
15 J.C. Schabort, Intern. Symp. on Control of the Human Environment, p. 62, Johannes-
    burg (1969) (Abstracts).
16 J.C. Schabort and D.C. Wilkins, J.S. Afr. Chem. Inst., 22 (1969) 59.
17 J.C. Schabort and D.J.J. Potgieter, Biochim. **Biophys.** Acta, 250 (1971) 329.
18 J.C. Schabort, D.C. Wilkins, C.W. Holzapfel, D.J.J. Potgieter and A.W. Neitz, Bio-
    chim. Biophys. Acta, 250 (1971) 311.
19 P.S. Steyn, J. Chromatog., 45 (1969) 471.
20 J.C. Schabort and H.A. Roberts, Biochem. Pharmacol., 20 (1971) 243.
21 P. Stern, Iugoslav. Physiol. Pharmacol. Acta, 7 (1971) 187.
22 B.J. Wilson, T. Hoekman and W.-D. Dettbarn, Brain Res., 40 (1972) 540.

Chapter 8

PENICILLIUM RUBRUM — RUBRATOXINS*

PAUL M. NEWBERNE

Department of Nutrition and Food Science, Massachusetts Institute of Technology, Cambridge, Mass. (U.S.A.)

The rubratoxins are members of a group of mould metabolites described collectively as mycotoxins. Large quantities of rubratoxins produced by some strains of *Penicillium rubrum* growing on feeds are a hazard to livestock consuming such feeds as well as a potential public health hazard to man. During the past decade, a large body of information has developed which confirms that numerous spoilage moulds can produce mycotoxins when they grow on grains and other feedstuffs with serious consequences if ingested. Twenty years ago, Sippel *et al.* [1] described a disease of cattle and swine which followed the ingestion of mouldy corn. In this outbreak, 13 cultures of moulds were isolated from the toxic corn but only 2 were found to cause illness and death when fed under appropriate conditions to experimental animals. One isolate each of *Aspergillus flavus* and *Penicillium rubrum* caused illness in animals; these two were used for further experimental investigations to show that fresh corn contaminated with the toxic isolate of *Penicillium rubrum* was considerably more toxic than corn contaminated with the toxic isolate of *Aspergillus flavus*. In fact, a total dose of 7 to 8 pounds of corn contaminated with *Aspergillus flavus* and fed over a period of about 5 days caused the death of pigs but a single dose of only 0.5 pound of corn contaminated with *Penicillium rubrum* resulted in death within one day. In another outbreak of mouldy corn poisoning in cattle, Burnside *et al.* [2] isolated from toxic corn two organisms, *Aspergillus flavus*, Link, and *Penicillium rubrum*, Stoll, which produced toxic substances when grown on natural or semisynthetic media. The material produced by these two cultures was lethal to pigs, horses, mice and chick embryos. Several additional investigations have confirmed the toxicity of diets contaminated with *Penicillium rubrum* and its metabolites [3—5]; controlled acute and chronic toxicity studies using the purified toxin have further defined the nature of this form of mycotoxicosis [6].

Evidence from field observations and from experimentally induced toxicity of *Penicillium rubrum* was sufficiently impressive to strongly impli-

* Contribution No. 2153 from the Department of Nutrition and Food Science, Massachusetts Institute of Technology, Cambridge, Mass. 02138 (U.S.A.)

cate it in the aetiology of mouldy corn toxicosis in farm animals. Mouldy corn toxicosis, however, is a general description of a toxicologic disease and this illness may be due to toxins produced by a number of moulds growing on corn products, particularly species of the genera Fusarium, Aspergillus, and Penicillium. In fact, it is very likely that the natural disease, mouldy corn toxicosis, results from an interaction of several toxins produced by a number of different moulds.

The toxicity of mouldy feeds for farm animals was recognized as early as the turn of the century [7]. It required, however, a more dramatic example represented by the outbreak of a disease in turkey poults in England in 1960, to focus sufficient attention on the problem to bring about an international effort to define the disease complex [8—10]. Along with the various episodes of mouldy corn poisoning in livestock and turkey "X" disease in poults, chicks and ducks, another disease in which mould toxins were implicated was described by Seibold and Bailey [11], a form of liver toxicosis in dogs designated hepatitis "X". Following some detailed experimental studies in dogs, Newberne *et al.* [12] and Bailey and Groth [13] concluded that the liver disease hepatitis "X" in dogs and mouldy corn toxicosis of pigs shared the same aetiology. Experiments utilizing purified aflatoxins in dogs revealed the similarity of the experimentally induced disease to that observed in field cases and further demonstrated that although aflatoxin B_1 elicited symptoms and lesions similar to hepatitis "X" [14,15], the synergism between aflatoxin B_1 and rubratoxin most closely represented the symptoms and lesions observed in field cases [14]. There can now be little doubt that many of the symptoms and lesions of disease association with mouldy corn poisoning may be ascribed to the action of aflatoxins. However, some of the gross haemorrhagic phenomena observed in farm animals poisoned on contaminated feeds are not reproduced in laboratory animals dosed with pure aflatoxins which suggests that other toxic substances in many of the mycotoxicoses have a synergistic effect with aflatoxin. Studies now in progress may help in the elucidation of these phenomena and point out the complex nature of mycotoxicoses which are the clinical manifestations of a number of different toxins present in feed-stuffs.

In another disease problem with another species, Forgacs and Carll [16] and Forgacs *et al.* [3] described a haemorrhagic disease of poultry and related it to a contaminated feed and further showed that *Penicillium rubrum* and the closely related species *Penicillium purpurogenum* caused good grain to become toxic to chicks and reproduced the haemorrhagic disease seen in the field.

Wilson and Wilson [17], in attempting to identify the toxic material produced by *Penicillium rubrum*, partially purified a substance that produced congestive, haemorrhagic, and degenerative lesions in mice similar to those produced by the intact mould cultures. Investigations following upon these reports eventually resulted in the isolation and purification of toxic compounds from cultures of the organism [18—20]. More recently the

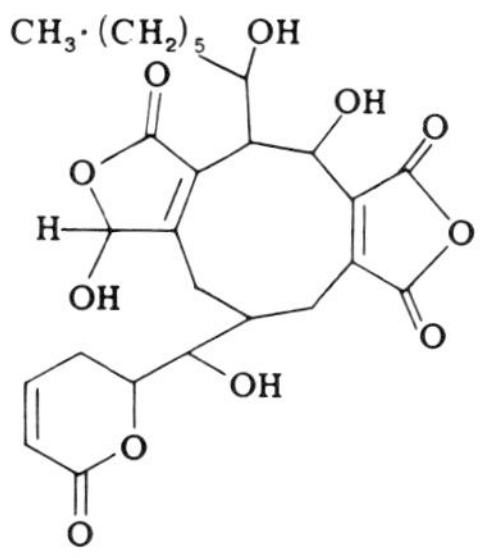

Fig. 1. Structural formula of rubratoxin A. 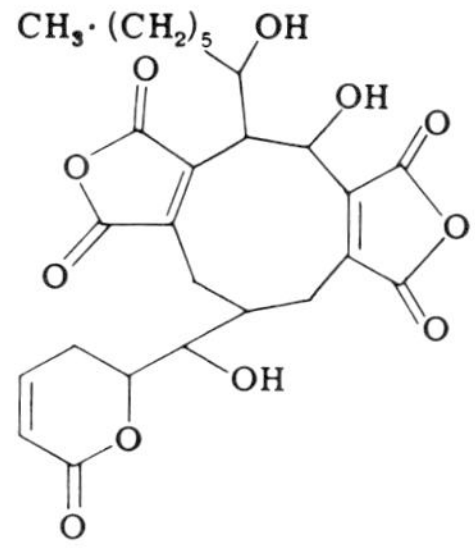   Fig. 2. Structural formula of rubratoxin B.

principal toxins, rubratoxin A and rubratoxin B (Figs. 1 and 2) have been chemically characterized and their structures proposed [18,21,22]. It seems clear now that the principal toxin produced by toxigenic strains of *Penicillium rubrum* is rubratoxin B, but there is little information available concerning the distribution of rubratoxin B in animal feeds or in human feedstuffs and except for the report of Wogan *et al.* [6], little information exists on the toxicity of purified compounds studied under controlled conditions.

Rose and Moss [23] have shown that modifications of structural configurations of the polyfunctional rubratoxin B molecule reduce or eliminate completely toxicity for mice. Edwards and Wogan [24] reported the results of experiments dealing with comparative acute and subacute toxicity of rubratoxin B to several animal species and Hayes and Wilson [25] described the effects of the toxin on liver composition and metabolism in mice. Although it is difficult at this point to assess the role of *Penicillium rubrum* in naturally occurring diseases, there is little doubt that strains of *Penicillium rubrum* do produce large quantities of toxic metabolites under controlled laboratory conditions and that the mould is often isolated from feeds collected at the site of animal disease outbreaks. There is no longer any doubt about contamination of natural feeds which results in animal toxicoses, but the magnitude of the problem is unmeasured.

Among others, the reports of Scott [26] and of Carlton *et al.* [27] illustrate through the isolation of species of toxigenic fungi from cereal and legume products and a number of species of Penicillium from the corn, that staple feeds are excellent substrates for growth of the moulds. One gains the impression, however, that the rubratoxins, while contributing to the toxicity of feeds to animals probably because of their relatively low toxicity, participate more in a spectrum of toxicologic diseases by potentiating the activity of more dangerous compounds such as the aflatoxins [20,24].

Growth of *Penicillium rubrum* and production of toxins

Pigments ranging in colour from yellow through orange to deep red and purple are produced by many of the Penicillia species grouped in the series of Biverticillata-symmetricea [28] . Both *Penicillium rubrum* and *Penicillium purpurogenum* characteristically produce red or purple-red pigments. Furthermore, these two species are very closely related and it is often difficult, if not impossible, to assign a particular isolate to one species or the other. In fact, the close relationship between the two is confirmed by the report that *Penicillium purpurogenum* may also produce rubratoxins [29]. Both species of mould occur widely in nature particularly in a range of organic matter, soil, and dead plant material. Examples include the isolates from legume and cereal products [26], kernels and peanut pods [30,31] and from sunflower seeds, bran and maize [32]. As has been shown with the various strains of *Aspergillus flavus* [20] colonies of isolates of *Penicillium rubrum* often exhibit wide variations when subcultured for any length of time on laboratory media. For example, Moss [33] points out that on Czapek-Dox agar, cultures can vary from thin grey-green heavily sporing colonies to thick colonies in the form of a dense felt with a regularly pigmented area of mycelium. Moreover, using the same medium the production of pigment on the under surface of the mycelium and its discharge into the surrounding agar may vary widely from one isolate to another. Some colonies will produce an intensely deep red pigment that imparts a near black colour to the undersurface of the colonies while others produce yellow or orange pigments. It is pointed out, however, that in spite of colonial variation, the microscopic features on the conidial structure are rather uniform with smooth conidiophore ends in a verticel of about 3 to 5 metolae which measure from 8—11 by 2 μ, each in turn bearing a verticel of a similar number of sterigmata measuring 8—14 by 1.8—2 μ. The conidia vary in dimensions and may be subglobose or even elliptical but are uniformly smooth walled; these structures usually measure about 2 to 4 μ.

The growth and toxin production of *Penicillium rubrum* varies considerably depending upon the type of medium that is used. When complex substrates are used both growth and toxic production are considerably greater than when synthetic media are used. In the earlier work of Wilson and Wilson [34], a crude toxic preparation was obtained from cultures of the mould grown on cracked hard corn moistened with 1% sucrose. The crude material was characterized to some degree and it induced symptoms in experimental animals similar to those observed in natural outbreaks of mouldy corn toxicoses [17]. Townsend *et al.* [18] isolated a crude toxic preparation similar to that of Wilson and Wilson from static cultures of *Penicillium rubrum* grown on a Raulin—Thom medium enriched with malt extract. Wogan and Mateles [20] obtained good yields of toxins when the mould was grown on a medium containing yeast extract.

The extraction and isolation methods used in the earlier work were complex and resulted in products that were rather difficult to isolate in pure form. A low yield of two crystalline metabolites was obtained from the crude amorphous powder developed by Townsend *et al.* [18] while Moss *et al.* [36] obtained pure crystalline toxins from the amorphous crude material by column chromatography using cellulose powder and eluting with benzene

which contained increasing concentrations of ethyl acetate; crystalline rubratoxin B came off the column when the concentration of ethyl acetate
reached about 10%.

It is interesting and significant that the toxins produced by *Penicillium
rubrum* are excreted into the medium and not retained by the mycelium.
Washed, dried mycelium of *Penicillium rubrum* is reported to be non-toxic
[33]. Moss and Hill [36] have described the relationship that exists between
toxin production, pH of the medium, growth, and the visible appearance of
the static culture. They have shown that toxin production is associated with
the active growth period of *Penicillium rubrum*, a fact of importance in
distinguishing the toxins from secondary metabolites which are typically
produced at maximal rates during the later phases of growth. During the
later phases of growth, the pH of the culture medium decreases rapidly and
sharply and remains at 2.5 through 2.7 until the culture begins to undergo
autolysis. In the static culture, the production of orange pigments on the
surface first and later the appearance of red pigments at the base of the
mycelial felt occurs only when the logarithmic growth phase has given way
to a period of decreasing growth rate. Thus, pigments of *Penicillium rubrum*
appear to conform with the pattern of production of typical secondary
metabolites and the structures of three of the pigments are known [37,38].
These include the red bis-*p*-quinone, phoenicin and 2 orange pigments,
mitorubrin and mitorubrinol.

The two crystalline toxic metabolites of importance were isolated from
crude preparations derived from cultures of *Penicillium rubrum* by Townsend
et al.[18]. These investigators assigned the names rubratoxin A and rubratoxin
B to the two crystalline metabolites, the titles reflecting the source of the
toxins, not their colour. Although both toxins are regularly isolated from
different strains of *Penicillium rubrum*, it now seems clear that the major
metabolite is rubratoxin B. While Hayes and Wilson [19] have been unable
to confirm the presence of rubratoxin A in the cultures they have worked
with, Wogan and Mateles [20] have obtained significant amounts or rubratoxin A from a single preparation which contained little or no concentrations of rubratoxin B.

Rubratoxin B is the bisanhydride, $C_{26}H_{30}O_{11}$; rubratoxin A,
$C_{26}H_{32}O_{11}$, probably has essentially the same structure with one of the
anhydride groups reduced to the lactol (Figs. 1 and 2). The rubratoxins
possess relatively stable anhydride functions, a feature which they share with
the nonadrides volcanic acid and glauconic acid produced by *Penicillium
purpurogenum*. These and other metabolites are described in the literature
by Barton and Sutherland [39].

Physical and chemical properties of rubratoxins A and B

The pure rubratoxins are very soluble in acetone, moderately soluble in
alcohol and esters and only sparingly soluble in water. They are completely
insoluble in non-polar solvents. The two toxins differ somewhat in their

solubility in the various solvents. For example, rubratoxin A is much more soluble in ethyl alcohol than is rubratoxin B and rubratoxin B is considerably more soluble in ethyl acetate than is rubratoxin A. Knowledge of these differences in solubility has been of considerable significance in the fractionation of mixtures of the two toxins. Furthermore, the toxins vary in the types of physical structures they form depending upon the solvent. For example, pure rubratoxin B crystallizes from mixtures of benzene and ethyl acetate as long lathes with the geometry of hexagonal plates extended along one axis in the plane of the hexagon. Pure rubratoxin B crystallizes from diethyl ether as rosettes of needles and it may crystallize as regular hexagonal plates from solvents such as amyl acetate. Some of the properties of the two rubratoxins are listed in Table I. Both rubratoxins decompose on melting and the pyrolytic decomposition of rubratoxin B results in the loss of a molecule of carbon dioxide. This accounts for the lack of a parent ion in the mass spectra of the rubratoxins. The mass spectrum of rubratoxin A is a peak at m/e 458 which corresponds to the ion ($C_{25}H_{30}O_8{}^+$) and the loss of both carbon dioxide and water from the parent molecule. The highest peak in the mass spectrum of rubratoxin B occurs at m/e 474 corresponding to ($C_{25}H_{30}O_9{}^+$).

The ultraviolet spectra of the rubratoxins, measured in non-hydroxylic solvents reveals a well-defined absorption at 250 nm, which is associated with the dye substituted maleic and hydride function. The absorption rapidly decreases in intensity in the presence of hydroxylic solvents pointing out the ease with which the anhydride rings equilibrate with the open forms.

Infra-red spectra reveal strong complex absorption in the carbonyl region which appears to be related to the presence of the anhydride and lactone functions. Strong absorption at about 3500 cm^{-1} and differences in the finger print region (700—1100 cm^{-1}) are related to the hydroxyl groups and these differences have been useful in confirming purity [40].

Rubratoxin A is unstable in alkaline solution and such solutions develop a yellow colour and a slightly pungent smell which are shown by Moss *et al.* [35] to be a result of a formation of a mixture of volatile aldehydes, mainly

TABLE I[a]

COMPARISON OF SOME PROPERTIES OF RUBRATOXIN

Property	Rubratoxin A	Rubratoxin B
Formula	$C_{26}H_{32}O_{11}$	$C_{26}H_{30}O_{11}$
Molecular weight	520	518
Melting point	210—214° (dec.)	168—170° (dec.)
$[\alpha]_D^{20}$ (c = 2 acetone)	+84°	+67°
λ_{max} (CH_3CN)	252 nm (ϵ 4430)	251 nm (ϵ 9700)
LD$_{50}$ (i.p. in propylene glycol to mice)	6.6 mg/kg	3.0 mg/kg

[a] Table from ref. 33 by permission of author and publisher.

heptaldehyde and acetaldehyde. Further, Moss and Hill [36] have suggested that the colour formed by the action of alkali on rubratoxin A might form the basis of an assay for this compound in mixtures of the two toxins. Rubratoxin B can be recovered in a very high yield unchanged by acidifying solutions of the toxin in excess alkali.

Thin-layer chromatography has been used to study rubratoxins A and B. Moss and Hill [36] observed that a chromatogram of a mixture of rubratoxin A and B could give a pattern of 4 spots instead of two and these investigators considered the artifacts probably due to hydrolysis of the anhydride groups. R_F values quoted by Moss and Hill do not agree with those of Hayes and Wilson [19]. Moss and Hill observed R_F values for rubratoxin A, 0.7 and rubratoxin B, 0.56. Hayes and Wilson, using a mixture of glacial acetic acid—methanol—chloroform (2:20:80 v/v/v) as the developing solvent, found that rubratoxin B gave a single spot at R_F 0.54 to 0.59. The differences in the results obtained by these two investigators are understandable if not explained, since chromatography of polar material such as the rubratoxins is quite sensitive to a number of factors including temperature, relative humidity and others.

For more detailed description of the chemical and the physical properties of the rubratoxins and related metabolites, the reader is referred to ref. 33.

Biosynthesis

Knowledge of the biosynthesis of the rubratoxins is incomplete; that which is known is described in detail by Moss and the reader is referred to the review in ref. 33. Progress toward an understanding of biosynthetic pathways of the rubratoxins is proceeding by way of degradation of the parent molecules. It is not difficult to prepare several derivatives of rubratoxin B. Acetylation with acetic anhydride and pyridine produces the triacetate (melting point 188°) atmospheric pressure yields the dihydroderivative (melting point 155°). Treatment of rubratoxin B with chromic acid in acetone at 0° results in a high yield of the monoketone derivative (melting point 165°) despite the three secondary hydroxyl groups in the parent molecule.

When the dihydro toxin is hydrolyzed the α-diol is formed which on periodate oxidation loses the δ-lactone moiety. Glutamic acid hemialdehyde, and likely the aldehyde, form the products of periodate oxidation.

Rubratoxin B is a substituted higher homologue of byssochlamic in which the *n*-propyl substituent is replaced by a C_7 chain and the ethyl group substituted by a C_6 chain containing the functional groups of the unsaturated δ-lactone.

It has been postulated that the rubratoxins which are C_{26} compounds, are produced by the head-to-tail, head-to-tail coupling of two C_{13} units, derived from a derivative of decanoic acid and oxaloacetic acid. The stage at which the oxygen functions are introduced is unknown. Thus, the complex nature of the biosynthesis of these interesting toxins awaits further investigation.

Analytical method

Although a routine analytical methodology for the rubratoxins has not been standardized and published, the several laboratories active in rubratoxin research have based their assays on the principle of solubility. Both rubratoxins A and B are highly soluble in acetone, moderately soluble in alcohol and esters, sparingly soluble in water and insoluble in non-polar solvents.

A reliable method is that described by Wogan and Mateles [20] in which the toxin-containing material is extracted with ethyl acetate, evaporate the ethyl acetate solution to dryness, extract the solids with petroleum ether, weighing and dissolving the residue in dioxane for determination of the infrared spectrum (700—1100 cm^{-1}, ref. 40).

A less accurate procedure but adequate for most purposes consists of three successive extractions with cold ethyl acetate, equilibration of the ethyl acetate with 0.1 volume of saturated sodium chloride solution, drying the organic layer with anhydrous magnesium sulphate and evaporation under vacuum at a temperature less than 45°. The resulting oil is evaporated to dryness, washed with hot petroleum ether and then diethyl ether and the tarry solid dissolved in and crystallized from acetonitrile. Separation of rubratoxins A and B can then be accomplished by thin-layer chromatography developed in a mixture of glacial acetic acid—methanol—chloroform (2:20:80 v/v/v) R_F values being 0.56 and 0.7 for rubratoxins A and B respectively.

For more accurate assessment of purity, melting point, IR, nuclear magnetic resonance (NMR) and mass spectra, thin-layer chromatography and UV molar extinction coefficient should be determined [33].

Biological activity

It is very likely that the outbreaks of hepatitis "X" in dogs in the early 1950s [11—13] were due in part to the interactions of rubratoxins, aflatoxins, and possibly other mould metabolites. This hypothesis was strengthened by the observations using purified aflatoxin and rubratoxin in controlled experiments with dogs [14]. Although aflatoxin alone induced many of the symptoms and lesions observed in field cases, it was the combined action of both aflatoxin and rubratoxin that induced the most characteristic clinical pathologic responses. The disease described by Sippel *et al.* [1] in pigs and cattle following the consumption of mouldy corn was probably a result of consuming corn containing rubratoxin perhaps along with other toxins, since Burnside and colleagues were able to reproduce the disease observed in spontaneous outbreaks by using a culture of *Aspergillus flavus* and a culture of *Penicillium rubrum* to contaminate fresh corn which was fed to animals under controlled experimental conditions. The work of Burnside *et al.* [2] demonstrated that corn infected with *Penicillium rubrum* was toxic to a male goat and to pigs, mice and horses. Clinical pathologic examinations of animals dosed with sublethal amounts of toxic feed revealed no specific changes other than an increase in vitamin A level in the serum. Pigs that were fed toxic doses of the contaminated corn were lethargic and exhibited salivation, pain, belligerent behaviour, and erythema of the central abdominal skin surface. Horses developed depression, anorexia and a general incoordination. At necropsy there was congestion of most of the tissues in the pig, which was particularly intense in the stomach and often accompanied by haemorrhage. The liver and kidneys of mice were pale and icteric, while in the horse, haemorrhage of the large and small intestine and stomach was a common finding. Further, the liver of the horse was described as grey-green to yellow, indicating a marked injury and fatty infiltration. Histopathology showed that in all animals the liver and kidney tissues suffered the most prominent lesions, but in some of the horses, the central nervous derangement was accompanied by lesions in the brain.

Forgacs *et al.* [3] described haemorrhage and congestion of the musculature of the breast, legs, thighs, lung, kidney, spleen, small intestine and caecum when chicks were fed grain contamined with *Penicillium rubrum* and *Penicillium purpurogenum*. The liver of mice, rats, guinea pigs and ducklings were capable of regenerating over a period of 7 days following a sublethal dose of rubratoxin, a concentration that was sufficient to produce haemorrhage and necrotic lesions in the liver when animals were killed and examined only 24 h after dosing.

Extracts from *Penicillium rubrum* grown on cracked corn moistened with sucrose were lethal to dogs, rabbits, guinea pigs and mice and the symptoms described by Wilson and Wilson parallel those described by other investigators using mouldy feeds. The route of administration has a significant influence on the relative toxicity of the crude material with oral dosing exhibiting a much lower index of toxicity than either intraperitoneal (i.p.) or intravenous (i.v.) administrations. The oral toxicity, for example, reported by Wilson and Wilson is 200 mg/kg while the i.v. and i.p. doses were 15 mg/kg and 7.5 mg/kg respectively. The work of Townsend [18] followed a similar pattern with oral LD_{50} 120 mg/kg and i.v. and i.p. LD_{50}s 6.5 and 4 mg/kg respectively. In each case, the solvent used as a vehicle was propylene glycol. For a realistic assessment of rubratoxin, in view of the oral toxicity, one must assume that the metabolites of *Penicillium rubrum* are not very toxic when studied in the laboratory in pure form and that this important route of exposure and its low level of toxicity implies that in field cases, the mould is capable of producing very large quantities of the metabolites. It should be pointed out also that the response of different animal species and even different strains of the same species of animals to any one group of toxins usually varies considerably and the variability can be extreme in some cases.

Rose and Moss [23] obtained a quantitative value for the acute toxicity of the rubratoxins in mice by using pure toxins and derivatives which they prepared during the determination of structures. In their studies, rubratoxin B was more toxic than any of its derivatives and it is particularly interesting to observe the decrease in toxicity associated with saturation or removal of the α—β unsaturated lactone ring. This is of some interest because Dickens and Jones [41] have pointed out that some of their previous studies suggested the presence of α—β unsaturated bond in a 5 or 6 member lactone ring as a permanent feature in some groups of carcinogens. The absence of α—β unsaturation or the opening of the lactone ring itself gives rise to a loss of biological activity. It must be stressed here also that the rubratoxins have not been demonstrated to be carcinogenic but there is little doubt of the functional groups present in the rubratoxin molecules, the intact α—β unsaturated lactone contributes to the acute toxicities.

Rose and Moss [23] in conducting toxicity studies, used propylene glycol as the solvent; given in appropriate doses, this solvent is considered to be non-toxic and also to allow for stability of the toxins. It has been observed that rubratoxin B has a far higher toxicity when dissolved in dimethylsulphoxide (DMSO), with an i.p. LD $_{50}$ dose for mice about 0.37 mg/kg. It should be remembered, however, that DMSO has a demonstrated activity of its own in biological systems and NMR studies indicate that the solvent interacts strongly with the rubratoxin molecule. Thus, it is important to consider the type of solvent one uses in determining biological activity of selected compounds.

Neubert and Merker [42] injected Sprague-Dawley rats with a sublethal dose of crude rubratoxin and observed a severe fatty infiltration of the liver. They reported visible damage to the mitochondria although *in vitro* tests failed to support the presence of apparent visible damage, and indicated that the toxin had no direct significant effect on the activity of the mitochondria. If the rats were cold-adapted by keeping them for 60 days at a temperature of 4° before the toxin was administered, the mitochondria were not damaged despite the fact that cold-adapted rats reacted to the toxin in a manner identical to normal rats. These investigators concluded that the toxic symptoms associated with *Penicillium rubrum* are not induced by fatty liver and mitochondrial injury.

Townsend *et al.* [18] administered doses of toxins which were small enough to result in microscopic lesions and demonstrated that the metabolic processes of the liver of mice were adversely influenced by this dose level. These investigators based their assumptions on the measurement of sleeping time after a standard dose of phenobarbitone, with and without a previous dose of toxin.

Based on reports in the literature, the liver appears to be the major site of injury, an observation that is not surprising in view of its central role in detoxication mechanisms. The other generalized effects of the toxins in most species of animals appears to be haemorrhage throughout a wide range of organs and tissues. More recently, Wogan *et al.* [6] have investigated the effects of rubratoxin B on several species of animals when the toxin was administered by different routes. The rubratoxin B used in these experiments was produced by Wogan and Mateles [20]. The *Penicillium rubrum* isolate (P-13) that was used was provided by Dr. B.J. Wilson, Vanderbilt University and was a subculture of that originally isolated by Burnside *et al.* in the natural field outbreaks [2].

In the studies of Wogan *et al.* a further attempt was made to identify the interaction, if any, between rubratoxin B and aflatoxin B_1. Animal species used included weanling Fischer rats of both sexes, adult Carworth mice of both sexes, adult Hartley guinea pigs, adult cats of both sexes, male mongrel dogs and male white leghorn chickens. The rats and mice were fed a purified agar-gel diet [43]; all other species were fed commercial rations of appropriate types, housed individually, and provided with food and water *ad libitum*. The rubratoxin B was dissolved in either DMSO or in propylene glycol; dosing was by i.p. injection or by stomach tube. The aflatoxin B_1 was isolated and purified from a mixture of aflatoxins by methods described previously [43]. The results of these studies are listed in Table II, which takes into account several different animal species, different sexes, and different routes of administration with two vehicles. When rubratoxin B was administered i.p. to rats, the toxin had an LD_{50} of about 0.35 mg/kg and the value was similar in both sexes. Furthermore, the vehicle appeared to have no influence on the response. On the other hand, when the compound was administered orally, there was a sharp contrast between the values obtained by this route of administration and those obtained by the i.p. route. The lethality of the compound was greatly reduced by oral administration as has been reported by other investigators.

TABLE II[e]

LETHAL POTENCY OF RUBRATOXIN B IN SEVERAL SPECIES

Species	Sex	Weight	Number of animals	Dosing route	Vehicle	LD_{50} (mg/kg)
Rat	M	58 g	60	i.p.	PG[a]	0.36 (0.27—0.49)[b]
Rat	F	60 g	50	i.p.	PG	0.36 (0.28—0.46)
Rat	F	59 g	70	i.p.	DMSO[c]	0.35 (0.28—0.45)
Rat	M	60 g	25	p.o.[d]	DMSO	*ca.* 400
Rat	F	58 g	25	p.o.	DMSO	*ca.* 450
Mouse	F	25 g	25	i.p.	DMSO	0.27 (0.22—0.34)
Mouse	F	25 g	30	i.p.	PG	2.6 (2.0—3.1)
Guinea pig	M	565 g	18	i.p.	DMSO	0.48 (0.41—0.56)
Cat	M	3 kg	3	i.p.	DMSO	*ca.* 0.2
Cat	M,F	3 kg	8	i.p.	PG	1.0—1.5
Dog	M	3 kg	7	i.p.	PG	>5.0
Chicken	M	500 g	6	i.p.	PG	>4.0

[a] Propylene glycol.

[b] 95% confidence interval; calculated by the method of Litchfield and Wilcoxon (1949).

[c] Dimethylsulphoxide.

[d] By stomach tube.

[e] Table from ref. 6 with permission of authors and publisher.

References p. 179

Perusal of Table II indicates that the mouse, guinea pig and cat all had LD_{50} values of less than 1.0 mg/kg of rubratoxin B when the toxin was injected in the DMSO vehicle. In the mouse, there was a significant influence of the vehicle with an LD_{50} value for propylene glycol solutions much higher than that for DMSO solutions. There was a similar but smaller vehicle effect observed in the cat. The dog and chicken were less susceptible than other species with LD_{50} values of 4 mg/kg or greater when rubratoxin B was dissolved and injected i.p.

In these experiments, all animals that died or were killed at intervals after dosing were subjected to complete necropsy and microscopic examination of all tissues and organs and the most consistent histologic findings, common to all species, included congestion, haemorrhage, and degeneration, particularly in the liver and spleen. The liver lesion originated in the mid-zonal area and spread peripherally in the lobule to the periportal zone (Fig. 3). Haemorrhagic and necrotic changes in mice were accompanied by the presence of polymorphonuclear leukocytes of the necrotic zones (Fig. 4). Depletion of lymphocytes and necrosis was a common finding in the spleen (Fig. 5). Extreme congestion and frequent haemorrhage was observed in some cases (Fig. 6). Cats appeared to be extremely sensitive to the effects of rubratoxin and responded with severe reactions; in the case of this species, in addition to microscopic lesions of liver and spleen similar to those observed in other species, the cat developed haemorrhagic and congestive lesions of lymph nodes, sub-endocardial haemorrhage, marked oedema of the

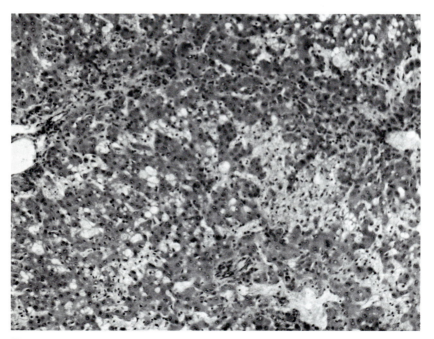

Fig. 3. Liver from guinea pig poisoned with rubratoxin. H. and E. × 125.

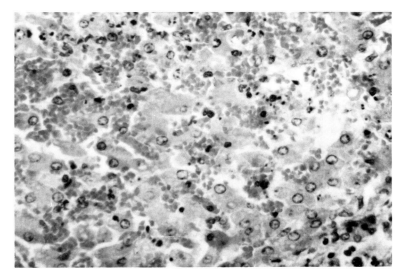

Fig. 4. Haemorrhage and necrosis in liver of mouse with acute rubratoxin B toxicity. H. and E. × 240.

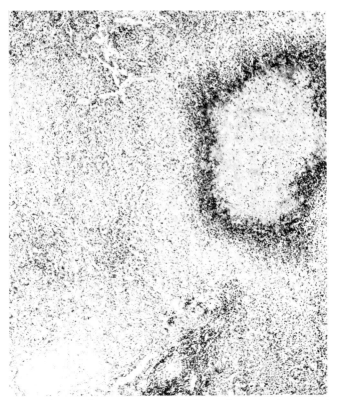

Fig. 5. Spleen from rat given and LD$_{50}$ dose of rubratoxin B. Note haemorrhagic necrosis and depletion of spleen lymphocytes. H. and E. × 28.

References p. 179

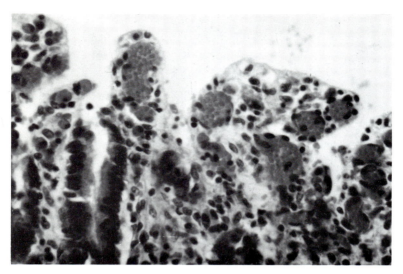

Fig. 6. Extreme congestion and necrosis of villous epithelium from rat poisoned with LD_{50} dose of rubratoxin B. H. and E. × 180.

serosal surface of the gall bladder, adrenal cortical congestion, and massive ascites. Guinea pigs were the only other species that developed ascites but in this species it was much less severe than in the cat.

Mild renal damage was observed in an occasional animal, but the finding was not consistent within a species or in any one group. No other gross or microscopic effects of toxins were observed.

An additional series of experiments was performed to characterize the chronic toxicity of rubratoxin B and to determine whether chronic oral treatment would reveal any evidence of carcinogenic activity. After a series of dose ranging studies that extended over 4 weeks of trial, it was found that only the two lowest doses (5 and 10 mg/kg body weight, 3 times per week by stomach tube) caused neither mortality nor significant growth depression. Therefore, these dose levels were selected for chronic toxicity studies and the long-term effects are listed in Table III. The rats that were treated and died during the 60-week treatment period exhibited a characteristic toxic syndrome with histologic changes of a remarkably similar pattern. Typically, the animals that died appeared healthy and showed normal weight gain until shortly before the onset of signs of toxicity. Generally, within a very short period, usually about 3 days, the animals developed anorexia, diarrhoea and a porphyrin discharge from the eyes, ears and nose. Rapid weight loss amounting to 20 to 30% body weight preceded death.

Histopathologic examination of such animals revealed extensive liver damage consisting mainly of massive haemorrhagic necrosis with entire sections of the organ destroyed. The degenerative process appeared to begin in the centrilobular region and spread rapidly throughout and beyond the lobule so that all architectural organization was lost. The sudden appearance

TABLE III[a]

CHRONIC TOXICITY TO RUBRATOXIN B IN RATS DURING 60 WEEKS OF TREATMENT

Compound	Treatment regimen	Total dose per rat		Sex	Mortality during treatment		Survived treatment	Killed after (wk)
					Killed	Died		
DMSO	0.05 ml 3 × /wk	9.0	ml	M	2/10	2/8	6/8	87
DMSO	0.05 ml 3 × /wk	9.0	ml	F	1/5	0/4	4/4	87
Rubra B	10 mg/kg 3 × /wk	616	mg	M	1/5	1/4	3/4	87
Rubra B	10 mg/kg 3 × /wk	458	mg	F	1/5	2/4	2/4	87
Rubra B	5 mg/kg 3 × /wk	307	mg	M	2/20	7/18	11/18	82

[a] Table from ref. 6 with permission of authors and publisher.

and rapid development of the lesion was further evidenced by the absence of regenerative changes and a phagocytic component that is generally attracted to such necrotic zones within 24 to 30 h. On the other hand, significant histologic lesions were not observed in the liver or other tissues among 4 of 30 rats that were killed during the treatment period and that did not show prior clinical symptoms. Negative findings were also obtained from 7 to 26 rats that died or were killed between the end of the treatment period (60 weeks) and 87 weeks at which time they were sacrificed. Furthermore, from the 19 or 26 toxin-treated animals that survived throughout the experimental period and were killed after 82 to 87 weeks, there were no significant lesions observed. From these studies then it was concluded that chronic treatment with rubratoxin B using the dosing regimens listed, resulted in the death of about 30% of rats within a 16—60-week period. Repeated exposures at frequent intervals appeared to be required for precipitation of the toxic response, as shown by absence of liver pathology of animals killed during treatment or at the end of the experimental period. Thus, it seems that a threshold must be reached with repeated challenges beyond which the liver is unable to withstand further insult and massive degeneration following in a very short period of time. These studies further indicated that rubratoxin B was not carcinogenic to rats.

The combined effect of rubratoxin B and aflatoxin B_1 was studied in rats by administering the toxins singly or together. Table IV lists the results of this experiment. In the case of rubratoxin B, the dose level used was that known from previous experience to cause slight depression of weight gain but no mortality. The total aflatoxin dose and regimen shown previously to result in only a small incidence of liver tumours was chosen to reveal any interaction of the two toxins that might exist [43]. The fourth group of rats were thus treated with both toxins simultaneously for 5 weeks and aflatoxin for 1 additional week thereafter. With reference to lethality, the two toxins acted synergistically. Neither agent alone caused mortality but when administered simultaneously, 9/20 animals died during the course of the ex-

TABLE IV[a]

TOXICITY TO MALE RATS EXPOSED SIMULTANEOUSLY TO RUBRATOXIN B AND AFLATOXIN B_1

Treatment regimen	Total toxin (mg/rat)	Body wt. at end of dosing (% of control)	Mortality	Preneoplastic liver lesions at 70—80 weeks[b]
Controls (DMSO 3 × /wk for 5 wk)	—	100	0/10	0/7
Rubratoxin B (25 mg/kg, 3 × /wk for 5 wk)	39.7	95	0/10	0/7
Aflatoxin B_1 (0.2 ppm in diet for 6 wk)	0.11	102	0/10	6/7
Rubratoxin B and aflatoxin B_1	—	86	9/20	5/8

[a] Table from ref. 6 by permission of authors and publisher.
[b] See Wogan and Newberne [43].

periment. Based on these observations there was not an interaction between the two toxins in terms of carcinogenicity since animals treated with both aflatoxin B_1 and rubratoxin B and those treated with aflatoxin alone had approximately the same incidence of focal parenchymal cell hyperplasia, a characteristic preneoplastic lesion associated with aflatoxin hepatocarcinogensis [43].

The observations in the rat studies illustrated clearly that the route of administration greatly influenced the lethal potency of the toxin with the oral LD_{50} value much higher than the i.p. LD_{50} value. This is of some concern and significance in view of the fact that animal and human populations would ordinarily be exposed to the toxins by the oral route. These findings are essentially in agreement with those of Townsend [18] and Hayes and Wilson [25] where mice and ducklings were used for the investigations. It is likely that the reduced toxicity associated with oral routes of administration is at least partially a result of the chemical instability of the compound in the acidic environment of the stomach. It was also interesting that the solvent that was used as a vehicle influenced toxicity. This was particularly evident in the mouse in which there was a 10-fold difference in the LD_{50} value between animals treated i.p. with toxin dissolved in DMSO compared to those treated with propylene glycol solutions; a similar but smaller solvent effect was observed in the cat. Somewhat analogous results had been reported by Townsend [18].

The LD_{50} values associated with i.p. injection of rubratoxin B dissolved in DMSO given to rats, mice, guinea pigs, and cats were in the range of 0.2 to 0.5 mg/kg. Comparable values using propylene glycol as the vehicle in mice, cats, dogs, and chickens are in the range of 1 mg/kg to more than 5 mg/kg body weight. These values indicate the wide species susceptibility to poisoning by the toxin and in general agree with the values reported for mice and ducklings [16,23,25].

The chronic toxicity studies in this investigation yielded information of considerable importance — namely, that none of the survivors on the 60-week treatment regimen at a maximum tolerated level had any evidence of neoplastic or preneoplastic lesions even after 87 weeks. This provided very strong evidence that rubratoxin B is not carcinogenic to rats. Furthermore, a total of 11 of 18 male rats survived a total dose of 307 mg/animal over a 60-week period and 3 or 4 tolerated 616 mg/animal over the same period. The clinical syndrome was an acute one in those animals that died and the histopathologic changes in liver and spleen were similar to those treated with a single dose. Lesions were absent in animals where an acute toxic response was not precipitated.

The fact that there is synergism between rubratoxin B and aflatoxin, as illustrated in the rat studies, is an important observation from the standpoint of public health and has obvious implications with respect to hazards presented by foodstuffs contaminated by both toxin-producing moulds or by both mycotoxins. There is little doubt that such multiple contamination does occur under natural conditions and with some degree of frequency. This may well contribute to the high mortality in some field outbreaks of mycotoxicosis in animals. These studies provide important information on the toxicity and metabolic effects of rubratoxin B but do not elucidate the mechanism of action of this toxin. It is particularly important to bear in mind the distinct possibility that rubratoxins, not particularly toxic in their own right, may be extremely important in potentiating the activity of more dangerous compounds such as the aflatoxins. Further studies along these lines may help elucidate the enormously complex and enigmatic responses to the mycotoxins in field cases.

References

1 W.L. Sippel, J.E. Burnside, M.B. Atwood, Proc. 90th Am. Meeting, Am. Vet. Med. Assoc., Toronto, 1953, p. 174.
2 J.E. Burnside, W.L. Forgacs, J. Carll, W.T. Atwood and E.R. Doll, Am. J. Vet. Res., 18 (1957) 817.
3 J. Forgacs, H. Koch, W.T. Carll and R.H. White-Stevens, Am. J. Vet. Res., 19 (1958) 744.
4 K. Madhavikutti and E.R.B. Shanmugasundaram, Proc. Indian Acad. Sci. B., 68 (1968) 261.
5 D. Blevins, M.W. Glenn, A.H. Hamdy, R.F. Brodasky and R.A. Evans, Am. J. Vet. Res., 154 (1969) 1043.
6 G.N. Wogan, G.S. Edwards and P.M. Newberne, Toxicol. Appl. Pharmacol., 19 (1971) 712.
7 T. Butler, Am. Vet. Rev., 26, 748; quoted by Udall in The Practice of Veterinary Medicine, Cornell University, Ithaca, N.Y., 1954, p. 307.
8 W.P. Blount, Turkeys, 9 (1961) 52.
9 F.D. Asplin and R.B.A. Carnaghan, Vet. Rec., 73 (1961) 1215.
10 W.D. Salmon and P.M. Newberne, Cancer Res., 23 (1963) 571.

11 H.R. Seibold and W.S. Bailey, J. Am. Vet. Med. Assoc., 121 (1952) 201.
12 J.W. Newberne, W.S. Bailey and H.R. Seibold, J. Am. Vet. Med. Assoc., 127 (1955) 59.
13 W.S. Bailey and A.H. Groth, J. Am. Vet. Med. Assoc., 134 (1959) 514.
14 P.M. Newberne, R. Russo and G.N. Wogan, Pathol. Vet., 3 (1966) 331.
15 B.J. Wilson, P.A. Teer, G.H. Barney and F.R. Blood, Am. J. Vet. Res., 28 (1967) 1217.
16 J. Forgacs and W.T. Carll, Vet. Med., 50 (1955) 172.
17 B.J. Wilson and C.H. Wilson, J. Bacteriol., 84 (1962) 283.
18 R.J. Townsend, M.O. Moss and H.M. Peck, J. Pharm. Pharmacol., 18 (1966) 471.
19 A.W. Hayes and B.J. Wilson, Appl. Microbiol., 16 (1968) 1163.
20 G.N. Wogan and R.I. Mateles, Prog. Ind. Microbiol., 7 (1968) 149.
21 M.O. Moss, F.V. Robinson, A.B. Wood, H.M. Paisley and J. Feeney, Nature, 720 (1968) 767.
22 M.O. Moss, F.V. Roginson and A.B. Wood, Tetrahedron Letters, 5 (1969) 367.
23 H.M. Rose and M.O. Moss, Biochem. Pharmacol., 19 (1970) 612.
24 G.S. Edwards and G.N. Wogan, Federation Proc. Fed. Am. Soc. Exptl. Biol., 27 (1968) 551.
25 A.W. Hayes and B.J. Wilson, Toxicol. Appl. Pharmacol., 17 (1970) 481.
26 de B. Scott, Mycopathol. Mycol. Appl., 25 (1965) 213.
27 W.W. Carlton, J. Tuite and P. Mislivec, Toxicol. Appl. Pharmacol., 13 (1968) 372.
28 K.B. Roper and C. Thom, A Manual of the Penicillia, Williams and Wilkins, Baltimore, Md., 1949.
29 S. Natori, S. Sakaki, H. Kurata, S. Udagawa, M. Ichinoe, M. Saito, M. Umeda and K. Ohtsubo, Appl. Microbiol., 19 (1970) 613.
30 C.R. Jackson, Plant Soil, 23 (1965) 203.
31 A.Z. Joffe and S.Y. Borut, Mycologia, 58 (1966) 629.
32 G. Cantini and J.G. Scurti, Allionia II, Turin, (1965) 29.
33 M.O. Moss, in A. Ciegler, S. Kadis, and S.J. Ajl (Eds.), Microbial Toxins, Vol. VI, Academic Press, New York, 1971, pp. 383—385.
34 B.J. Wilson and C.H. Wilson, J. Bacteriol., 83 (1962) 693.
35 M.O. Moss, R.V. Robinson, A.B. Wood and A. Morrison, Chem. and Ind. (London), (1967) 755.
36 M.O. Moss and I.W. Hill, Mycopathol. Mycol. Appl., 40 (1970) 81.
37 T.P. Curtin, G. Fitzgerald and J. Reilly, Biochem. J., 34 (1940) 1605.
38 G. Buchi, J.D. White and G.N. Wogan, J. Am. Chem. Soc., 87 (1965) 3484.
39 D.H.R. Barton and J.K. Sutherland, J. Chem. Soc., 87 (1965) 1769.
40 A.B. Wood and B. Buckingham, Spectrochim. Acta, 26A (1970) 465.
41 F. Dickens and H.E.H Jones, Brit. J. Cancer, 19 (1965) 392.
42 D. Neubert and H.J. Merker, in H.W. Ravdanat (Ed.), Symposium on Recent Advances in the Pharmacology of Toxins, Prague, Czechoslovakia, Proc. Intern. Pharmacol. Meeting 2, MacMillan, New York, 1965, p. 9.
43 G.N. Wogan and P.M. Newberne, Cancer Res., 27 (1967) 2370.

Chapter 9

STACHYBOTRYS AND STACHYBOTRYOTOXICOSIS

JOSEPH V. RODRICKS and ROBERT M. EPPLEY

Division of Chemistry and Physics, Bureau of Foods, Food and Drug Administration, U.S. Department of Health, Education and Welfare, Washington, D.C. (U.S.A.)

In early 1939 former Soviet Premier Nikita Kruschev arrived in the Ukraine with orders from Premier Stalin to organize and improve the region's agricultural operations. In his recently published memoirs [1], Kruschev recalls that his first task was to appoint a number of scientific commissions to investigate a frequently fatal disease of unknown origin which was at that time, and had been since 1931, severely affecting the Ukrainian horse population. Within a relatively short time, reports Kruschev, the cause of the disease had been located. The chronicle of those early investigations can be found in a series of communications from the head of one of the investigating groups, Professor V.G. Drobotko [2—4], and in a number of summary accounts [5,6] of the pioneering work of Drobotko and other Soviet investigators. In 1938, 7 years* after the first outbreak of the disease, the Soviet workers had confirmed that it was associated with fungus-contaminated feed and had classified it as a mycotoxicosis; in particular, the disease could be reasonably designated as stachybotryotoxicosis since it could be demonstrated that it was caused by unknown toxic agents elaborated by certain strains of the saprophytic mould, *Stachybotrys alternans*. Stachybotryotoxicosis has been studied and, as recently as 1972, reviewed in great depth by Forgacs [6—8]. The present survey leans heavily on the reviews of Forgacs insofar as it deals with the clinical and pathological characteristics of this mycotoxicosis, but adds appreciable information regarding the chemistry of the involved microorganism. Recent studies in our laboratories provide strong albeit indirect evidence [9] that the compounds responsible for stachybotryotoxicosis belong to the series of sesquiterpenes known as 12,13-epoxy-Δ^9-trichothecenes.

Stachybotryotoxicosis in farm animals and man

Stachybotryotoxicosis was first reported in the Ukraine in 1931; it affected horses and to some extent, it appears, human beings as well [2,10].

---

* Kruschev is perhaps in error in his recollection that he ordered the study groups into action in 1939. Of course, there had been considerable research on the matter before Kruschev arrived on the scene.

The investigation into the aetiology of the disease, which had been found to be non-infectious, proceeded on the assumption that the horses were receiving a toxic chemical as a result of some type of feed contamination. That the toxic agent was of fungal origin was suggested by the discovery, on a farm where the disease was enzootic, of hay heavily contaminated, indeed blackened, with *Stachybotrys alternans*. Subsequently, naturally infected hay, artificially inoculated hay, and pure cultures of *S. alternans* were each fed to healthy horses and in all instances the clinical characteristics of the illness were reproduced [2,7]. These studies clearly demonstrated that the illness in horses was mould-related and was most probably due to some toxic metabolite of *S. alternans*.

The illness, or intoxication, presents itself clinically in horses in one of two forms: the typical form, which seems to proceed in three stages, and the atypical form, which has but one stage. The latter form seems to occur only after the animal has eaten large quantities of fungus-contaminated hay or straw; the signs of illness appear within 72 h of ingestion. The typical form develops after long ingestion of smaller amounts of toxic fodder. Both forms of the illness can be and often are fatal. The major clinical characteristics of equine stachybotryotoxicosis are summarized in Table I.

At autopsy haemorrhage is observed in almost all organs, and most especially on the mucous and serous membranes of the gastrointestinal tract and in the subcutis and skeletal musculature. Necrotic ulcers are found on the mucous membranes of the mouth, throat, stomach and intestines. The lymph nodes are enlarged and filled with blood. The lungs are often haemorrhagic and are always oedematous. These and other pathologic observations are common to both the typical and atypical forms of equine stachybotryotoxicosis. A clinical and pathological account of this disease is related in full by Forgacs [7].

Stachybotryotoxicosis has been diagnosed in other animals and in man. It is thought that either inhalation of aerosols generated during the handling of infected hay or direct skin contact with the hay, or both, can induce stachybotryotoxicosis in man. Drobotko [2,3] and Vertinskii [10] report that in areas where equine stachybotryotoxicosis was common a similar illness was observed among humans handling mouldy animal fodder. In humans a skin rash developed which frequently became a moist dermatitis; there was inflammation and pain in the mouth, burning sensations in the nose and sometimes nasal bleeding. Very often there was chest pain and a rise in body temperature. Drobotko also reports human stachybotryotoxicosis in areas far removed from regions where horses were diseased, especially in those homes where infected straw was used for bedding. Fungi isolated from such straw were cultured and their extracts tested on shaved rabbit skin for dermatitis production; only isolates of *Stachybotrys alternans* provoked a reaction. Members of the scientific teams investigating the illness developed a skin rash after coming in contact with naturally contaminated and artificially inoculated hay. These skin reactions in humans are similar to those which have been described by laboratory workers who have handled mem-

TABLE I

MAJOR CLINICAL CHARACTERISTICS OF EQUINE STACHYBOTRYOTOXICOSIS
AS REVIEWED BY DROBOTKO [2, 3] AND BY FORGACS [7]

Typical form	Atypical form
Stage I	Loss of reflex response and vision.
Usually lasts 8 to 12 days.	Poor control of movement.
Irritation of mouth, throat, nose and lips.	Hyperirritability.
Swelling and soreness of glands.	Appetite depressed.
	Body temperature elevated.
Stage II	Cardiac activity initially intensified,
Lasts 15 to 20 days.	followed by weakening.
As in Stage I, with leukopenia appearing gradually.	Prolonged agonal period accompanied by tremors.
Stage III	Shock symptoms occasionally.
	Blood abnormality absent.
Lasts 1 to 6 days with death common.	Death from respiratory failure.
Fever up to 41.5°.	
Necrotic ulcers on mucous surfaces of mouth and throat.	
Leukopenia persists with agranulocytosis developing.	
Fresh necrotic areas develop on mucous membranes of cheeks, gums, frenum, lips.	
Pulse weak, diarrhoea frequent.	

bers of the 12,13-epoxy-Δ^9-trichothecene series of fungal metabolites [11,12]; observations such as these provide support to the hypothesis which will be developed later that certain 12,13-epoxy-Δ^9-trichothecenes are among the compounds responsible for stachybotryotoxicosis [9].

In addition to the reports cited above, field cases of stachybotryotoxicosis have been reported for cattle, poultry, swine, a hippopotamus, and a bison. In 1958 and 1959, again in the Ukraine, the illness struck the horse population but also spread to include thousands of head of cattle [13]. The form of the toxicosis in cows resembled atypical equine stachybotryotoxicosis. A number of other outbreaks of stachybotryotoxicosis in cattle have been reported from the Soviet Union [14—16]. Sarkisov [5] has described a fatal disease of hens during the period 1940—1946 and has attributed the illnesses to ingestion of *S. alternans*-contaminated feedstuffs. The deaths of a hippopotamus and of a bison in a zoo are associated with the ingestion of *S. alternans*-infected fodder [17]. Palyusik [18] has recently reported outbreaks of stachybotryotoxicosis among poultry, horses and swine in Hungary. There have been other outbreaks of stachybotryotoxicosis in farm animals in Eastern Europe in recent years, but these reports have not been available to us [19,20].

Biological tests for toxic Stachybotrys

Fungi of the genus Stachybotrys are in the class Fungi Imperfecti, order Moniliales, family Dematiaceae, subfamily Macronemiae; they are typical saprophytes which are distributed throughout the world. They have been

References p. 196

isolated not only from straw and hay, as has been mentioned, but also from soil, dead plant tissues, and a variety of cellulose-rich materials. *Stachybotrys alternans* has been isolated from tissues of animals that died of stachybotryotoxicosis; however, the experimental evidence argues that this fungus is not a primary pathogen. Rather, animals suffering from stachybotryotoxicosis are very susceptible to attack by a variety of microorganisms, including *S. alternans* [2,7,10].

Although the Soviet investigators of stachybotryotoxicosis identify *S. alternans* as the involved organism, a number of investigators consider this species as synonymous with *S. atra* [18,21]. Korpinen and Ylimäki [22] have isolated toxigenic strains of *S. chartarum* and consider *S. atra* and *S. alternans* synonymous with this species. Although he has reproduced the clinical and pathological characteristics of stachybotryotoxicosis in experimental animals using isolates of *S. atra*, Forgacs [7] insists that there has been no comparison of *S. alternans* as used by Soviet scientists and those *S. atra* strains which have been the subject of much investigation by Western scientists. In keeping with Forgacs' recommendation, we shall refer to the various species of Stachybotrys isolates which have been implicated in stachybotryotoxicosis as identified by the authors, realizing that these identifications may not be valid.

(A) Dermal toxicity

Following the establishment of the relationship between a well defined disease and certain strains of Stachybotrys and the finding that some metabolite or combination of metabolites of this fungus were most probably involved, investigators began to search for the toxin(s). Soviet research workers screened many fungal isolates and found that diethyl ether extracts of toxic hay or of certain strains of *S. alternans* produced dermal toxicity when applied to the shaved skin of rabbits [23,24]. The rabbit dermal toxicity test has come to be a standard test for differentiating toxic from non-toxic isolates of Stachybotrys. Using this test as a guide, Soviet investigators describe 2 important strains of *S. alternans*; a toxic strain (var. *jateli*) which is common in the southern regions of Eastern Europe and a non-toxic strain (var. *atoxica*) which predominates in the north.

40 strains of *S. atra*, grown under laboratory conditions, were studied by Forgacs and co-workers [25] in the United States, using the rabbit dermal toxicity test as the indicator of toxin production. These investigators found that the toxic fraction was ether-soluble, that the amount of ether-soluble extract varied greatly, and that the intensity of the dermal inflammation was not related to the total weight of ether extract [25]. Of the 40 strains tested, 26 produced intradermal reactions and 15 of these produced severe inflammation. Extracts from 2 of these strains also proved to be dermally toxic to cattle and horses. Forgacs *et al.* demonstrated that as little as 0.00175 μg of a crude ether extract of some of the more potent strains of *S. atra* could bring about hyperaemia when applied topically to rabbit skin.

(B) Animal feeding studies

Although the history of the investigations into the chemistry of the metabolites of species of Stachybotrys will be dealt with in the next section, it will clarify the development of this review to introduce some of the early work on the isolation of toxic Stachybotrys metabolites. Palyusik [18] and Pidoplichko and Bilai [24] outline work done in 1949 by Fialkov and Serebriani in which two compounds, designated as stachybotryotoxin A and stachybotryotoxin B, each presumably dermally active, were isolated from the ether extracts of toxic strains of *S. alternans*. Stachybotryotoxin A was precipitated from ether by the addition of petroleum ether and compound B crystallized from ether solution upon evaporation of the solvent. It now appears that these two isolates are most probably complex mixtures containing small amounts of highly potent dermal toxins and, perhaps, other toxic compounds as well [9]. In carrying out animal feeding studies some workers used mouldy feed, others used ether extracts of toxic feed or of cultures of Stachybotrys, and others used the isolates designated by Fialkov and Serebriani as stachybotryotoxins A and B. The language surrounding the experimental study of stachybotryotoxicosis is somewhat confused because some investigators use the term stachybotryotoxin when referring to an obviously crude extract, but use it at other times to refer to what is thought to be purified compound. Where possible, the term stachybotryotoxin will not be used in the following outline of animal feeding studies. An attempt will be made to describe the technique used by each investigator to prepare toxic isolate for study. In the section on chemistry, we will return to these questions to attempt to clarify the matter further.

Mice. Korneev [26] reports that oats inoculated with toxin-producing strains of *S. alternans*, fed at the rate of 2 g per day to mice, produced symptoms of intoxication within 48 h and death within 76 h. Necropsy revealed haemorrhage in the stomach. Both white Swiss mice and grey mice were susceptible.

Using a toxic strain of *S. atra* cultured for 3 weeks on oats and feeding white Swiss mice *ad libitum*, Forgacs *et al.* [25] observed the development of atony, depression, incoordination, and loss of appetite after 4 days, and the continued development of these symptoms until the eighth day, when death occurred. Pathologic examination revealed mottled and congested livers, haemorrhage and congestion in the lungs, stomach, and heart, and leukocytic infiltration of the stomach, liver, and small intestine.

Palyusik [18] prepared an ether extract from oat cultures and from agar cultures of *S. alternans*. A toxic material was precipitated from the ether solution by the addition of petroleum ether, and the precipitate was collected by filtration and washed with petroleum ether. The precipitate produced dermal toxicity in rabbits; the LD_{50} (i.p.) for albino mice was 51.6 mg per kg of body weight. The lack of proof of chemical homogeneity places doubt on the usefulness of this LD_{50} value.

References p. 196

3 strains of *S. chartarum* isolated in Scandinavia yielded toxins which produced the clinical and pathologic characteristics normally associated with stachybotryotoxicosis [22]. 2 of the strains were found in samples of mouldy grain and the third was isolated from a commercial feed which was suspected of causing abortions in pigs. Each of the strains was cultured on a grain mixture and, after 2 months of cultivation at room temperature, was fed to a test group of two white mice, one male and one female; the grain mixture comprised the entire diet. A control group received the same grain mixture, sterilized but not inoculated with the fungal strains. All 3 strains produced toxic effects. The first signs of disease appeared after 3 days; all the mice in the test groups died within 10 days. Depression and apathy, incoordination, shivering, and peritoneal pain developed in all animals of the test group. Ears, nose, and paws became cyanotic; the nose became swollen and lost hair. Pathologic examination revealed haemorrhages in the stomach, small intestines, and the cerebral meninges. Necrosis was observed all along the digestive tract and in the mucous membrane. Though only a small number of animals was used, these observations together with some chemical evidence to be cited later provide strong evidence that these isolates of *S. chartarum* were indeed producers of some of the toxins associated with stachybotryotoxicosis.

Guinea pigs. Guinea pigs fed 1.5 to 2.0 g daily of toxic oats prepared by inoculation with *S. alternans* were observed to develop congestion of the mouth and necrosis of the lower lip after 3 days [25]. In 10 to 16 days leukopenia, thrombocytopenia and agranulocytosis developed; body temperature increased and death occurred in 25 to 40 days. Gross necropsy revealed haemorrhages and other manifestations typical of equine stachybotryotoxicosis.

The LD_{50} (i.p.) for the guinea pig of a petroleum ether precipitate from an ethyl ether extract of oat cultures and agar cultures of toxic *S. alternans* has been reported to be 62.4 mg/kg. The chemical composition of the precipitate was unknown [18].

Spores of *S. alternans* were introduced into the lungs of guinea pigs by spraying, as a test of the idea that local or systemic effects in man might be produced by inhalation of toxin-containing aerosols generated during the handling of feed contaminated with toxic strains of *S. alternans* [27]. Necropsy findings on animals receiving the spray disclosed the presence of lung inflammation and congestion, focal bronchopneumonia, emphysema, and a number of other degenerative changes in the heart, liver, spleen, and kidney. Many of these effects resembled those seen in human and animal stachybotryotoxicosis. No fungal growth was observed, indicating that some toxic metabolite(s) of the fungus can be absorbed in the lungs and act systemically [7].

Cattle, sheep and swine. Forgacs [7] and Forgacs *et al.* [25] have described in detail studies in which strains of toxic *S. atra* were cultivated on sterile wheat straw for 60 days at room temperature and fed to a 250 lb calf, a 130 lb ewe, and a 125 lb pig. All animals eventually died, the ewe 25 h after a single feeding of 6 oz. of toxic straw, the pig 12 h after a single dose of 3.5 oz. of toxic straw, and the calf 3 days after receiving a total of 6 lbs of toxic straw over a 14-day period; the clinical symptoms observed in these animals and the findings of gross necropsy and histology reveal an intoxication very similar to equine stachybotryotoxicosis and to that reported for Soviet field cases of stachybotryotoxicosis in cattle [13].

Stachybotryotoxicosis has been induced in hogs by feeding straw infected with *S. alternans* [28]. Sows receiving the toxic feed had dry necroses on the teat tegument and suckling piglets developed necroses on the mouth, snout and mucous membrane. Continuous feeding of the infected toxic straw resulted in the deaths of 57—65% of the animals receiving the fodder [28].

Poultry. A number of studies have demonstrated that chicks are susceptible to stachybotryotoxicosis. Schumaier *et al.* [29] fed an ether extract of wheat, on which a toxic strain of *S. atra* had been cultured, to chicks for 13 days and observed depression, loss of appetite and reduced growth rates. There was a watery discharge from the mouth, and necropsy revealed necrotic lesions in the mouth and crop. When the chicks received contaminated wheat directly the same effects were seen. Palyusik [30] has reported that 5- to 7-day-old chicks develop stachybotryotoxicosis when fed oats infected with *S. atra* or an ether extract of such oats, or a material obtained as a precipitate when petroleum ether is added to the ether extract. Birds exhibited normal appetites early in the feeding trial, but the appearance of oral lesions after 24 h resulted in decreased food consumption. Oral lesions became very severe in 2 to 3 days, and the majority of deaths occurred in 6—12 days. Palyusik points out that stachybotryotoxicosis in chicks may be mistakenly diagnosed as the diphtheroid form of fowl pox.

The petroleum ether precipitate described above, which is of unknown chemical composition, was added to the chick diet at the level of one part per thousand and fed continuously; from this part of the study an LD_{50} for 6-day-old chicks was determined to be 795.3 mg/kg body weight [30]. An ethanol solution of the precipitate was prepared, diluted with water, and the ethanol was evaporated. The remaining aqueous solution was administered to the chicks and an LD_{50} was determined. In this case the LD_{50} was found to be 188.7 mg/kg body weight; interestingly, the chicks died without developing the oral lesions typical of stachybotryotoxicosis.

Palyusik *et al.* [31] have also produced stachybotryotoxicosis in 1-day-old goslings; these authors speculate that a widespread, unknown disease of goslings which had been diagnosed by some as "goose influenza" could very well be a mycotoxicosis and may well be, in part, stachybotryotoxicosis. The experimental stachybotryotoxicosis induced in 1-day-old goslings showed clinical and pathological manifestations very similar to those seen in chicks.

Other animals. In addition to those animals discussed in the foregoing, stachybotryotoxicosis has been induced experimentally by feeding Stachy-botrys-contaminated materials to dogs [26], horses [2,7,10] and rats [18].

(C) Other types of biological activity

Bodon and Palyusik [32] have studied the cytotoxic effects of mate-rials isolated by ether extraction of oat cultures of S. alternans. The addition of petroleum ether to the ether extract resulted in the precipitation of an almost colourless powder. It was this material, of unknown chemical compo-sition, which was examined by the tissue culture technique. Using calf-kidney epithelial cells in monolayer cell cultures and administering an aque-ous colloidal suspension of the precipitate at levels of 23.5 μg/ml and at six serial 10-fold dilutions of this level, these workers observed cytotoxic effects characterized by a decrease in the amount of cellular RNA, focal cell de-struction, and shrunken cells showing granulation of the nuclear chromatin; in some cases the nuclear membrane was absent but in others it remained normal. Cell cultures inoculated with some of the high dilution solutions contained many bi- or multinuclear cells. The authors recommend this test as a very sensitive bioassay for toxins of S. alternans; however, it was not demonstrated that the isolate tested was a single chemical compound or even a mixture of closely related chemical types.

Phytotoxicity is elicited by some of the metabolites obtained from cultures of S. alternans [33]. 20-day-old cultures of the mould were ex-tracted with methylene chloride, followed by chloroform, methanol, and petroleum ether extractions. Column chromatography gave a material which, at a concentration of 1 mg/ml, produced the characteristic inflammation of the rabbit ear skin, and at concentrations of 10—1000 μg/ml decreased the rate of germination of rye and radish seeds and the growth of roots of lettuce and rye seedlings. The physical and chemical data given for the isolate tested indicate that the material is not a pure chemical compound.

A precipitate produced by the addition of petroleum ether to an ether extract of S. alternans was found to be lethal to the protozoon Paramecium caudatum; the lowest concentration of this chemically uncharacterized sub-stance which produced the test was 20 μg/ml [18].

Eppley and Bailey [9] have found that the common brine shrimp (Artemia salina) is very sensitive to some of the toxic metabolites of S. atra and have used this organism to monitor the isolation of a variety of pure, toxic materials.

Metabolites of Stachybotrys

The major features of the biological effects of Stachybotrys toxins have now been outlined; the work of Forgacs [7] can be consulted if greater detail is required. We have sought to present this background before filling in the chemical details, since it seems certain, as we have already noted, that there has been no toxicological evaluation of any chemically homogeneous metabolite of any of the species of Stachybotrys implicated in stachybotryo-toxicosis. The details of some key Russian work have been available only through secondary sources, but a careful reading of these sources has led to the conclusion that the substances described by the Russian workers were incompletely defined and of questionable purity. Prior to very recent investigations in our laboratories the major attempts to deal with the chemical nature of Stachybotrys metabolites were those of Fialkov and Serebriani; these investigations have been described by Pidoplichko and Bilai [24] and by Palyusik [18]. Using cultures of *S. alternans* grown on wort agar, Fialkov and Serebriani isolated a colourless, amorphous powder by ether extraction of the cultures followed by the addition of petroleum ether to the extract. This material, which was named stachybotryotoxin A, demonstrated a positive rabbit skin toxicity test, and is the same type of material which a number of investigators have used for toxicological studies (see p. 184). A combustion analysis of stachybotryotoxin A revealed the possibility of two molecular formulas: $C_{25}H_{34}O_6$ or $C_{26}H_{38}O_6$. This work was done in 1949 when spectroscopic capability was limited. The investigators determined that the material designated as stachybotryotoxin A possessed a steroid nucleus, at least four hydroxy groups, and an unsaturated lactone of undetermined ring size. There is no evidence that stachybotryotoxin A was a pure compound; the evidence does indicate that this isolate was impure.

Using the isolation technique described by Fialkov and Serebriani, Pashevich prepared stachybotryotoxin A but found an empirical formula of $C_{24}H_{44}O_4$, and presented evidence for carboxylic acid functionality; this report has not been available for study and has come to us only through a secondary source [18]. Pashevich's work has not, to our knowledge, been followed by complete characterization of the isolate. This finding casts further doubt on the purity of stachybotryotoxin A. Fialkov and Serebriani also described stachybotryotoxin B, a crystalline material obtained when an ether extract of cultures of *S. alternans* is allowed to evaporate. There is some evidence that stachybotryotoxin B is a chemical relative of stachy-botryotoxin A but it is reported to be less toxic and less soluble in organic solvents. Nothing else is known of the chemical characteristics of stachy-botryotoxin B.

A material was isolated from cultures of *S. alternans* by Seredyuk and co-workers [33] and given the trivial name of stachybotryotoxin. The method of isolation used by these workers makes it unlikely that this material is equivalent to either stachybotryotoxin A or B, but it may be one component of the isolates which have been called stachybotryotoxins A and B. Stachybotryotoxin exhibits an empirical formula of $C_{16}H_{27}O_5$, melting point 62—69° and an experimentally determined molecular weight (Rast) of 302.8 ($C_{16}H_{27}O_5$ has a calculated molecular weight of 299). A stachy-botryotoxin concentration of 1 mg/ml produced the characteristic rabbit skin inflammation. There are no data regarding the purity of this material or its chemical constitution.

Recent work has resulted in some clarification of the chemistry of stachybotryotoxins. In all of the fractionation work outlined below, brine shrimp were used to follow the biological activity. This organism is reliable for monitoring the isolation of toxic Stachybotrys metabolites in that compounds demonstrated to be active against the brine shrimp also give a positive rabbit skin toxicity test. The brine shrimp bioassay has the advantages of being relatively rapid (16 h per test) and requiring very small amounts of test materials [34]. A strain of *Stachybotrys atra* which had been demonstrated as capable of producing stachybotryotoxicosis (strain number M-1126, Food and Drug Administration, Department of Health, Education and Welfare; Strain Number QM-1533, Quartermaster Corps, U.S. Army; Strain Number 1006, Harvard University, Cambridge, Mass.) was cultured on sterilized oats [9]. After 1 month the cultures were air-dried and extracted with diethyl ether in a Soxhlet apparatus for 20 h. The ether extract was concentrated and hexane added to the concentrate. The precipitate which formed was collected and dried and both the filtrate and the precipitate were toxic to brine shrimp. When examined by thin-layer chromatography (TLC) the toxic precipitate showed the presence of a large number of compounds (developing solvent: isopropanol/chloroform, 3/97.) This precipitate should be equivalent to stachybotryotoxin A. Although it might be argued that the *S. atra* as used in our laboratories is not equivalent to the *S. alternans* used by the Russian workers, it has been postulated, and accepted by some, that *S. atra* and *S. alternans* are synonymous species and in any event the described toxicology for the strain used by us closely resembles the toxicology described by the Russians.

Yus'kiv [35] has described a colour test for Stachybotrys toxins which is based on the reaction of the toxin with resorcinol in acidic solution; this test has been considered a reliable indicator of toxic components of materials isolated from Stachybotrys cultures [18,22]. The stachybotryotoxin A isolated in our laboratories gave a positive test with resorcinol, but when the resorcinol reagent described by Yus'kiv was sprayed on the TLC plate on which our isolate had been spotted and developed, three red spots appeared; however, in subsequent fractionation of our stachybotryotoxin A, all three of these compounds were absent from fractions toxic to the brine shrimp. Those compounds which do exhibit toxicity to brine shrimp are almost certainly responsible for at least part of the toxic effects of *S. atra* (especially the dermal inflammation, the effects on the mouth and gastrointestinal tract, and the haemorrhages) in larger animals and man; thus, the resorcinol test must be used with caution because it detects only compounds usually associated with the Stachybotrys toxins. Of course, the resorcinol-positive fraction may contain compounds which are toxic but which do not produce an effect on the brine shrimp.

Our stachybotryotoxin A, a complex mixture, was dissolved in chloroform and the solution added to hexane (1/10, chloroform/hexane) to give a non-toxic precipitate. The filtrate was evaporated to near dryness and applied to a silica gel column. The column was eluted with isopropanol/chloroform (1/99, V/V) and the course of the fractionation followed using TLC and the brine shrimp bioassay.

Two compounds were isolated during the chromatographic fractionation, one of which proved to be the well known steroid ergosterol, and the other was found to be the compound known as mellein (Fig. 1). The latter is a known metabolite of *Aspergillus ochraceus* [36]. Mellein showed no toxicity to the brine shrimp, but did produce a mild skin reaction when applied to rabbit ears. However, the dermal response elicited by mellein was very mild compared to that produced by other metabolites of *S. atra*, and for this reason it can be concluded that mellein is of little importance as a Stachybotrys toxin.

The column chromatographic fractionation procedure, coupled with the use of preparative TLC, resulted in the isolation of five pure compounds which were highly toxic to brine shrimp; compounds giving positive brine shrimp responses also gave positive rabbit skin toxicity tests [34]. These five compounds were designated as satratoxins C, D, F, G and H, according to their descending order on TLC plates. Satratoxin D was soon identified as the known fungal metabolite roridin E (Fig. 2) [37], and for this reason the term satratoxin D was dropped. That satratoxin D is equivalent to roridin E was established by direct comparison (UV, infrared, mass and nuclear magnetic resonance (NMR) spectra and TLC characteristics) with an authentic sample of roridin E [38]. Other compounds present in the mixture seen on TLC plates (*i.e.*, compounds "A", "B", "E") were not toxic to the brine shrimp.

Satratoxins C, F, G, and H, and roridin E do not possess the steroid characteristics ascribed to stachybotryotoxin A by Fialkov and Serebriani. The characteristics of stachybotryotoxin A described by the Russian workers may be explained in at least two ways, if it is accepted that their material was a mixture: (*1*) a steroidal, unsaturated lactone may exist as a major component of the mixture, but fail to give a positive rabbit skin or brine shrimp toxicity test and would thus not be detected by our method; or (*2*) the presence of large amounts of ergosterol in stachybotryotoxin A mixture could be responsible for the observation of a steroidal nucleus. Other chemical features observed by the Russian workers could be due to some of the

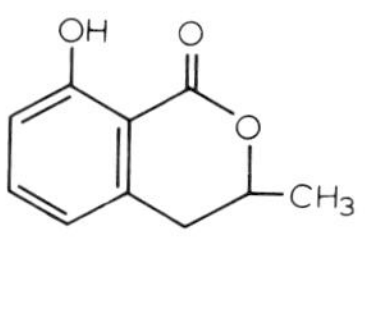

MELLEIN

Fig. 1.

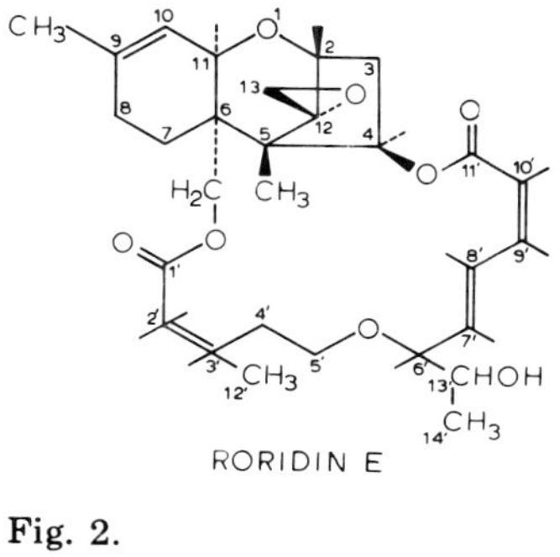

RORIDIN E

Fig. 2.

other compounds present. The compounds which we have called satratoxins C, F, G, and H are produced in very small amounts (see Table II) by the isolate of *S. atra* used in our work, and are thus quantitatively minor components of the stachybotryotoxin A mixture. Nevertheless, these compounds appear to represent the major toxic components of the stachybotryotoxin A mixture [9]. In Table II are listed some of the properties of these *S. atra* metabolites.

Roridin E (Fig. 2) is a metabolite of the soil fungus *Myrothecium verrucaria* [37], and is one of a series of complex sesquiterpenoids classified as 12,13-epoxy-Δ^9-trichothecenes [12,39].There are well over twenty naturally occurring compounds in this series of sesquiterpenoids; all are fungal metabolites and have in common the tetracyclic ring system exemplified by the roridin E hydrolysis product, verrucarol (Fig. 3). It is the nature of the ring substituents which distinguishes one member of the series from another. Some members are similar to roridin E in that a macrocyclic ring system is fused to the tetracyclic terpenoid (*e.g.*, roridins A [40], and D [41], and verrucarins A [42], B [43], and J [44]; other members of the series do not possess a macrocyclic ring system (*e.g.*, diacetoxyscirpenol [45], T-2 toxin [46], fusarenone [47], trichothecin [48], nivalenol [49], *etc.*). A recent and very comprehensive review by Bamburg and Strong [12] should be consulted for further details regarding these metabolites.

Satratoxins G and H have been isolated in sufficient quantities for further investigation. Upon hydrolysis with methanolic KOH both of these compounds yield the known compound verrucarol (Fig. 3) [38]; thus both of these substances are members of the 12,13-epoxy-Δ^9-trichothecene group. Satratoxins G and H have been studied using spectroscopy, but the nature of the substituents attached to the verrucarol nucleus have not as yet been determined. Molecular weight characteristics, coupled with the UV spectral features of these compounds (Table II), support the hypothesis that satratoxins G and H both possess macrocyclic ring systems having the unsaturated ester moieties present in roridin E but differing from this compound in other ways. The low resolution mass spectra of satratoxins G and H are shown in Fig. 4. Satratoxin F has not been available in sufficient quantities for characterization by chemical studies; however, UV, NMR, and mass spectra indicate that this compound is also a member of the series of 12,13-epoxy-Δ^9-trichothecenes. Structural work is continuing on all four of these materials but is slow because of the severely limited amounts of compounds available. None of these compounds appears to be equivalent to any presently known member of the trichothecene series.

Thus *Stachybotrys atra* can be added to the already large list of fungi which are capable of elaborating 12,13-epoxy-Δ^9-trichothecenes, and which now includes *Myrothecium verrucaria, M. roridum, Trichothecium roseum, Fusarium equiseti, F. scirpi, F. nivale, F. concolor, F. tricinctum* and *Trichoderma lignorum* [12]. Because of the similarity between the known toxic effects of a number of 12,13-epoxy-Δ^9-trichothecenes and the major clinical and pathological effects observed in animals suffering from stachybotryo-

TABLE II

PROPERTIES OF TOXIC COMPOUNDS ISOLATED FROM OAT CULTURES OF *STACHYBOTRYS ATRA* [9]

Compound	Yield from 6 kg oat culture (mg)	R_F on silica gel TLC[a]	Melting point, (°C)	Molecular weight	Molecular formula[b]	UV spectra (methanol)	Significant infrared absorptions, (cm^{-1})	LD_{50}[c]
Satratoxin C	2	0.95	—	484	—	—	—	150
Roridin E[d] (Satratoxin D)	3	0.50	170–176	514	$C_{29}H_{38}O_8$	222 nm ($\epsilon = 19\,600$) 260 nm ($\epsilon = 18\,700$)	3400–3500, 1710, 1645, 1600	130
Satratoxin F	2	0.42	164–168	542	—	250 nm ($\epsilon = 5\,900$)	3400–3500, 1750, 1715, 1625, 1600	150
Satratoxin G	8	0.28	167–170	544	$C_{28}H_{32}O_{11}$	256 nm ($\epsilon = 6500$)	3400–3500, 1750, 1710, 1625, 1598	150
Satratoxin H	12	0.20	162–166	528	$C_{29}H_{36}O_9$	225 nm ($\epsilon = 14\,700$) 255 nm ($\epsilon = 10\,400$)	3400–3500, 1770, 1650, 1595	200

a Isopropanol-CHCl$_3$ (3:97, v/v). When tested with a variety of solvent systems and spray reagents, all compounds showed one spot on TLC plates.

b Based on high resolution mass spectroscopy.

c ng of compound per 0.5 ml of test medium to kill 50% of the brine shrimp (*Artemia salina*) [34].

d Known metabolite of *Myrothecium verrucaria* (see text) [37].

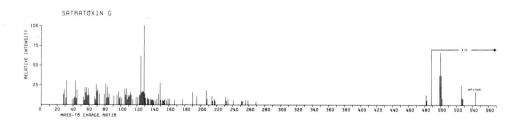

VERRUCAROL

Fig. 3.

toxicosis, Bamburg and Strong [12] had proposed that this mycotoxicosis might be partially attributable to the presence of these types of compounds. Our findings support this hypothesis.

Toxic effects of 12,13-epoxy-Δ^9-trichothecenes

Since purified epoxytrichothecenes derived from *S. atra* have been tested only on rabbit skin and brine shrimp, it will be useful to note some of the animal feeding studies which have been carried out with another member of this series of compounds. T-2 toxin, an epoxytrichothecene elaborated by the fungus *Fusarium tricinctum*, has been administered orally to experimental animals [50]. At subacute dosage levels symptoms such as haematuria, vomiting, loss of weight, thirst, and ataxia appeared in the animals. Following an initial phase during which there was degeneration of myelocytes in bone marrow, a leukocytosis appeared; this was in turn followed by severe leukopenia. Pathological findings were severe inflammation of the gastrointestinal tract: lesions in the lymph nodes, thymus, and testes; and degeneration of nerve cells. In cattle, feeding of T-2 toxin produced massive haemorrhage of the large intestine [12].

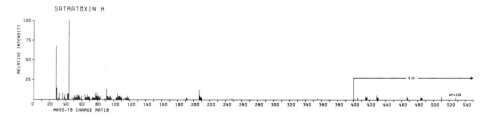

Fig. 4. Low resolution mass spectra of satratoxins G and H.

The simple description of the facts derived from animal feeding studies given in the last paragraph, coupled with the facts that many members of this group of compounds are known potent dermal toxins, both in animals and in humans [12], allow the tentative conclusion that the roridin E and other related 12,13-epoxy-Δ^9-trichothecenes (satratoxins C, F, G, and H) are at least partially responsible for stachybotryotoxicosis.

Conclusion

It has been impossible to proceed, as have many of the other chapters in this book, by first discussing a mould and its properties, the chemistry of its metabolites, the analytical methods for those metabolites, the conditions for toxin production, and so on. Since the chemistry of the toxic metabolites of *Stachybotrys atra* is only beginning to emerge, it has been impossible, until now, to study the conditions for the production of these toxins except in those crude ways which have been outlined in this chapter and presented in greater detail in the work of Forgacs [7]. Likewise, no direct analytical methods are available and biosynthetic studies do not exist. Obviously, the field is ripe for investigation by mycologists, analytical chemists, organic chemists, biochemists, and toxicologists. The excellent work on other 12,13-epoxy-Δ^9-trichothecenes which has come out of the laboratories of J. Bamburg, F.M. Strong and co-workers at the University of Wisconsin, Prof. Ch. Tamm and co-workers at the University of Basel, and others, should provide a sound background for future studies of the complex 12,13-epoxy-Δ^9-trichothecenes which now seem to be involved in stachybotryotoxicosis.

Stachybotryotoxicosis is perhaps just one of many mycotoxicoses in which 12,13-epoxy-Δ^9-trichothecenes may be involved. Bamburg and Strong [12] have already made a case for the idea that 12,13-epoxy-Δ^9-trichothecenes are likely suspects as the causative agents for alimentary toxic aleukia, a mycotoxicosis responsible for widespread human disease and death [51]. These authors have also presented evidence, largely based on clinical and pathological data, that these compounds may be involved in mouldy corn toxicosis of poultry, pigs, and cattle, fescue foot, pink rot of fruit, and "scab" barley disease.

Stachybotryotoxicosis seems preventable, as are most mycotoxicoses, if conditions conducive to mould growth in foods and feedstuffs are avoided; Forgacs [7] has described the sanitary practices required to avoid *S. atra* contamination of forage. What seems critical at this stage is the need for evaluating the chronic toxic effects of all 12,13-epoxy-Δ^9-trichothecenes, regardless of fungal source, to learn which are potential human health hazards, and the development of assay techniques which may be used to locate these compounds in food supplies.

References

1 N. Kruschev, Kruschev Remembers, with Introduction. Commentary and Notes by Edward Crankshaw, Little, Brown, Boston, 1970, pp. 111—114.
2 V.G. Drobotko, Stachybotryotoxicosis, a New Disease of Horses and Humans, Report presented at the Academy of Science, U.S.S.R., 1942.
3 V.G. Drobotko, Am. Rev. Sov. Med., 2 (1945) 238.
4 V.G. Drobotko, P.E. Marushenko, B.E. Aizeman, N.G. Kolesnik, P.D. Iatel and V.D. Melnichenko, Vrach. Delo, 26 (1946) 125.
5 A.K. Sarkisov, Veterinariya, 24 (1947) 25.
6 J. Forgacs and W.T. Carll, Advan. Vet. Sci., 7 (1962) 278.
7 J. Forgacs, Stachybotryotoxicosis, in S. Kadis, A. Ciegler and S.J. Ajl (Eds.), Microbial Toxins, Vol. VIII, Academic Press, New York, 1972, pp. 95—126.
8 J. Forgacs, Stachybotryotoxicosis and moldy corn toxicosis, in G.M. Wogan (Ed.), Mycotoxins in Foodstuffs, M.I.T. Press, Cambridge, Mass., 1965, pp. 87—93.
9 R.M. Eppley and W.J. Bailey, Science, 181 (1973) 758. (Much of this work is taken from R.M. Eppley, Ph. D. Thesis, University of Maryland, College Park, Md.,).
10 K.I. Vertinskii, Veterinariya, 17 (1940) 61.
11 J.R. Bamburg, Ph. D. Thesis, University of Wisconsin, Madison, Wis., 1969.
12 J.R. Bamburg and F.M. Strong, 12,13-Epoxytrichothecenes, in S. Kadis, A. Ciegler and S.J. Ajl (Eds.), Microbial Toxins, Vol. VII, Academic Press, New York, 1971, pp. 207—292.
13 V.A. Fortuskny, A.M. Govrov, I.Z. Tebybenko, A.S. Biochenko and E.T. Kalitenko, Veterinariya, 36 (1959) 67.
14 I.G. Levenberg, L.I. Ivantsvov and M.P. Prostakov, Veterinariya, 38 (1961) 38.
15 I.A. Kurmanov, Veterinariya, 38 (1961) 41.
16 I.A. Ismailov and B.F. Moroshkin, Veterinariya, 4 (1962) 27.
17 A.F. Tkachenko, Comparative data on stachybotryotoxicosis of the bison and the hippopotamus, in V.I. Bilai (Ed.), Mycotoxicoses of Man and Agricultural Animals (English Transl.), Office of Tech. Ser., U.S. Dept. of Commerce, Washington, 1960, pp. 187—195.
18 M. Palyusik, Acta Vet. Hung., 20 (1970) 57.
19 G. Danko and J. Tanyi, Magy. Allatorv. Lap., 23 (1967) 225.
20 P. Pavlov, G. Dimitrov, H. Stankoushev and K. Surtmadjiev, Vet. Med. Nauki, Sofija, 4 (1967) 49.
21 G.R. Bisby, Brit. Mycol. Soc. Trans., 26 (1943) 133.
22 E-L. Korpinen and A. Ylimäki, Experientia, 28 (1972) 108.
23 A.K. Sarkisov and V.N. Orshanskaiya, Veterinariya, 21 (1944) 38.
24 N.M. Pidoplichko and V.I. Bilai, Toxic fungi which develop in food products and fodder, in V.I. Bilai (Ed.), Mycotoxicoses of Man and Agricultural Animals (English Transl.), Office of Tech. Ser., U.S. Dept. of Commerce, Washington, 1960, pp. 3—37.
25 J. Forgacs, W.T. Carll, A.S. Herring and W.R. Hinshaw, Trans. N.Y. Acad. Sci., 20 (1958) 787.
26 N.E. Korneev, Veterinariya, 25 (1948) 36.
27 P.F. Samsonov and A.P. Samsonov, The respiratory mycotoxicoses, experimentally, in V.I. Bilai (Ed.), Mycotoxicoses of Man and Agricultural Animals (English Transl.), Office of Tech. Ser., U.S. Dept. of Commerce, Washington, 1960, pp. 140—151.
28 V. Volintir, I. Popescu, I. Jivănescu, R. Mogaminzat, M.P. Vlah and S. Constantinescu, Rev. Zool. Med. Vet., 1 (1971) 68.
29 G. Schumaier, H.M. DeVolt, N.C. Laffer and R.D. Creek, Poultry Sci., 62 (1963) 70.
30 M. Palyusik, Sabouraudia, 8 (1970) 4.
31 M. Palyusik, I. Szép and F. Szöke, Acta Vet. Hung., 18 (1968) 363.
32 L. Bodon and M. Palyusik, Acta Vet. Hung., 20 (1970) 289.
33 L.S. Seredyuk, A.A. Svishcuk and P.A. Moroz, Fiziol. Biokhim. Osnovy Vzaimod. Rast. Fitotsenozakh, (1971) 133; Chemical Abstracts, 76 (1972) 137.

34 R.M. Eppley, J. Ass. Off. Anal. Chem., 57 (1974) 618.
35 R.V. Yus'kiv, Mikrobiol. Zh., 28 (1966) 68.
36 T. Yabuta and Y. Sumiki, J. Agric. Chem. Soc. (Japan), 9 (1933) 1264.
37 W. Zürcher and Ch. Tamm, Helv. Chim. Acta, 49 (1966) 2594.
38 Professor Ch. Tamm, Institut für Organische Chemie der Universität Basel, Basel, Switzerland kindly provided samples of roridin E and verrucarol.
39 V. Herout, Biochemistry of sesquiterpenoids, in T.W. Goodwin (Ed.), Aspects of Terpenoid Chemistry and Biochemistry, Academic Press, New York, 1971, pp. 80—82.
40 B. Böhner and Ch. Tamm, Helv. Chim. Acta, 49 (1966) 2527.
41 B. Böhner and Ch. Tamm, Helv. Chim. Acta, 49 (1966) 2547.
42 J. Gutzwiller and Ch. Tamm, Helv. Chim. Acta, 48 (1965) 157.
43 J. Gutzwiller and Ch. Tamm, Helv. Chim. Acta, 48 (1965) 177.
44 E. Fetz, B. Böhner and Ch. Tamm, Helv. Chim. Acta, 48 (1965) 1669.
45 E. Flury, R. Mäuli and H.P. Sigg, Chem. Commun., (1965) 26.
46 J.R. Bamburg, N.V. Riggs and F.M. Strong, Tetrahedron, 24 (1968) 3329.
47 J.F. Grove, Chem. Commun., (1969) 1266.
48 W.O. Godfredsen and S. Vangedal, Proc. Chem. Soc., (1964) 188.
49 T. Tatsuno, M. Saito, M. Enomoto and H. Tsunoda, Chem. Pharm. Bull., 16 (1968) 2519.
50 W.F.O. Marasas, J.R. Bamburg, E.B. Smalley, F.M. Strong, W.L. Ragland and P.E. Degurse, Toxicol. Appl. Pharmacol., 15 (1969) 471.
51 A.Z. Joffe, Alimentary toxic aleukia, in S. Kadis, A. Ciegler and S.J. Ajl (Eds.), Microbial Toxins, Vol. VII, Academic Press, New York, 1971, pp. 139—189.

Chapter 10

TOXIC TRICHOTHECENES*

EUGENE B. SMALLEY[a] and FRANK M. STRONG[b]

[a]*Department of Plant Pathology and* [b]*Department of Biochemistry, University of Wisconsin, Madison, Wis. 53607 (U.S.A.)*

The trichothecenes are a chemically related group of biologically active fungal metabolites produced in culture by various species of Fusarium, Cephalosporium, Myrothecium, Trichothecium, Trichoderma, and Stachybotrys. The fungi which produce the more toxic of these trichothecenes are associated with a wide variety of human and animal intoxications in temperate regions of the world. Because of unresolved analytical problems, however, only one of the trichothecenes has been chemically identified in toxic mouldy feeds [1].

A comprehensive review of this subject by Bamburg and Strong [2] covered the literature available through the first half of 1970. At that time, 22 naturally occurring members of this group had been discovered. The subsequent 3-year period has been one of considerable activity that has seen a number of additional laboratories taking up an active interest in the trichothecenes. Various Japanese investigators have been particularly active (see p. 263). Special progress has been made in the discovery of five additional natural trichothecenes, in further clarifying the biosynthetic pathways, and in the development of improved analytical and bioassay screening methods. As presently understood, low temperature mouldy corn toxicosis (haemorrhagic disease) [1,3], mouldy cereal emesis [4,5], "akakabibo" toxicosis [6], alimentary toxic aleukia [7,8], mouldy bean hull intoxication [9], stachybotryotoxicosis [10], dendrodochiotoxicosis (myrotheciotoxicosis) [11,12], and fusariotoxicosis [7,8] all probably can be considered primarily trichothecene intoxications.

Chemical nature

All of the natural trichothecenes have the basic tetracyclic sesquiterpene structure shown in Fig. 1. The skeletal structure (all R groups = H) includes a 6-membered oxygen-containing (oxane) ring, a 12,13-epoxide group, a 9,10-olefinic bond, tertiary methyl groups at positions 5 and 6, and a vinyl methyl at 9. All but one of the naturally occurring members of the group bear oxygen-containing substituents located at one or more of the positions 3, 4, 7, 8, and 15. These substituents may be hydroxyl, esterified hydroxyl, keto (position 8 only), epoxide (position 7, 8 only), or combinations there-

---

* Literature reviewed in this chapter covers the period up to July 1st, 1973.

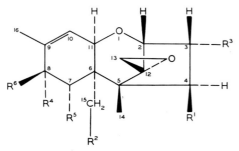

Fig. 1. 12,13-Epoxy-Δ^9-trichothecene (scirpene, all R groups = H).

of. The verrucarins and roridins constitute a special subgroup characterized, respectively, by a macrocyclic ester, or ester—ether, bridge between positions 4 and 15.

The naturally occurring trichothecenes which have been reported to date are listed in Table I. The parent compound itself, 12,13-epoxy-Δ^9-trichothecene (scirpene*) has very recently been reported [13]. In general the substances with higher numbers of substituents on the skeletal ring possess more pronounced biological activity.

The verrucarins and roridins have the structures shown in the latter portion of Table I and are all very closely related. These compounds are particularly potent toxins.

Most of the natural substances in Table I have at least one and usually more than one ester group. Saponification of the esters readily gives rise to the parent alcohols, of which there are ten. Only trichodermol (identical with roridin C), scirpen-4,8-diol, trichothecolone, and nivalenol have been reported to occur naturally. The structures of these four are indicated in Table I and the structures of the remaining six are evident also from Table I viz., crotocol (from crotocin), scirpen-3,15-, and 4,15-diols, scirpentriol, scirpentetraol (T-2 tetraol), and scirpenpentaol.

Production

(A) Trichothecene-producing fungi

All of the fungi known to produce trichothecene are members of the order Hyphomycetales in the Deuteromycotina (Fungi Imperfecti). Where their sexual stages are known, these are mainly found in the order Hypocreales of the Ascomycotina class — Pyrenomycetes. One or two perfect stages are also known from the related order Sphaeriales. Of the five imperfect genera having perfect stages in the Hypocreales, three are species of Nectria and another belongs in the closely related genus Hypomyces. Clearly

*For convenience the term "scirpene" is used in this chapter in place of 12,13-epoxy-Δ^9-trichothecene and as a basis for informative, relatively short common names for several natural trichothecenes, e.g., diacetoxyscirpenol.

the production of trichothecenes by fungi has important taxonomic implications, and suggests close natural relationships among these otherwise diverse species of imperfect fungi.

Most of the species of imperfect fungi which produce trichothecenes are members of the family Moniliaceae. This family is characterized by having hyaline or brightly coloured conidia with conidiophores not grouped into sporodochia or coremia [19]. However, two important trichothecene-producing genera (Myrothecium and Stachybotrys) have quite a different morphology, having either dark conidia or conidiophores and these belong to the family Dematiaceae [20].

(1) Dematiaceus trichothecene producers

In the genus Myrothecium 2 of the 13 recognized species [11], *M. roridum* and *M. verrucaria*, produce trichothecenes and have been intensively studied relative to their toxin-producing capabilities [2,17,18,21—23]. In the other dematiaceous genus Stachybotrys, only one of the eight recognized species (*S. atra = S. alteranans*) [20] has been shown to produce toxic trichothecenes [10]. These genera also differ from the other known trichothecene producers in that their toxins (called verrucarins and roridins) possess a unique "bridge structure" as a part of the basic trichothecene moiety [2] (Table I).

Taxonomically, the conidiogenous cells in Myrothecium are phialides, borne in sessile sporodochia and the conidia they produce are hyaline to pale olive and non-septate [11]. In Stachybotrys the conidiogenous cells are also phialidic, but borne in groups at the apex of each hyphal stipe or branch. The conidia are non-septate, hyaline to pale olive and aggregate in large slimy often black and glistening heads [20,24,26]. The perfect stage, and the only one known for Myrothecium is in the genus Nectria of Hypocreales (*N. bactridioides*) [11]. In the genus Stachybotrys the only described perfect stage was identified as Melanopsamma in the Sphaerales (*M. pomiformis*) [25,27].

Members of the genus Myrothecium are commonly isolated from soil and especially from soils high in organic matter [20]. One species (*M. inundatum*) is found only on dead mushrooms (*Russula ajusta*), while most of the common species have been found on dead and dying leaves of various plants [11]. Most of the species carry on a saprophytic existence, but one species (*M. roridum*) is a plant pathogen causing serious leaf spots and diebacks on a number of hosts in temperate and tropical regions. Their association with mycotoxicoses is discussed in detail in a later section (see p. 223).

The toxic trichothecenes produced by Myrothecium (verrucarins and roridins) have been studied mainly from the point of view of antibiotic production and cytotoxicity. Toxic fungi for these studies have not generally come from foods or feeds involved in animal intoxication, but rather from corporation screening trials using mainly fungus isolates from type culture collections.

Most of the species of Stachybotrys known are found naturally in moist

TABLE I

NATURALLY OCCURRING TRICHOTHECENES

Common name[a]	Source[b]	Substituent[c] at position[d]					Ref.
		3	4	7	8	15	
1. Scirpene	4	—	—	—	—	—	13
2. Trichodermol (Roridin C)	2	—	β OH	—	—	—	2
3. Trichodermin	5	—	β OAc	—	—	—	2
4. Scirpen-4,8-diol	4	—	β OH	—	α OH	—	13
5. Trichothecolone	4	—	β OH	—	=O	—	2
6. Trichothecin	4	—	β iso-Crot	—	=O	—	2
7. Crotocin	4, 6	—	β iso-Crot	-epoxide-	—	—	2, 45
8. Diacetylverrucarol	1	α OAc	β OAc	—	—	OAc	2
9. Deacetylcalonectrin	10	α OAc	—	—	—	OH	14
10. Calonectrin	10	α OH	—	—	—	OAc	14
11. Acetoxyscirpendiol	7	α OH	(β OAc)[e]	—	—	(OAc)[e]	2
12. Diacetoxyscirpenol	7, 8, 9, 11, 13	α OH	β OAc	—	—	OAc	2
13. Neosolaniol	7, 8, 9, 13	α OH	β OAc	—	α OH	OAc	9
14. HT-2 toxin	8, 9	α OH	β OH	—	α iso-Val	OAc	2
15. T-2 toxin	7, 8, 9, 11, 13	α OH	β OAc	—	α iso-Val	OAc	2
16. Nivalenol	10, 12	α OH	β OH	α OH	=O	OH	2
17. Fusarenon-X	7, 8, 10, 12, 14	α OH	β OAc	α OH	=O	OH	15
18. Diacetylnivalenol	10, 14	α OH	β OAc	α OH	=O	OAc	2, 16
19. Triacetoxyscirpendiol	7	α OH	β OAc	α OH	α OAc	OAc	2

Bridge structure[f]

Common name[a]	Source[b]	Bridge structure[f]	Ref.
20. Verrucarin A	1, 2	$-\overset{O}{\overset{\parallel}{C}}CHOHCHMeCH_2CH_2O\overset{O}{\overset{\parallel}{C}}CH{=}CHCH{=}CH\overset{O}{\overset{\parallel}{C}}O-$	2
21. Dehydroverrucarin A	2	$-\overset{O}{\overset{\parallel}{C}}\overset{O}{\overset{\parallel}{C}}CHMeCH_2CH_2O\overset{O}{\overset{\parallel}{C}}CH{=}CHCH{=}CH\overset{O}{\overset{\parallel}{C}}O-$	2

TABLE I (continued)

Common name[a]	Source[b]	Bridge structure[f]	Ref.
22. Verrucarin B	1, 2	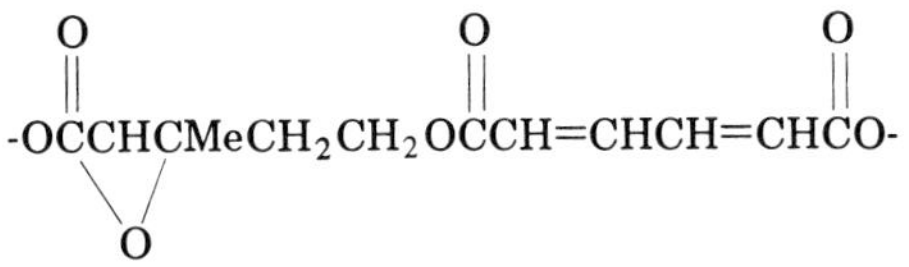 -OCCHCHMeCH$_2$CH$_2$OCCH=CHCH=CHCO- (epoxide O)	2
23. Verrucarin J	1	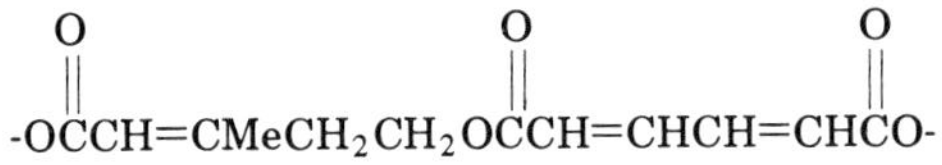 -OCCH=CMeCH$_2$CH$_2$OCCH=CHCH=CHCO-	2
24. Roridin A	1, 2	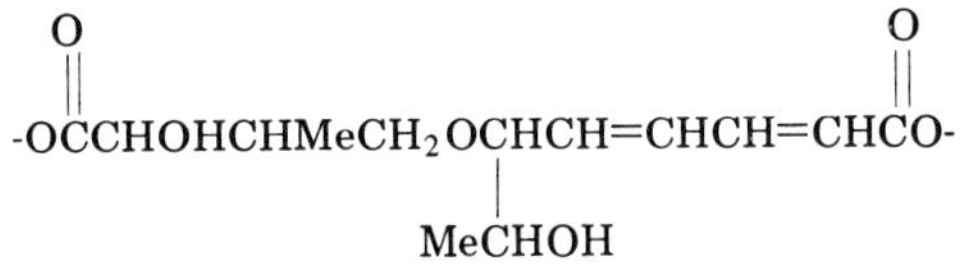 -OCCHOHCHMeCH$_2$OCHCH=CHCH=CHCO- MeCHOH	2
25. Roridin D	1	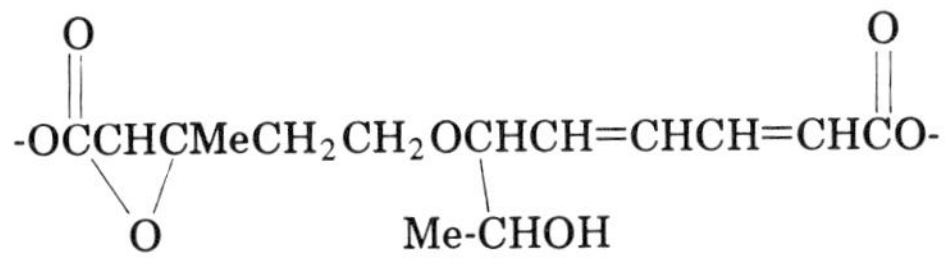 -OCCHCHMeCH$_2$CH$_2$OCHCH=CHCH=CHCO- (epoxide O) Me-CHOH	2
26. Roridin E	1, 3	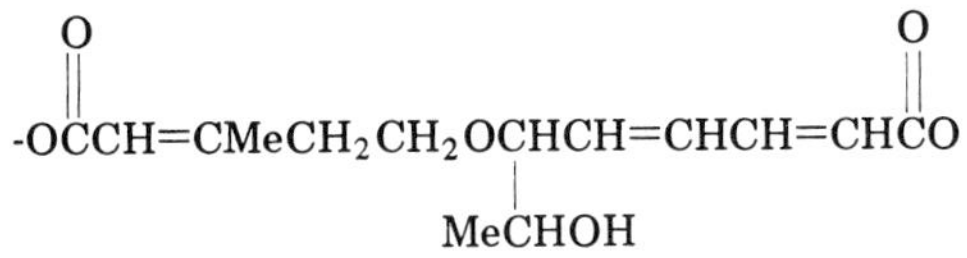 -OCCH=CMeCH$_2$CH$_2$OCHCH=CHCH=CHCO MeCHOH	17
27. Roridin H	1	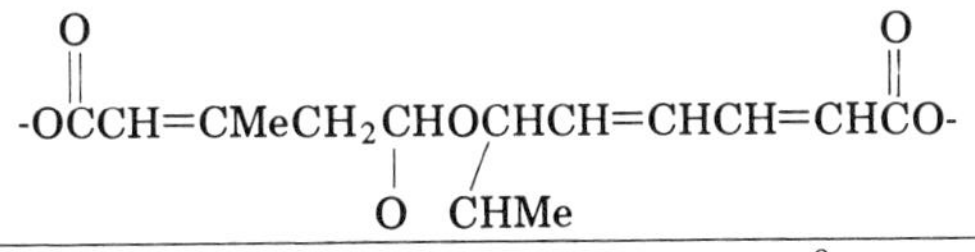 -OCCH=CMeCH$_2$CHOCHCH=CHCH=CHCO- O CHMe	18

a As suggested by the original discoverer of the substance; "scirpene" used to designate 12, 13-epoxy-Δ^9-trichothecene.
b Key to fungus sources: 1, *Myrothecium verrucaria*; 2, *M. roridum*; 3, *Stachybotrys atra*; 4, *Trichothecium roseum*; 5, *Trichoderma viride*; 6, *Cephalosporium crotocinigenum*; 7, *Fusarium roseum* (syn. *F. concolor, F. graminearum, F. equiseti, F. scirpi*); 8, *F. solani*; 9, *F. tricinctum* (syn. *F. poae, F. sporotrichioides*); 10, *F. nivale* (perfect stage - Calonectria); 11, *F. lateritium*; 12, *F. episphaeria*; 13, *F. rigidiusculum*; 14, *F. oxysporum*. c — = H; OAc = acetoxy; iso-Crot = isocrotonyloxy; iso-Val = iso-valeroxy. d Numbering as in Fig. 1. e One acetoxy group either at 4 or 15. f From position 15 (left side in Fig. 1) to position 4; no other substituents.

References p. 224

situations on decaying tree branches in the tropics (*S. theobromae, S. neph-rospora, S. cylindrospora, S. kampalensis, S. dichroa* [20]. *S. parvispora* has been found on decaying leaves of various tropical plants and the type species *S. atra* is found widely on paper, seeds, soil, textiles, dead grass, straw and forage [20,27]. No plant parasitic capabilities have apparently been discovered. A detailed review of Stachybotrys and its well known toxicosis is found elsewhere in this text [10] (see p. 182) and includes the discovery of the trichothecene nature of its toxin. Other members of the dematiaceous hyphomycetes may also be found to produce trichothecenes. We have tested toxic strains of Nigrospora and Epicoccum whose extracts produce dermal responses in rats similar to responses obtained with pure trichothecenes [3]. Keyl [28] in 1967 also reported that extracts from cultures of *Epicoccum nigrum* (2 strains), and *Cladosporium cladosporioides* (2 strains) both gave strong skin responses when tested on rabbits.

(2) Moniliaceous trichothecene producers

The remainder of the genera which produce trichothecenes without the "bridge structure" are members of the family Moniliaceae (Table I). These fungi produce hyaline, lightly coloured (never dark) conidia. The presently recognized trichothecene producing members of this family are all species of Fusarium, Cephalosporium, Trichoderma, or Trichothecium.

In the genus Fusarium, 8 of the 9 recognized species (under the Snyder and Hanson system [29]) have been reported to produce trichothecenes (*F. tricinctum, F. solani, F. roseum, F. nivale, F. oxysporum, F. lateritium, F. rigidiusculum* and *F. episphaeria*). However, major problems in identification of Fusaria exist. These problems and the various proposed taxonomic systems are discussed by Snyder and Hanson [30—33], Toussoun and Nelson [29], Raillo [34], Gordon [35], Messiaen and Cassini [36], Bilai [37], and most recently by Booth [38]. Confusion in the nomenclature of certain of these toxic species has resulted in somewhat inappropriate trivial chemical names in several cases.

Taxonomically, the conidiogenous cells in Fusarium are either simple- or polyphialidic-slime spores or enteroblastic. Conidia may occur as 0—1 septate, pyriform, fusoid to oval microconidia through to straight or curved 0—10 or more septate macroconidia. The presence or absence of microconidia and chlamydospores together with the shape of both the macro and microconidia are used to establish sections or groups or species [29,31, 32,38]. Where perfect stages are recognized in the 9 Fusarium species [31,32], 3 are Gibberella stages (*F. moniliforme, F. lateritium,* and *F. roseum*), 2 Calonectria (*F. rigidiusculum,* and *F. nivale*), while the remaining 2 either form Hypomyces (*F. solani*) or Nectria (*F. episphaeria*) stages [29]. A somewhat different view of Fusarium nomenclature with certain changes in the names of the perfect stages, and drastic changes in the imperfect species names is proposed by Booth [38].

The Fusaria occur widely in nature on many hosts and substrates and are among the most commonly occurring of all the fungi. Many of the species are parasitic on higher plants and cause serious economic losses. These aspects are broadly discussed by Booth [38] and others. Their associations with problems of mycotoxicoses are well known. These relationships have previously been reviewed in detail by Bamburg and Strong [2], Mirocha *et al.* [39], Joffe [40], Saito and Tatsuna [41], Smalley *et al.* [3], Balai [8], and others.

In the genus Cephalosporium, only 1 of approximately 20—70 species [27,42,43] has been shown to produce toxic trichothecenes. *C. crotocinigenum* [44,45] is unique in producing a trichothecene having a second epoxide in the basic structure (Table I). Since many other species of Cephalosporium have been shown to produce antibiotics [46], other toxic trichothecene producers may ultimately be found in nature. Problems of nomenclature also exist in this genus. Certain mycologists consider Acremonium to be the correct name for this genus [42]. In his treatment of the genus in 1963, Durrell [43] states that the variation of single spore strains of any "species" (he includes 57 in his study) of Cephalosporium are as great as the variation within the genus. He suggests that we could readily return to the single original species (*C. acremonium*) unless perfect stages are associated. Perfect stages of Cephalosporium (Acremonium) species which have been described are Ascomycetes either in the family Eurotiales (Emericellopsis) [43,47] or in the Hypocreales genus Nectria [42].

Taxonomically, conidiogenous cells in Cephalosporium are monophialidic with the phialides arising directly and singly from the vegetative hyphae and giving rise to non-septate hyaline conidia borne in balls [27]. Members of this genus are seen widely in nature, both as saprophytes and occasionally as parasites on plants and animals.

In the genus Trichoderma, of the approximately 20 species recognized [42], only 1 or 2 are known to produce trichothecenes (*T. viride, T. lignorum*). In a recent monograph, however, Rifai [48] has reduced the number of acceptable genera to only 9, and only *T. viride* can be considered to be a trichothecene producer.

Members of the genus Trichoderma are commonly isolated from soil, decaying wood, and other moist vegetable debris [48]. *T. viride* can frequently be isolated from mouldy corn or other mouldy feeds and is often found in association with mycotoxin problems [49]. Members of this genus are mainly saprophytic, but *T. viride* apparently has limited parasitic capabilities [50].

Taxonomically, the conidiogenous cells in Trichoderma are monophialidic with hyaline to green non-septate conidia which gather in moist balls at the mouth of the phialide [48]. Most of the accepted species of Trichoderma have perfect stage associations, and all of those known belong to the genus Hypocrea of the Hypocreales [48]. These include *H. pilulifera* (*T. pilulififerum*), *H. semiorbis* (*T. hamatum*) *H. aureoviridis* (*T. aureoviride*), *H. schweinitzii* (*T. longbrachiatum*), and *H. rufa* (*T. viride*).

In the genus Trichothecium only 1 of approximately 4 species is known to produce toxic trichothecenes. A monograph on this small genus has been published by Rifai and Cooke [51]. Taxonomically, conidia arise as blown-out ends of the conidiophore, murogenously, and after the first conidium has been cut off by the septum, but before it is mature, the next conidium is blown out and grows obliquely at one side of the first spore. Mature conidia are one-septate, hyaline and become pinkish in mass. The major species in this genus, *Trichothecium roseum* is world-wide in distribution usually occurring as a saprophyte on decaying vegetable matter, but occasionally it becomes a weak or secondary parasite on stored fruits and vegetables [50]. Other species include *T. luteum* whose colonies form a dirty yellow colour, and *T. parvum* whose colonies are white, and the Trichotheceum species associated with the perfect stage, *Hypomyces trichothecioides* [52].

(B) Trichothecene production in culture

From the literature it is not always clear from the point of view of animal intoxication, which of the many toxic trichothecenes are major toxicants produced in nature and which play only minor roles. Many reports indicate loss of toxin production in older cultures and problems relating to maintenance of physiological characters over long periods of culture are well known. Thus, one becomes concerned about the reality of much of the published toxic production studies. For example, the strain of *Fusarium nivale* (*Calonectria nivali*) used to produce calonectrin, the most recently described "new trichothecene" (CMI 14764) [14], was originally isolated in 1926 from Triticum by G.H. Cunningham in New Zealand [38]. One can only imagine the metabolic or genetic changes taking place during those 46 years. The isolate of *Gibberella intricans* (along with *F. scirpi* and *F. equiseti*) used by Brian *et al.* [53] in their discovery of diacetoxyscirpenol was an original Wollenweber strain described 30 years previously [54]. Although many of the described toxins have been derived from more recently isolated strains, the problems of interpretation remain confused.

(1) Production of macrocyclic esters

Most of the strains of *M. verrucaria* and *M. roridum* used for the various studies on production of verrucarins and roridins commonly produce verrucarins A and B and roridin A in culture [18,21,22,55] (Table I). Most of these investigations, however, have been directed primarily towards isolation and chemical characterization of trichothecenes rather than towards maximum or natural toxin production. From the available information it would appear, then, that the major trichothecene metabolites to be expected in Myrothecium-associated mycotoxin problems would be verrucarin A (from *M. verrucaria* contamination) and roridin A (from *M. roridum* contamination), and because of its phytopathogenic capabilities, *M. roridum* would appear to be potentially most important as a cause of mycotoxicoses. Mortimer *et al.* [12] and Vertinskii *et al.* [56] have isolated toxic strains of *Myrothecium verrucaria* and *M. roridum* from forage, and Mortimer *et al.* [12] identified verrucarin A and roridin A from his cultures. The recent discovery that roridin E is produced by toxic strains of *Stachybotrys atra* suggests that it too may belong in this category of major metabolites [10].

(2) Tetracyclic trichothecenes

The majority of the investigations concerning epoxytrichothecenes without the "bridge structure" have been associated with mycotoxin intoxications. As a result, somewhat more attention has been given to strain differences and conditions necessary for optimum toxin production. Most of the work on these metabolites to date suggests, as in the case of the verrucarins and roridins, that only a few of the compounds described are actually potential causes of mycotoxicoses. Those major compounds, which appear to be the most likely causes of mycotoxicoses are diacetoxyscirpenol, T-2 toxin, HT-2 toxin, fusarenon-X, and neosolaniol. These compounds are produced almost exclusively by species of Fusarium. Diacetoxyscirpenol is now known to be produced by *Fusarium tricintum* [2] (synonymous with *F. poae*, *F. sporotrichioides*), *F. roseum* [53] (synonymous with *F. sambucum*, *F. graminearum*, *F. equisti*, *F. scirpi*, *Gibberella zeae*, *Gibberella intricans*), *F. solani* [57], and *F. lateritium* [58]. T-2 toxin has been reported to be produced by *F. tricinctum* [2], *F. roseum* [59], *F. solani** [57], *F. lateritium* [59], and *Trichoderma viride* [2]. Fusarenone X has been identified from cultures of *F. nivale* [4], strain** Fn 2B, *F. oxysporum* and *F. episphaeria* (synonymous with *F. dimerum* and *F. merismoides* [38]. Neosolaniol is produced by *F. roseum*, *F. solani*, *F. tricinctum*, and *F. rigidiusculum* [63] (Table I).

Methods and conditions for *in vitro* production of 12,13-epoxytrichothecenes have been quite varied and have been dependent to a large extent upon the specific research interest of the investigator. Laboratories whose research goals and interests were mainly the production of antibiotics have tended to use comparatively simple and usually semichemically defined liquid media [17,18,21,22,53,55,61,64,65]. Incubation conditions in these cases were frequently at laboratory temperatures with large-scale fermentations accomplished by submerged-aerated culture techniques. Most workers used various modifications of Czapeks—Dox mineral liquid medium [66] with various additional carbon sources added, depending upon the requirements of the fungus. More frequently these mineral media were simplified to include only the minimum essential salts, carbohydrates, with additional yeast extract and vitamins. Using such techniques, maximum yields of 83, 220, 125, 10, 80, and 214 mg/l were obtained respectively for verrucarin A (*M. verrucaria*) [21], roridin A (*M. roridum*) [21], diacetoxyscirpenol (*F. scirpi*) [53], neosolaniol (*F. solani*) [57], fusarenon-X [15] and T-2 toxin [67].

In general, research workers whose main interest is in mycotoxins and

---

*The identity of the isolate is questioned by Marasas [60]. His examination of the culture suggests that it is actually a strain of *F. tricinctum*.

**Our examination of this strain clearly indicates that it is not *F. nivale*, but most probably an unusual variant of *F. tricinctum*. Ellis [62] indicates similar possibilities.

References p. 224

mycotoxicoses have utilized more complex culture media. These often have included extracts or preparations from original host plants, feed sources, or seeds, with growing conditions similar to those encountered in nature. Thus, by using a sterile corn seed medium with low incubation temperatures Ikediobi et al. [68] obtained yields of 1515 mg T-2 toxin per kg of dried corn when F. tricinctum was incubated at 8° for 30 days. Bamburg [69] achieved 900 mg T-2 toxin per kg of dried corn in F. tricinctum cultures incubated for 2 weeks at 8°. Similar yields were obtained by Burmeister [70] for T-2 toxins produced on moist, white corn grits. Unfortunately, no studies have been carried out to compare the ability of various fungi to produce trichothecene toxins using the same culture media, conditions of aeration and temperature.

Temperature apparently plays a critical role in the production of some toxic trichothecenes. T-2 toxin production is favoured by long-duration low-temperature incubation, but the level of production varies with the medium. In most studies highest yields were obtained at temperatures of 15° or below [68,70—72]. The highest levels were apparently produced at 8° (1515 mg/kg dry wt. of corn culture). The diacetoxyscirpenol-producing isolate of F. tricinctum also had low temperature optima for toxin production (as indicated by bioassay) [71], with peak production at 8° after 4 weeks on Gregory's liquid medium. Higher temperatures were apparently not detrimental to diacetoxyscirpenol production, since most of the original work [61] is based on production on liquid culture media at 24°. T-2 toxin production, however, declines at temperatures above 20° [70]. Isolates which produce T-2 toxin at low temperatures produce another trichothecene (HT-2 toxin) at higher temperatures [69].

In contrast the highest levels (based on bioassay) of fusarenon-X are produced by strain Fn 2B of F. nivale at 27°, and for production purposes the fungus was routinely grown at 25—27° [73,74]. Most of the other trichothecene-producing fungi have been grown for toxin production at 24° (or laboratory temperature) and critical studies on the effects of temperature are lacking.

(C) Chemical synthesis

Most of the extensive chemical studies on the trichothecenes have been on the proof of chemical structures, study of rearrangements and interconversions and the preparation of various derivatives. The most recent characterization studies have been those regarding roridin E [17] and roridin H [18]. Schumacher et al. [75] have also prepared from trichodermol, verrucarol and trichothecolone the isomeric 2β, 13-oxido-Δ^9-apotrichothecenes (Fig. 2), which possess the skeletal ring structure originally, but erroneously, proposed for the trichothecenes themselves. The biological effects of these synthetic products were not reported.

The first complete chemical synthesis of a trichothecene was accomplished by the preparation of ± trichodermin [76]. The difficult sixteen step

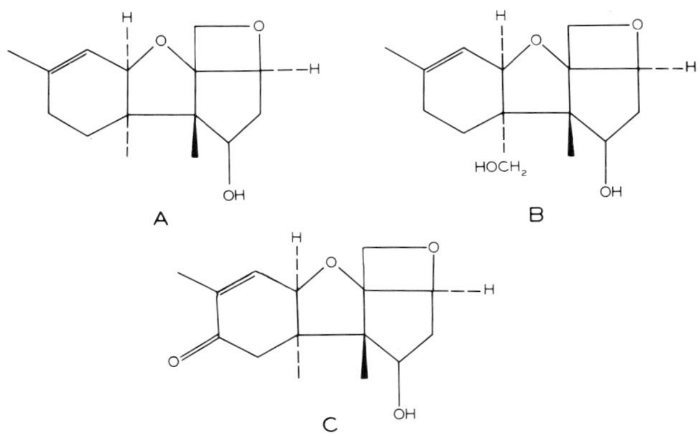

Fig. 2. Isomeric 2B,13-oxido-Δ^9-apotrichothecenes prepared from trichodermol (A), verrucarol (B) and trichothecolone (C).

procedure resulted in a crystalline product identical with the natural compound in all respects except optical rotation. This elegant synthesis provides confirmation of the structure but does not provide a practical source of material. An alternative synthesis of one of the key intermediates ("lactone 2") has also been worked out [77].

(D) Biosynthesis

It has been well established that the trichothecene skeleton is formed from three molecules of mevalonate *via* the usual pathway of lipid biosynthesis involving isopentenyl-, geranyl-, and farnesyl-pyrophosphates [78–80]. Little agreement has been reached, however, on the detailed mechanism by which the open-chain farnesyl skeleton

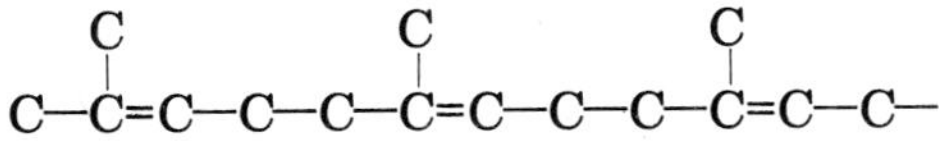

$$C \atop | \\ C-C=C-C-C-C=C-C-C-C=C-C-$$

is cyclized. There have been several suggestions that a β- or γ-bisabolene-type intermediate [81] may be involved. However, this possibility has been disproved since the (4R)-[4-^{3}H]mevalonoid hydrogen of the centre unit in farnesyl-pyrophosphate is retained in trichothecin and trichodermol [82,83]. The appropriate carbon atom in the proposed bisabolene intermediate carries no hydrogen atom.

Another compound which has been suggested [78,84] as a probable intermediate is the hydrocarbon trichodiene (Fig. 3). In a very significant development, Nozoe and Machida [85] have recently isolated this particular substance as a naturally occurring metabolite in cultures of *Trichothecium roseum*. They also obtained from the same cultures a closely related epoxy-alcohol, trichodiol [86] (Fig. 3). Machida and Nozoe have prepared tritiated trichodiene [87] and shown that, when it was fed to *T. roseum* cultures, the tritium label was incorporated into trichodiol as well as into scirpene and trichothecolone. The same authors also fed [4(R)-4-^{3}H, 2-^{14}C]mevalonate to *T. roseum*, as Achilladelis *et al.* [88] had done previously, and found it be incorporated into both trichodiol and trichothecolone [13]. Suitable degra-

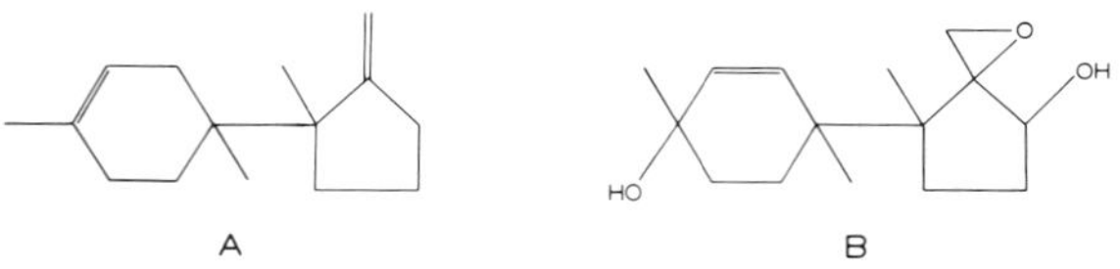

Fig. 3. Probable intermediates in trichothecene biosynthesis recently isolated by Nozoe and Machida [85] from pure cultures of *Trichothecium roseum*. Trichodiene, $C_{15}H_{24}$ (A) and trichodiol, $C_{15}H_{24}O_3$ (B).

dation procedures permitted them to conclude that C-8 and C-10 of the trichothecolone molecule are derived from carbons 2 and 4, respectively, of mevalonate, contrary to the early results of Jones and Lowe [78], but in agreement with Achilladelis *et al.* [88].

Other recent biosynthetic studies have shown that the 4-hydroxyl group of trichodermol [89] and of verrucarol [90] replaces a "pro-2(R)" mevalonoid hydrogen, which implies that hydroxylation proceeds with over-all retention of configuration at C-4, that the starter unit methyl groups of farnesyl pyrophosphate retain their individuality in the final trichothecene [89], and that a ^3H label from the terminal mevalonate unit probably migrates from C-8 to C-7. It has also been established that trichodermol may serve as a precursor of trichothecin, perhaps *via* crotocin or a similar epoxide [89].

Thus it is clear that the mechanism by which the trichothecenes are generated *in vivo* from farnesyl pyrophosphate must involve a number of intermediate substances which probably include at least trichodiene and very possibly trichodiol as well. Several additional details regarding sites of incorporation of labelled atoms from mevalonate precursor molecules have been established but the available information has not as yet been integrated and formulated into an over-all biogenetic pathway. It is not yet clear, for example, whether oxygen substitution on the scirpene nucleus is usually accomplished before or after the skeletal ring system is formed, although the identification of scirpene itself as a natural metabolite [13] obviously shows that the former can occur. Oxy-substitution at position 8 probably is subsequent to that at 4 inasmuch as trichodermin can serve as a precursor for trichothecin. No information is available as to the mechanism by which oxygen-containing functional groups are introduced into positions 3, 7, or 15.

The question as to whether some of the "naturally occurring" trichothecenes listed in Table I are to be regarded as **artefacts** of isolation resulting from hydrolysis of ester groups during the isolation process is perhaps rather pointless since the parent alcohol probably is generated first during biosynthesis and subsequently esterified. Therefore both the esters and alcohols might well be regarded as natural products.

Physical and chemical properties

(A) Spectra

Of the trichothecene group only the roridens and verrucarins possess a conjugated diene system which results in strong ultraviolet absorption. The most prominent UV band appears at about 258 to 262 nm and has a molecular extinction value of about 40 000. The tetracyclic trichothecenes such as T-2 toxin and trichothecolone have no UV absorption and do not show fluorescence. However, following treatment with 20% H_2SO_4 on thin-layer chromatography (TLC) plates T-2 toxin, diacetoxyscirpenol, neosolaniol, HT-2 toxin, nivalenol, fusarenon X and diacetylnivalenol fluoresence a sky-blue under long-wave UV light [63].

As previously emphasized the nuclear magnetic resonance (NMR) spectrum is characteristic of the entire group particularly with reference to the epoxide proton signals appearing in the region of δ 2.9. All of the compounds give characteristic mass spectra but the individual members of the group give spectra which result from different fragmentation patterns and therefore are more useful for characterization of individual trichothecenes than for detection or analysis of the group as a whole.

(B) Chemical properties

The solubility chracteristics of the natural trichothecenes are those of neutral lipid-like materials in that they are soluble generally in organic solvents of moderate polarity although relatively insoluble in very non-polar solvents such as hexane. As might be expected, the parent alcohols, especially those having 3 to 5 free hydroxyl groups, show a much more polar character and accordingly are more soluble in polar-type solvents.

As a class the trichothecenes show quite remarkable stability to a variety of environmental conditions including moderate variations in temperature, exposure to light and air and to moderate variations in pH. Although hydrolized by strong alkalis and rearranged by strong acids they are stable in the pH range of most normal foods. The stability properties, therefore, are such that the trichothecenes are not destroyed under conditions to which foods or feeds are normally exposed either on storage or during cooking [2].

Biological properties

(A) General toxicity

The biological activity of the naturally occurring trichothecenes in many respects is quite similar. All of them are capable of inducing dermal reactions consisting of severe local irritation, inflammation, and desquamation when applied to skins of experimental animals [2,28,69,91,92]. These responses are relatively non-specific and develop on humans as well as on laboratory animals [2,12,69]. This response has been used semiquantitatively in purification studies [2,69] with T-2 toxin and in attempts to detect such components in naturally toxic foods and feeds [1,3].

With the exception of trichothecin, trichodermin and crotocin, most of the major naturally occurring trichothecenes have acute oral or intraperitoneal (i.p.) LD_{50} in young mice or rats below 10 mg/kg body weight [2]. Oral LD_{50} as low (0.6 mg/kg), however, have been reported in the case of T-2 toxin in non-ruminant calves, but these levels were non-lethal in older ruminant animals [93]. Similar levels (LD_{50} 0.5 to 0.75) were also reported [94] for verrucarin A and J, injected i.p. However, in different trials the intravenous (i.v.) LD_{50} for verrucarin A in mice has been reported to be 1 mg/kg [21]. In the case of several trichothecenes, however, the route of application does not apparently greatly affect the LD_{50} dose level. For example, fusarenon-X has an oral LD_{50} of 4.4 mg/kg body weight and an i.p. LD_{50} of 3.4 mg/kg. Kosuri *et al.* [95] found the oral LD_{50} for T-2 toxin was 3.8 mg/kg in young rats, while Yates *et al.* [96] found the intraperitoneal LD_{50} for T-2 toxin was 3.04 mg/kg in mice. Obviously, good comparative studies using standard animals and standard methods of application are lacking for these important toxic compounds.

Most of the trichothecenes studied are cytotoxic when tested against various plant and animal tissue culture systems, but few comparative studies are available. Perlman *et al.* [97] contrasted several of the naturally occurring trichothecenes including T-2 toxin, diacetoxyscirpenol, trichothecin, crotocin, trichodermal and some of their derivatives for cytotoxic effects against KB cells (human epidermal carcinoma). Their results indicated inhibition zones ranging from 22 or 30 mm for trichothecolone and trichodermol to 49 mm for T-2 toxin. Concentrations used were 1 mg/ml. However, in the case of T-2 toxin, measurable inhibition zones were seen at concentrations as low as 0.55 μg/ml. Their tests did not include such active compounds as fusarenon X, neosolaniol, verrucarin A, or roridin A. Their cytotoxicity, however, has been reported elsewhere [9,21,98,99]. Used in agar culture media of tobacco callus tissue cultures, T-2 toxin gave 50% growth inhibition (ED_{50}) at concentrations of $6 \cdot 10^{-3}$ μM [100,101]. The standard growth inhibitors actinomycin D or actidione (cycloheximide) in the same test gave 50% inhibition at approx. 10^{-1} μM. Helgeson *et al.* [100] concluded that T-2 toxin was the most active agent yet tested which inhibited cell division in tobacco callus tissue cultures. Fusarenon-X is also highly cytotoxic and inhibited growth of HeLa S-3 cells at 0.5 μg/ml [99]. Diacetoxyscirpenol inhibited the multiplication of KB cells at $7 \cdot 10^{-3}$ μg/ml (ED_{50}) [102].

It is interesting to note that crotocin, trichodermin and trichothecin (or their parent alcohols) which are only slightly cytotoxic [97,103,104] are also only moderately toxic when administered to test animals. Where LD_{50} values have been determined, these range between 100 and 1000 mg/kg. They are in fact 100- to 1000-fold less toxic then verrucarin A or T-2 toxin.

There appears to be good correlation between dermal toxicity, cytotoxicity and phytotoxicity to higher plants among the various trichothecenes. The original purification study with diacetoxyscirpenol in fact was accomplished on the basis of a phytotoxicity assay [53]. Similar effects have been reported for T-2 toxin and Marasas *et al.* [105], Burmeister *et al.* [106,107] and others [4] have suggested the use of these plant systems to detect low order trichothecene contamination in feeds and foods. As might be expected from the cytotoxicity information, the germinating seeds of higher plants are also extremely sensitive to the action of trichothecenes. Trichothecin has been reported to inhibit virus local lesion development in tobacco necrosis virus on inoculated beans (at dose levels between 20 and 100 mg/l), but it is suggested that these effects are mainly on host plant susceptibility to the virus rather than acting directly on the virus particles [108].

Although T-2 toxin and diacetoxyscirpenol are inactive as a contact insecticide against Drosophila [58,109], Kishaba et al. [110] found that culture filtrates of Myrothecium roridum inhibited feeding and resulted in the death of Mexican bean beetles, Epilachna varivestis, Mulsant. Using this as a bioassay, two compounds were obtained in pure form which are reported to be identical with verrucarin A and roridin A [2,58] although they have not been compared with these pure compounds. More recently, Cole and Robinson [58] reported on the larvicidal activities of diacetoxyscirpenol purified from Fusarium lateritium against 48-h first instar larvae of Lucilia sericata (LC$_{50}$ 7.5 mg/ml). In work with culture filtrates he demonstrated similar activity for culture extracts from trichothecene producing strains of M. verrucaria, M. roridum, Cephalosporium crotocinigenum and Trichothecium roseum, and suggested that these effects were due respectively to the larvicidal activity of verrucarins, roridins, crotocin and trichothecin. They found that pure trichothecin and trichothecolone had an LC$_{50}$ of about 100 mg/l against L. sericata larvae.

Concentrations of 50 mg of fusarenon-X per ml or higher inhibited all cell division of the protozoa Tetrahymena pyriformis in exponentially growing mass cultures [111]. Later work by Ueno et al. [112] indicated complete inhibition of synchronous division of Tetrahymena at 0.05 mg/ml for diacetoxyscirpenol, 5 mg/ml for fusarenon-X and 25 mg/ml for nivalenol. The brine shrimp (Artemia salina) has been reported to be very sensitive to the toxic metabolites of Stachybotrys atra [10]. Pure roridin E isolated from this fungus had a brine shrimp LD$_{50}$ of 130 μg/0.5 ml of test media.

A few of the trichothecenes have been studied primarily because of their antibiotic or cytotoxic properties, and, in fact, these may be considered antibiotics rather than toxins [2]. None of them have appreciable antibacterial activity [2] except verrucarin A which shows slight inhibitory effects against some gram negative bacteria at 50 μg/ml [21] and only a few have really high level antifungal activity [104].

Trichothecin has been studied intensively in the U.S.S.R. and elsewhere for its usefulness in controlling plant diseases [113—121]. To a lesser extent, trichodermin has also been evaluated against plant diseases [122]. Trichothecin apparently has systemic fungicidal activity in plants and has been used experimentally to control fusarium and verticillium wilts of cotton [115, 119], fusarium wilt of pea and chickpea [118], fusarium wilt of flax [116], fusariosis of maize [120], wilt smut of wheat [121], and downy mildew of tobacco [117]. In a recent study, Burmeister [107] found that T-2 toxin was not bacteriostatic (50 μg/assay disc) to a broad range of bacteria, but it was slightly toxic to 6 of 11 fungi tested. This activity was most apparent against yeast, although Penicillium digitatum was also actively inhibited. Filter paper discs with 4 μg or more of T-2 toxin caused were fungistatic against Rhodotorula rubra, the most responsive organism studied. The other trichothecenes with antifungal properties (verrucarin A and crotocin) also appear to have the greatest activity against yeasts although other fungi are also affected [21,104,123]. Although neither nivalenol nor fusarenon-X was shown to be active against bacteria, fungi or prophage, in more complex tests Ueno et al. [4] demonstrated mutagenic effects of fusarenon-X to RD-mutant cells of the yeast, *Saccharomyces cerevisiae*.

(B) Toxicology

Only limited experimental comparisons have been made between the toxicological or histopathological effects of various toxic trichothecenes in animals. The available results suggest that the nature of the response to the various toxic trichothecenes is similar, although the severity varies with the toxin and the species of animal [2,4,9,41,53,95,124,125]. Although the skin necrotizing effects of these compounds have been emphasized, this is not the usual route of exposure and a more realistic appraisal of their toxicity has come from oral dosage. In general, animals administered these toxins orally become listless or inactive, often with the hairs of their coats erect. Respiration rates increase and in later stages diarrhoea develops which may be bloody. In later stages at lethal doses body temperatures may fall and respiration becomes shallow and difficult. At sublethal doses in chronic feeding studies, severe sores or necrotic lesions often develop in the mouth which make eating difficult [126—128].

Internally, erosion of the mucosal epithelium of the stomach and small intestines commonly takes place resulting in severe haemorrhagic gastroenteritis [4,95]. When fed to rainbow trout, T-2 toxin caused extensive sloughing of the intestinal mucosa which was shed into the water [129]. Such symptoms, however, were not reported in fusarenon-X feeding trials to mosquito fish (*Oryzas latipes*) [4]. At lethal dose levels animals usually die within 12 to 72 h after toxin administration [4,12,95].

Various investigations have suggested that the character of the trichothecene-induced lesions in actively dividing cells is very similar to those induced by radiation [4,41,125,130,131]. Pathologically, cell degeneration is observed in the bone marrow, lymph nodes, intestines, and other organs. Postmortem examinations frequently show engorgement of capillaries around the body wall and intestinal tract with accompanying ecchymotic haemorrhages in the peritoneal surfaces of the small and large intestines. Often lungs are congested and fatty degeneration is present in the liver [4,12,41,95]. Frequently in larger animals massive haemorrhages are seen in the lumen of the small and large intestines.

In addition to generalized toxicoses, emesis is induced by several of the more toxic trichothecenes in certain animals [4,5]. In Japan, Ueno et al. [4] have observed that fusarenon-X induced vomiting in cats and 10-day-old Pekin ducks at doses of 5 mg/kg administered subcutaneously (s.c.) (orally in ducks) and that the s.c. minimum dose for vomiting was 0.4 to 0.5 mg/kg. Crude toxins from cultures of *Fusarium tricinctum* (producing T-2 toxin) and *Fusarium solani* (producing T-2 toxin and neosolaniol) also induced vomiting [4]. Ellison and Kotsonis [5] obtained similar results when pure T-2 toxin was administered intravenously or orally to pigeons. They found that the oral ED_{50} for emesis in pigeons was 1 to 2 mg/kg. I.v. emesis was obtained at 0.1 mg/kg or less. Ueno *et al.* [4] and Ellison and Kotsonis [5] concluded that emesis associated with the ingestion of scabby grains in early reports resulted from intake of toxic trichothecenes.

(C) Metabolism and excretion

The mechanism of action of trichothecenes is not c ear nor has the route of metabolism or the method of excretion been explained. In the most extensive study on this subject to date Ueno *et al.* [4] found that radioactivity of ^{3}H-labelled fusarenon-X administered s.c. to mice (3.6 · 10^5 cpm/animal) after 24 h was present in the urine (89 000 cpm) and faeces (3490). After 30 min the maximum radioactivity (9000 cpm) was found in the liver with smaller amounts in the kidney (3500), small intestines (2300), large intestines (2300), and the faeces (2980). After 3 h radioactivity was only present in the kidneys (1200 cpm), the spleen (50), the bile (100), with most of the activity in the urine (16 300) and faeces (3540). After 12 h no radioactivity was present in any of the organs. No radioactivity was ever found in the heart, brain, or testes. From this data he concluded that the toxin was eliminated rapidly through the kidneys, although there was an early unexplained concentration in the liver. The breakdown products in the urine (which were shown not to be fusarenon-X) were not identified, but Ueno and associates suggested that fusarenon-X was metabolized to more polar compounds.

All of the studies so far suggest that the toxic trichothecenes act to prevent cell division. These effects have been shown in tissue cultures of both plant and animal cells [4,9,100,101] as well as in whole animals [4] and in germinating plant seeds [4,67,105,107]. Helgeson *et al.* [100] found that at a concentration of 5 · 10^{-8} M T-2 toxin completely prevented tissue growth of tobacco callus tissue culture. Increased toxin concentrations progressively decreased log growth rates. Furthermore, when tissues were removed from the toxin medium and transferred to fresh medium lacking toxin, they began to grow again at normal (control) rates. His studies also demonstrated that carbohydrate metabolism is not severely disrupted by the action of this toxin. Biochemically, trichothecenes inhibit DNA and protein synthesis in a variety of systems [4,9] (HeLa cells, Ehrlich ascites tumours, mouse tissue, rabbit reticulocytes, and protozoa).

Analytical methods

(A) Biological methods

Although the number of toxic trichothecenes known to be produced in culture by fungi associated with cases of apparent food- or feed-related intoxications becomes larger each year, their confirmed presence in these foods or feeds has hardly been demonstrated at all [1]. The lack of characteristic absorptions in the UV, visible or infrared regions and of PMR or mass spectral peaks suitable for routine analytical exploration has so far prevented the development of spectral methods for determining trichothecenes in food or feed samples. Several methods, however, have been proposed for their detection in cultures [1,2,68,106,107]. Similarly, no colorimetric chemical methods are known or seem likely to be developed because of the lack of characteristic colour reactions that could be made the basis of specific qualitative or quantitative measurements [2].

Qualitative decisions as to whether or not mouldy feed or food samples contain appreciable amounts of toxic trichothecenes can at the present time best be made on the basis of the biological screening test. Once this has been accomplished for a given sample, further qualitative identification of the individual compound(s) present and their concentrations in the sample will require a rather extensive preliminary cleanup procedure followed by analysis of the purified material by specialized techniques as described.

Although trichothecenes produce many biological effects with varying degrees of specificity, few of these appear at low trichothecene concentrations. Gross symptomatology of reacting animals in the field can be indicative of trichothecene intoxication, but used alone can also be misleading. The haemorrhagic syndrome, an obvious last stage of toxicosis, is not specific for trichothecenes [132]. Early symptomatology is usually diffuse and can resemble many other problems. Veterinarians attempting to treat animals having early stage symptoms by injections with antibiotics or vitamins, usually fail to obtain dramatic responses. Such negative therapeutic effects, however, may in themselves be somewhat diagnostic for trichothecene toxicosis. The specific pathology in the organs of reacting animals as described in Section A (see p. 211) may be a somewhat better indicator of trichothecene action, but here again it in itself is not confirmatory.

A usual first step in any biological analysis of toxic feed samples is a qualitative and/or quantitative determination of the presence of known trichothecene-producing fungi in the sample. Such methods have been described in detail by Marasas [67,133], Hsu et al. [1], and others. The detection of living trichothecene-producing fungi in the suspect sample only indicates that these compounds may be present in the sample. In the late stages of mould development, however, non-toxin-producing fungi may dominate and thus make the detection of trichothecene-producing fungi difficult or impossible. The non-detection of trichothecene-producing species may likewise only indicate that the original toxin-producing fungi have died.

Biological analysis of extracts from toxic feeds may also be useful in determining the general nature of the problem, but do not identify the toxin. A one-step extraction of the sample to concentrate toxic ingredients prior to bioassay is usually necessary and ethyl ether or ethyl acetate have proven to be the most useful solvents for toxic trichothecenes [3]. The general levels of toxicity of the concentrates can be determined by oral or parenteral administration to laboratory animals. Rats, mice, and guinea pigs have been commonly used for this purpose, but recently brine shrimps [10], insect larvae [58] and chick embryos [4] have been shown to be extremely sensitive to toxic trichothecenes. For more specific information, tests can be devised utilizing the skin necrotizing, cytotoxic, protein synthesis inhibition, seed germination inhibition, or the emetic responses of these compounds. Ueno *et al.* [63], suggest a combination biological test using the rabbit reticulocyte [4] assay combined with a TLC system in which plates are examined under UV light after being sprayed with 20% H_2SO_4 and heated at 100° for 10 to 20 min. All trichothecenes tested in this system exhibited a sky-blue fluorescence. Although this system has obvious uses in testing unknown pure cultures for trichothecene production, it has not been tested on naturally contaminated foods or feeds. Hsu *et al.* [1] discusses the extreme difficulty in analysis of trichothecenes from contaminated natural products.

(B) Physical methods

Two techniques have so far proven to be most useful in the chemical analysis of trichothecenes, *viz.*, TLC and gas-liquid chromatography (GLC). To date TLC has been used primarily for research purposes such as following the progress of isolation procedures and checking individual trichothecene samples for purity. Spots can be located on the TLC plates by spraying with H_2SO_4 and charring or by exposure to iodine vapours — both very general methods for visualization of organic substances. Spots due to verrucarins and roridins also can be seen under UV light, and the other (tetracyclic) trichothecenes by quenching of a fluorescence indicator. Rhodanine [110] and 3-hydroxypyrene-5,8,10-trisulphonic acid sodium salt [18] have been used for this purpose. Solvent systems, adsorbents, and the corresponding R_F values for individual trichothecenes have been tabulated [2].

Since TLC has only moderate resolving power, qualitative identifications based on R_F values alone, even when measured with several different solvent systems, are hazardous and frequently unreliable. GLC is far more superior in this respect and furthermore sensitive to smaller amounts of material and capable of yielding quantitative results of much higher precision. A GLC method was successfully applied to 13 of the simpler trichothecenes, using the trimethylsilyl (TMS) derivatives of those having free hydroxyl groups [68]. Closely reproduceable methylene unit (MU) values, established for each compound studied, should be useful for qualitative identification purposes. A combination of preliminary clean-up, GLC, and mass spectroscopy was used to prove the presence of T-2 toxin in a sample of mouldy corn associated with poisoning of dairy cattle in Wisconsin [1].

As far as we are aware none of the roridins or verrucarins have been successfully chromatographed by GLC. These important, highly toxic members of the trichothecene group would, therefore, probably not be detected by this method. Furthermore, the multiplicity of closely related naturally occurring trichothecenes obviously complicates the problem of detecting the presence and quantity of each individual member in mouldy feeds and foods. For these reasons it has appeared desirable to simplify the analytical problem by alkaline treatment of the partially purified sample extracts in order to convert the naturally occurring trichothecene esters to their corresponding parent alcohols. As noted above, this reduces the number of possible individual substances to be considered (as of this writing) from 27 to 10. A start on this approach has been made in our laboratories [134] and has so far been extended to eight of the alcohols concerned [135].

The sample (mouldy corn, silage, fungal cultures, *etc.*) was dried, ground, and twice extracted with ethyl acetate (2 × 250 ml/100 g sample). The combined extract was shaken out with one-half volume of water, the organic layer evaporated, the residue dissolved in 80 ml methanol plus 20 ml water, the solution shaken out twice with 80 ml portions of Skelly-solve B (or hexane), and the aqueous-methanol layer diluted with 60 ml of water. This solution was extracted twice with 160 ml portions of chloroform:ethyl acetate (1:1, v/v) and the organic layer evaporated to give an oily residue amounting usually to 0.1–1 g/kg of dry sample.

A portion, usually 100 mg (or an amount estimated to contain at least 0.2 mg of a trichothecene), was dissolved in 5 ml of 0.3 N sodium hydroxide in ethanol:water (90:10, v/v) and allowed to stand 12–24 h at room temperature. The mixture was then adjusted to pH 8–9 (Hydrion paper) with 0.2 N ethanolic HCl, evaporated to dryness at 40° and the residue taken up in tetrahydrofuran:benzene (1:3, v/v). After filtering through a sintered glass filter into a small vial and removing the solvent in a stream of nitrogen, the remaining material was silylated and chromatographed as previously described [68].

If, as was frequently the case, the saponified material obtained as above (usually *ca.* 5–20 mg) contained too many interfering substances for successful GLC analysis, it was further purified by TLC on a 5 × 20 cm preparative plate. The sample (*ca.* 20 mg) was applied as a solution in 0.2–0.3 ml of tetrahydrofuran:benzene (1:3, v/v) and the plate developed three successive times with tetrahydrofuran:methanol (15:85, v/v) in which the R_F values of trichothecene alcohols with two or less oxy-substituents (including trichothecolone and crotocol) are in the range 0.05–0.30. Adsorbent from the appropriate areas of the plate was then removed, eluted with tetrahydrofuran:methanol (1:1, v/v) and the solvent evaporated to leave a residue which was suitable for silylation and GLC analysis.

Trichothecene alcohols with more than two oxy-substituents do not move from the origin when subjected to TLC as above described [134]. The preparative plate as left from the above procedure was therefore further developed with ethyl acetate:acetone:absolute ethanol (4:4:1, v/v) (in which the R_F values of the higher alcohols so far investigated are about 0.4–0.5), the appropriate areas of the plate eluted, and the eluate evaporated and silylated as before. With samples prepared in this manner, little difficulty has been encountered in separating and identifying most of the trichothecene alcohols so far studied.

Suitable control experiments with known trichothecenes and their parent alcohols have shown that this procedure, although unfortunately rather lengthy, should be capable of detecting and quantitatively estimating trichothecenes in field samples at concentrations of *ca.* 0.2 ppm or higher. Further studies will, of course, be necessary to ascertain which of the natural esters may have been present in the original sample. This general approach contributed to the recent discovery that the toxic metabolites of *Stachybotrys atra* belong to the roridin group [10].

Practical importance

As the chemical nature, biology and toxicology of the trichothecenes has become better understood, it is apparent that these toxins may have been underrated as causal agents of serious food or feed related intoxication in man and animals. This results from the fact that toxic trichothecenes with the exception of T-2 toxin have not been identified chemically from toxic foods or feeds causing toxicoses. With this exception, the equivalent of Koch's postulates has not established a "proof of cause" for any of the other trichothecenes. Thus, these compounds are known only as fungal metabolites in the laboratory. Most of the more active toxin producers, however, have been isolated and identified from toxic foods or feeds, and frequently extracts from these toxic feeds have induced identical symptoms to those produced in experimental animals by pure toxins.

The lack of evidence for the occurrence of trichothecenes in toxic foods or feeds primarily results from the fact that no adequate practical method for their chemical detection has yet been devised. Even so, the widespread nature of the trichothecene-producing fungi in foods and feeds suggests that they may come to be recognized as the most important of the acutely toxic mycotoxins. Certainly their potential as inducers of mycotoxicoses appears to be much greater than that of the more widely investigated aflatoxins [136].

(A) Natural occurrence and pathogenicity of trichothecene-producing fungi

The fungi which produce toxic trichothecenes occur widely in nature both as saprophytes and as plant parasites. *Myrothecium roridum* which produces the toxic roridins and verrucarins (Table I) is parasitic on a wide variety of host plants. It has been shown to attack gardenia [137], tomato fruits [138], cotton [139], pansy [140,141], gloxinia [142], snapdragon [140], cantaloupe [143], cocoa [144], coffee [145,146], and red clover [147]. Diseases caused by Myrothecium result in characteristic target-shaped stem or leaf lesions which often result in blighted foliage under conditions of high humidity [146]. *M. verrucaria*, a saprophytic species, is widespread in nature and is very commonly found on dead and dying plant leaves [11]. In New Zealand *Myrothecium* spp. are reported to be a normal part of the mycoflora of pasture plant leaves [12]. Mortimer *et al.* [12] suggest that the occurrence of seasonal unthriftiness in sheep and cattle on apparently nutritious dry autumn pastures may in part be due to intake of Myrothecium toxins. Vertinskii *et al.* [56] report *M. verrucaria* intoxication in sheep eating contaminated rye straw. The potential for animal intoxication by trichothecenes produced by *Myrothecium* spp. seems very large because of the many host plants involved which might easily find their way into animal or human diets. The parasitic activities of *M. roridin* on coffee berries are of great interest as a possible mode of trichothecene contamination of human food [145,146].

Many of the other trichothecene-producing fungi are also plant parasites. Several important plant diseases are induced by species of Cephalosporium (Acremonium) including brown spot of celery (*C. apii*) [148], balsam fir canker (*C. album*) [149], persimmon wilt (*C. diospyri*) [148], brown stem rot of soybean (*C. gregatum*) [150], sugar cane wilt (*C. sacchari*) [151] and Cephalosporium leaf wilt of Barbasco (*C. deformans*) [152]. A few other species are also pathogenic to man and include *C. infestans, C. kileense, C. modurae, C. recifei* and *C. spinosum* [153]. Apparently none of the parasitic Cephalosporium species has been examined for production of trichothecenes.

Trichoderma viride and *Trichothecium roseum* both have limited plant pathogenic capabilities. *T. viride* has been reported to be the causal agent of stored product rots in both garlic and lemons, but it apparently does not commonly attack other stored fruits and vegetables [50]. *T. roseum* is occasionally parasitic on apple fruits, and it also causes a pink rot of stored celery [50].

Many of the trichothecene producing species of Fusarium are well known plant pathogens, although, in general, their parasitic abilities have not generally been related directly to production of phytotoxic trichothecenes [38]. Bilai [8,37] and Seemuller [154] have both discussed in detail the phytopathological and taxonomic significance of the Fusarium representatives of the section Sporotrichiella, which are recognized as the major trichothecene producing fungi (particularly *F. tricinctum*). Seemuller [154] in fact uses pathogenicity on conifers, wheat, maize, carnations, peas and lupines as a major character to differentiate 5 species in this section (*vs.* a single species (*F. tricinctum*) in the Snyder and Hanson system). Bilai [8] observed that *F. sporotrichiella* (= *F. tricinctum*) is widely distributed in the world and its varieties are found on a variety of cultivated plants (wheat, oats, forage grasses, sugar beets, stone fruits, citrus fruits, maize, soybean, peas, lupine, tomato and carnation). Although definitive studies are not available the phytotoxic trichothecenes appear to play a major role in the process of pathogenesis in the various diseases.

(B) Role of trichothecene-producing fungi in mycotoxicoses

(1) T-2 toxicoses

Toxicological problems in farm animals associated with consumption of mouldy wheat, oats and particularly corn have long been recognized and cases are well documented [2,3,7,155,156]. The causes of these problems are complex and probably depend on the environmental conditions prevailing during the development of the mouldy corn or grains. Mouldy corn toxicoses have been particularly important in the North Central United States in recent years, and major outbreaks have occurred in association with the 1962, 1964, 1965, 1970, and 1972 corn crops [1,3]. Such problems have been associated with corn late to mature or high in moisture at the time of the first killing frost. When stored in conventional cribs without artificial drying, this corn may be quite mouldy by early spring. Symptoms in animals following ingestion of mouldy corn are frequently quite varied and range from feed refusal to general digestive disorders accompanied by bloody diarrhoea and lack of weight gain or death accompanied by haemorrhagic lesions in the stomach, heart, intestines, lungs, bladder and kidneys. These latter symptoms have been called the haemorrhagic syndrome [156]. This general problem has been illustrated and discussed in detail by Smalley *et al.* [3].

The fungal flora and resulting toxic metabolites developing in mouldy corn are closely linked to prevailing environmental conditions. In warm areas, the development of *Aspergillus flavus* and the production of aflatoxins in mouldy corn can be quite common [155,157]. More recently, *Fusarium moniliforme* producing unspecified toxins has been associated with equine leucoencephalomalacia from mouldy corn in Egypt and South Africa [158– 160]. Other toxic strains of *F. moniliforme* have been isolated from Southern leaf blight damaged corn seed in Georgia [161]. Symptoms induced by South African strains of the fungus included subcutaneous oedema and icterus. The gross pathological lesions consisted of severe cardiac haemorrhages, petechiae and ecchymoses of major organs, oedema, icterus and liver damage. These authors failed to reproduce leucoencephalomalacia as obtained by Wilson in older equines [158]. In cooler north temperate regions, mouldy corn is less likely to be contaminated with *Aspergillus flavus*. Instead, fungi are present which develop at low temperatures. Toxin-producing strains of *Trichothecium roseum*, *Nigrospora*, *Epicoccum nigrum*, *Alternaria tenuis*, *Fusarium roseum*, *Fusarium moniliforme*, *Fusarium tricinctum*, and various Penicillia are commonly isolated [3].

Clearly, *Fusarium tricinctum* is consistently the most toxic fungus isolated from mouldy corn in low temperature storage and its presence in mouldy corn samples is correlated with farm outbreaks of mouldy corn poisoning [1,3]. In fact, experiments aimed at duplicating the field conditions leading to development of toxic corn have usually been unsuccessful in the absence of *F. tricinctum* contamination.

In an attempt to establish a causal relationship between northern mouldy corn toxicosis and the toxic trichothecene metabolites of *F. tricinctum*, a typical case was studied in detail in Wisconsin. Over a 5-month period in the winter of 1970—71, 20% of a herd of lactating Holstein cows died after prolonged ingestion of ground mouldy corn [1]. Postmortem examinations of affected animals revealed extensive internal haemorrhages typical of those observed in earlier cases of mouldy corn poisoning and experimentally produced by administration of T-2 toxin. The major fungi present were also typical of those isolated in previous cases of mouldy corn poisoning. Dilution plate counts made from the mouldy corn samples indicated a population of $2 \cdot 10^5$ propagules of *F. tricinctum*/g of mouldy corn. Ethyl acetate extracts of this corn gave a very toxic response in the rat dermal assay and 1 ml of this residue intubated into 60-g rats resulted in death in 3 h.

The toxin from this mouldy corn sample was purified following procedures described by Bamburg [69] and Ikediobi [68]. These techniques indicated the presence of T-2 toxin, but positive identification was not possible because of interfering products. However, with additional purification, gas chromatographic, mass spectral and biological properties of the highly purified material confirmed its identity as T-2 toxin [1]. The approximate concentration of T-2 toxin in the original feed was estimated from peak areas of GLC tracings to be 2 mg/kg of mouldy corn, although this figure was probably low because of unavoidable losses during the purification procedures.

Diagnosis of T-2 toxicoses by chemical means is still a difficult problem and procedures have not been perfected to the stage where routine feed sample analysis such as can be done with aflatoxins is possible. Clearly, analysis of contaminated feeds presents many more difficulties than the procedures based on analysis of fewer *F. tricinctum* corn cultures. GLC techniques for determining final concentrations can no doubt be improved to make such procedures for routine analysis a possibility.

(2) Feed rejection and emesis

Rejection and emesis have frequently been reported in association with mouldy cereal grains and particularly in association with *F. roseum* (*Gibberella zeae* ear rots) [162,163]. Although causal relationships between this fungus and toxic mouldy cereal grains has not been clearly established, *G. zeae* is the major fungus present in such mouldy corn.

Curtin and Tuite [162] have suggested that feed refusal factors in the corn are possibly associated with smells or tastes and differ from those factors causing emesis or uterine hypertrophy. Recent outbreaks of this problem in the Midwestern United States in the fall of 1972 has stimulated new interest in this problem, which is particularly important since it frequently is a forerunner to later development of the haemorrhagic syndrome associated with T-2 toxicosis.

Early investigations concerning the emesis factor have been reviewed by Prentice and Dickson [163]. More recently the findings that trichothecenes, particularly T-2 toxin and fusarenon-X, induce emesis in reactive animals suggest their possible role in this problem [4,5]. Ueno *et al.* [4], in fact, conclude that earlier reports of emesis resulting from ingestion of scabby wheat, barley and mouldy corn actually resulted from the intake of toxic trichothecenes. These findings need to be confirmed by positive detection of trichothecenes in emetic cereals or corn.

(3) Alimentary toxic aleukia

Various investigators have reviewed early Russian studies concerning the serious problem of human intoxication associated primarily with ingestion of mouldy cereals contaminated with the fungus *F. tricinctum* (synonymous with *F. poae*, *F. sporotrichioides* and *F. sporotrichiella*). The symptoms induced in experimental animals in the Russian studies suggest similarities to controlled trichothecene intoxications induced with chemically defined products [8,40]. The toxins identified by Russian scientists as being probably responsible for the disease, have been called poaefusarin and sporofusarin. Chemical structure of these steroidal toxins has been described by Olifson [164,165]. Recently, Mirocha and Pathre [7] have chemically analyzed an authentic sample of poaefusarin received from Soviet scientists. They could not confirm the presence of a steroid structure in the product, but found that the sample contained 2.5% of T-2 toxin. In addition, they found that both T-2 tetraol and zearalenone (F-2) were present in the sample. Toxicity tests with this product suggested that the dermal necrosis induced by skin application was due to the presence of the T-2 toxin. These findings tend to confirm conclusions proposed by various investigators that alimentary toxic aleukia is a trichothecene induced problem and more specifically identical to T-2 toxicosis [2—4].

(4) Stachybotryotoxicosis and dendrodochiotoxicosis

Symptoms reported by Russian scientists in animals ingesting mouldy forage contaminated with *Stachybotrys atra* resemble those of mouldy corn toxicoses and suggest a possible trichothecene origin [156]. This problem is described in detail in Chapter 9 of this text [10]. These authors' discovery of the presence of roridin E and possibly other 12,13-epoxytrichothecenes in extracts from pure cultures of toxic *S. atra* isolates suggests the possible major role of trichothecenes in Stachybotrys toxicosis as previously proposed by Bamburg and Strong [2].

Still another type of animal intoxication reported in Russian mycotoxin literature also resembles trichothecene intoxication in its animal symptomatology [166]. This disease called dendrodochiotoxicosis mainly results from ingestion of *Dendrodochium toxicum* infested forage by horses, sheep and pigs. Tullock's [11] recent finding that the fungus *D. toxicum* is synonymous with the fungus *Myrothecium roridum* would strongly suggest that the roridin and verrucarin trichothecenes are the actual toxicants in these mycotoxicoses. Recently Mortimer *et al.* [12] have described experimentally induced disease and death in rabbits, sheep, and clover resulting from administration of verrucarin A and roridin A isolated from 18 strains of *M. roridin* and *M. verrucaria*. They also suggest the possible relationship to dendrodochiotoxicosis.

It would thus seem reasonable to assume that both stachybotryotoxicoses and dendrodochiotoxicosis can be ascribed to the activities of species of Myrothecium and most probably due to the verrucarin and roridin mycotoxins.

References

1 I.C. Hsu, E.B. Smalley, F.M. Strong and W.E. Ribelin, Appl. Microbiol., 24 (1972) 684.
2 J.R. Bamburg and F.M. Strong, 12,13-Epoxytrichothecenes, in S. Kadis, A. Ciegler and S.J. Ajl (Eds.), Microbial Toxins, Vol. 7, Academic Press, New York, 1971, pp. 207—292.
3 E.B. Smalley, W.F.O. Marasas, F.M. Strong, J.R. Bamburg, R.E. Nichols and N.R. Kosuri, Mycotoxicoses associated with moldy corn, in M. Herzberg (Ed.), Toxic Microorganisms, Unnumbered Publ. U.S. Dept. of Interior and U.J.N.R. Panels on Toxic Microorganisms, Washington, D.C., 1970, pp. 163—173.
4 Y. Ueno, I. Ueno, Y. Iitoi, H. Tsunoda, M. Enomoto and K. Ohtsubo, Jap. J. Exptl. Med., 41 (1971) 521.
5 R.A. Ellison and F.N. Kotsonis, Appl. Microbiol., 26 (1973) 540.
6 Y. Ueno, Y. Ishikawa, M. Nakajima, K. Sakai, K. Ishii, H. Tsunoda, M. Saito, M. Enomoto, K. Ohtsubo and M. Umeda, Jap. J. Exptl. Med., 41 (1971) 257.
7 C.J. Mirocha and S. Pathre, Appl. Microbiol., 26 (1973) 719.
8 V.I. Bilai, Ann. Acad. Sci. Fenn. A. IV Biol., 168 (1970) 19.
9 Y. Ueno, K. Ishii, K. Sakai, S. Kanaeda, H. Tsunoda, T. Tanaka and M. Enomoto, Jap. J. Exptl. Med., 42 (1972) 187.
10 J.V. Rodricks and R.M. Eppley, Stachybotrys and Stachybotryotoxicosis, in I.F.H. Purchase (Ed.), Mycotoxins, Chapter 9, Elsevier, Amsterdam, 1974.
11 M. Tullock, C.M.I. Mycological Papers, No. 130, Commonwealth Mycological Institute, Kew, Surrey, 1972, 42 p.
12 P.H. Mortimer, J. Campbell, M.E. DiMenna and E.P. White, Res. Vet. Sci., 12 (1971) 508.
13 Y. Machida and S. Nozoe, Tetrahedron, 28 (1972) 5113.
14 D. Gardner, A.T. Glen and W.B. Turner, J. Chem. Soc. Perkin. Trans. I, 18 (1972) 2576.
15 Y. Ueno, I. Ueno, K. Amakai, Y. Ishikawa, H. Tsunoda, K. Okubo, M. Saito and M. Enomoto, Jap. J. Exptl. Med., 41 (1971) 507.
16 T. Tatsuno, Y. Morita, H. Tsunoda and M. Umeda, Chem. Pharm. Bull., 18 (1970) 1485.
17 P. Traxler, W. Zurcher and Ch. Tamm, Helv. Chim. Acta, 53 (1970) 2071.
18 P. Traxler and Ch. Tamm, Helv. Chim. Acta, 53 (1970) 1846.
19 G.C. Ainsworth, Dictionary of the Fungi, Commonwealth Mycological Institute, Kew, Surrey, 1961.
20 M.B. Ellis, Dematiaceous Hyphomycetes, Commonwealth Mycological Institute, Kew, Surrey, 1971, p. 522, 540.
21 E. Härri, W. Loeffler, H.P. Sigg, H. Stähelin, Ch. Stoll, Ch. Tamm and D. Wiesinger, Helv. Chim. Acta, 45 (1962) 839.
22 B. Bohner, E. Fetz, E. Härri, H.P. Sigg, Ch. Stoll and Ch. Tamm, Helv. Chim. Acta, 48 (1965) 1079.
23 W. Gams and K.H. Domsch, Pilze aus Agrarböden, Stuttgart, 1970, p. 88.
24 G.R. Bisby, Trans. Brit. Mycol. Soc., 28 (1945) 11.
25 C. Booth, C.M.I. Mycological Papers, No. 68, Commonwealth Mycological Institute, Kew, Surrey, 1957, 22 pp.
26 O. Verona and G. Mazzuchetti, I generi Stachybotrys e Memnoniella, Publ. Ente Naz. Cellul. Carta, 1968.
27 G.L. Barron, The Genera of Hyphomycetes from Soil, Williams and Wilkins, Baltimore, 1968, p. 286.
28 A.C. Keyl, J.C. Lewis, J.J. Yates, S.G. Yates and H.L. Tookey, Mycopathol. Mycol. Appl., 31 (1967) 327.
29 T.A. Toussoun and P.E. Nelson, Identification of Fusarium species, Pennsylvania State University Press, University Park, Pa., 1968, 51 pp.
30 W.C. Snyder and H.N. Hansen, Am. J. Bot., 27 (1940) 64.
31 W.C. Snyder and H.N. Hansen, Am. J. Bot., 28 (1941) 738.

32 W.C. Snyder and H.N. Hansen, Am. J. Bot., 32 (1945) 657.

33 W.C. Snyder and H.N. Hansen, Ann. N.Y. Acad. Sci., 60 (1954) 16.

34 A.I. Raillo, Fungi of the Genus Fusarium, Government Publishers of Agricultural Literature, Moscow, 1950, 415 pp.

35 W.L. Gordon, Can. J. Bot., 30 (1952) 209.

36 C.M. Messiaen and R. Cassini, Ann. Epiphyt., 19 (1968) 387.

37 V.I. Bilai, Ann. Acad. Sci. Fenn. A IV Biol., 168 (1970) 7.

38 C. Booth, The Genus Fusarium, Commonwealth Mycological Institute, Kew, Surrey, 1971, 237 pp.

39 C.J. Mirocha, C.M. Christensen and G.H. Nelson, F-2 (Zearalenone) estrogenic mycotoxin from Fusarium, in S. Kadis, A. Ciegler and S.J. Ajl (Eds.), Microbial Toxins, Vol. 7, Academic Press, New York, 1971, pp. 107—138.

40 A.Z. Joffe, Alimentary toxic aleukia, in S. Kadis, A. Ciegler and S.J. Ajl (Eds.), Microbial Toxins, Vol. 7, Academic Press, New York, 1971, pp. 139—206.

41 M. Saito and T. Tatsuno, Toxins of *Fusarium nivale*, in S. Kadis, A. Ciegler and S.J. Ajl (Eds.), Microbial Toxins, Vol. 7, Academic Press, New York, 1971, pp. 293—316.

42 J.A. von Arx, The Genera of Fungi Sporulating in Pure Culture, Cramer, Lehre, 1970, p. 177.

43 L.W. Durrell, Notes on Cephalosporium Species, Colorado State University, Fort Collins, 1963, 28 pp.

44 M.B. Schol-Schwarz, Trans. Brit. Mycol. Soc., 48 (1965) 51.

45 B. Achilladelis and J.R. Hanson, Phytochemistry, 8 (1969) 765.

46 J.M. Roberts, Mycologia, 44 (1952) 292.

47 M.P. Backus and P.A. Orpurt, Mycologia, 53 (1961) 64.

48 M.A. Rifai, C.M.I. Mycological Paper No. 116, Commonwealth Mycological Institute, Kew, Surrey, 1969, 56 pp.

49 E.B. Smalley, personal observation, 1973.

50 Anon., Index of Plant Diseases in the United States, U.S.D.A. Agricultural Handbook No. 165, 1960, 531 pp.

51 M.A. Rifai and R.C. Cooke, Trans, Brit. Mycol. Soc., 49 (1966) 147.

52 K. Tubaki, Nagao, 9 (1960) 29.

53 P.W. Brian, A.W. Dawkins, J.F. Grove, H.G. Hemming, D. Lowe and G.L.F. Norris, J. Exptl. Bot., 12 (1961) 1.

54 H.W. Wollenweber, Z. Parasit. Kde., 3 (1931) 269.

55 Ch. Tamm, Angew. Chem., 78 (1966) 496.

56 K.T. Vertinski, Kh.A. Dzhilavyan and U.P. Koroboa, Byul. Vses. Inst. Vet., 2 (1967) 86.

57 K. Ishii, K. Sakai, Y. Ueno, H. Tsunoda and M. Enomoto, Appl. Microbiol., 22 (1971) 718.

58 M. Cole and G.N. Robinson, Appl. Microbiol., 24 (1972) 660.

59 H.R. Burmeister, J.J. Ellis and C.W. Hesseltine, Appl. Microbiol., 23 (1972) 1165.

60 W.F.O. Marasas, personal communication, 1973.

61 P.W. Brian and H.G. Hemming, J. Gen. Microbiol., 1 (1947) 158.

62 J.J. Ellis, personal communication, 1973.

63 Y. Ueno, N. Sato, K. Ishii, K. Sakai, H. Tsunoda and M. Enomoto, Appl. Microbiol., 25 (1973) 699.

64 J. Gyimesi, Acta Chim. Acad. Sci. Hung., 45 (1965) 323.

65 G.G. Freeman, J. Gen. Microbiol., 12 (1955) 213.

66 J. Tuite, Plant Pathological Methods, Burgess, Minneapolis, 1969, p. 27.

67 W.F.O. Marasas, Mycoflora, Toxicity, and Nutrient Value of Mouldy Corn, Ph. D. Thesis, Univ. of Wisconsin, Madison, 1969.

68 C.O. Ikediobi, I.C. Hsu, J.R. Bamburg and F.M. Strong, Anal. Biochem., 43 (1971) 327.

69 J.R. Bamburg, Mycotoxins of the Trichothecane Family Produced by Cereal Molds, Ph. D. Thesis, Univ. of Wisconsin, Madison, 1969.

70 H.R. Burmeister, Appl. Microbiol., 21 (1971) 739.

71 M.W. Gilgan, Studies on the Chemistry and Metabolism of the Toxins from *Fusarium tricinctum* and *Lathyrus odoratus*, Ph. D. Thesis, Univ. of Wisconsin, Madison, 1965, pp. 2—23, 98.
72 J.R. Bamburg, N.V. Riggs and F.M. Strong, Tetrahedron, 24 (1968) 3329.
73 Y. Ueno, I. Ueno, T. Tatsuno, K. Ohokubo and H. Tsunoda, Experimentia, 25 (1969) 1062.
74 Y. Ueno, Y. Ishikawa, K. Saito-Amakai and H. Tsunoda, Chem. Pharm. Bull., 18 (1970) 304.
75 R. Schumacher, J. Gutzwiller and Ch. Tamm, Helv. Chim. Acta, 54 (1971) 2080.
76 E.W. Colvin, R.A. Raphael and J.S. Roberts, Chem. Commun., (1971) 858.
77 S.C. Welch and R.Y. Wong, Tetrahedron Letters, 19 (1972) 1853.
78 E.R.H. Jones and G. Lowe, J. Chem. Soc., (1960) 3959.
79 A.W. Dawkins, J. Chem. Soc., (1966) 116.
80 H.P. Sigg, R. Mauli, E. Flury and D. Hauser, Helv. Chim. Acta, 48 (1965) 962.
81 W.B. Turner, Fungal Metabolites, Academic Press, London, 1971, pp. 219—223.
82 P.M. Adams and J.R. Hanson, Chem. Commun., (1971) 1414.
83 J.M. Forrester and T. Money, Can. J. Chem., 50 (1972) 3310.
84 R. Bentley and I.M. Campbell, in M. Florkin and E.H. Stotz (Eds.), Comprehensive Biochemistry, Vol. 20, Elsevier, Amsterdam, 1968, p. 452.
85 S. Nozoe and Y. Machida, Tetrahedron Letters, (1970) 2671.
86 S. Nozoe and Y. Machida, Tetrahedron, 28 (1972) 5105.
87 Y. Machida and S. Nozoe, Tetrahedron Letters, 19 (1972) 1969.
88 B. Achilladelis, P.M. Adams and J.R. Hanson, Chem. Commun., (1970) 511.
89 P.M. Adams and J.R. Hanson, Chem. Commun., (1970) 1569.
90 R. Achini, B. Müller and Ch. Tamm, Chem. Commun., (1971) 404.
91 R. Wei, E.B. Smalley and F.M. Strong, Appl. Microbiol., 23 (1972) 1029.
92 Y. Ueno, Y. Ishikawa, K. Amakai, M. Nakajima, M. Saito, M. Enomoto and K. Ohtsubo, Jap. J. Exptl. Med., 40 (1970) 33.
93 V. Rac, personal communication, 1973.
94 A.M. Guarino, A.B. Mendillo and J.J. DeFeo, Biotechnol. Bioeng., 10 (1968) 457.
95 N.R. Kosuri, E.B. Smalley and R.E. Nichols, Am. J. Vet. Res., 32 (1971) 1843.
96 S.G. Yates, H.L. Tookey, J.J. Ellis and H.J. Burkhardt, Phytochemistry, 7 (1968) 139.
97 D. Perlman, W.L. Lummis and H.J. Geiersbach, J. Pharm. Sci., 58 (1969) 633.
98 Y. Ueno, M. Hosoya, Y. Moria, I. Ueno and T. Tatsuno, J. Biochem. (Tokyo), 64 (1968) 479.
99 K. Ohtsuba and M. Saito, Jap. J. Med. Sci. Biol., 23 (1970) 217.
100 J.P. Helgeson and G.T. Haberlach, Plant Physiol., 52 (1973) 660.
101 C. Stahl, L.N. Vanderhoef and N. Siegel, Plant Physiol., 52 (1973) 663.
102 H. Stahelin, M.E. Kalberer-Rusch, E. Signer and S. Lazary, Arzneimittel-Forsch., 18 (1968) 989.
103 E.T. Glaz, E. Csanyi and J. Gyimesi, Nature, 212 (1966) 617.
104 E.T. Glaz, E. Scheiber and K. Jarfas, Acta Physiol. Acad. Sci. Hung., 18 (1960) 225.
105 W.F.O. Marasas, E.B. Smalley, J.R. Bamburg and F.M. Strong, Phytopathology, 61 (1971) 1488.
106 H.R. Burmeister, J.J. Ellis and S.G. Yates, Appl. Microbiol., 21 (1971) 673.
107 H.R. Burmeister and C.W. Hesseltine, Appl. Microbiol., 20 (1970) 437.
108 F.C. Bawden and G.G. Freeman, J. Gen. Microbiol., 7 (1952) 154.
109 E.B. Smalley, Unpublished research, 1973.
110 A.N. Kishaba, D.L. Shankland, R.W. Curtis and M.C. Wilson, J. Econ. Entomol., 55 (1962) 211.
111 N. Nakano, Jap. J. Med. Sci. Biol., 21 (1968) 351.
112 Y. Ueno and H. Yamakawa, Jap. J. Exptl. Med., 40 (1970) 385.
113 D.S. Ahkmedov, Zashch. Rast., Moskva, 12 (1967) 29; RAM, 47 (1968) 584.

114 S. Askarova and R.Ya. Joffe, Khlapkovodstva, 12 (1962) 59; RAM, 41 (1962) 713.
115 F.A. Babaeo, Study of trichothecin in control of cotton wilt, in Trans. of the all-union symposium on cotton wilt control, Tashkent Iydet Uzbekistan, 1964, pp. 165—167; RAM, 47 (1966) 807W.
116 E.S. Gurinovich and M.E. Kobyyrea, Action of various antibiotics on *F. lini.* Fiziol. Asoben. Kultini Rast. Minsk, (1964) 124—131.
117 R.I. Keryukliena, Zashch. Rast. Moskva, 15 (1970) 51; RPP, 49 (1970) 2174.
118 G. Kuzmina, Zashch. Rast. Moskva, 11 (1966) 31.
119 G.M. Kublanovskaya and N.V. Sukhacheva, Zashch. Rast. Moskva, 6 (1961) 25; RAM, 41 (1962) 458.
120 A.V. Nechaeva, Zashch. Rast. Moskva, 8 (1969) 922; RPP, 49 (1970) 134.
121 A.N. Silyanova, Nauch. Dokl. Ryssh. Shk. Biol. Nauki, 13 (1970) 102; RPP, 50 (1971) 74.
122 Kh. Tillaev, Khlapkovodstva, 14 (1964) 48; RAM 43 (1964) 1655.
123 E.T. Glaz, E. Scheiber, J. Gyimesi, I. Horvath, K. Steczek, A. Szëntirmai and G. Bohus, Nature, 184 (1959) 908.
124 S.G. Yates, Toxin producing fungi from fescue pasture, in S. Kadis, A. Ciegler and S.J. Ajl (Eds.), Microbial Toxins, Vol. 7, Academic Press, New York, 1971, pp. 191—206.
125 Y. Ueno, N. Saito, K. Ishii, K. Sakai and M. Enomoto, Jap. J. Exptl. Med., 42 (1972) 461.
126 R.D. Wyatt, B.A. Weeks, P.B. Hamilton and H.R. Burmeister, Appl. Microbiol., 24 (1972) 251.
127 C.M. Christensen, R.A. Meronuck, G.H. Nelson and J.C. Behrens, Appl. Microbiol., 23 (1972) 177.
128 W.F.O. Marasas, J.R. Bamburg, E.B. Smalley, F.M. Strong, W.L. Ragland and P.E. Degurse, Toxicol. Appl. Pharmacol., 15 (1969) 471.
129 W.F.O. Marasas, E.B. Smalley, P.E. Degurse, J.R. Bamburg and R.E. Nichols, Nature, 214 (1967) 817.
130 M. Saito, M. Enomoto and T. Tatsuno, Gann, 60 (1969) 599.
131 T. Tatsuno, Cancer Res., 28 (1968) 2393.
132 S.J. Cysewski, A.C. Pier, G.W. Engstrom, G.W. Richard, J.L. Daugherty and J.R. Thurston, Am. J. Vet. Res., 29 (1968) 1577.
133 W.F.O. Marasas and E.B. Smalley, Onderstepoort J. Vet. Res., 39 (1972) 1.
134 I.C. Hsu, Studies on Trichothecene Mycotoxins, Ph. D. Thesis, Univ. of Wisconsin, Madison (1972).
135 I.C. Hsu, E.B. Smalley and F.M. Strong, (1973) Unpublished results.
136 L.A. Goldblatt, Aflatoxin, Academic Press, New York, 1969, p. 472.
137 C.L. Fergus, Mycologia, 49 (1957) 124.
138 J.A. Stevenson and L.P. McColloch, Plant Dis. Reptr., 31 (1947) 147.
139 M. Cognee and L.S. Bird, Phytopathology, 54 (1964) 621.
140 N.C. Preston, Trans. Brit. Mycol. Soc., 26 (1943) 158.
141 W.R. Orchard, Proc. Can. Phytopathol. Soc., 21 (1953) 15.
142 R.H. Luttrell, Plant Disease Reptr., 49 (1965) 78.
143 D.M. McLean and S. Bailey, Plant Disease Reptr., 45 (1961) 728.
144 F.C. Deighton, Plant Pathology Section, Rept. Dept. Agr. Sierra Leone, (1955) 33.
145 T.R. Nag Raj and K.V. George, Ind. Phytopathol., 11 (1958) 153.
146 E. Schieber and G.A. Zentmyer, Plant Disease Reptr., 52 (1968) 115.
147 B.M. Cunfer and F.L. Lukegic, Phytopathology, 60 (1970) 341.
148 M.A. Smith and G.B. Ramsey, Bot. Gaz., 112 (1951) 394.
149 G.H. Hepting, Diseases of Forest and Shade Trees of the United States, USDA, Forest Service, Ag. Handbook 386, 1971, p. 8, 156.
150 J.G. Dickson, Diseases of Field Crops, McGraw-Hill, New York, 1956, 392 pp.
151 C.W. Edgerton, Sugar Cane and its Diseases, Louisiana State Univ. Press, Baton Rouge, 1959, 113 pp.

152 B.S. Crandall, Phytopathology, 40 (1950) 34.
153 Anon., The American Type Culture Collection, 9th ed., Catalogue of Strains, 1970, 101 pp.
154 E. Seemüller, Nutr. Biol. Bun. Anst. Ld-Forstu., Berlin-Dahlem, 127 (1968) 93 pp.
155 J.E. Burnside, W.L. Sippel, J. Forgacs, W.T. Carll, M.B. Atwood and E.R. Doll, Am. J. Vet. Res., 18 (1957) 817.
156 J. Forgacs. Stachybotryotoxicosis and moldy corn toxicosis, in G.N. Wogan (Ed.), Mycotoxins in Foodstuffs, M.I.T. Press, Cambridge, 1964, p. 87.
157 V.L. Diener and N.O. Davis, in L.N. Goldblatt (Ed.), Aflatoxin, Academic Press, New York, 1969, p. 13.
158 B.J. Wilson and R.R. Maronpot, Vet. Rec., May 8 (1971) 484.
159 L. Badioli, M.H. Abow-Youssef, A.I. Radwan, F.M. Hamdy and P.K. Hildebrandt, Am. J. Vet. Res., 29 (1968) 2029.
160 T.S. Kellerman, W.F.O. Marasas, J.G. Pienaar and T.W. Maude, Onderstepoort J. Vet. Res., 39 (1972) 205.
161 R.J. Cole, J.W. Kirksey, H.G. Cutter, B.L. Doupnik and J.C. Peckham, Science, 179 (1973) 1324.
162 T.A. Curtin and J. Tuite, Life Sci., 5 (1966) 1937.
163 N. Prentice and A.D. Dickson, Biotech. Bioeng., 10 (1968) 413.
164 L.E. Olifson, Chemical and Biological Properties of Toxic Substances from Grains Infected with *Fusarium sporotrichiella* Fungi, Ph. D. Thesis, I.T. of Industrial Nutrition, Moscow (1965).
165 L.E. Olifson, Proc. Symp. Mycotoxins, Acad. of Sci., Ukrainian S.S.R., Kiev, Oct. 3—9 (1972) 12.
166 V.I. Bilai and N.M. Pidoplencko, Toksinobrazjyushchie Mikroskopicheskie Griby, Akad. Nauk Ukrainskoj S. S. R., Kiev, 1970, 289 pp.

Chapter 11

TOXICITY OF *FUSARIUM POAE* AND *F. SPOROTRICHIOIDES* AND ITS RELATION TO ALIMENTARY TOXIC ALEUKIA

A.Z. JOFFE

Laboratory of Mycology and Mycotoxicology, Department of Botany, The Hebrew University of Jerusalem (Israel)

The Fusarium species *F. poae* and *F. sporotrichioides* achieved notoriety during the closing years of World War II as the cause of septic angina or Alimentary Toxic Aleukia (ATA) in some districts of the U.S.S.R., especially in the Orenburg District. ATA, which occurred widely between longitudes of approximately 40° to 140° east and latitudes 50° to 60° north, is a very serious and in most cases lethal disease, accompanied by typical spots on the skin, by necrotic angina, extreme leukopenia and multiple haemorrhages, sepsis and exhaustion of the bone marrow. The large-scale poisonings of rural populations due chiefly to these species rank among the most pernicious effects ever recorded as a result of consumption of mouldy grain which has overwintered in the field [1—19].

Earlier research on the conditions attending outbreaks of ATA has previously been reviewed [13—21]. This research has more recently been followed up by detailed studies of the histological and pathological effects of the toxins concerned on animals, of the phytotoxic effects on a variety of plants, and on the possible relationship between effects of the toxins on animals and plants.

An overall survey of the present state of knowledge regarding toxins produced by *F. poae* and *F. sporotrichioides* will be presented here, and this will be supplemented by a review of experience gained in studies on aetiology, epidemiology and treatment of ATA.

Characterisation of the fungi

The fungi *F. poae* and *F. sporotrichioides* belong to the section Sporotrichiella Wr. em. Joffe [22]. The characteristics both these species have in common are: macroconidia sparse, small, oblong, narrowly fusoid to falcate, pedicellate, formed in aerial mycelium or in sporodochia: chlamydospores intercalary, terminal, in chains or knots, occasionally with plectenchymatous sclerotia; perithecial states absent.

The characteristics that distinguish the two species from one another are detailed below:

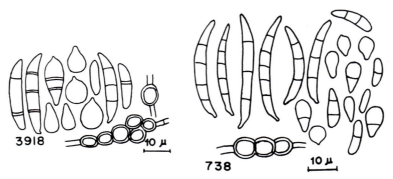

Fig. 1. Macrospores, microspores and chlamydospores of: *Fusarium poae* (No. 3918) (left); *F. sporotrichioides* (No. 738) (right).

F. poae (Peck.) Wr. (Fig. 1, No. 3918) — Cultures white, yellow or red-brown. Microconidia in the aerial mycelium abundant, in relation to macroconidia, oval, spherical with basal papilla, spindle-ellipsoid or elongate, rarely pear-shaped. Macroconidia sparse, curved, falcate, in aerial mycelium, sporodochia absent, pionnotes absent or very rare.

0—sept., oval	6.8 — 9	× 4.5 — 8.2
0—sept., spindle-ellipsoid shaped	10 — 15	× 2.6 — 4.2
1—sept., oval, pear-shaped	10.5 — 16	× 4.5 — 7.4
1—sept., spindle-ellipsoid shaped	12 — 22.5	× 3.6 — 4.5
3—sept., curved falcate	19 — 38	× 3.5 — 6.8

Chlamydospores intercalary, single or in pairs, knots, or chains. Growth rate 7.2 cm.

F. sporotrichioides Sherb. (Fig. 1, No. 738) — Cultures white, white-rose or red. Microconidia globose, pyriform, ellipsoid, elongate or slightly falcate, dispersed in aerial mycelium. Macroconidia formed in aerial mycelium or in sporodochia, falcate to curved, with foot-cell.

0—sept., pear-shaped	6 — 9	× 5.5 — 7.0
0—sept., ellipsoidal	7 — 11	× 4.0 — 7.8
0—sept., elongate or sickle shaped	8 — 11	× 2.8 — 3.8
0—sept., spindle-shaped	9 — 16	× 2.6 — 3.8
1—sept., pear-shaped	8 — 15	× 5.2 — 7.0
1—sept., ellipsoidal	9 — 20	× 3.8 — 8.0
1—sept., elongate or sickle-shaped	12.5 — 17	× 3.0 — 4.2
1—sept., splindle-shaped	15 — 26	× 2.5 — 4.2
3—sept., sickle or spindle-shaped	22 — 35	× 3.6 — 4.7
4—sept., sickle or spindle-shaped	33 — 40	× 3.8 — 4.5
5—sept., sickle or spindle-shaped	37 — 45	× 4.0 — 5.2

Chlamydospores intercalary, single or in pairs, knots or chains. Growth rate 3.5 cm.

Methods of testing toxicity

For this purpose pure cultures of *F. poae* and *F. sporotrichioides* were grown on various agar and liquid media or natural sterilized media (millet, wheat, barley, and others). Our observations had shown that accumulation of toxin in overwintered cereals increased with sharp temperature fluctuations. Therefore, after being grown at temperatures of $-5°$ to $+8°$ for 25 to 70 days, cultures were subjected to successive freezing and thawing. In some cases the Fusarium was grown on wheat for a period of 40 days at temperatures ranging from $7°$ to $24°$ (3 days at $24°$, followed by 21 days at $7°$, 10 days at $18°$ and again 6 days at $7°$).

Cultures were then autoclaved for 20 min at $100°$, thallus and substrate were removed from the flasks, dried at $60°$ for 24 h and ground. Subsequently the toxins were extracted in 70% or 96% ethyl alcohol or in ether, shaken for 12 h and then evaporated to dryness using a vacuum rotary evaporator.

The toxic properties of the cultures were assessed by skin tests on rabbits using the method adapted by the author [16,17]. Only rabbits with non-pigmented skin and weighing at least 1.5 kg were used.

Each rabbit was kept in a separate cage and was maintained on laboratory feed. On the back and sides of each rabbit, squares of skin measuring 3×3 to 4×5 cm were carefully cleared of hair so that five rows with five, seven or eight squares each were obtained, and the skin looked like a chessboard (Fig. 2). Extract of the cultures was applied twice, either by micropipette or by platinum loop, to the skin of the rabbit, with an interval of 24 h.

Two squares on each rabbit served for control and were treated with alcohol or ether extract of autoclaved wheat grains not infected by fungi. No

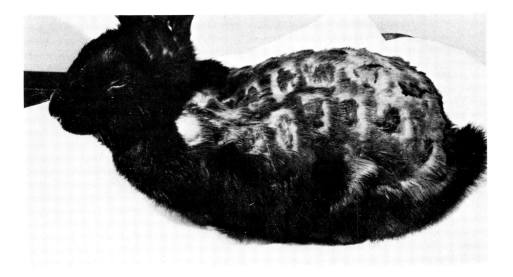

Fig. 2. Rabbit subjected to skin test of toxins. The white squares on the left are controls.

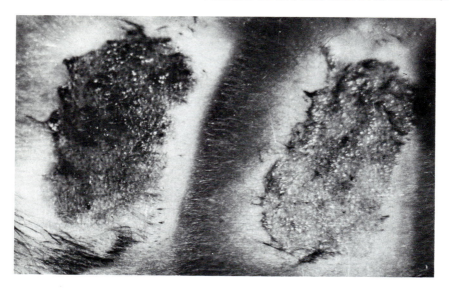

Fig. 3. Close-up of rabbit skin reaction to toxins: strong necrotic reaction (left); leuko-cytic and oedemohaemorrhagic reaction (right).

changes on the skin were observed in the control. Toxic reactions were recorded for 48 h and the rabbits were kept under observation for at least 8 to 10 days after the first application.

Skin reactions are of two types — leukocytic and oedemohaemorrhagic-necrotic (Fig. 3). The former is characterized by the formation of the skin surface of a whitish, easily detachable film which consists of a mass of leukocytes in the horny layer of the epithelium. The latter type of skin reaction involves an acute oedema, haemorrhage and necrosis; in this case, there is no leukocytic film. The estimation of the reaction was based on its principal components: leukocytic film, oedema, haemorrhage, and necrosis. The intensity of necrosis was determined on the 8th day; the other components were recorded on the 3rd day.

The intensity of the leukocytic component was estimated according to massiveness of the superficial leukocytic film, the intensity of oedema was assessed by the thickness of the skin fold, the intensity of haemorrhage by the quantity of visible extravasations, and necrosis by the massiveness of the scab and the time of its shedding. The presence of oedema, haemorrhage, and necrosis was regarded as evidence of marked toxicity of the fungus. A pronounced leukocytic film, unaccompanied by any of the other components, was also considered as an indication of the toxicity of the experimental material, and fungi inducing this reaction were included in the toxic group; a reaction represented by a thin leukocytic film, scattered vesicles, reddening, and desquamation with or without slight oedema of the skin was assessed as weak and doubtful, and the associated fungi were termed mildly toxic.

Intensity of the skin reaction produced by each culture of Fusarium was assessed on the scale of grades described by Joffe [16,17,33].

The toxins

(a) Composition of the toxins

Numerous investigations have been reported of the chemistry of toxins from the *F. poae* and *F. sporotrichioides*. Much attention has been paid also to the study of the structure and chemical properties of toxins from over-wintered grains.

Gubarev and Gubareva [23] isolated two fractions from ether extracts of toxic grains. One fraction contained derivatives of fatty acids that caused the local skin inflammatory reaction in rabbits. The second was a fraction of non-saponifiable material which produced a necrotic skin reaction.

Kretovitz and Bundel [24] found that the toxins are present in over-wintered grains as oxidation products of unsaturated fatty acids. Kretovitz and Sosedov [25] found that toxic grains contain more non-protein nitrogen and amino nitrogen and less starch than non-toxic grains. Toxic grains show-ed increased activity of dextrinogenic amylase and decreased activity of peroxidase and oxidase.

Gabel [26] concluded that the toxic fractions isolated from overwinter-ed grains were saponins, and Svoyskaya [37] found that these toxins had acidic properties, but also believed that there were other toxins in the form of neutral lactones and compound ethers.

Okuniev [28] ascribed the toxicity to the presence of hydroxy acids and steroid components resembling vitamin E and an oxy-derivative of coumarol.

It is obvious from the many theories that there was no unanimous opinion on the chemical composition of Fusarium toxins isolated from over-wintered toxic cereals because no one has yet succeeded in isolating a pure toxin and determining its structure.

Many details have been added by the studies by Olifson [29—32] who found a method for purification and isolation of toxins from overwin-tered grains and from normal grains previously autoclaved and then infected with toxic fungi (*F. poae* and *F. sporotrichioides*) of varying degrees of toxicity. Olifson's basic assumption was that the toxin is found in the lipid fraction of the grains. When such a lipid extract is applied to the skin of rabbits, it causes a strong inflammatory reaction. Cats, mice and guinea pigs which were fed on a diet mixed with this lipid fraction died. Olifson [32] also determined the physicochemical constants for these lipids. The analysis indicated an increase of acid value from a normal of 14.1—17.8 to 121.4 for toxic lipids and also an increase in peroxide value from a normal of 1.3—1.8 to 6.8—8.4 in the non-saponifiable residue. On the other hand the refractive index decreased and the iodine value decreased from 132.4 to 66.6 for toxic lipids. If these constants for free and bound lipids are compared, it is evident that free lipids yield a higher acid value (121.4) and a lower peroxide value (6.8—8.4) than bound lipids whose acid value and peroxide value are 15.8 and 29.1, respectively. It thus seems that a big difference exists in the chemical composition of free and bound lipids extracted from prosomillet grains.

A comparison was also made between the chemical composition of prosomillet grains intentionally infected with toxic cultures of *F. poae* and *F. sporotrichioides*, and overwintered grains which became toxic in the field. The grains infected with pure cultures had been grown for some days at —10° to —15° and then transferred to +3° to +5° for one month. The cultures were then autoclaved and after the infected grains had completely dried in the air, tests for humidity, ash content, proteins, cellulose, acid value and iodine value were performed. The infected grains showed an increased ash content and acid value and a decreased iodine value.

The lipid content of prosomillet grains which had been infected with *F. poae* and *F. sporotrichioides* differed considerably from that of uninfected grains. The grains of each culture were separated into two fractions — a neutral fatty fraction and an acidic fatty fraction. Olifson [32] found that the toxicity of the neutral fraction was more marked for *F. poae* and *F. sporotrichioides* than for some other toxigenic fungi, *e.g.* species of Cladosporium.

Olifson [30—32] isolated from prosomillet which had been infected with *F. sporotrichioides* a saponinic fraction, which he named sporofusarin whose empirical formula is $C_{65}H_{96}O_{25}$ and melting point 246—248°. Hydrolysis for 4 h with 5% H_2SO_4 yielded a sapogenin whose empirical formula is $C_{24}H_{31}O_4$ and which was given the name sporofusariogenin.

Olifson isolated a monoglycoside from *F. poae* named poaefusarin whose empirical formula is $C_{35}H_{39}O_{12}$. This glucoside contains xylose and a steroidic aglucone. The aglucone was designated poaefusariogenin $(C_{24}H_{28}O_5)$ which differs from sporofusariogenin by the presence of an aldehyde group instead of a methyl group. These two derivatives were tested on cats and compared with the effect of a lipid material, called lipotoxol by Olifson, which was isolated from the lipid fraction of toxic overwintered prosomillet.

All three derivatives gave a similar syndrome in cats, characterised mainly by a constant leukopenia. The oral lethal dose was 0.5 mg and all animals died within 15—16 days. In cats and dogs lipotoxol inhibits the normal action of the heart. The lethal dose for mice was 0.06—0.07 mg of 1% lipotoxol extracted from the fatty fraction of prosomillet. On the skin of rabbits lipotoxol caused a typical haemorrhagic-oedematous reaction. Lipotoxol resembles in structure and properties both sporofusariogenin and poaefusariogenin which are both steroids and since the syndrome produced in cats was very much like ATA in man, Olifson concluded that the cause of the disease was the toxin secreted by the two fungi *F. sporotrichioides* and *F. poae* in overwintered grains.

It is also possible that intermediate products of these final derivatives

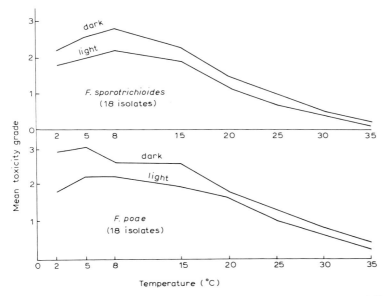

Fig. 4. Effect of temperature and light on toxicity of isolates of *F. poae* and *F. sporotrichioides*.

such as unsaturated fatty acids, oxyacids and sterols further contribute to the syndrome of ATA*.

(b) Conditions of toxin formation

(i) Effects of temperature and light

The overall toxicity produced by extracts from wheat cultures grown in light and darkness was determined at 7 temperatures from 2 to 35°.

F. poae and *F. sporotrichioides* produced the highest toxicity both in light and darkness, at 5° and 8°. Darkness clearly favoured development of toxicity of the 2 fungi at all temperatures (Fig. 4).

Closely similar results regarding the temperature and light effects on production of toxicity by the species were obtained in additional test series, in which the fungi were grown on liquid substrate [33].

The effects of alternate freezing and thawing on toxin formation by *F. poae* was tested under laboratory conditions [18]. Cultures grown on potato-dextrose agar, potato-dextrose acid agar and carbohydrate-peptone media were alternately kept at room temperature (18°) and at various freezing temperatures down to −15°, and this was repeated 4—5 times. Results clearly showed that sharp fluctuations of temperature greatly increased the toxicity of extracts from these cultures. Application of such extracts to the skin of rabbits resulted in acute oedema, haemorrhage and necrosis.

---

* See Note added in proof, p. 262.

TABLE I

EFFECT OF TEMPERATURE AND OF THE DEVELOPMENTAL STAGES OF *F. POAE* AND *F. SPOROTRICHIOIDES* ON TOXIN FORMATION AS EXPRESSED BY REACTION OF MICE AND RABBITS

Developmental stage of fungus	Effect of subcutaneous injection on mice (A, alive, D, dead)					
	F. poae			*F. sporotrichioides*		
	0.2 ml	0.5 ml	1.0 ml	0.2 ml	0.5 ml	1.0 ml
(1) Cultured at 23 to 25° for 40 days						
Before sporulation	A	A	A	A	A	A
Sporulation	A	A	A	A	A	D
Senescence	A	A	A	A	A	A
(2) Cultured at +1 to −2° for 40 days						
Before sporulation	A	A	A	A	A	A
Sporulation	D	D	D	D	D	D
Senescence	A	A	A	A	A	A
(3) Cultured at 0 to 5° for 10 days, then at −7 to −15° for 30 days, continued at 0° to +5° for 4 days						
Before sporulation	A	A	A	A	A	D
Sporulation	D	D	D	D	D	D
Senescence	A	A	A	A	A	A
(4) Cultured at −7° to −10° for 40 days						
Before sporulation	D	D	D	A	A	A
Sporulation	D	D	D	D	D	D
Senescence	A	A	A	A	A	A

a Nature of rabbit skin reactions: L, leukocytic; O, oedematous; N, necrotic.

(ii) Temperature and developmental stage of the fungus

Pure cultures of *F. poae* and *F. sporotrichioides*, sown on liquid medium (synthetic with starch and carbohydrate-peptone) and grown under different temperature conditions, were assayed for toxicity at each of the three stages of development, *i.e.* prior to sporulation, at the time of abundant sporulation, and at the stage of senescence. Assays were carried out with other extracts of the liquid media and extracts of the fungal mass by skin tests on 22 rabbits. The liquid substrates were passed through Seitz filters, and the sterile filtrates obtained in this way were tested for toxicity on white

Effects of application to rabbit skin[a]
(+ slightly toxic, ++ toxic, +++ highly toxic)

Ether extract from liquid						Fungal mass					
F. poae			F. sporotrichioides			F. poae			F. sporotrichioides		
L	O	N	L	O	N	L	O	N	L	O	N
	+										
	++	+	+	+	+	+			+	+	
	+	+	+								
	+										
	+++	++	++	++	+		++		+	+	
	+										
			+	+							
	+++	+++	++	+++	+		++		+	++	+
+	++	+		+		+	+				
	++	+									
	+++	+++	++	++	++		++		++	+	
	++	+		+		+	++				

mice by subcutaneous injections of 0.2, 0.5 and 1.0 ml.

Results are presented in Table I. They show that injection of filtrates of cultures obtained at different stages of development at the temperature of 23—25° did not cause death of white mice [21].

In the case of extracts of cultures maintained at low temperatures, as well as those kept at 0°—5° with intervening freezing, death of the mice occurred within 12—48 h, depending on the dosage and the fungal species. Death was due to systemic toxic action. *Post mortem* examinations disclosed in every case necroses in the digestive system and other organs. The greatest toxicity was displayed by filtrates of Fusarium obtained during the stage of abundant sporulation from cultures grown at the temperature of $-2°$ to $-7°$ and prior to sporulation from the $-7°$ to $-10°$ series, while extracts obtained at an advanced stage of senescence were considerably less toxic.

Application of liquid ether extracts of *F. poae* and *F. sporotrichioides* to the skin of rabbits invariably produced a distinct skin reaction when the fungus was at the sporulation stage, even in the case of cultures grown at incubator temperatures of $23°-25°$. Sporulation seems to be connected with high concentrations of the toxin concerned. It should, however, be mentioned that in every instance cultures grown at low temperature produced more pronounced reactions than incubator-reared cultures. Greatest toxicity was associated with material obtained at the stage of abundant spore formation.

The data in Table I bring out yet an other interesting characteristic, namely that the application of ether extracts of the liquid substrate to rabbit skin produced a stronger reaction than did extracts of the fungal film of *F. poae* and *F. sporotrichioides* cultures. This indicates that the toxins of *F. poae* and *F. sporotrichioides* are excreted into the surrounding medium, and thus act as exotoxins.

(iii) Substrates

Effects of substrates on the production of overall toxicity were first studied by growing the fungi at the above temperatures on 3 grain substrates (wheat, barley, millet) and on 3 liquid substrates: carbohydrate-peptone (30 g sucrose, 10 g glucose, 1 g peptone, 1000 ml distilled water); Czapek's (2 g $NaNO_3$, 1 g K_2HPO_4, 0.5 g HCl, 0.5 g $MgSO_4$, $7H_2O$, traces of $FeSO_4$, 30 g sucrose, 1000 ml distilled water) and starch (minerals as in Czapek's, but 25 g soluble starch per 1000 ml distilled water substituted for the sugar).

In the case of the liquid cultures toxicity was determined separately for the mycelial mass grown on the liquid (thallus) and for the substrate after removal of that mass. The data summarized in Table II show that toxicity derived from the liquid substrate was always higher than that from the thallus grown on that substrate. Overall toxin production was constantly strongest on the starch substrate, and in the case of *F. poae* it was somewhat stronger on Czapek's than on the carbohydrate peptone substrate.

Among the grain substrates, which were tested in light and darkness, barley yielded growth with higher overall toxicity of *F. poae* and *F. sporotrichioides* both in light and in darkness. The favourable effect of darkness on toxin production was evident on all three of these substrates.

(iv) Vigour of growth and toxigenicity

Is there any relation between the growth of these fungi *in vitro* and between the degree of toxicity of their culture extracts to rabbit skin? The following conclusions can be drawn from results obtained with 5 isolates of *F. poae* and 6 of *F. sporotrichioides* showing strong toxicity. The peak for growth of both species was at $18-24°$, where toxicity was moderate. But toxicity was strongest at $6-12°$, where growth was limited. At temperatures

TABLE II

EFFECTS OF SUBSTRATES ON PRODUCTION OF TOXINS
Mean of results obtained at 7 temperatures in 144 cultures.

	Mean grade of toxicity per isolate			
	F. poae 9 isolates		F. sporotrichioides 9 isolates	
	liquid	thallus	liquid	thallus
Liquid substrates				
Carbohydrate-peptone	1.2	0.5	1.4	0.8
Czapek's	1.4	0.7	1.5	0.9
Starch	1.6	0.8	1.8	1.0
Mean	1.4	0.7	1.6	0.9
	18 isolates		18 isolates	
	light	dark	light	dark
Grain substrates				
Prosomillet	1.3	2.2	1.5	2.0
Wheat	1.3	1.8	1.3	1.9
Barley	1.7	2.3	1.7	2.1
Mean	1.4	2.1	1.5	2.0

of 30—35° growth was moderate and toxicity weak. Thus there seems to be no relation between vigour of growth at any of the above temperatures and the toxigenicity of these fungi. This is clearly illustrated in Fig. 5.

(c) Persistence of toxins

In the course of nearly 7 years, 40 samples of overwintered cereals responsible for cases of ATA, from which toxic fungal cultures had been isolated in 1943—1944, were tested again for toxicity and were subjected to mycological analysis. It was established that these samples did not lose their toxicity after 6—7 years of storage under laboratory conditions. However, concurrent mycological investigations of these samples failed to isolate *F. poae* and *F. sporotrichioides* which had obviously perished. These findings may account for the difficulties encountered in attempts to isolate toxic fungi from cereals after a prolonged period of storage.

(d) Comparative toxicity of the 2 species to animals

Isolates of *F. poae* and *F. sporotrichioides* made in U.S.S.R. from grain

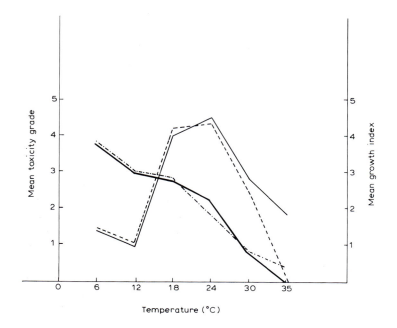

Fig. 5. Relation between growth *in vitro* and grade of overall toxicity of *F. poae* (5 isolates) and *F. sporotrichioides* (6 isolates) at various temperatures. ———, *F. poae*, growth; ------, *F. sporotrichioides*, growth; — · — ·, *F. poae*, toxicity; ━━━━, *F. sporotrichioides*, toxicity.

that had actually caused ATA with lethal effects were cultured on wheat kernels at 6 temperatures. Extracts from these cultures were repeatedly applied to rabbits, and the skin reaction was graded from ± (very slight) to ++++ (very strong).

TABLE III

COMPARATIVE TOXICITY OF *F. POAE* AND *F. SPOROTRICHIOIDES* TO RABBIT SKIN, AFTER CULTURING AT 6 TEMPERATURES

Temperature °C	Type and grade of rabbit skin reaction			
	F. poae		*F. sporotrichioides*	
	Oedema	Necrosis	Oedema	Necrosis
6	++++	+++	+++	++
12	+++	+++	+++	++
18	+++	++	++	++
24	+++	+	++	+
30	+	+	+	±
35	+	±	±	0

The results obtained, as shown in Table III, show that *F. poae* was more toxic than *F. sporotrichioides* at almost all temperatures.

These results corroborate those obtained in tests on effects of temperature and developmental stage of the two fungi on their production of toxins (Table I). Further confirmation for the greater toxicity of *F. poae* to animals has been obtained in feeding tests on horses, as will be mentioned later.

Effects of the toxins on animals, man and plants

(a) Effects on animals

The toxins of Fusarium cultures have both a localized and general toxic effect. The localized effect is first apparent in an inflammatory reaction and is accompanied by subsequent skin necrosis at the site of toxin application. The general effect is apparent in defective haemopoiesis and acute degeneration in the internal organs, as well as in extreme hyperaemia especially of the digestive tract.

The symptoms in animals vary according to the potency and quantity of the toxin, the route of administration and the sensitivity of the animal. General effects of the toxins have been studied in a variety of animals and results of experiments carried out by us and others will be briefly summarized.

(i) Protozoa

The effect of dilute (1:1000) alcohol extracts of millet grain experimentally infected with *F. poae* and *F. sporotrichioides* was studied on certain Protozoa and a toxic effect demonstrated in *Paramecium caudatum*, *Stylonychia mytiles*, *Opalina ranarum* and *Nyctotheras cordiformis* [34,35].

(ii) Frogs

Frogs were tested in order to establish their susceptibility to *F. poae*. Following repeated feeding with the dry fungi the animals died within 4 to 14 days depending on the dose. The effect of the toxin was cumulative. *Post mortem* dissection showed extreme hyperaemia, haemorrhages and oedema of the digestive tract.

(iii) Mice and rats

The toxic properties of *F. poae* and *F. sporotrichioides* were also tested on white mice. The mice died 2—7 days after they had been fed cultures of the two Fusarium species, Fusaria extract or dry fungi and their liquid filtrates. Subcutaneous injections of fungus filtrates prepared during their abundant sporulation produced a lethal effect after 13—24 h [16].

In recent work, Schoental and Joffe [36] have studied long-term effects of *F. poae* and *F. sporotrichioides* on mice and rats.

Crude concentrated alcoholic culture extracts were applied to rodents by various routes. Extracts from both fungi appeared to produce similar

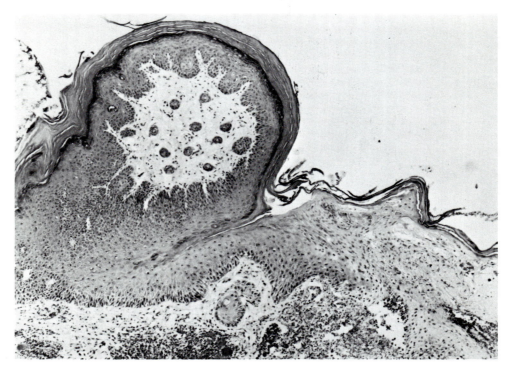

Fig. 6. Small papilloma on the skin of a mouse after application of extract of *F. sporo-trichioides*. × 76.

effects in mice and in rats. Single doses, of the order of 0.1 ml, were toxic to young animals, which died within a few days. Smaller doses allowed the animals to survive longer so that dosage could be repeated at various time intervals. Chronic lesions developed in animals that survived several weeks or months, suggesting that the fusarial metabolites may be carcinogenic and immunosuppressive.

High doses, which kill the animals within one or more days, cause depletion of the lymphoid tissues, as a consequence of which infections develop in various organs, including the heart and the kidneys.

Smaller doses applied locally to the skin or by intragastric intubation to the oesophagus and stomach have local cytotoxic effects which are followed by regeneration and basal cell hyperplasia of the squamous epithelium.

Fig. 6 shows a small papilloma in the skin of a mouse, which developed in the course of 10 weeks after a few applications of *F. sporotrichioides*. Fig. 7 shows the hyperplasia and an invagination of the oesophagus in a rat that received *F. poae* extract by intragastric tube. The regeneration of the basal cells suggests that the lesions may become neoplastic in due course.

The effects of the extract on the thymus, spleen and lymph glands, and the development of infective foci in various tissues including the kidneys, strongly suggest immunosuppressive action. These experiments are being continued.

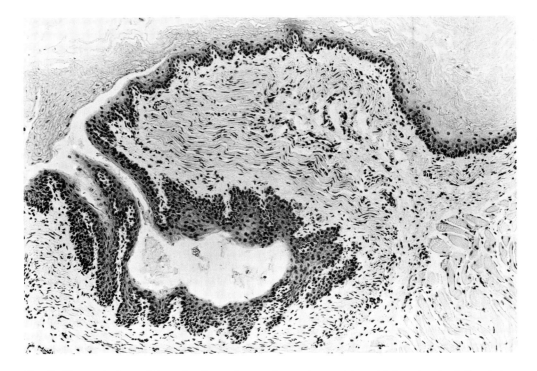

Fig. 7. Hyperplasia and invagination of oesophagus in a rat to which extract of *F. poae* was applied by intragastric tube. × 100.

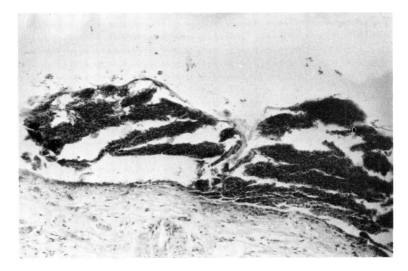

Fig. 8. General necrosis of the epidermis following the application of toxic *F. poae* extract. Intraepidermal vesiculation filled with an exudate containing white blood cells. In the dermis oedema and leukocytic infiltration. × 135.

References p. 260

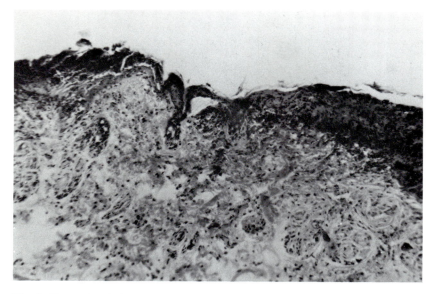

Fig. 9. Necrosis of the epidermis with an infiltration of leukocytes in the upper layers of the dermis. Here necrosis is found accompanied by oedema and leukocyte infiltration following the application of toxic *F. sporotrichioides* extract. × 135.

(iv) Guinea pigs and rabbits

The histological effects of application of toxins of *F. poae* and *F. sporotrichioides* to rabbit skin are shown in Figs. 8 and 9.

In feeding tests with guinea pigs and rabbits, these were given liquid filtrates or dry mycelium of toxic *F. poae* and *F. sporotrichioides*. The guinea pigs died within 5—21 days of receiving 1.0—2.0 ml of liquid culture of 0.05—0.07 g of dry fungus. The animals which received the toxic fungi in liquid form were more severely affected. The rabbits received 3—5 ml filtrate or 0.1—0.15 g mycelium. Death occurred on the 8th—24th day after feeding of toxic cultures and autopsy showed dilated blood vessels and haemorrhages in the walls of the intestines and many other organs and tissues. After subcutanous injection of rabbits with *F. poae* extract marked migration of leukocytes through the walls of small blood vessels was noted (Fig. 10).

In both mice and guinea pigs to which toxic overwintered grains had been infused, Alisova and Mironov [13] and Alisova [14] found a leukopenia which was not constant. The number of leukocytes declined to 2900 per mm^3 in mice and to 500 per mm^3 in guinea pigs. A similar leukopenia was obtained even after the toxin had been autoclaved.

Bilai [37] fed small rabbits weighing 600—900 g with food infected with *F. poae* and *F. sporotrichioides* and found that the effect of the toxin depended on the weight of the animals and that rabbits were more sensitive than guinea pigs and mice. *Post mortem* examinations of the rabbits revealed hyperaemia of the subepithelial tissues and gastrointestinal system as well as diffuse haemorrhages in the liver.

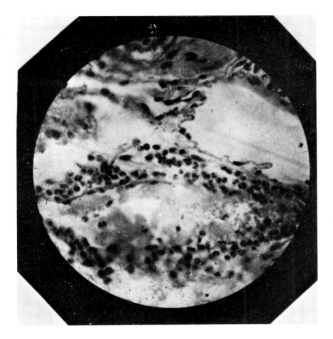

Fig. 10. Leukocytes migrating through walls of small blood vessels after subcutaneous injection of *F. poae* into rabbits.

Maisuradge [38] found that when rabbits were fed germinating oats infected with *F. sporotrichioides*, they lost weight, the body temperature rose, the pulse became quicker and the breathing heavy and slow. The leukocyte count rose to 37 000 per mm^3 in the middle of the experimental period and later fell to 600—200 per mm^3. A decrease in the red cell count was also evident.

(v) Dogs

The effect of different quantities of agar cultures of *F. poae* on the motor activity of dogs has also been studied by the author [39]. *F. poae* cultures were fed orally and through a gastric fistula under various conditions. The fungus culture had been mixed with 200 g of meat, and gastric motility was registered by a kymograph. It was found that introduction of large doses of fungus (1 g and more) caused poisoning of the animal and cessation of stomach motor activity.

When dogs were fed oats infected with *F. poae* the following symptoms were detected by Maisuradge [38]: The body temperature rose to 40° and the pulse to 120/min. The leukocyte count decreased to 800 per mm^3. Stomatitis and gastrointestinal haemorrhages were evident, as well as degeneration of the liver and kidneys and changes in the epicardium and nervous system.

References p. 260

(vi) Chickens

Sergiev [40] stated that chickens die after feeding on toxic grains but studies in our laboratory [13] and independent studies by Sarkisov [41] suggest that chickens are resistant to the toxin.

(vii) Horses

Detailed studies were made on two horses fed with *F. poae* cultures [42]. Two series of agar cultures were used. The first series was grown at room temperature without being subjected to freezing and subsequent thawing. The second series was grown at low temperatures and found to be more toxic than the first; 40 g of fungus culture of the second series produced an acute toxicosis and the horse died 36 h after the administration of the culture. The clinical and pathological findings were those of a haemorrhagic diathesis.

Our observations proved that *F. poae* is apparently more toxic than *F. sporotrichioides*. Thus according to Sarkisov *et al.* [43] the feeding of a horse with 16.4 kg cereals infected by *F. sporotrichioides* caused the development of stomatitis and gingivitis only, while our experiment showed that 40 g of a culture of *F. poae* resulted in the death of the horse within 36 h.

Horses which had fed on herbage or whole grains of oats which were toxic showed the following symptoms according to Maisuradge [38]: 2 days after ingestion of toxic food, hyperaemia was already evident in the mouth with swelling and splitting of the lips. After 7 days a foul-smelling slough appeared in the oral mucosa. Deformation of the head of the horses was observed and the animals lost weight because they refused to eat the toxic food after a while. Although one horse ate 23 kg of toxic food and another 18 kg, they recovered slowly and after a very long period returned to normal health.

(vii) Cats

Of all laboratory animals the best results were obtained in *cats* as these developed all the clinical symptoms which occur in ATA in man. We studied the effect of *F. poae* and *F. sporotrichioides* in the form of agar cultures, millet cultures, dry mass and culture liquid in 26 cats. The daily dose of agar and millet cultures was 0.05—0.12 g and that of the liquid substrate 0.5—1.0 ml. In all the forms administered to cats, both Fusarium species were lethal after variable periods of time, depending on the daily dose of fungus and on individual properties of the respective organism. The death of the cats followed a failure in blood production. A fall in haemoglobin, red cells, leukocytes and neutrophils was observed with a relative increase in lymphocytes. The lowest leukocyte count found was 100 per mm^3. The symptoms included vomiting, a haemorrhagic diathesis and neurological disturbances.

In the majority of cases the cats died on the 6th—12th day. Autopsy revealed marked hyperaemia of internal organs, especially of the digestive tract and kidneys and extreme changes in the adrenal glands. Histological examinations of organs of cats which had died after infection of *F. poae* and

F. sporotrichioides revealed changes in the blood-producing tissue, which were similar to those seen in ATA. It should be added that cats were found to be the best model on which the whole clinical picture of ATA could be reproduced.

Sarkisov [44] fed 7 cats with doses of 0.1—2 g infected millet. The lethal dose was 0.4—16.5 g, and death occurred within 2—34 days. The body temperature rose to 41° and a progressive leukopenia appeared, the white cell count falling to 50—200 cells/mm^3.

Rubinstein and Lass [45] fed millet infected with *F. sporotrichioides* to cats and monkeys which developed symptoms similar to those of ATA in man.

(b) Effects on man

In order to establish a diagnosis of ATA, the recent dietary history of the patients should be carefully investigated. The quantity and duration of feeding on overwintered toxic grains or their products must be determined.

The chemical findings have been described by Chilikin [46], Manburg [47], Manburg and Rachalski [48], Romanova [49] and Yefremov [4].

(i) The four stages of ATA

The clinical features of ATA are usually divided into four stages. If the disease is diagnosed during the first stage and even at the transition from the second to third stages, early hospitalization may still permit the patient's life to be saved. If, however, the disease is detected only during the third stage, the patient's condition is usually desperate and in most cases death cannot be prevented. Only very few patients survive the third stage.

The first stage. The symptoms characteristic of this stage appear shortly after ingestion of the toxic grains. They may appear after a single meal of overwintered toxic grains, and disappear completely even if the patient continues to eat the grains. These symptoms include primary changes in the buccal cavity and gastrointestinal tract. Shortly after eating food prepared from toxic grain, the patient feels a burning sensation in the mouth, tongue, throat, palate, oesophagus and stomach as a result of the action of the toxin on the mucous membranes. The tongue may feel swollen and stiff, and the mucosa of the oral cavity may be hyperaemic. Inflammation of the gastric and intestinal mucosa results in vomiting, diarrhoea and abdominal pain. In most cases excessive salivation, headache, dizziness, weakness, fatigue and tachycardia accompany this stage, and there may be fever and sweating. The leukocyte count may already decrease in this stage to levels of 2000/mm^3, with relative lymphocytosis and there may be an increased erythrocyte sedimentation rate [4].

This stage may not always be detected because it appears and disappears relatively quickly; the patient may be become accustomed to the toxin and a quiescent period follows while the effects of the toxin accumulate and the patient enters the second stage. The first stage may last from 3 to 9 days.

The second stage. This is often called the latent stage [46] because the patient feels well and is capable of normal activity. Sometimes it is also called the leukopenic stage [48,49] because its main features are disturbances in the haemopoietic system characterized by a progressive leukopenia, a granulopenia and a relative lymphocytosis. In addition there is anaemia and a decrease in the platelet count. The decrease in leukocytes lowers the resistance of the body to bacterial infection. In addition to changes in the haemopoietic system, there are also disturbances in the central and autonomic nervous systems. Weakness, headache, palpitations and mild asthmatic symptoms may occur. The skin and mucous membranes may be icteric, the pupils dilated, the pulse soft and labile, and the blood pressure decreased. The body temperature does not exceed 38° and the patient may even be afebrile. There may be diarrhoea or constipation.

This stage normally lasts from 3 to 4 weeks but it may extend over a period of 2 to 8 weeks. If consumption of toxic grain continues, the symptoms of the third stage rapidly develop.

The third stage. The transition from the second to the third stage is sudden. The patient's resistance is now already low, and violent symptoms

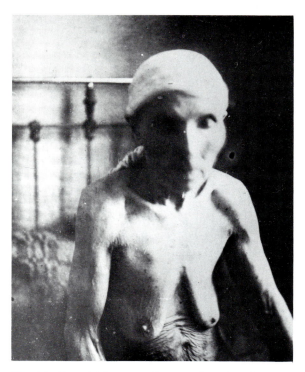

Fig. 11. Petechial spots, first small and red, later blue, caused by intradermal or submucous haemorrhage on the chest.

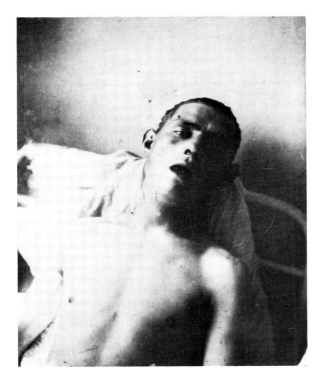

Fig. 12. General view of a patient with a severe form of ATA. Nosebleed (epistaxis). Respiratory distress. Haemorrhage on the left arm.

may be present, especially under the influence of stress, associated with physical exertion and fatigue.

The first visible signs of this stage are petechial haemorrhages on the skin of the trunk, in the axillary and inguinal areas, on the lateral surfaces of the arms and thighs, on the chest (Fig. 11) and in serious cases, also on the face and head. The petechial haemorrhages vary from a few millimeters to larger areas a few centimeters in diameter.

As a result of increased capillary fragility any slight trauma may cause the haemorrhages to increase in size. Haemorrhages may also be found on the mucous membranes of the mouth and tongue, and on the soft palate and tonsils. Nasal, gastric and intestinal haemorrhages may occur (Fig. 12).

Necrotic changes soon appear in the throat, with difficulty and pain in swallowing. The necrotic lesions may extend to the uvula, gums, buccal mucosa, larynx and vocal cords, and are usually contaminated with a variety of avirulent bacteria. The necrotic areas are an excellent medium for bacterial infection which is possible because of the lowered resistance of the body due to the damage to the haemopoietic and reticuloendothelial systems. Bacterial infection causes an unpleasant odour from the mouth due to the enzymatic activity of the bacteria on proteins.

Areas of necrosis may also appear on the lips and on the skin of the nose, jaws and eyes [21].

The regional lymph nodes are frequently enlarged. The submandibular and cervical lymph nodes may become so large and the adjoining connective tissue so oedematous that the patient experiences difficulty in opening his mouth. Oesophageal lesions may occur and involvement of the epiglottis may cause laryngeal oedema. In such cases, death may occur by strangulation. Death of about 30% of the patients was directly related to stenosis of the glottis.

The blood abnormalities observed initially in the first and second stages become intensified during the third stage. The leukopenia increases to counts of 100 or even fewer leukocytes per mm^3. The lymphocytes may constitute 90% of the white cells present, the number of thrombocytes decreases below 5000 per mm^3 and the erythrocytes below 1 million per mm^3.

The blood sedimentation rate is increased. The prothrombin time ranges between 20 to 56 sec, and the clotting time is usually not much prolonged. There may be a deficiency in fibrinogen in severe cases.

Some investigators found that patients suffer from an acute parenchymatous hepatitis accompanied by jaundice. Bronchopneumonia, pulmonary haemorrhages and lung abscesses are frequent complications.

The fourth stage. This is the stage of convalescence and its course and duration depend on the intensity of the toxicosis. Usually 2 months or more elapse until the blood-forming capacity of the bone marrow returns to normal.

(ii) Pathological findings

Toxic materials from overwintered grains have a local as well as a general effect on the tissues of man or animals. The local action is manifested by clinically burning sensations in the mouth, palate, pharynx, oesophagus and stomach. The tongue becomes stiff and vomiting occurs. These phenomena are usually transient.

According to Tomina [50] toxins from overwintered cereals which are absorbed in the stomach and intestine possess an accumulative effect on various organs and tissues.

The most severe effects are on the haemopoietic system and result in depression of leukopoiesis, eythropoiesis and thrombopoiesis [51].

An important contribution to the pathogenesis of ATA was that of Strukov and Tishchenko [52] who showed that disturbances of the haemopoietic system develop although the bone marrow remains viable. ATA causes a temporary depression and disturbance of the haemopoietic system which is reversible and not a destruction of bone marrow.

Strukov [53] and Aleshin *et al.* [54] thought that toxins of overwintered cereals did not act primarily on the bone marrow but on an extramedullary apparatus which regulates the haemopoietic, autonomic nervous and endocrine systems.

Haemorrhages and thrombi are observed in the blood vessels of many organs and necrotic foci are present along the entire gastrointestinal tract. Haemorrhages are also present in the adrenal glands, lungs, pleura and kidneys.

Various investigators have observed changes in the nervous system in ATA [55—57], such as impaired nervous reflexes, meningismus, general hyperaesthesia, encephalitis, cerebral haemorrhages and destructive lesions in the nervous and sympathetic ganglia.

(c) Effects on plants

Toxic effects of *F. poae* and *F. sporotrichioides* are manifested chiefly where the toxins act on tender tissues of young plants [16,18,20,22,58,59] and on over-mature tissue of plants in storage or in the soil [17,60].

(i) Effects of media on which P. poae or F. sporotrichioides have grown

The effects of culture filtrates on seed germination was determined on pea, bean, wheat and barley seed. Cultures of 6 isolates each of *F. poae* and *F. sporotrichioides* were kept for 5 days on 2 liquid substrates at $25°$, for 40 days at $10°$ and for 75 days at $4°$. After sterilization at $100°$ for 15 min, equal amounts of filtrate were applied to filter papers, in large petri dishes and the seeds were placed on them (10 seeds each in 10 petri dishes for each plant/fungus/temperature combination). Control seeds were sown on filter paper wetted with tap water.

All filtrates greatly reduced germination of seeds of all 4 plants from the 89—91% germination of the controls. The reduction increased as the temperature at which cultures were kept decreased; at $25°$, germination ranged from 23 to 59%, at $10°$ from 11 to 21% and filtrates from cultures kept at $4°$ suppressed germination almost entirely. Beans were most and wheat least susceptible to these effects.

In another test, drops of such culture filtrates were applied to leaf blades or leaf axils of young pea, bean and tomato plants. At the site of application strongly necrotic brown-black spots appeared after 9—24 h.

To determine the role of substrates on phytotoxicity, isolates were grown in Erlenmeyer flasks containing 250 ml of each of 3 liquid substrates: *Substrate I* — 2 g $KHNO_3$, 1 g KH_2PO_4, 0.5 g KCl, 0.5 g $MgSO_4 \cdot 7 H_2O$, traces $FeSO_4$, 10 g sucrose, 1000 ml distilled water.
Substrate II — 2 g $KHNO_3$, 1 g KH_2PO_4, 0.5 g KCl, 0.5 g $MgSO_4 \cdot 7 H_2O$, traces of $FeSO_4$, 5 g soluble starch, 2.5 g dextrose, 2.5 g sucrose, 1000 ml distilled water.
Substrate III — 25 g malt extract, 1000 ml distilled water.

After 12 days' culturing at $24°$ the fungal growth was killed by autoclaving. Branches of the 3 test plants were then introduced into each flask and the rate at which they wilted was recorded. Each isolate was tested on each plant and substrate in 4 flasks, so that the test comprised 72 flasks for each substrate. The results showed that phytotoxicity was least on substrate

TABLE IV

EFFECTS OF SEEDLING INOCULATION WITH *F. POAE* AND *F. SPOROTRICHIOIDES* IN ISRAEL

Cultured at (°C)	Plants inoculated										
	Bean	Cucumber	Watermelon	Tomato	Eggplant	Pepper	Onion	Cotton	Wheat	Maize	All hosts
F. poae	(1) Number of seedlings killed (out of 18 for each host-isolate combination)										
6	15	13	9	18	14	16	17	9	2	4	117
12	13	11	8	17	13	14	16	8	1	3	104
18	11	10	6	15	10	11	13	6	0	1	83
24	9	5	4	11	7	6	8	5	0	0	55
30	3	3	2	6	3	3	5	3	0	0	28
35	0	2	0	2	1	1	2	1	0	0	10
F. sporotrichioides											
6	9	16	14	20	18	9	16	11	5	1	119
12	8	13	11	17	15	8	14	9	3	0	98
18	5	11	7	14	11	6	12	7	2	0	75
24	3	6	5	11	7	4	9	4	0	0	49
30	1	2	1	3	3	2	5	2	0	0	19
35	0	0	0	1	1	0	3	1	0	0	6
	(2) Total number of seedlings killed by 8—9 isolates (out of 18 for each host-isolate combination) and number of isolates causing mortality on each host [a].										
F. poae, 8 isolates	31/4	57/6	48/6	61/7	55/8	38/5	71/5	50/6	15/2	6/2	432
F. sporotrichioides, 9 isolates	61/6	33/5	35/6	57/6	42/8	49/7	62/7	57/6	0	0	396

[a] In each pair of figures, the first indicates the total number of the seedlings of each host killed by all the isolates, the second indicates the number of isolates that affected this host. Thus, 31/4 indicates that 4 isolates (out of the 8 tested) caused mortality of 31 seedlings.

III, strongest for *F. poae* on substrate I, and for *F. sporotrichioides* on substrate II.

(ii) Inoculation tests with seedlings

Such tests were carried out in Israel [61] and Germany [58].

In Israel phytotoxicity of 8 isolates of *F. poae* and 9 of *F. sporotrichioides* was tested in inoculation tests with seedlings of 10 field and vegetable crops.

Each isolate was sown on two petri dishes and incubated for 7 days at 24°. In addition one isolate of each species was cultured for 7 days at 12°, 18°, 24°, 30°, 35° or for 46 days at 6° (in two replicates). The contents of uncontaminated dishes were then introduced into a blender and 200 cc. water were added. The resultant suspension was divided over 3—4 beakers, into which 5—8-day-old seedlings were dipped for 1 min. The seedlings were subsequently planted into previously prepared holes in unsterilized soil held in plastic trays. After the seedlings were planted, the soil was well clamped down. There were 18 seedlings of each test crop. Results were recorded after 21 days, but if any one plant of the uninoculated series of a crop wilted, the inoculated series of that crop was disregarded.

The results obtained in these tests are summarized in Table IV. They show that both fungi affected a wide range of plants in their seedling stage, especially tomato, eggplant, pepper, onion, cucumber, cotton and bean, while wheat and maize seedlings were rarely affected. Cultures grown at 6° developed the highest phytotoxicity, which decreased with the rise in temperature and was slight at 30—35°. A large number of isolates affected plants of many species, often as many as 7—9 out of the 10 species tested (Fig. 13).

In Germany, Seemüller [58] has carried out comprehensive tests with *F. poae* and *F. sporotrichioides* on seedlings of wheat, pea and lupine.

The results, as presented in Table V, show that germination of wheat was not affected at all, while that of lupines and peas was affected strongly by *F. sporotrichioides*, though not by *F. poae*. Seemüller [58] similarly found that 4-week-old seedlings of *Pinus sylvestris*, *P. nigra* and *Picea abies* were strongly affected by soil inoculation with *F. sporotrichioides*, but not by *F. poae*.

(iii) Effects on plants past the seedling stage

Studies on effects of *F. poae* and *F. sporotrichioides* on growing plants past the seedling stage have been reported by Seemüller [58]. After he had isolated *F. sporotrichioides* from lesions on a dead branch of *Pinus sylvestris*, 15—26 isolates of this fungus were inoculated into branches of this pine in 3 years' tests. All isolates proved clearly pathogenic, causing wilting of branches from the top downwards, and their ultimate death. Similar symptoms of disease have been caused on other conifers by this fungus in experiments carried out by Gerlach [58].

The white rot of corn-cobs has been found in Germany to be caused by *F. poae* [62]. Seemüller [58] did not generally succeed in inoculations of

Fig. 13. Seedling wilt induced by *F. poae* on onions (top left) and tomato (bottom left) and by *F. sporotrichioides* on cotton (top right) and cucumber (bottom right). Seedlings in control containers are shown in each picture.

TABLE V

EFFECTS OF INOCULATION WITH *F. POAE* AND *F. SPOROTRICHIOIDES* ON GERMINATION OF SEEDLINGS (FROM REF. 58)

	Wheat	Pea	Lupine
F. poae			
Number of isolates tested	30	27	22
Number of seeds sown	1584	1600	1050
% germination	98.4	94.1	83.1
F. sporotrichioides			
Number of isolates tested	26	26	26
Number of seeds sown	1800	1600	1300
% germination	96.7	36.7	17.3
Control			
Number of seeds sown	312	240	240
% germination	97.1	95.0	85.1

TABLE VI

RELATION BETWEEN GRADE OF TOXICITY TO RABBIT SKIN AND SEVERITY OF SEEDLING MORTALITY IN ISOLATES OF F. POAE AND F. SPOROTRICHIOIDES

Species	Grade of toxicity[a]	Number of isolates	Mean % seedling mortality	Mean number of hosts affected
F. poae	7	4	50.5	9.3
	5	1	18.3	7
	3	1	16.1	6
	1	1	4.4	2
	0	1	1.1	2
F. sporotrichioides	7	5	35.3	7.8
	6	1	28.9	8
	5	1	22.8	5
	0	3	0.2	1

[a] Grades of toxicity: 0, no skin reaction; 1-2, slight reddening and/or incipient oedema; 3-4, oedema and/or some leukocytarrhoea; 5, more extensive leukocytarrhoea or slight necrosis; 6, pronounced leukocytarrhoea or necrosis; 7, severe leukocytarrhoea or necrosis.

cobs by *F. poae* unless the cobs had been injured but *F. sporotrichioides* infected uninjured cobs as well. Inoculations were most successful when cobs were in full flower.

A rot of the central bud of carnations due to *F. poae* has been described by Cooper [63]. Seemüller [58] found this fungus capable of causing such rot only in autumn or winter, and then only on buds formed on weak shoots. This author has also reported successful wound inoculation of apple fruits by *F. sporotrichioides*. This fungus, when inoculated by spore suspensions to wheat at the stage of earing, further caused blanching of the ear, shrinking of kernels, and sometimes death of the ear. *F. poae* did not affect the wheat in these tests.

(d) Relation between toxicity to animals and to plants

In order to elucidate the relation, if any, of toxicity of *F. poae* and *F. sporotrichioides* to animals and plants, a tabulation of results obtained with the same isolates in both cases has been made. For this purpose phytotoxicity was expressed in terms of seedling mortality, and toxicity to animals in terms of severity of symptoms provoked on rabbit skin. These data are presented in Table VI. They clearly show a positive relationship. The higher the grade of toxicity to animals, the higher also the mortality rate of seedlings and the larger, in general, the number of plant hosts affected. Seedling mortality caused by isolates non-toxic to rabbits was always negligible.

References p. 260

Aetiology, epidemiology and treatment of ATA

(a) Sources of toxins causing ATA

Large-scale investigations were carried out from 1943 to 1949 on samples from various plants and plant organs and from soil, all collected from experimental cereal fields in the district in which ATA outbreaks in the U.S.S.R. were heaviest [19,21]. Results of these investigations are presented separately for 1943/44, when the outbreak was disastrous and for the remaining years of lighter outbreaks (Table VII).

The data show the enormous rise in the percentage of toxic samples in the year in which outbreaks of the disease was most severe. In the subsequent years of diminished disease severity, samples of wheat were more frequently affected than those of millet or barley. Grains were always more severely toxic than vegetative parts; in soil the percentage of slightly toxic samples was particularly high.

(b) Relation between toxicity of cereals and incidence of ATA

The correlation between the incidence of ATA and the toxicity of overwintered cereals, in a number of years, is presented in Fig. 14. In 1943 the percentage of toxic samples was high but was not paralleled by high incidence of disease: this is due to the fact that in 1943 food in Russia was not yet so scarce as to force large numbers of people to collect overwintered grains. However, in 1944 the high prevalence of toxicity in the samples coincided with extreme scarcity of food, and large parts of the population

TABLE VII

TOXICITY OF SAMPLES OF VARIOUS CEREALS, PLANT ORGANS AND SOIL IN THE YEAR OF HEAVIEST ATA OUTBREAKS (1943/44) AND IN 5 SUBSEQUENT YEARS WITH LOWER INCIDENCE OF DISEASE

Source of samples	Number of samples examined	Percentage of samples		
		Toxic	Slightly toxic	Non-toxic
(1) *Species of cereal*				
1943/44 millet	243	60.8	15.2	27.9
1944/49 millet	420	2.6	5.2	92.2
wheat	415	4.6	4.6	90.8
barley	255	1.9	5.5	92.6
(2) *Plant parts and soil*				
1943/44 vegetative parts	134	39.6	20.1	40.3
grains	109	78.0	9.2	12.8
1944/49 vegetative parts	510	1.9	4.3	93.8
grains	539	4.6	8.1	89.2
soil	285	2.8	8.4	88.8

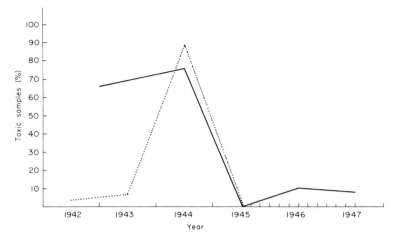

Fig. 14. The incidence of ATA as compared with the percentage of toxicity of samples collected from overwintered cereal crops in 1942—1947. Solid line, percentage of samples of overwintered cereals found to be toxic; broken line, percentage of consumers of over-wintered cereals affected with ATA.

could only subsist by searching for overwintered grain; this resulted in a very high incidence of disease in that year. In the following years, the toxicity of samples decreased and the food situation improved, so that incidence of the disease was greatly reduced.

(c) Weather and the incidence of ATA

A comprehensive analysis of the connection between various weather conditions prevailing in autumn and winter and the incidence of ATA in the 50 countries of the Orenburg district of U.S.S.R. was carried out over the 7 years 1941/42—1947/48 (Table VIII).

These data show that the year of highest disease incidence, 1943/44, was characterized by the following partly interrelated weather factors: (*i*) temperatures in January and February were considerably higher than in most other years; (*ii*) the thickness of the snow-cover in March far exceeded that in all other years; (*iii*) the depth to which the soil was frozen was less in autumn and was especially slight in March.

The higher January—February temperatures coupled with the heavy snow cover in March evidently prevented the soil from freezing to its usual depth, which in February—March usually goes down to 80—120 cm. As described in the section on conditions of toxin formation, temperatures in the −5—10° range and alternate freezing and thawing are exceedingly favourable to toxin production. The weather prevailing in 1943/44 thus greatly favoured accumulation of toxin in the cereals overwintering under the snow, and caused the disastrous outbreaks of ATA in 1944.

During the 30 years preceding the outbreaks of ATA in 1943/44, simi-

TABLE VIII

RELATION BETWEEN AUTUMN AND WINTER WEATHER AND THE INCIDENCE OF ATA IN THE ORENBURG DISTRICT OF U.S.S.R., 1941/42–1947/48

Weather		1941/42	1942/43	1943/44	1944/45	1945/46	1946/47	1947/48
Temp. (°C):	January	−17.7	−18.8	−8.3	−16.2	−12.4	−16.6	−7.0
	February	−18.7	−15.7	−8.0	−20.8	−5.4	−15.1	−12.6
Snow cover (in cm)	December	14	40	29	5	11	4	8
	January	22	53	50	5	29	6	15
	February	23	50	79	6	50	25	21
	March	25	43	108	17	70	18	28
Depth to which soil was frozen (cm)	November	52	21	12	25	40	33	n.r.
	December	n.r.[a]	36	14	30	50	75	41
	January	n.r.	n.r.	27	n.r.	50	106	68
	February	100	n.r.	29	80	n.r.	123	118
	March	n.r.	n.r.	10	80	91	n.r.	124

Incidence of ATA in 50 counties

Number of counties in which the population was affected by ATA at the following rates per 10 000 head:

	1941/42	1942/43	1943/44	1944/45	1945/46	1946/47	1947/48
No disease	31	20	3	36	42	38	0
0–50 cases	12	17	3	14	8	11	0
50–500 cases	7	13	19	0	0	0	0
500–1000 cases	0	0	16	0	0	0	0
more than 1000 cases	0	0	9	0	0	0	0

[a] n.r., not recorded.

lar combinations of low September—October rainfall (up to 25 mm) and relatively mild January—February temperatures (−9.5—12.5°) were recorded in the Orenburg district only in 1924/25 and 1934/35. It is of interest to note that, according to Geminov [7], cases of poisoning occurred in considerable numbers in the rural population of this district in these 2 years.

(d) Summary of factors affecting outbreaks of ATA

The principal factors contributing to the outbreaks of ATA in human populations may be summarized as follows:

(i) Toxicity of the food
The *quantity* of overwintered grain ingested: usually the disease appeared after eating at least 2 kg overwintered grain and death occurred after eating 6 kg.

The *kind* of cereal ingested: millet and wheat were the most toxic.

The *period* over which toxic grain was eaten: symptoms usually appeared 2—3 weeks after toxic grain was eaten, but death was generally caused only by repeated ingestion of grain over several weeks; lethal effects then appeared 6—8 weeks after the first symptoms.

Concentration of toxin in the food: Great variations were found in the toxicity of overwintered grain, even within samples of one field. Families who fed on non-toxic overwintered grain were not affected. Entire families were found who ingested highly toxic grains and were all affected by ATA.

(ii) Sensitivity of the individual
Age: Babies less than 1 year old were not affected. Disease occurred most frequently and caused highest mortality between the ages of 8 and 50.

Sex: There was no difference in the incidence of ATA in men and women.

Nutritional status: Undernourished persons were affected more severely, as their diet consisted almost entirely of cereals, which caused avitaminosis.

(iii) Environmental and agricultural factors
The *season* of harvesting: grains harvested during the spring thaws were toxic, those harvested in autumn were not.

Weather: relatively high January—February temperatures, deep snow covers, preventing the soil from freezing to a great depth, and freezing and thawing of the soil in spring were the weather factors most favourable to ATA outbreaks.

Altitude: the disease has not been found in the U.S.S.R. at altitudes above 350—400 m, probably because of the lower winter temperatures.

(e) Prophylaxis and treatment

The most important prophylactic measure is to refrain from eating

overwintered toxic grains. Primary preventive measures thus consist in edu-
cating the rural population as to the aetiology and clinical symptoms of
ATA, and such measures have greatly reduced outbreaks of ATA. When the
outbreaks of the disease were first reported, medical teams were sent to the
affected areas, and the population was examined clinically and haematologi-
cally.

Grain samples should be examined for toxicity by skin tests. At the
same time, toxic cereals should be replaced with normal grain. When ATA
was detected in the second stage, treatment included blood transfusion and
the administration of nucleic acid and calcium preparations, sulphonamides,
and vitamins C and K. When the number of leukocytes declined below
$3000/mm^3$, hospitalization was recommended. During the third stage, the
same measures are indicated as in the second stage, but treatment should be
more intensive. Following recovery, a rich diet must be given for 1 month,
and the patient should remain under periodic haematological surveillance.

References

1 I.B. Davydovski and A. Kestner, Arch. Pathol. Anat. Pathol. Physiol., 1 (1935) 11.
2 I.P. Vasiliev and N.A. Abragimov, in Alimentary Toxic Aleukia, Publ. Tartar SSR,
   Kazan, 1945, p. 7.
3 L.S. Lass, Vitebsk. Med. Inst., (1940) 3.
4 V.V. Yefremov, Publ. Med. Lit. Moskow, (1948) 1—118.
5 A.P. Onegov and B.A. Naumov, Acta Kirovsk. Zootech. Vet. Inst., 5 (1943) 110.
6 B.A. Talaev, B.I. Mugunov and E.N. Sharbe, Nutrit. Probl., 5 (1936) 27.
7 H.B. Geminov, in Septic Angina and its Medical Treatment, USSR Min. of Health,
   Kuibyshev, 1945 p. 3.
8 B.A. Riazanov, Lectures on the Republic Conference on Alimentary Toxic Aleukia,
   Moscow, 1947.
9 B.A. Riazanov, Lectures on the Republic Conference on Alimentary Toxic Aleukia,
   Moscow, 1948.
10 G.N. Beletzki, Hyg. Sanit., 3 (1943) 22.
11 L.N. Karlic, in A.E. Nestierow A.N. Sysin and L.N. Karlic (Eds.), Alimentary Toxic
   Aleukia (Septic Angina), Medgiz, Moscow, 1945, p. 12.
12 J. Forgacs and W.T. Carll, Advan. Vet. Sci., 7 (1962) 273.
13 Z. Alisova and S. Mironov, in Data on Septic Angina, Vol. 1, Kharkov Med. Inst. and
   Chkalov Inst. Epidemiol. Microbiol., 1944, p. 17.
14 Z. Alisova, Acta Chkalov Inst. Epidemiol. Microbiol., Second Commun., 2 (1947)
   104.
15 A.Z. Joffe and S.G. Minonov, Acta Chkalov Inst. Epidemiol. Microbiol., 2 (1947) 35.
16 A.Z. Joffe, Bull. Res. Counc. Israel, 8D (1960) 81.
17 A.Z. Joffe, Bull. Res. Counc. Israel, 9D (1960) 101.
18 A.Z. Joffe, Mycopathol. Mycol. Appl., 11 (1962) 201.
19 A.Z. Joffe, Plant Soil, 18 (1963) 31.
20 A.Z. Joffe, in G.N. Wogan (Ed.), Mycotoxins in Foodstuffs, M.I.T., Cambridge, Mass.,
   1965, p. 77.
21 A.Z. Joffe, in Microbial Toxins, Vol. 7, Academic Press, New York, 1971, p. 139.
22 A.Z. Joffe, Mycopathol. Mycol. Appl., Vol. dedicated to Honorato Verona, 1974, pp.
   1—29.
23 E.M. Gubarev and N.A. Gubareva, Biochemistry, 10 (1945) 199.

24 B.L. Kretovitz and A.A. Bundel, Biochemistry, 10 (1945) 216.

25 B.L. Kretovitz and N.I. Sosedov, Biochemistry, 10 (1945) 279.

26 U.O. Gabel, Acta Chkalov Inst. Epidemiol. Microbiol., 2 (1947) 42.

27 B.D. Svoyskaya, Acta Chkalov Inst. Epidemiol. Microbiol., 2 (1947) 45.

28 N.V. Okuniev, Lectures on the Republic Conference on Alimentary Toxic Aleukia, Moscow, 1948.

29 L.E. Olifson, Publ. Acad. Sci. Ukr. S.S.R., Kiev, (1956) 21.

30 L.E. Olifson, Bull. Chkalov Sect. D.E. Mendeliev All Soviet Union Chem. Assoc., 7 (1957a) 21.

31 L.E. Olifson, Bull. Chkalov Sect. D.E. Mendeliev All Soviet Union Chem. Assoc., 7 (1957b) 37.

32 L.E. Olifson, Ph.D. Diss., I.T. of Industrial Nutrition, Moscow, 1965 pp. 1—36. (Abstr.)

33 A.Z. Joffe, Mycopathol. Mycol. Appl., 53 (1974) in press.

34 B.S. Drabkin and A.Z. Joffe, Acta Chkalov Med. Inst., 2 (1950) 92.

35 B.S. Drabkin and A.Z. Joffe, Microbiol., 21 (1952) 700.

36 R. Schoental and A.Z. Joffe, J. Pathol., 112(1) (1974) 37.

37 V.I. Bilai, Publ. Acad. Sci. Ukr. S.S.R., Kiev, (1953) 1.

38 G.I. Maisuradge, Cand. Diss., Moscow, 1953 pp. 1—15. (Abstr.)

39 E.T. Hrootski and A.Z. Joffe, Acta Chkalov Agr. Inst., 6 (1953) 59.

40 P.G. Sergiev, in A.E. Nestierow, A.N. Sysin and L.N. Karlic (Eds.), Alimentary Toxic Aleukia (Septic Angina), Medgiz, Moscow, 1945, p. 7.

41 A.Kh. Sarkisov, Ed. Min. Agr. U.S.S.R., Moscow, (1948) 22.

42 N.A. Antonov, G.S. Belkin, A.Z. Joffe, A.Y. Lukin and I.N. Simonov, Acta Chkalov Agr. Inst., 4 (1951) 47.

43 A.Kh. Sarkisov, H.E. Korneev and E.S. Kvashnina, Ed. Min. Agr. U.S.S.R., Moscow, (1948) 54.

44 A.Kh. Sarkisov, U.S.S.R. Agr. Min., Moscow, (1954) 1—216.

45 Y.I. Rubinstein and L.S. Lass, Hyg. Sanit., 7 (1948) 33.

46 V.I. Chilikin, Acta Chkalov Inst. Epidemiol. Microbiol., 2 (1947) 147.

47 E.M. Manburg, in Data on Septic Angina, Vol. 1 Chkalov, 1944, p. 85.

48 E.M. Manburg and E.A. Rachalski, Acta Chkalov Inst. Epidemiol. Microbiol., 2 (1947) 152.

49 E.D. Romanova, Acta Chkalov Inst. Epidemiol. Microbiol., 2 (1947) 164.

50 M.V. Tomina, Republic Conference on Alimentary Toxic Aleukia, Moscow, 1948. (Abstracts).

51 M.A. Koza, I.A. Leontiev and P.Y. Yasnitzki, Alimentary Toxic Aleukia, Moscow, 1944, p. 3.

52 A.I. Strukov and M.A. Tishchenko, Acta Chkalov Inst. Epidemiol. Microbiol., 2 (1947) 120.

53 A.I. Strukov, Acta Chkalov Inst. Epidemiol. Microbiol., 2 (1947) 117.

54 B.V. Aleshin, Sh.A. Burstein and B.I. Cherniak, Acta Chkalov Inst. Epidemiol. Microbiol., 2 (1947) 125.

55 Z.A. Gurewitch, in Data on Septic Angina, Vol. 1 Chkalov, 1944 p. 103.

56 A.S. Poznanski, Acta Chkalov Inst. Epidemiol. Microbiol., 2 (1947) 176.

57 E.N. Kovalev, Neuropathol. Psychiat. (U.S.S.R.), 13 (1944) 75.

58 E. Seemüller, Arb. Biol. Bund Anst. Land Forstw., 127 (1968) 1—93.

59 H. Tint, Phytopathology, 35 (1945) 421.

60 M.W. Gilgan, E.B. Smalley and F.M. Strong, Arch. Biochem. Biophys., 114 (1966) 1—3.

61 A.Z. Joffe, Z. Pflanzenkrankh. Pflanzensch., 80(2) (1973) 92.

62 A. Focke und W. Kühnel, Nachrichtenbl. Deut. Pflanzenschutzd., 18 (1964) 116.

63 K.W. Cooper, Phytopathology, 30 (1940) 853.

64 J.R. Bamburg, N.V. Riggs and F.M. Strong, Tetrahedron, 24 (1968) 3329.

65 E.B. Smalley, W.F.O. Marasas, E.M. Strong, J.R. Bamburg, R.E. Nichols and N.R. Kosuri, in M. Herzberg (Ed.), Toxic Micro-Organisms, U.S. Dept. of Interior, Washington, D.C., 1968, pp. 163—173.
66 J.R. Bamburg and F.M. Strong, in S. Kadis, A. Ciegler, and S.J. Ajl (Eds.), Microbial Toxins, Vol. 7, Academic Press, New York, 1971, pp. 207—292.
67 C.J. Mirocha and S. Pathre, Appl. Microbiol., 26 (1973) 719.
68 L.E. Olifson, Proceedings of the Symposium on Mycotoxins, Academy of Sciences of Ukrainian S.S.R., Kiev, Oct. 3—9, 1972, p. 12.
69 L.E. Olifson, S.M. Kenina and V.I. Kartashova, Orenburg Regional Administration of D.I. Mendeliev, All-Union Chemical Association, 1972, pp. 3—8.
70 Y. Ueno, N. Sato, K. Ishii, K. Sakai and M. Enomoto, Jap. J. Exptl. Med., 42 (1972) 461.

Note added in proof

However, recent studies carried out mostly in the U.S.A. [64—67] throw doubt on the role of steroid materials as responsible for the toxic action of *F. sporotrichioides* and *F. poae*. These studies were carried out with toxins of these fungi obtained from the U.S.S.R., and with American Strains determined as *F. tricinctum* (Corda) Sacc. emend. Snyder and Hansen, which we have reason to assume belong to what, by mycological standards, should more properly be called *F. poae* (Peck.) Wr. and *F. sporotrichioides* Sherb. In this work the toxins were found to be diacetoxyscirpenol and T-2 toxin, which are sesquiterpenoids and derivatives of the trichothecene group. Bamburg and Strong [66] have concluded that ATA is more likely due to trichothecene toxins rather than to the steroidal toxins reported by Olifson [29—32,68] and Olifson *et al.* [69].

In Japan, too, Ueno *et al.* [70] showed by chemical analysis that in *F. pcae*, and *F. sporotrichioides* trichothecene-type toxins (T-2 toxin, neosolaniol and traces of HT-toxin) were present.

Chapter 12

TRICHOTHECENE TOXINS OF FUSARIUM SPECIES

M. SAITO and K. OHTSUBO

The Institute of Medical Science, The University of Tokyo, Shirokanedai, Minato-ku, Tokyo-108 (Japan)

Poisoning due to scab grains has a rather long history. In 1891 Woronin reported that in European Russia and Eastern Siberia "scabby grains", as they were then called, were toxic to men and animals [64,93]. In Japan Akakabi-byo (red mould disease) or Kokuten-byo (black spot disease) of wheat have been noticed since the last century. Diseased oats reduced the growth and breeding of military horses and *F. saubinetti* was identified as the causative agent [28]. Fusarium toxicosis was also reported in association with toxic corn disease of cows in the U.S.A. [5,56], fescue foot disease, alimentary toxic aleukia in the U.S.S.R. (p. 229) and stachybotryo-toxicosis (p. 181), but in all of these cases the mycotoxins concerned have not yet been directly isolated from the toxic feed.

This blight was called "wheat scab" in England, where the disease was first described [57]. In America it is variously called Fusarium blight, Fusarial head blight or ear blight. In Germany a similar disease of wheat, rye, oats, barley and maize was called "Schorfkrankheit" [49]. The blighting of wheat is mostly attributed to a fungus, *Gibberella zeae* (Schw.) Petch or *Fusarium graminearum* Schwabe, its asexual or conidial phase.

As described below, over 20 closely related substances with the common basic structure of 12,13-epoxytrichothecene have been isolated as the metabolites of several families of imperfect fungi; Fusarium, Trichothecium, Myrothecium and Cephalosporium. All these fungi belong to field fungi from the viewpoint of food hygiene. [10]. Among them Fusarium spp. are most important because they most frequently destroy wheat, rice, and other cereals and beans. According to extensive research by Ishii *et al.* [25] and Nishikado [38], the "red mould" disease is prevalent in the regions of the Japan Islands facing the Pacific Ocean. This is probably due to the coincidence of the rainy season of these districts and the blooming of wheat [24,40,60]. The causative fungus was found to be *F. graminearum* in most cases [38] and rarely *F. nivale*, *F. poe* and *F. oxysporium** [72].

In 1963, owing to a large outbreak of akakabi-byo throughout the western part of Japan, the yearly harvest of wheat in this district was one-

---

* Microbial classification of Fusaria or Gibberella is still controversial. We adopted here the classification of Tsunoda [75] based on that of Snyder and Hansen [58].

half of that of a normal year [72]. From the scabbed wheat grains at Kumamoto Prefectural Laboratory of Agriculture in Kyushu, strains of *F. graminearum*, *F. nivale etc.* were isolated. When grown on rice most of them were more or less toxic to mice, but *F. nivale* Fn 2b was found to produce the highest mortality [65,75]. Rapidly proliferating cells of the animals, such as crypt cells of the intestinal mucosa and bone marrow cells, were selectively affected. Extracts of this isolate were cytotoxic to the HeLa cells [91]. These radiomimetic effects were used as a guide in the isolation of a new toxic substance, nivalenol, from the moulded rice grains [63,66] and thus *F. nivale* Fn 2b became a standard strain for our toxicological study. The colony of this strain was floccose, pink coloured and a purple pigment eluted into the plate. Crescent- or spindle-shaped conidia (3.5 to 4.5 · 10 to 31 μ with 1 to 3 septa) attach to the top of 1 to 3 sterigmata extending from twig-like branches of conidiophores [75].

Outbreaks of disease attributable to the fungus

In 1928 an extensive outbreak of barley blight occurred in the U.S.A. Mouldy barley was exported to Europe and caused wide-spread poisoning of pigs [49]. In northern Germany the imported barley was found to be heavily contaminated with *F. roseum* accompanied by rose yeast, *Cladosporium hearbarum* and Alternaria spp. [33]

In 1955 intoxication due to ingestion of deteriorated rice occurred in Tochigi Prefecture and Kochi Prefecture suburbs of Tokyo. More than 40 persons were affected with nausea and vomiting and diarrhoea occurred in some cases. Fusarium sp. was isolated from more than 50% of the rice sampled in these cases. 2 out of 4 strains of fungi isolated showed marked toxicity, mice dying in 2 weeks after receiving a diet containing 50% powdered rice which was artificially infected with the two strains [73]. Crude toxins extracted from the rice mentioned above or from the culture broth (Czapek solution supplemented with 0.5% peptone) were also fatal to rats [74].

In October and November 1956 food poisoning affecting more than 100 persons occurred at a small institute for agricultural training in Hokkaido. The patients, consisting of about 80% of the trainees, developed headache (18%), abdominal pain, nausea (91%), vomiting (70%), diarrhoea (4%), chills (30%) and fever (0.4%) 5 to 30 min after ingestion of the suspected diet [41]. There were no fatal cases. In both accidents the noodles were found to be made from flour contaminated with Fusarium sp.

Similar cases were encountered several times in the post-World War II period in Hokkaido [61] and in 1946 in Tokyo due to imported wheat flour [22]. In all of these cases *F. graminearum* was isolated but it was not proved that the isolated fungi actually produced toxic metabolite(s).

The earlier reports of outbreaks of disease in Japan were concerned mainly with livestock and laboratory animals [23,51,62]. Intestinal catarrh in guinea pigs and similar symptoms with diarrhoea in horses were observed when these animals were fed scab wheat grains [29,70].

Toxic metabolites of the fungus

The major toxic metabolites (mycotoxins) of Fusaria are butenolide, zearalenone and 12,13-epoxytrichothecene derivatives (Fig. 1). Various derivatives of trichothecenes, so far more than 20 compounds, have been isolated from Trichothecium, Cephalosporium, Myrothecium and Fusarium [4,26,34]. The first trichothecene compounds, verrucarrins, were isolated in the search for antibiotics in 1946 [8]. Diacetoxyscirpenol was the earliest isolate of phytotoxic trichothecenes [9].

In Japan Tatsuno and co-workers isolated nivalenol from cultures of *F. nivale* on rice grains [63,66]. Fusarenon-X was purified from a culture broth of *F. nivale* Fn 2b [77]. Fusarenon, isolated by Morooka [34,75] from cultures of the same straim of *F. nivale* as that used by Tatsuno with a slightly different procedure, was recently confirmed to be identical with

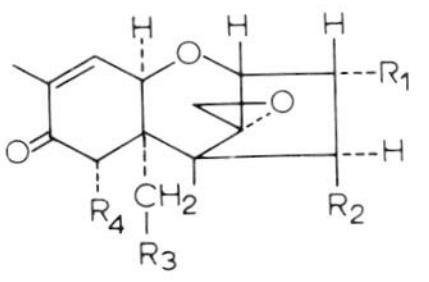

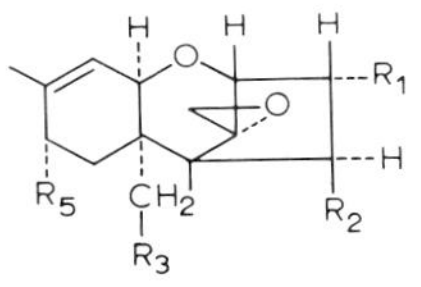

Fig. 1. Chemical structures of naturally occurring trichothecenes.

(I)

	R_1	R_2		R_3	R_4	References to isolation and chemical properties
Nivalenol	OH	OH		OH	OH	65, 66
Fusarenon-X (Fusarenon)	OH	OAC		OH	OH	35, 65, 75, 76, 84
Trichothecolone	H	OH		H	H	1
Trichothecin	H	OOCCH=CHMe		H	H	12, 13, 17

(II)

	R_1	R_2	R_3	R_5	
Trichodermol	H	OH	H	H	21
Diacetoxyscirpenol	OH	OAC	OAC	H	5, 9, 11, 55
T-2 toxin	OH	OAC	OAC	$OOCCH_2 CHMe_2$	6, 86, 87
HT-2 toxin	OH	OH	OAC	$OOCCH_2 CHMe_2$	4
Diacetylverrucarrol	H	OAC	OAC	H	48
Trichodermin	H	OAC	H	H	17
Neosolaniol	OH	OAC	OAC	OH	26, 86, 87

References p. 279

fusarenon-X [35]. From *F. solani*-contaminated bean hulls Ueno and his coworkers isolated neosolaniol and T-2 toxin [26,86]. The former is assumed to be the hydrolyzed product of the latter. Other trichothecenes isolated in this country are diacetoxyscirpenol [86,87], 4,15-diacetoxy-3,7-dihydroxyscirpenone and diacetylnivalenol [19,68]. All of these compounds have in common the chemical structure 4-β-hydroxy(acyloxy)-12,13-epoxy-Δ^9-trichothecene that characterizes naturally occurring toxins of this group [4] (Fig. 1). Recently Morooka and associates [94] isolated dehydroxynivalenol, a derivative without a hydroxyl group at position 4, directly from wheat naturally contaminated with *F. roseum*. The 3-mono-acetate was also isolated from a culture of a strain of *F. roseum* as were hydroxynivalenol and butenolide [94].

During his extensive screening of trichothecene-producing Fusaria [76,83,86,87] Ueno noticed that 8-enol and 8-keto derivatives are produced by different strains (species) of Fusaria. For example, *F. nivale* Fn 2b produced nivalenol and fusarenon-X, but not T-2 toxin or diacetoxyscirpenol. On the other hand, *F. solani* M-1-1 produced T-2 toxin, neosolaniol and diacetoxyscirpenol and HT-2 toxin, but not nivalenol and fusarenon-X [26,86,87]. In general the fungal strains producing the former type of trichothecenes were isolated from food originating from colder districts and those producing the latter groups of toxins from the rather warm areas of the world. He considers that the synthesis of 8-ketone or 8-ol derivatives has taxonomic significance.

Chemistry of nivalenol, fusarenon-X and neosolaniol

The basic chemical structure of the compounds of this group, formally called sesquiterpenoids, has been systematically named trichothecane [18]. Of the natural products of *F. nivale*, nivalenol is the simplest 8-keto derivative and neosolaniol the simplest 8-hydroxy derivative (Fig. 1). Physico-chemical properties of nivalenol and fusarenon-X isolated from *F. nivale* and of neosolaniol [26], the most recently identified substance of this group, are tabulated in Table I. For the chemistry of other trichothecenes the detailed review by Bamburg and Strong [4] should be consulted.

Purification procedures of representative trichothecene compounds of Fusarium developed in our laboratories are as follows.

Nivalenol [67]. *F. nivale* Fn 2b was cultivated on sterile rice at 28° for 3 weeks. The lipids and non-toxic pigments were removed from the mouldy rice by acetone extraction and then the residue was extracted with aqueous alcohol. The toxic materials in the alcohol solution were adsorbed onto activated charcoal, eluted with methanol, concentrated and non-toxic metabolites were precipitated by adding chloroform. The methanol—chloroform-soluble materials were chromatographed on a Kieselgel column using chloroform—methanol as eluant. From the most toxic fraction, the crystals of nivalenol were obtained.

TABLE I

PHYSICOCHEMICAL PROPERTIES OF TOXINS ISOLATED FROM *FUSARIUM NIVALE* AND *F. SOLANI* [26, 67].

Property	Nivalenol		Fusarenon-X		Neosolaniol	
m.p.	$222-223°$		$91-92°$		$171-172$	
R_F on a TLC[a]	0.05		0.31		0.15	
M^+ ion m/e	312		354		382	
Molecule formation	$C_{15}H_{20}O_7$		$C_{17}H_{22}O_8$		$C_{19}H_{26}O_8$	
Elemental analysis	Found,	Calc.	Found,	Calc.	Found,	Calc.
C:	56.29	57.68	56.42	57.62	59.78	59.65
H:	6.01	6.49	6.06	6.26	6.66	6.85
O:	37.70	35.86	37.52	36.12	33.56	33.49
UV_{max} (mμ)	218 (CH_3OH)		220 (CH_3OH)			
IR VKBr (cm^{-1}) OH	3500—3200		3400		3450	
OCH$_3$	—		1720		1735	1250
C=O	1680—1610		1680		—	
$[\alpha_D]^{24}$(C_2H_5OH)	$+21.5°$		$+56.1°$			

a Kieselgel G-ethylacetate: *n*-hexane (3:1).

Fusarenon-X [84]. *F. nivale* Fn 2b was cultured on a stationary peptone-supplemented Czapek medium (PSC-medium) for 2 weeks at 25° to 27°. The yellow-coloured filtrate was passed through a charcoal column, eluted with methanol and dried under reduced pressure. The residue (a yellow powder) was dissolved in hot methanol and the deep yellow methanol solution was mixed with 5 vol. of chloroform and the supernatant was decanted and dried *in vacuo*.

This "crude" toxin was chromatographed using chloroform—methanol as eluant and a Kieselgel column as adsorbent. From one toxic fraction nivalenol was isolated and the other toxic fraction was re-chromatographed using chloroform—acetone as eluant. The most toxic fraction collected from this column was dissolved in dichloromethane, followed by drop-wise addition of *n*-pentane. This solution was kept at 10° overnight to obtain transparent bispyramids, fusarenon-X. The yield was 150 to 200 mg from 6 g of the "crude" toxin or 10 l of the culture filtrate.

Neosolaniol [26]. *F. solani* M-1-1 was grown in PSC-medium in a stationary flask at 25° for 12 days.

Crude toxin, 13 g from 20 l of the original filtrate, was prepared from the culture filtrate, as described in the case of fusarenon-X, and chromatographed successively on a silicagel column with *n*-hexane-ethyl acetate, ethyl acetate, ethyl acetate-methanol and methanol to obtain 7 fractions. From 2 fractions T-2 toxin and diacetoxyscirpenol were identified. A third toxic fraction was similarly rechromatographed with acetone-*n*-hexane. The puri-fied fraction gave rise to 200 mg of crystals from ethyl acetate-*n*-hexane. This new substance was named solaniol, but this name was later [86] altered to neosolaniol to avoid confusion with a quinoid known as "(+)-solaniol" [2].

Biosynthesis

[1-^{14}C]acetate and [2-^{14}C]mevalonolactone were incorporated into trichothecin [30]. On the basis of these data Sigg *et al.* [55] proposed a pathway of biosynthesis.

In his studies of biosynthesis of trichothecolon using *Trichothecium roseum*, Nozoe [39] has identified, as intermediate products, compound Q (12,13-epoxytrichothecene) and compound R (4,8-dihydroxy-12,13-epoxy-trichothecene). Both samples were toxic to HeLa cells but about 100 times less toxic than fusarenon-X [91]. The complete pathway of biosynthesis, however, has not been elucidated [39].

Conditions of mycotoxin production

As described above, many strains of Fusaria which had been isolated in Japan produced either fusarenon-X, nivalenol and diacetylnivalenol or neo-solaniol, T-2 toxin and diacetoxyscirpenol. Butenolide was occasionally de-tected in association with the latter group [87].

Yields of fusarenon-X and nivalenol depended on the substrate on which *F. nivale* Fn 2b was cultured. On rice this strain produces much more nivale-nol than fusarenon-X, whereas with the PSC medium the same fungus syn-thesizes more fusarenon-X than nivalenol [82]. *F. nivale* Fn 2b was most toxic when cultured on rice. The toxicity of mouldy kernels, in descending order of toxicity, is: wheat, corn, rye and barley [75,82]. Of the liquid media, Czapek medium supplemented with yeast extract or peptone supported both mycelial growth and toxin production of the same strain of the fungus while Raulin-Thom and Sabouraud medium supported mycelial growth only.

Production of fusarenon-X by *F. nivale* Fn 2b in PSC medium was higher when cultured at 27° than at 10° or 20°. Mycelial growth, on the other hand, was most rapid at 10° [82]. As for production of T-2 toxin by *F. tricinctum* Bamburg *et al.* [6] reported that the toxin was only produced at temperatures as low as 4° to 8° and not at 20°, although the fungal growth was faster at the higher temperature. This was also the case with diacetoxyscirpenol production by a different strain of *F. tricinctum* [14]. These facts indicate the important influence of temperature on mycotoxin production.

Physical and biological assay of trichothecenes

Gas chromatography has been the most sensitive and reliable method for chemical detection of trichothecenes. A detailed gas chromatographic analysis of T-2 toxin was reported by Bamburg [3].

Tatsuno succeeded in separating several trichothecenes [69] using a column containing 1.5% OV-17 with shimalite W as its support. Trichothecenes having free hydroxyl groups were at first trimethyl-silylated with 10% bis-trimethylsilylacetamide and 5% trimethylsilylchloride in benzene at 60° for 15 min. Each trichothecene compound chromatographed at 220°–240° on the column mentioned above showed a single peak and they were well separated from each other giving retention times of 3.55, 3.95, 8.95, 5.95 and 7.50 min for nivalenol, fusarenon-X, diacetylnivalenol, diacetoxyscirpenol and neosolaniol, respectively. Trichothecin and T-2 toxin were not separated on this column (r.t. 8.55 min), but on an SE-30 column, there was a 3.5 min difference between their retention times.

A 5 g sample of powdered feed for laboratory animals containing 40 ppm of fusarenon-X was assayed using the following technique. The sample was extracted with a mixture of 10% aqueous NaCl and 10% methanol, the lipids removed with n-hexane, fusarenon-X was transferred to chloroform, dried and finally trimethylsilylated before gas chromatography. There was an 86% recovery of fusarenon-X. This method, therefore, can be used as a quantitative assay for trichothecenes in foodstuffs or cereals. The method is capable of detecting 2 or 3 μg, as long as the final extract contains 1 mg/ml.

Thin-layer chromatography is one of the most useful methods for the purification and identification of trichothecenes. R_F values of many trichothecenes in various solvent systems and criticism of the methods are described by Bamburg and Strong [4] and Ueno et al. [88].

Ueno has investigated the detection of mycotoxins produced by Fusaria [76,88]. On a Kieselgel G plate trichothecenes, butenolide and zearalenone were clearly separated by different R_F values using chloroform—methanol (95:5), ethylacetate—n-hexane (3:1) (Table II), benzene—acetone (3:2) and ethylacetate—toluene (3:1) as developers. Trichothecenes yield no visible or fluorescent spots by ultraviolet light but are detected by spraying the plate with 20% H_2SO_4 and heating to 100° when they appear as brown (nivalenol, fusarenon-X, diacetylnivalenol) or grey (neosolaniol, diacetoxyscirpenol, T-2 toxin) spots. After this treatment the latter 3 and butenolide give rise to sky-blue fluorescence under ultraviolet light.

Various biological methods for detection of trichothecenes have been reported. Inflammatory reaction of the skin of rats [56] rabbits [81] and guinea pigs after application of trichothecenes is varied, but usable for T-2 toxin, HT-2 toxin, diacetoxyscirpenol and fusarenon-X. Phytotoxicity assays using pea seedling growth inhibition and growth inhibition of pea internode [9] are sensitive methods for detection of diacetoxyscirpenol and T-2 toxin [32]. Antibiotic methods are of little value except for trichothecin [13]. None of these methods were specific for trichothecenes but some of them can be used as a qualitative and semiquantitative guide during isolation [56].

In this connection inhibition of protein synthesis in the rabbit reticulocyte is of particular value because of its specificity as a screening method for

TABLE II

COMPARATIVE TOXICITY OF TRICHOTHECENES [42, 44, 45, 76, 80, 81, 85, 86, 88, 91, 94]

Trichothecenes (fungal strains)		Acute toxicity to male mice (LD_{50} mg/kg i.p.)	Irritation on rabbit skin		Cytotoxicity to HeLa cells at		Inhibition of protein synthesis	
			10 μg	1.0 μg	3.0 μg/ml	0.3 μg/ml	In HeLa cells (ID_{50} μg/ml)	In rabbit reticulocytes (ID_{50} μg/ml)
Nivalenol	(*F. nivale*)	4.0	+	—	4	3	0.4	2.5
Fusarenon-X	(*F. nivale, F. episharia etc.*)	3.4			4	3.5	0.1	0.25
Diacetylnivalenol	(*F. nivale*)	3.5						0.30
Diacetoxyscirpenol	(*F. solani*)	23.0	++	+	4	4	about 0.05	0.03
Neosolaniol	(*F. solani*)	14.5			4	4	about 0.1	0.25
T-2 toxin	(*F. solani, etc.*)	5.2		+	4	4		0.03
Trichothecin	(*Trichothecium roseum*)				4	2		0.30
4-Dehydroxynivalenol	(*F. roseum*)	70.0			4	2.5		
3-Acetyl-4-dehydroxy-nivalenol	(*F. roseum*)	49.4			3	0		
Dihydronivalenol		about 18			13.5	0.5	2.0	

trichothecenes [76,79]. With the method developed by Ueno, incorporation of radioactive amino acid, usually [^{14}C]leucine, into acid-insoluble fractions of reticulocytes was specifically inhibited by trichothecene compounds but not or only slightly with other mycotoxins such as butenolide, luteoskyrin, cyclochlorotine, citrinin and aflatoxin [76,80].

As we shall see, cultured cells are highly susceptible to most of the compounds of this group and could be used for screening in the same way as they have been for other mycotoxins [53]. With the use of the plastic panel technique, originally designed by Toplin to screen antitumour antibiotics [71], bioassay of mycotoxins has become very easy and practical, requiring only small amounts of test samples. If the cells are cultured on a cover slip they may be directly fixed and stained for histological examination. Then, besides the lethal and growth-inhibitory effect, characteristic morphological changes [90] enable diagnosis of the specific cytotoxicity caused by trichothecenes or other mycotoxins.

Biological activity of the toxins

(1) Acute toxicity of trichothecene compounds to mice and other animals

The acute LD_{50} values of various trichothecenes in male mice are shown in Table II [52,76,85,86,88,94]. The acute signs and symptoms in rats and mice injected with lethal doses of nivalenol or fusarenon-X appeared rapidly, within half an hour, and were weakness, tachycardia, low body temperature, and then diarrhoea after several hours. In cats and Peking ducklings vomiting was the first and predominant symptom [76,85]. Nervous symptoms were absent. With T-2 toxin tachycardia and low blood pressure were observed in rats [56].

The mice injected with doses more than several times the LD_{50} of fusarenon-X survived about 10 h, although the symptoms appeared very rapidly. With lower doses survival time was longer, up to 3 or 4 days. In the early phase the animals died in shock. When the animals survived for a few days, lesions in the intestinal tracts (haemorrhages, erosions and ulcerations with massive fluid content) were responsible for death. These may be compared with acute death after whole-body irradiation [52]. The number of peripheral blood cells, especially of lymphocytes, increased markedly (2 to 5 times the normal value) during the 3 to 6 h after injection of nivalenol, fusarenon-X and neosolaniol [52,89]. With sublethal doses white blood cell counts almost recovered to the initial level after 24 h. The red cell counts were affected little throughout the experimental period. This leukemoid-like reaction did occur with lower doses after which bone marrow histology revealed no pathological change. In a subacute experiment with rats, leukopenia and anaemia were observed in 4 weeks after daily intravenous injections of 0.54 μg/kg of diacetoxyscirpenol [59].

Histopathological changes induced in mice are common with all 12,13-epoxytrichothecene derivatives examined. As repeatedly described [26,50,52,84,87], the proliferating cells of the tissues were selectively damaged. Fusarenon-X, for example [85], causes necrosis of the epithelium of the digestive system, especially of the small intestine (Fig. 2), karyorrhexis of the undifferentiated lymphoid cells of the follicles in the spleen, lymph nodes and other lymph apparatus, of the thymus, of the haemopoietic cells of the bone marrow and spleen. Partial atrophy of seminiferous tubules and destruction of the granulosa cells of the ovary are occasionally noted. The responses of the animals were almost similar irrespective of the route of administration. As far as histological changes are concerned, there are no differences between male and female and between old and young.

Histological findings in other species of animals administered fusarenon-X were very similar to those observed in the mouse, though the susceptibility to the toxin differed rather markedly. For example the acute LD_{50} values of fusarenon-X in various animals are as follows [76,85]: male and female mice, 3.4 mg/kg, new-born mice, about 0.1 mg/kg, male and female Wistar rats, 4.4 and 4.0 mg/kg, respectively, guinea pigs about 0.5 mg/kg and cats about 1 mg/kg. Horses seem to be the most susceptible. 1 to 2 g of fusarenon-X or T-2 toxin is sufficient to kill a horse weighing about 500 kg [47].

Distribution of fusarenon-X in the tissues of mice was investigated by Ueno et al. [85] using tritium-labelled material. The radioactivity was found 30 min after intraperitoneal administration in the liver (3% of the injected radioactivity), kidneys (1%), intestines (1.5%), stomach, spleen, bile and plasma. Tritiated fusarenon-X was rapidly and for the most part excreted in the urine, almost no radioactivity remaining in the tissues after 3 h. Analysis of a thin-layer chromatogram of the urine revealed that the major counts

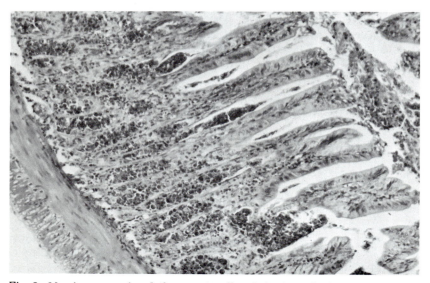

Fig. 2. Massive necrosis of the crypt cells of the intestinal mucosa of a mouse injected intraperitoneally with 150 μg/10 g of fusarenon-X and sacrificed after 3 h. (H.E. stain, × 160)

were detected at the original spot on the plate and the fraction correspond-ing to fusarenon-X contained only negligible counts. This indicated that fusarenon-X is metabolized into more polar compounds. Further study of the absorption, detoxication and excretion of trichothecenes in the animal has not been carried out.

The skin-irritating effect of trichothecene compounds was already no-ticed when Brian and co-workers [8] first isolated verrucarins. The skin test in rats has been an excellent guide during isolation of diacetoxyscirpenol from *F. equiseti* [9] and *F. tricinctum* [15]. T-2 toxin was also isolated using this skin-irritating effect as a bioassay method [6]. Concentrated cul-ture broth or crude extract was applied to the rat skin causing an erythema-tous lesion, eschar formation, and in most toxic cases, death of the animals [6,32]. The skin of various kinds of animals other than rats has also been shown to be sensitive to most of the trichothecenes [76,81,87]. The results on the rabbit skin are listed in Table II.

Human skin is also susceptible to trichothecenes. Bamburg [3] reported that the accidental contact of laboratory workers with crude extract contain-ing T-2 toxin (*ca.* 100 mg/l) caused severe irritation, loss of sensitivity and desquamation of the skin of the hands. Ueno and Tatsuno informed us that splashing of the crude toxin solution caused irritating dermatitis of the hands and face in the students and laboratory assistants who were engaged in isolation of T-2 toxin and fusarenon-X. Identical symptoms were reported in laboratories working with large-scale cultures of *F. tricinctum*, strain T-2 [5] and of *Myrothecium roridium* [37].

(2) Chronic toxicity of Fusaria and fusarenon-X to mice and rats [54]

2 strains of Fusaria, *F. nivale* Fn 2b and *F. graminearum* NRRL 2830, were selected for long-term feeding tests on rats (Donryu, male) and mice (DDD, male) with artificially infected rice. The mouldy rice, prepared at the Food Research Institute, Tokyo, was powdered and mixed with commercial powder feed (CE-II). The concentrations of the mouldy rice (between 10 and 60%) were determined by a preliminary test and it was fed from 8 weeks of age throughout the life span of the animals except for rats fed with *F. nivale*-grown rice, in which the feeding period was 200 days.

In the rats fed with *F. nivale*-rice, 23 out of 30 died between 120—240 days and the rest survived over 1 year. The prominent pathological changes in the rats which died earlier were hypoplasia of the bone marrow, atrophy of the thymus, spleen and testes. Hyperplasia of the intrahepatic bile ducts and atypical hyperplasia in the intestinal mucosa were observed in the rats which survived longer. No tumours occurred. In the mice fed with *F. nivale*-rice, 19 of 37 survived over 1 year. The pathological lesions of note were atrophy of the thymus, spleen and testes, atypical hyperplasia in the gastric and intestinal mucosa and intrahepatic bile ducts, as observed in the rat. In the mice, however, malignant tumours occurred: one case each of hepatoma, myeloic and lymphocytic leukemia, subcutaneous sarcoma and adenocar-cinoma of the jejunum (Fig. 3).

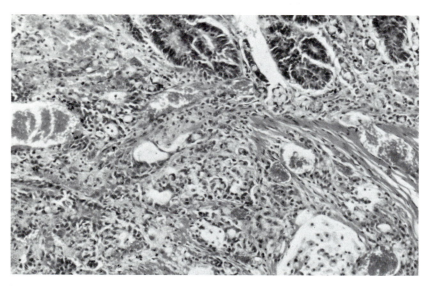

Fig. 3. Mucocellular adenocarcinoma infiltrating the upper ileum occurring in a mouse fed mouldy rice of *F. nivale* Fn 2b for 481 days. The animal died of tumour intussusception of the ileum. (H. E. stain, × 320)

All of 18 rats fed with *F. graminearum*-grown rice survived over 1 year and died at random intervals thereafter, showing pathology similar to those of the *F. nivale* group. Of the mice fed with *F. graminearum*-grown rice 21 out of 43 died within 70 days, mostly showing diarrhoea. The rest survived over 1 year and died of general weakness. The pathological changes were similar to those of the mice of the *F. nivale*-group, but hypoplasia of the bone marrow in this group was more prominent. One case each of hemangio-sarcoma of the liver and lymphocytic leukemia was observed.

Repeated subcutaneous or oral administration of fusarenon-X was carried out using 38 rats and 34 mice. In a group of 20 rats 40 μg/100 g body weight of fusarenon-X was administered weekly by a gastric tube. 60% of the rats survived 50 doses and *post mortem* examination revealed intrahepatic bile duct hyperplasia, atypical hyperplasia in the gastric and intestinal mucosa in several animals and hepatoma in one rat. Hypoplasia and atrophy of the bone marrow, thymus and spleen were observed in one half of the animals. The other group of rats was injected weekly with 40 μg/100 g body weight of fusarenon-X subcutaneously. Most of the rats surviving the 22 injections survived more than 1 year. Similar lesions were observed in them (Fig. 4) with one case of lung adenoma. Two groups of 16 and 18 mice received 10 and 20 weekly subcutaneous injections, respectively, of 25 μg of fusarenon-X per 10 g body weight. Most of them survived repeated administration. Local alopecia at the injection site was seen but the hair regenerated in a few months. Pathological changes in the animals were scarce except for moderate atrophy of the thymus. One case of leukemia was observed.

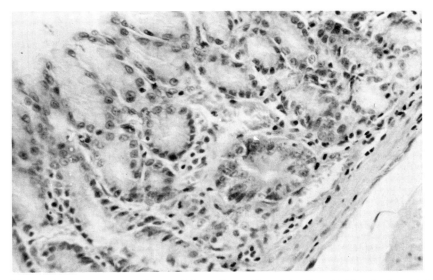

Fig. 4. Gastric mucosa of a rat injected weekly with 40 μg/100 g body weight of fusarenon-X 22 times and sacrificed on the 350th day of experiment. Glands with typical epithelial cells are seen in the antrum. (H.E. stain, × 320)

In control animals (10 rats and 11 mice) no tumour occurred in the experimental period of over 400 days. Atrophy of the organs was mild and observed in only a few cases.

In conclusion, long-term feeding of Fusarium toxins caused a generalized atrophy or hypoplasia of haemopoietic tissues [59]. At the same time regenerative hyperplastic reaction may occur in the haemopoietic tissue and in the gastrointestinal tract, such as atypical proliferation of glands of the stomach and small intestine. Malignant tumours of various organs were observed, though their incidences were in general very low. Loss of hair at the injection site of fusarenon-X can be compared with the skin-test reaction. However, contrary to the results of skin test there were scarcely any inflammatory reactions in the injected subcutaneous regions. Negative results of tumourigenicity were reported by Marasas *et al.* [32] in rats and rainbow trout.

(3) Cytotoxicity to cultured cells

All of the 1(2),13-epoxytrichothecene derivatives are highly toxic to cultured cells. The doses that produce complete cytolysis (grade 4* in Table II) of HeLa cells when cultured for 3 days are 1 μg/ml for fusarenon-X and nivalenol, 0.3 μg/ml for neosolaniol and 0.03 μg/ml for diacetoxyscirpenol and T-2 toxin. The values for dehydroxynivalenol [42] and 9,10-dihydro-

---

* The degree of cytotoxicity was estimated on a scale ranging 0 (no cellular damage) through 4 (complete cytolysis) [71].

nivalenol [44] are somewhat higher (30 and 10 μg/ml, respectively) indicating the dependence of toxicity on the chemical structure. Inhibition of cell growth (grade 1 to 2 in Table II) occurred with about one tenth of the above-mentioned dose [42]. With diacetoxyscirpenol, growth inhibition of L-cells, P-815 mastocytoma cells and KB cells was produced at concentrations of 0.024, 0.0017, 0.007 μg/ml (ED_{50}), respectively [59], and that of HEp^2 cells and BHK cells at 0.005 and 0.0015 μg/ml (lowest toxic doses), respectively [20].

The morphological appearance of the cells was similar when the cells were treated with the compounds at the doses causing complete growth inhibition. After 24 h (Fig. 5b) the cells were slightly enlarged, the pale enlarged nuclei had a darkened membrane, granularly dispersed chromatin and normal nucleoli. After 72 h (Fig. 5c) the cells were shrunken, the nuclei pyknotic with large, round and eosinophilic nucleoli. Mitotic figures were scarce after 24 h.

Nivalenol and fusarenon-X were proved to affect HeLa cells in every phase of the growth cycle. By means of autoradiography nivalenol, at concentrations higher than 0.5 μg/ml, evoked a G1 block about 2 h prior to the beginning of the S phase, and a G2 block just before mitosis [45]. Inhibition of DNA synthesis was also demonstrated by incorporation study [45].

Effects of fusarenon-X on the DNA synthesis of mouse fibroblasts, L cells, were examined by means of quantitative cytophotometry of DNA [43]. Fusarenon-X was added to the culture medium of synchronized cells in G1 and S phase (Fig. 6) at a final concentration of 0.5 μg/ml. As shown on the left side of the figure (a, b and c), the cells synthesized DNA synchronously. By adding fusarenon-X to the cells in G1 phase (d) the cells were prevented from entering the S phase and DNA synthesis was inhibited. The cells in S phase (e) were also sensitive to the toxin and synthesized little new DNA. When the toxin was added after 16 h of inoculation the pattern of the histogram was essentially unchanged after 8 h, although no mitosis was seen, indicating the cells did not complete the first cycle after inoculation in this

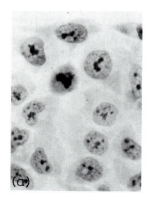 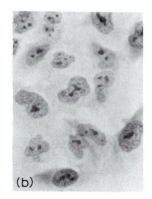 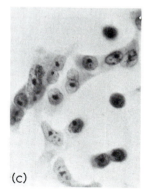

Fig. 5. HeLa cells. (a) Control; (b) treated with 1 μg/ml of fusarenon-X for 24 h; and (c) for 72 h. (H.E. stain, $\times$ 640)

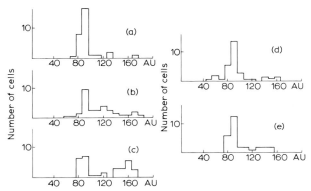

Fig. 6. Cytophotometry of synchronized L-cells. The cells were synchronized in a nutritionally deficient medium and inoculated in Leighton tubes in replicate. After 4 and 16 h, *i.e.* in G1 and middle S phase, fusarenon-X, 0.5 µg/ml in final concentration, was added and the cells were further incubated for 12 (d) and 8 (e) h, respectively. The control cultures were fixed after 4 (a), 16 (b) and 24 (c) h. The fixed cells were stained with Feulgen and their DNA content was cytophotometrically measured using an integrating microdensitometer. Abscissa, absorbance at 570 nm in arbitrary units; ordinate, number of cells. 30 cell nuclei on each slide were measured [43].

period. When the asynchronously growing L cells were incubated with fusarenon-X for 6 h, the histogram of DNA content of the cells differed little from that of the control cells. Thereafter the mitotic rate fell rapidly reaching 0.7% after 6 h, compared with 3.5% in the untreated control. These results indicate that most of the cells neither enter into, nor proceed through S phase.

(4) Antimicrobial effects

With doses of up to 125 µg/ml, fusarenon-X showed no bacteriocidal effect on *B. subtilis*, *E. coli*, etc. by the cup method [85]. As for other trichothecene compounds, trichodermin, crotocin and trichothecin are reported to have antibiotic activity. According to Glaz *et al.* [16] the sensitivity of the cup-plate method is 3 µg/ml for trichodermin and 10 µg/ml for crotocin.

Trichothecin inhibited the spore germination of *Penicillium digitatum* with doses of less than 0.5 µg/ml [13]; germination of the same fungi was inhibited by 100 µg/ml of diacetoxyscirpenol [9]. Mycelial growth of *Aspergillus oryzae*, *A. niger*, *P. islandicum*, *F. nivale* and Mucor sp. was not influenced by concentrations of fusarenon-X up to 100 µg/ml [85].

It is of interest that fusarenon-X induced a respiratory-deficient mutant of the yeast, *Saccharomyces cervisiae* [85]. When the cells were cultured in a medium containing 250 µg of fusarenon-X per ml at 27° for 1 week and then plated on agar containing trypan blue and eosin, white colonies consisting of 4% of the total were detected.

Multiplication of *Tetrahymena pyriformis* was completely inhibited by 50 µg/ml of fusarenon [36] and synchronous division after heat treatment was markedly delayed by adding 2.5 µg/ml of fusarenon-X [78].

References p. 279

(5) Phytotoxic effect

Expansion of bean-leaf disks was prevented by diacetoxyscirpenol at concentrations of 0.1 μg/ml and above [9]. Germination of *Brassica oleracea* was retarded when 10 to 100 μg of fusarenon-X per ml was added to the agar plate [85]. These observations suggest that proliferating plant cells are also affected by trichothecenes, probably by the same mechanism of action.

Mode of action of nivalenol and fusarenon-X

Biochemical mode of action of trichothecenes has been studied in detail with fusarenon-X and nivalenol.

As reported in our earlier study [45] on the inhibitory effect of nivalenol on macromolecule synthesis by HeLa cells, the toxin inhibited the incorporation of [³H]leucine and [³H]thymidine into protein and DNA, respectively, at concentrations ranging from 0.1 to 5 μg/ml, while it had little effect on the incorporation of [³H]uridine into RNA. This dual inhibition of protein and DNA synthesis was observed in various strains of cultured cells treated with various 12,13-epoxytrichothecenes (fusarenon-X [44], diacetoxyscirpenol [59], neosolaniol [42]). Stähelin *et al.* [59] reported that diacetoxyscirpenol inhibited protein synthesis in cultured cells (P-815 mastocytoma cells). DNA and protein synthesis by Ehrlich ascites tumour cells are also inhibited *in vitro* [76,77].

In Tetrahymena a similar pattern of inhibition of macromolecular synthesis was reported [36,78]. In the study by Ueno and coworkers [79], using the cell-free system of rabbit reticulocytes, nivalenol (0.5 μg/ml) inhibited poly U-dependent synthesis of polyphenylalanine. The activation reaction of amino acids was not affected. In the cell-free ribosomal system of rat liver, the polyphenylalanine was inhibited by synthesis of 0.1 to 10 μg/ml of fusarenon-X, nivalenol and T-2 toxin [76]. Polyribosomes of L-cells disappeared rapidly when treated with 2 to 3 μg/ml of fusarenon-X for 5 min [46]. This effect was almost completely inhibited by pretreatment with cyclohexamide, suggesting that the early detachment of ribosomes had occurred or reassembly of ribosomes had been blocked by fusarenon-X. Interestingly, there was no increase in the number of subunit particles. This mode of action seems to be common to all of the 12,13-epoxytrichothecenes, because several compounds, as stated above, have an identical pattern of inhibition of macromolecular synthesis.

The mechanism(s) of inhibition of DNA synthesis by these compounds has not been elucidated in the same way as have other inhibitors of protein synthesis [7,31,92]. Activities of thymidine kinase and DNA polymerase were not influenced by nivalenol in Ehrlich ascites tumour cells [77]. Under the influence of fusarenon-X, the thymidine nucleotide pool of L-cells is not decreased but rather increased. The activity of DNA polymerase, however, was not influenced, but alkaline sucrose gradient sedimentation of replicating DNA revealed some delay of polimerization [43].

Acknowledgements

The authors thank professors M. Enomoto of St. Marianna Medical College, Kawashaki, and Y. Ueno of Tokyo Science University, Tokyo, for their kind advice and discussion in preparing the manuscript.

References

1 B. Achilladelis and J.R. Hansen, Phytochemistry, 8 (1969) 765.
2 G.P. Arsenault, Tetrahedron, 24 (1968) 4745.
3 I.R. Bamburg, Cited in 4.
4 J.R. Bamburg and F.M. Strong, 12,13-Epoxytrichothecenes, in S. Kadis, A. Ciegler and S.J. Ajl (Eds.), Microbial Toxins, Vol. 7, Academic Press, New York, 1971, pp. 207—292.
5 J.R. Bamburg, W.F. Marasas, N.V. Riggs, E.B. Smalley and F.M. Strong, Biotechnol. Bioeng., 10 (1968) 445.
6 J.R. Bamburg, N.V. Riggs and F.M. Strong, Tetrahedron, 24 (1968) 3329.
7 L.L. Bennett Jr., D. Smithers and C.T. Worl, Biochim. Biophys. Acta, 87 (1964) 60.
8 P.W. Brian and J.C. McGowan, Nature, 157 (1946) 334.
9 P.W. Brian, A.W. Dawkins, J.F. Grove, H.G. Hemming, D. Lowe and G.L.F. Norris, J. Exptl. Bot., 12 (1961) 1.
10 C.M. Christensen, in G.N. Wogan (Ed.), Mycotoxins in Foodstuffs, MIT, Cambridge, 1965, pp. 175—186.
11 A.W. Dawkins, J.F. Grove and B.K. Tidd, Chem. Commun., (1965) 27.
12 G.G. Freeman, J. Gen. Microbiol., 12 (1955) 213.
13 G.G. Freeman and R.I. Morrison, J. Gen. Microbiol., 3 (1949) 60.
14 M.W. Gilgan, Cited in 1.
15 M.W. Gilgan, E.B. Smalley and F.M. Strong, Arch. Biochem. Biophys., 114 (1966) 1.
16 E.T. Glaz, E. Scheiber, J. Gyimesi, I. Hovanth, K. Steczek, A. Szentirmai and G. Bohus, Nature, 184 (1959) 908.
17 W.O. Godtfredsen and S. Vangedal, Acta Chem. Scand., 19 (1965) 1088.
18 W.O. Godtfredsen, J.E. Grove and Ch. Tamm, Helv. Chim. Acta, 50 (1967) 1666.
19 J.F. Grove, J. Chem. Soc., 1 (1970) 375.
20 J.F. Grove and P.H. Mortimer, Biochem. Pharm., 18 (1969) 1473.
21 E. Härri, W. Loeffler, H.P. Sigg, H. Stähelin, C. Stoll, Ch. Tamm and D. Wiesinger, Helv. Chim. Acta, 45 (1962) 835.
22 A. Hirayama and M. Yamamoto, Bull. Natl. Hyg. Lab., 66 (1948) 85.
23 M. Ide, T. Niibayashi, J. Ohara and T. Yonemura, Bull. Natl. Inst. Anim. Health, 54 (1967) 34.
24 H. Ishii, Byogaichu Hassei Yosatsu Hokoku, 8 (1961) 1.
25 H. Ishii, Y. Kashiwagi and N. Sasaki, Tokushima Noshi Shiken Hokoku, 1 (1953) 10.
26 K. Ishii, K. Sakai, Y. Ueno, H. Tsunoda and M. Enomoto, Appl. Microbiol., 22 (1971) 718.
27 S. Ito, Hokkaido No-kaiho, 12 (1912) 1.
28 S. Ito, Hokkaido No-kaiho, 12 (1912) 51.
29 S. Ito, Hokkaido No-kaiho, 12 (1912) 133.
30 E.R.H. Jones and G. Lowe, J. Chem. Soc., (1960) 3959.
31 J.W. Littlefield and P.S. Jacobs, Biochim. Biophys Acta, 108 (1965) 625.
32 W.F.O. Marasas, J.R. Bamburg, E.B. Smalley, F.M. Strong, W.L. Ragland and P.E. Degurse, Toxicol. Appl. Pharmacol., 15 (1969) 471.
33 H. Miessner und G. Schoop, Deut. Tierärzt. Wochenschr., 16 (1929) 167.
34 N. Morooka and T. Tatsuno, Toxic substances (fusarenon and nivalenol) produced by *Fusarium nivale*, in M. Herzberg (Ed.), Proc. 1st U.S.—Japan Conf. on Toxic Micro-

organisms, UJNR and U.S. Department of the Interior, Washington, D.C., 1970, pp. 114—119.

35 N. Morooka, N. Nakano, T. Nakazawa and H. Tsunoda, J. Agr. Chem. (Tokyo), 45 (1971) 151.

36 N. Nakano, Jap. J. Med. Sci. Biol., 21 (1968) 351.

37 A. Nespiak, M. Koćor and A. Siewiński, Nature, 192 (1961) 138.

38 Y. Nishikado, Studies on the wheat scab, caused by *Gibberella zeae* (Schw.) Petch and its control, Jap. Ministry for Agriculture and Forestry, Tokyo, and University of Okayama, 1957.

39 S. Nozoe, Personal communication from Tatsuno.

40 K. Ogasawara, Hokkaido Eisei Kenkyushoho, 2 (1951) 35.

41 K. Ogasawara, J. Food Hyg. Soc. Japan., 6 (1965) 81.

42 K. Ohtsubo, Unpublished data.

43 K. Ohtsubo and C. Mittermayer, Unpublished data.

44 K. Ohtsubo and M. Saito, Jap. J. Med. Sci. Biol., 23 (1970) 217.

45 K. Ohtsubo, M. Yamada and M. Saito, Jap. J. Med. Sci. Biol., 21 (1968) 185.

46 K. Ohtsubo, P. Kaden and C. Mittermayer, Biochim. Biophys. Acta, 287 (1972) 520.

47 K. Okubo, Personal communication.

48 M. Okuchi, M. Itoh, Y. Kaneko and S. Doi, Agr. Biol. Chem. (Tokyo), 32 (1968) 394.

49 Oppermann und Dönecke, Deut. Tierärzt. Wochenschr., 16 (1929) 165.

50 M. Saito and K. Okubo, Studies on the target injuries in experimental animals with the mycotoxins of *Fusarium nivale*, in M. Herzberg (Ed.), Proc. 1st U.S.—Japan Conf. on Toxic Micro-organisms, UJNR and U.S. Department of the Interior, Washington, D.C., 1970, pp. 82—93.

51 M. Saito and T. Tatsuno, Toxins of *Fusarium nivale*, in A. Ciegler, S. Kadis and D.J. Ajl (Eds.), Microbial Toxins, Vol. 7, Academic Press, New York, 1971, pp. 293—316.

52 M. Saito, M. Enomoto and T. Tatsuno, Gann, 60 (1969) 599.

53 M. Saito, K. Ohtsubo, M. Umeda, M. Enomoto, H. Kurata, S. Udagawa, F. Sakabe and M. Ichinoe, Jap. J. Exptl. Med., 41 (1971) 1.

54 M. Saito, M. Enomoto, K. Ohtsubo and M. Murata, Trans. Soc. Pathol. Jap., 60 (1971) 79.

55 H.P. Sigg, R. Mauli, E. Flury and D. Hauser, Helv. Chim. Acta, 48 (1965) 962.

56 E.B. Smalley, W.F.O. Marasas, F.M. Strong, J.R. Bamburg, R.E. Nichols and N.R. Kosuri, Mycotoxicoses associated with moldy corn, in M. Herzberg (Ed.), Proc. 1st U.S.—Japan Conf. on Toxic Micro-organisms, UJNR and U.S. Department of the Interior, Washington, D.C., 1970. pp. 163—173.

57 W.G. Smith, Cited in 38.

58 W.C. Snyder and H.N. Hansen, Am. J. Bot., 32 (1945) 657.

59 H. Stähelin, M.E. Kalberer-Rüsch, E. Signer and S. Lazáry, Arzneimittelforsch., 18 (1968) 989.

60 R. Takakuwa, Personal communication through Udagawa.

61 S. Takeda and K. Ogasawara, Rept. Hokkaido Inst. Public Health, 2 (1951) 47.

62 Y. Tochiuchi, Byochu Zasshi, 20 (1933) 106.

63 T. Tatsuno, Cancer Res., 28 (1968) 2383.

64 T. Tatsuno, Seikagaku, 41 (1970) 153.

65 T. Tatsuno, N. Morooka, M. Saito, M. Enomoto, M. Umeda and K. Okubo, Folia Pharmacol. Jap., 62 (1966) 26.

66 T. Tatsuno, M. Saito, M. Enomoto and H. Tsunoda, Chem. Pharmacol. Bull., 16 (1968) 2519.

67 T. Tatsuno, Y. Fujimoto and Y. Morita, Tetrahedron, 33 (1969) 2823.

68 T. Tatsuno, Y. Morita, H. Tsunoda and M. Umeda, Chem. Pharm. Bull., 18 (1970) 1485.

69 T. Tatsuno, K. Ohtsubo and M. Saito, J. Appl. Chem., 351 (1973) 309.

70 U. Tokura, Byochu Zasshi, 20 (1933) 90.

71 I. Toplin, Cancer Res., 19 (1959) 959.
72 H. Tsunoda, Micro-organisms which deteriorate stored cereals and grains, in M. Herz-
    berg (Ed.), Proc. 1st U.S.—Japan Conf. on Toxic Micro-organisms, UJNR and U.S.
    Department of the Interior, Washington, D.C., 1970, pp. 143—162.
73 H. Tsunoda, O. Tsuruta, S. Matsunami and S. Ishii, Food Res. Inst. (Tokyo), 12
    (1957) 25.
74 H. Tsunoda, O. Tsuruta and S. Matsunami, Food Res. Inst. (Tokyo), 13 (1958) 25.
75 H. Tsunoda, M. Toyazaki, S. Morooka, N. Nakano, H. Yosiyama, K. Okubo and M.
    Isoda, Proc. Food Res. Inst., 23 (1968) 89.
76 Y. Ueno, Toxicological and biological properties of fusarenon-X, a cytotoxic myco-
    toxin of *Fusarium nivale* Fn 2b, in I.F. Purchase (Ed.), Mycotoxins in Human Health,
    Macmillan, London, 1971, pp. 163—178.
77 Y. Ueno and K. Fukushima, Experientia, 24 (1968) 1032.
78 Y. Ueno and H. Yamakawa, Jap. J. Exptl. Med., 40 (1970) 385.
79 Y. Ueno, M. Hosoya, Y. Morita, I. Ueno and T. Tatsuno, J. Biochem., 64 (1968) 479.
80 Y. Ueno, M. Hosoya and Y. Ishikawa, J. Biochem., 66 (1969) 419.
81 Y. Ueno, Y. Ishikawa, K. Amakai, M. Nakajima, M. Saito, M. Enomoto and K.
    Ohtsubo, Jap. J. Exptl. Med., 40 (1970) 33.
82 Y. Ueno, Y. Ishikawa, K. Saito-Amakai and H. Tsunoda, Chem. Pharm. Bull., 18
    (1970) 304.
83 Y. Ueno, Y. Ishikawa, N. Nakajima, K. Sakai, K. Ishii, H. Tsunoda, M. Saito, M.
    Enomoto, K. Ohtsubo and M. Umeda, Jap. J. Exptl. Med., 41 (1971) 257.
84 Y. Ueno, I. Ueno, K. Amakai, Y. Ishikawa, H. Tsunoda, K. Okubo, M. Saito and M.
    Enomoto, Jap. J. Exptl. Med., 41 (1971) 507.
85 Y. Ueno, I. Ueno, Y. Iitoi, H. Tsunoda, M. Enomoto and K. Ohtsubo, Jap. J. Exptl.
    Med., 41 (1971) 521.
86 Y. Ueno, K. Ishii, K. Sakai, S. Kanaeda, H. Tsunoda, T. Tanaka and M. Enomoto, Jap.
    J. Exptl. Med., 42 (1972) 187.
87 Y. Ueno, N. Sato, K. Ishii, K. Sakai and M. Enomoto, Jap. J. Exptl. Med., 42 (1972)
    461.
88 Y. Ueno, N. Sato, K. Ishii, K. Sakai, H. Tsunoda and M. Enomoto, Appl. Microbiol.,
    25 (1973) 699.
89 Y. Ueno, N. Sato, K. Ishii, N. Shimada, K. Tokita, M. Enomoto, M. Saito, K. Ohtsubo
    and I. Ueno, Jap. J. Pharmacol., 23 (Suppl.) (1973) 133.
90 M. Umeda, Jap. J. Exptl. Med., 41 (1971) 195.
91 M. Umeda, Personal communication.
92 B.G. Weiss, J. Cell Physiol., 73 (1969) 85.
93 M. Woronin, Botan. Z., 49 (1891) 81.
94 A. Yoshizawa, N. Uratsuji and N. Morooka, Proc. Ann. Cong. Jap. Assoc. Agr. Chem.,
    (1972) 291.

Chapter 13

CITREOVIRIDIN FROM *PENICILLIUM CITREO-VIRIDE* BIOURGE

YOSHIO UENO

Department of Chemical Microbiology, Faculty of Pharmaceutical Sciences, Tokyo University of Science, Ichigaya, Tokyo (Japan)

Acute cardiac beriberi, which is called "shoshin-kakke" in Japanese, has been prevalent during the last three centuries in rice-eating countries including Japan. This disease is characterized by violent and tragic symptoms such as ascending paralysis, convulsions and respiratory arrest. Numerous people investigating the disease presented many speculative theories about its cause and these theories may be classified into three categories, namely infection, avitaminosis and intoxication. Since the famous discovery of vitamin B_1 and the contemporaneous disappearance of the disease, the second theory of vitamin deficiency has been generally accepted as the explanation for the cause of beriberi including the acute type of cardiac beriberi. As for the last theory of intoxication, Sakaki [1] in 1891 succeeded in demonstrating that an ethanol extract from naturally mildewed rice grains contained neurotoxic compound(s) which caused convulsions, paralysis and death in frogs, rabbits and mice. This is the first experimental approach to the aetiology of cardiac beriberi from the standpoint of the intoxication theory, although both the causal fungi and the toxic principle were not identified at that time.

In 1918, the late professor I. Miyake started screening fungi for toxicity under the working hypothesis that a "rice-fungus-toxin" might be responsible for the disease. Subsequently he found two "yellow rice" samples invaded by Penicillium sp., one harvested in Taiwan (Formosa) and imported into Japan in 1936 and the other harvested in Japan between 1934 and 1939 [2]. In 1947, he [3] designated the fungus as *P. toxicarium* Miyake sp. nov., which was recently revised to *P. citreo-viride* Biourge by his collaborator, Naito [4].

Toxicological study of the metabolites of this fungus was started by Uraguchi with artificially moulded rice grains, so-called "yellow rice" (Ō-hen-mai in Japanese). He demonstrated that the ethanol-ether extract of the mouldy rice contained a neurotoxic metabolite which affected several kinds of animals and that the symptoms were progressive paralysis of the ascending type, lowering of body temperature, convulsions and respiratory failure, all of which were similar to the reported clinical manifestations of cardiac beriberi in man [5].

In 1947, a chemical study on the metabolite of *P. citreo-viride* was performed by Hirata and his collaborators [6,7] and they isolated a yellow pigment, citreoviridin, although at that time no toxicological study was undertaken. Recently, the author reexamined the chemical and biological properties of the toxic principle of *P. citreo-viride* in cooperation with Prof. Uraguchi, and the results were reported at the symposia on "Mycotoxins" at the 1st International Congress of Plant Pathology held in London in 1968 [8] and on "Mycotoxins in Human Health" held in Pretoria in 1970 [9].

In addition to these mycological and toxicological studies, Uraguchi [8] pointed out, on the basis of his statistical observations, that the rapid disappearance of fatal beriberi in Japan in the early 19th century coincided with the increase in the extent of rice inspection and that the discovery and medical application of vitamin B_1 were made somewhat later than the start of decrease in incidence.

Occurrence of cardiac beriberi

(I) Clinical symptoms

Beriberi (kakke) may be classified on clinical symptoms into several types. *Acute cardiac beriberi* (shoshin-kakke in Japanese), the most tragic type usually progressed very rapidly with vomiting, convulsions, ascending paralysis, lowering of body temperature and respiratory arrest as major symptoms. Once the disease started, the patient usually died within 3 days and no method of saving the patient was available because of the rapid progression of the disease. Other types were *atrophic beriberi* and *wet beriberi* which were characterized by atrophy, anaemia or dropsy, without severe paretic signs. The whole course of the disease extended over a 2—30-week period depending upon the degree of severity; in some cases motor disturbance remained in the legs 1—2 years after development.

Irisawa *et al.* [10] described (in Japanese) the clinical syndrome of fatal cases of beriberi; Uraguchi [8] translated this description as follows:

"On affection, it begins with palpitation, cardiac distress, tachypnoea, *etc.* Dyspnoea gets worse with nausea or vomiting. Within a few days, the condemned patient falls into indescribable agony, moaning in pain and struggling with jactitation, or sometimes with violent maniacal excitement, which presents a heart-rending sight.

"On clinical examination, the heart is usually dilated to the right, the first sound at the apex prolonged, and systolic murmur is audible. The second pulmonary sound is accentuated. Blood pressure is depressed; pulse is quickened, sometimes exceeding 120 per min, and softer with palpitation. Dyspnoea gradually increases, with cold and cyanotic extremities, anaemic and dry skin, and husky voice. The pulse becomes more enfeebled. The pupil is dilated. At last consciousness is disordered or almost lost, and respiration fades to final arrest."

Furthermore, Uraguchi [8] added the very lucid description by Wright [11], as follows:

"Within a varying time — from 5 to 72 h — nervous symptoms appear.

The kneejerks either become exaggerated, or more usually are found to be absent. The perineal and anterior tibial muscles rapidly weaken and become flaccid. Foot-drop follows, and in pronounced cases the flaccid paresis marches rapidly to the leg, thigh, hip, finger, wrist, arm and shoulder girdle muscles.

"The laryngeal and pharyngeal muscles may become involved. The tongue has been affected in some of my cases. The internal ocular muscles may not escape. All forms of sensation may be disturbed or fail entirely. The calf and arm muscles become tender and pressure of them is resented by the patient. The sensory disturbance may be coextensive with the motor paresis, or may be confined to the legs and hand.

"There are varying degrees of oedema. In my experience a slight pretibial oedema is always present. Beginning as such it may rapidly develop into a general anasarca — the wet beriberi of the older writers.

"The nervous mechanism of the circulatory system is early involved, perhaps as early as the sensory motor system. There is first cardiac irritability, then palpitation on the slightest exertion, with a low-tension, rapid, irregular pulse. The veins of the neck and the epigastrium pulsate largely. Signs of venous engorgement become apparent. Rapid dilatation of the right heart occurs. The precordia heaves alarmingly, contrasting markedly with the low-tension, weak pulse. Haemic murmurs are heard. There are reduplications of various sounds. Cardiac exhaustion becomes apparent. Delirium cordis soon dominates the situation. Dyspnoea frets the patient, he gasps for air, clutches his chest, rolls from side to side, while his face betrays a terrible anxiety. He at last passes to death in an agony distressful to behold. The mind is clear to the last. I have had patients give me a clear answer as to their feelings, then wildly stare for a moment and suddenly pass."

(II) Statistical evidence in Japan

Severe epidemics of shoshin-kakke prevailed in East Asia and other rice-eating countries including Japan which was heavily plagued with this disease. Therefore, numerous investigations were carried out into the epidemiology of the disease and many excellent reviews and books on kakke were published. In this report, the author quotes statistical data from the excellent publication by Toyama entitled "Aetiological Studies on Beriberi" [12], and from other publications.

(a) Social levels and beriberi

The incidence of beriberi was connected with the social class or the occupation of patients. From the Genroku Era (1600) to the early Meiji Era (1870), beriberi was reported to have prevailed among the people who belonged to the high social class of princes and generals; during the middle Meiji Era (1880—1900), high percentages of the patients were found among the middle class: bonzes and scholars (37%), merchants (29%), engineers and craftsmen (11%), government officials (7%) and peasants (5%). Further investigation revealed a close relationship between the development of fatal beriberi and the occupation of the patient. In the last decade of the Meiji Era (1900—1910), the high incidence moved from the middle class to lower classes; examples of the death rate per 1000 cases were as follows: soldiers and sailors (68), fishermen (64), shipping men (63), navy officers (61), mechanical engineers (57), merchants (17) and peasants (5).

These statistical data strongly support the theory that the epidemic of beriberi shifted with passage of time from the higher social classes to the lower classes of people. This is quite peculiar, because malignant epidemics such as cholera and other infectious diseases commonly occur among lower-class people, and a change in the social class of the majority of patients with passage of time is not usually demonstrable.

(b) Population increase and fatal beriberi

Acute cardiac beriberi appeared first in Edo (now Tokyo) in the late 17th century, when a rapid population increase and marked political and social changes took place. Similar cases were reported in big cities such as Osaka and Yokohama.

Table I shows yearly incidences of fatal beriberi and populations from 1899 to 1908 in Japan, and statistical data for three big cities are listed in Table II. The average death rate was around 1—2 per 10 000 but increased to 4—10 in Tokyo, Osaka and Yokohama, indicating clearly that the incidence of fatal beriberi in big cities was higher than in other locations. Uraguchi [8] stressed that social problems such as changes in dietary habits accompanied by improved techniques for rice preparation, storage and transportation in town existed concomitantly with the higher incidence in city dwellers.

(c) Seasonal change of beriberi incidence

Numerous studies have revealed that the incidence of beriberi increased in early summer and late autumn. According to Toyama [12], during

TABLE I

INCIDENCE OF FATAL BERIBERI CASES

Year	Number of deaths	Population (× 1000)	Death rate/10 000
1899	9034	44269	2.03
1900	6500	44831	1.45
1901	7180	45404	1.58
1902	11099	45990	2.41
1903	10783	46577	2.31
1904	9408	47108	1.93
1905	11703	47641	2.46
1906	7766	48187	1.61
1907	8767	48754	1.80
1908	10786	49319	2.19

TABLE II

INCIDENCE OF FATAL BERIBERI CASES IN THE THREE BIG CITIES OF JAPAN

Year	Tokyo		Osaka		Yokohama	
	Number of deaths	Death rate/10 000	Number of deaths	Death rate/10 000	Number of deaths	Death rate/10 000
1901	1450[a]	8.85				
1902	1990	11.65				
1903	—	—				
1904	1708	9.13				
1905	1016	5.16				
1906	917	4.44				
1907	942	4.40				
1908	1216	5.61				
1909	1038	4.77			174	4.27
1910	1121[b]	5.12	1377[b]	11.11	128[b]	3.05
1911	909	4.19	1315	10.32	137	3.09
1912	434	2.00	586	4.40	36	0.79
1913	575	2.65			73	1.84
1914					110	2.68

Population

a 1901 (Meiji 34) Tokyo   1,638,094
b 1910 (Meiji 43) Tokyo   1,805,800
                 Osaka   1,239,373
                 Yokohama   419,830

References p. 301

1911—1921, the Kanebo Spinning Industry, one of the most famous cotton spinning companies in Japan, improved the nutrition of several thousand weavers in order to reduce the prevalence of beriberi. The incidence of the disease, mostly atrophic and paralytic types of beriberi, was maximal during spring and summer. Statistical analysis by Uryu [13] revealed that the number of fatal beriberi cases during 1899—1938 formed a clear wave-like curve with the highest incidence in summer and the lowest in winter.

Besides the seasonal effect, several reports pointed out that climatic conditions, such as heavy rain and low temperature, strongly influenced the prevalence of beriberi.

Although gastrointestinal diseases are common in summer, the prevalence of beriberi at this time of the year, with its warm, humid conditions, is worthy of attention because of the environmental and ecological conditions which favour growth and toxin production by the fungus, as will be discussed later.

(d) Sex and age differences

In 1882, beriberi disease was very prevalent in Tokyo, and the Kakke Hospital recorded 555 cases; 450 outpatients and 105 inpatients. Statistical analysis disclosed the very interesting phenomenon that female outpatients and inpatients numbered only 20 and 5, respectively, and that the death rate was 21 cases, or 20% of all inpatients. Furthermore, the percentages of patients in various age groups were as follows; 2.19% below 14 years of age, 74.26% 14—25 years, 14.01% 26—35 years, 6.26% 36—45 years, 2.44% 46—55 years, 1.2% over 56. Several other reports also pointed out that young men were more susceptible to the acute type of cardiac beriberi than were old. These results indicate that men, especially young men, are highly susceptible to the tragic disease of beriberi.

According to Fukushima [14], the Hospital of Kyoto University recorded 300 cases of infant beriberi during 1921—1931. Of these patients 193 were male and 107 female; 19% of the cases were fatal. It was noteworthy that the incidence was high in early summer when heavy rain continued during the rainy season from May to June and furthermore that 50% of their mothers were also attacked by beriberi.

(e) Rice inspection and concomitant disappearance of cardiac beriberi

Shoshin-kakke was frequently encountered in Japan from the end of the 17th century until the beginning of the 18th century, but at the present time it is rarely seen in this country. As for the time of disappearance of the disease, data have been obtained from the recently published statistical study by Miyake *et al.* [15], which covered all the autopsied cases of cardiac beriberi during the preceding 80 years in the Department of Pathology at the University of Tokyo. In this study clinical observations as well as pathological changes, such as dilatation of the right ventricle of heart, were employed for diagnosis of cardiac beriberi. As is shown in Table III, a total of between 10 and 20 cases were reported annually up to 1909, and only a few cases per

TABLE III

NUMBER OF CARDIAC BERIBERI CASES AUTOPSIED IN THE DEPARTMENT OF PATHOLOGY, UNIVERSITY OF TOKYO

Year	Total autopsied cases	Cardiac beriberi	Year	Total autopsied cases	Cardiac beriberi
1883—1887	195	2	1911	327	9
1888	132	20	1912	549	4
1889	119	26	1913	483	1
1890	209	12	1914	576	9
1891	162	8	1915	357	7
1892	187	4	1916	437	19
1893	136	3	1917	337	8
1894	114	6	1918	371	4
1895	143	9	1919	328	9
1896	244	13	1920	275	1
1897	201	8	1921	260	4
1898	248	11	1922	273	8
1899	305	13	1923	257	4
1900	430	12	1924	219	2
1901	443	11	1925	230	3
1902	474	19	1926	213	0
1903	402	14	1927	179	0
1904	424	21	1928	191	0
1905	378	2	1929	209	0
1906	434	14	1930	201	1
1907	365	6	1931	238	0
1908	441	17	1932	221	2
1909	521	16	1933	261	0
1910	372	5	1934	275	2
			1935	262	0
			1936	251	0
			1937	263	0

year after 1910 except for the temporary incidence rise during World War I (1916—1917).

From the epidemiological data mentioned above and statistical analysis on rice inspection data, Uraguchi [8] pointed out that the decrease in beriberi was closely related to the increased frequency of rice inspection by the Japanese government. According to his data, official rice inspection to check damaged or mouldy rice covered only 30% of all prefectures until 1909 as against 60% of the whole country in 1911, which meant that rice-exporting prefectures were placed under government control. These data indicate that the more extensive the rice inspection, the fewer the cases of cardiac beriberi that were detected. This finding supports the view that cardiac beriberi was caused by the intake of damaged or mouldy rice.

If it is assumed that cardiac beriberi is caused by a mycotoxin, the shift of beriberi cases from the upper social class to the lower class, the high incidence of beriberi in people who lived in town or in confined places such as ships, and the seasonal variation of incidence are all explained by the following ecological considerations; (*1*) the technique for preparing polished rice was first utilized by upper class people and later spread to lower classes; (*2*) transportation or unsatisfactory storage of polished rice, which is more susceptible to fungal attack than unpolished rice, increased contamination by fungi, and (*3*) the mycotoxin-producing ability of fungi was influenced by seasonal or climatic conditions such as temperature and humidity.

(III) Occurrence

Several types of beriberi appeared in Asian countries such as Korea, Taiwan, Indonesia and India, and several cases were also reported in Hawaii, South America and Africa. The following data are quoted from the excellent review by W. Liao, entitled "Tokyo-Kakkebyo-Kenkyu (Studies on Oriental Kakke-diseases)" [16] and from other reports.

(*Korea*) Incidence of beriberi was high in towns; 41 cases in 1913, 63 cases in 1914 and 14 cases until the middle of 1915, and 92 male and 26 female out of the total 118 cases. Most of the patients (65%) contracted the disease from May to August, and 69% were young. Furthermore, in the same district of Korea, the incidence was higher among Japanese than among Koreans, indicating that food habits were one of the important factors in the incidence of beriberi.

(*Taiwan*) In 1877, a violent endemic, named Keelung-disease, appeared in Keelung, one of the big towns in Formosa, and 10 out of 12 patients were reported to have died from cardiac beriberi. In 1888, gold deposits were located in a suburban district of Keelung, and over 60% of workers suffered from the tragic disease with symptoms such as dropsy, paretic signs, acceleration of heart rate, respiratory failure and agony, and many died. In 1902–1903, 50% of the workers in the camphor industry were attacked by the disease with the same symptoms; but fatal cases were rare.

(*Malaya*) Malaya is one of the countries well known for a high incidence of beriberi. In Singapore, for example, fatal beriberi cases in 1911 were 40 per 10 000 persons, which was around ten times higher than the incidence in Tokyo.

(*Hawaii*) In 1912, there were 22 fatal cases of beriberi, among which were 17 Japanese and 5 Filipinos, and all of them consumed polished rice as major foodstuff. An interesting finding was that the Japanese people in Hawaii imported the polished rice from their home country at that time.

These examples indicate that acute cardiac beriberi as well as other types of beriberi occurred sporadically in rice-eating countries, and certain people such as fishermen, sailors, miners or those who habitually eat imported polished rice, were liable to be stricken by the epidemic. It is noticeable that the area with a high incidence of acute cardiac (or fatal) beriberi also had a high incidence of wet or atrophic beriberi at the same time. Toxicologically, this presumably means that several types of beriberi are aetiologically identical and the major difference probably results from differences in dose and duration of intake of the toxic agent.

At the present time, avitaminosis is generally accepted as the cause of beriberi. However, as far as acute cardiac beriberi or shoshin-kakke is concerned, the patient dies within a short period of 2—3 days and such a rapid course of the disease has not been demonstrated in vitamin B_1 -deficient animals. Neither dilation of the right ventricle nor hypertrophy of adrenal medulla, which are characteristic findings in cardiac beriberi cases, were demonstrable in the vitamin deficiency. Furthermore, according to Ogata [17], vitamin B_1 -deficient animals recovered from their avitaminosis when liver removed from typical shoshin-kakke patients was administered to them. This proved that the liver of patients dying from cardiac beriberi contained adequate amounts of the vitamin at the time of death.

In addition to the above experimental evidence, the history of vitamin B_1 discovery is not compatible with the avitaminosis theory of cardiac beriberi. Suzuki *et al.* [18] first reported a constituent which was effective in controlling beriberi in rice bran in 1910 and Funk [19] identified the vitamin in 1911. In the Hospital of Kyoto University, parenteral administration of vitamin B_1 to the patients started in 1922. However, according to Uraguchi [8], the rapid decrease of cardiac beriberi cases started in 1910, the year of vitamin B_1 discovery or about 10 years before its practical use as a medicine.

From the above considerations, Uraguchi [8,20] proposed the theory of the mycotoxic aetiology of cardiac beriberi.

Toxic metabolite of *P. citreo-viride*

Adopting an ethanol-ether extraction method, Sakaki [1] and Uraguchi [5] obtained from naturally or artificially moulded rice a crude material which killed experimental animals such as mice, cats, dogs and frogs. This finding indicated that the toxic metabolite was alcohol- and ether-soluble. Sakabe *et al.* [7] reported the isolation of a yellow pigment, citreoviridin, from the ethanol-soluble fraction of the mouldy rice.

Recently, employing lethal toxicity in mice as a bioassay tool, the author re-examined the isolation of the toxic metabolite from *P. citreo-viride*-moulded rice as follows [21].

(a) Culture

P. citreo-viride Biourge was inoculated on an Ushinsky agar slant (glycerin 40 ml, ammonium lactate 10 g, NaCl 5 g, aspartate 4 g, K_2HPO_4 2.5 g, $MgSO_4$ 0.4 g, $CaCl_2$ 0.1 g and water 1 l, pH 7) and incubated at 15—20° for 2 weeks. The substrate for the toxin production was 5% polished rice and before fermentation it was immersed in tap water for 30 min and autoclaved for 20 min. After inoculation with the fungal spores, the rice grains were fermented at 20° for 3—4 weeks. The yellow rice thus obtained was dried at 60—70° and powdered.

(b) Isolation of the toxic metabolite

The isolation procedure is schematically presented in Fig. 1. Each fraction was dissolved into olive oil and administered intraperitoneally (i.p.) or subcutaneously (s.c.) to mice and the toxicity was determined from the death rate, the time lapse before death and the symptoms of poisoned mice.

The mouldy rice powder (4 kg) was extracted several times with ethanol, and from the combined extracts a brown oily "ethanol extract" (20 g) was obtained. On dissolving the brown oil in hot benzene followed by standing overnight at 4°, a non-toxic brown precipitate was deposited. By dropwise addition of *n*-hexane to the yellow benzene solution, the "ethanol— crude toxin" (2 g) was precipitated as yellow powder. The toxicity test in mice revealed that, in a dose of 20—100 mg/kg i.p., the ethanol crude toxin caused paresis in the hind legs followed by respiratory failure 10—60 min after administration.

Column chromatographic fractionation of the crude toxin on silica gel with *n*-hexane—acetone (1:1) as solvent gave six fractions, of which only the 5th fraction (520 mg) was toxic. On crystallization from methanol, toxic yellow needles (400 mg) were obtained. Extraction of the mouldy rice powder in a preparative Soxhlet extractor first with *n*-hexane followed by ether yielded the "ether—crude toxin", from which the yellow crystalline mycotoxin was isolated.

From m.p., UV spectrum and other chemical properties, the crystalline product was identified as citreoviridin, as described below. The same com-

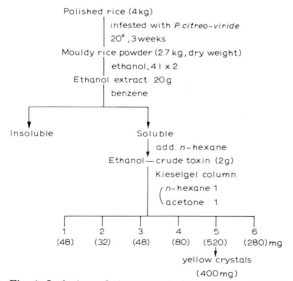

Fig. 1. Isolation of citreoviridin from the mouldy rice grains.

pound was isolated from *Eupenicillium ochrosalmoneum* Udagawa [21] and *P. pulvillorum* Turfitt [22].

Chemical properties of citreoviridin

Hirata and his collaborators started to work on fungal metabolites of *P. citreo-viride* in 1940 and they isolated citreoviridin [7]. In the experiment summarized above the author confirmed that the toxic metabolite of *P. citreo-viride* was identical with citreoviridin [21].

The compound, crystallized from methanol, has m.p. 107—111° and a molecular formula of $C_{23}H_{30}O_6 \cdot CH_3OH$, $[\alpha]_D^{20}$ —107° (C = 1 in CHCl$_3$). On storage under reduced pressure the methanol molecule is lost. The ethanol solution of the pigment exhibits maximum peaks at 388 nm (ϵ 48 000), 294 nm (27 000), 286 nm (24 600), 234 nm (10 200), 204 nm (17 000). The IR spectrum (KBr) gave absorptions at 3500, 1689, 1654, 1530, 1452, 1405, 1249, 1150, 1094, 1069, 999, 821 and 811 cm^{-1}. The toxin is soluble in ethanol, ether, benzene, chloroform and acetone, but insoluble in *n*-hexane and water.

On acetylation, it forms a monoacetate, (m.p. 99—101°) and on *p*-nitrobenzoylation, a mono- (m.p. 178°) and di-nitrobenzoate (m.p. 269—271°).

From the NMR spectrum and chemical degradation, the structure of citreoviridin was found to be composed of three components, an α-pyrone chromophore, a conjugated polyene moiety and a hydrofuran ring, as shown in Fig. 2 [7].

A photochemical reaction of citreoviridin in the presence of iodine yields isocitreoviridin [22]. On exposure to UV illumination (366 nm), the mycotoxin emits brilliant yellow fluorescence [21].

Production

(I) On natural media

The author [21] employed sterilized rice grains for the production of citreoviridin. Unpolished rice grains were first roughly ground in a mechanical mortar, and then immersed in tap water for 20—30 min. The damage to the surface of the grains and the high content of water allowed rapid invasion of the fungus.

A detailed study of the culture requirements of *P. citreo-viride* for

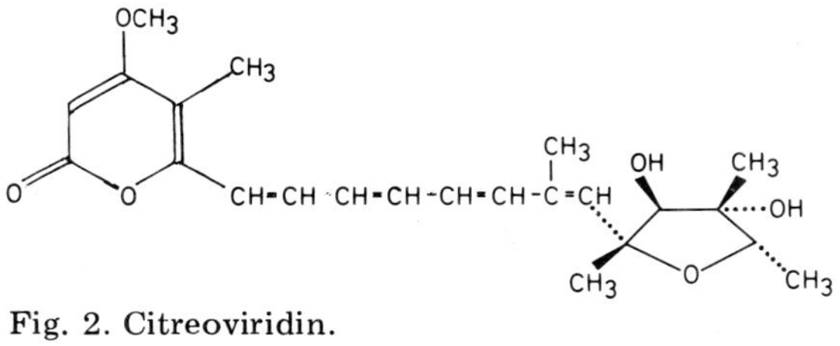

Fig. 2. Citreoviridin.

higher production of citreoviridin was undertaken by the author [21]. *P. citreo-viride* is liable to lose the mycotoxin-producing ability during storage of the strain, and furthermore, culture temperature and relative humidity have a marked influence on the yield of citreoviridin.

In order to prevent the decrease in toxin-producing ability, the fungal strain was cultured on Ushinsky-agar slants at 20° for 2—3 weeks and stored at 4° until used. In some cases, the fungal spores were inoculated on steril-ized rice grains, and after cultivation at 15—20° for 3 weeks the mouldy grains were stored at 4°. Special techniques such as single spore isolation are needed for selection of the mycotoxin-producing strain, when albino-type spores are detected on the culture slant.

As for environmental factors influencing production of citreoviridin, the author examined the effect of culture temperature on mycotoxin pro-duction on rice. Chemical analysis revealed that the relative yield of citreo-viridin on polished rice was high when the fungus was cultured at rather low temperatures (12—22°) as shown in Fig. 3 and the high water content in-creased the content of the mycotoxin.

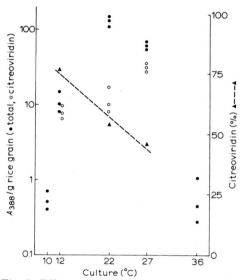

Fig. 3. Effect of culture temperature of the production of citreoviridin by *P. citreo-viride* on rice grains.

The toxicity test in mice also demonstrated that the infested rice grains were more toxic when cultured at lower temperatures such as 12°, as is shown in Table IV.

The yield of citreoviridin from moulded rice grains ranged from 0.2 g to 1.0 g from 1 kg of the dried powder. The author also demonstrated that *Eupenicillium ochrosalmoneum* Scott and Stolk (*P. ochrosalmoneum* Udagawa) produced citreoviridin on rice grains, but the yields were less than that by *P. citreo-viride*.

TABLE IV

EFFECTS OF CULTURE TEMPERATURE ON THE PRODUCTION AND THE LETHAL TOXICITY OF THE CRUDE TOXIN

Polished rice, 100 g, was inoculated with *P. citreo-viride* and incubated for 2 weeks at different temperatures. The mouldy rice was dried and extracted with ethanol.

Temperature (°C)	Ethanol extract (g/100 g rice)	Death rate of mice (mg/10 g s.c.)	
		5	10
12	0.94	2/2 (88, 55)[a]	2/2 (39, 46)
18	2.03	2/2 (60, 66)	2/2 (40, 60)
24	2.04	1/2 (120)	2/2 (71, 89)
27	2.57	0/2	2/2 (80, 100)
36	0.96	0/2	0/2

[a] Lifetime in minutes.

According to Nagel *et al.* [22], *P. pulvillosum* Turfitt produces high concentrations of citreoviridin on maize meal.

(II) On synthetic media

Synthetic media are preferred for the isolation of a mycotoxin as well as for study of its biogenetic system.

The author made several attempts to develop a suitable synthetic medium, and also examined environmental factors influencing the production of citreoviridin [23].

Of the five synthetic liquid media tested (Mannit, Czapek-Dox, Glycerin-Czapek-Dox, Waksman and Ushinsky media) a stationary culture with Ushinsky medium supported the highest level of production of both the mycelium and citreoviridin (Fig. 4).

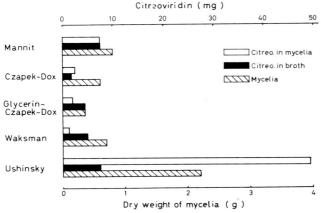

Fig. 4. Production of citreoviridin on the liquid medium. *P. citreo-viride* was grown in a stationary culture at 20° for 2 weeks in 500-ml Erlenmeyer flasks, each containing 150 ml of the culture medium indicated in the figure.

References p. 301

Photometric determination revealed that citreoviridin was recovered from the mycelium in a yield of 2.2% of its dry weight. On mass cultivation of *P. citreo-viride* with this medium at 20° for 3 weeks, 127 g of the dried mycelium yielded 1.1 g of the ethanol—crude toxin, from which 420 mg of the crystalline citreoviridin was isolated. With submerged cultivation, however, the yield was extremely low.

An important finding was that the culture temperature influenced the yield of citreoviridin in the same manner as in solid culture; at 20—24° the growth of the fungus was rather slower than at 27—37°, while the maximum yield of citreoviridin was obtained below 20° with a decrease in yield as the culture temperature increased. These findings suggest that citreoviridin production by *P. citreo-viride* is favoured by low temperature and high humidity.

(III) Ecology of P. citreo-viride

Miyake *et al.* [2] and Naito [4] reported that the growth of *P. toxicarium* Miyake (now *P. citreo-viride* Biourge) occurred when the water content of rice grains reached 14.6%. Other fungi require 1% higher water content for growth. Naito [4] observed that the fungal growth was greatest and the pigment formation highest in rice grains followed by corn and wheat.

This means that *P. citreo-viride* can grow preferentially on rice grains soon after the grains become wet under unsatisfactory storage conditions. Furthermore, Miyake *et al.* [2] stated that this fungus was detected frequently not only in imported rice but also in the domestic rice produced in the district facing the Japan Sea, where the weather is colder and less sunny than in the district facing the Pacific Ocean.

On the basis of these ecological considerations of the toxigenicity of *P. citreo-viride*, Uraguchi [8,20] suggested that the high prevalence of acute cardiac beriberi in the past was a consequence of the introduction of polished rice into diets of the major rice-producing areas and that unsatisfactory storage of this rice allowed mould damage to occur. Seasonal changes in the incidence and its acceleration by heavy rain are also presumed to be connected with the ecological behaviour of the fungus.

Analysis for citreoviridin

(I) Chemical methods

Citreoviridin is soluble in ethanol and benzene, and absorbs light in visible region; λ_{max}^{EtOH} 388 nm (48 000). These chemical features were applied by the author to the analytical procedure for citreoviridin in mouldy rice as well as in biological materials such as tissues and excreta [21]. The spectrophotometric method for determining citreoviridin was standardized as described below. All the experiments should be carried out under dim light.

The method described below is one that has been found to give optimum results. The activity of the plates and the choice of developing solvents appear to be critical for the success of this procedure.

Apparatus: Chromatoplates, 20×10 cm; UV lamp, principal wavelength 365 nm.

Reagents: *n*-hexane, benzene, acetone, all solvents should be of analytical-reagent grade; Kieselgel G, obtainable from E. Merck & Co. Inc.

Procedure: Prepare slurry by adding two parts by weight of water to one part of dry Kieselgel G and shaking the mixture vigorously in a stoppered flask for 1 min. Coat the plates with a layer of Kieselgel G of thickness 500 μ with a conventional spreader. Allow the plate to stand for 30 min and then activate it by drying at 110° for 1 h.

In case of cereal grains, extract 50 g of finely ground, homogenized powder with 250 ml of *n*-hexane in Waring blender for 5 min, and re-extract the residue twice with 250 ml of ethanol for 10 min. Transfer the ethanol solution to a coloured flask, and concentrate the solution to dryness at 50° under reduced pressure and dissolve the residue in 1 ml of benzene.

In the case of biological materials such as tissues and faeces of animals, homogenize the samples with 5 vol. of ethanol in a Potterglass homogenizer. Urine samples are acidified by adding HCl to a final concentration of 0.1 N, followed by extraction with 2 vol. of benzene. Concentrate the ethanol or benzene solutions to dryness and redissolve the residue in a small volume of acetone.

Apply a 0.2 ml portion of the acetone-soluble fraction of extracts in a line approx. 6 cm wide across a chromatoplate. Develop the chromatoplate and air-dry the plate to examine it briefly under UV light. Mark the brilliant yellow band due to citreoviridin, and then scrape the Kieselgel G containing the citreoviridin from the plate and extract it with ethanol for 5 min. Filter off the ethanol and wash the gel three times with ethanol, making the combined ethanol filtrate up to 5 ml. The R_F values are as follows:

0.45 (*n*-hexane—acetone, 1:1)
0.50 (ethylacetate—toluene, 1:1)
0.35 (chloroform—methanol, 9:1)

Record the UV absorption spectrum of the ethanolic solution in a 10-mm cell and calculate the amount of citreoviridin present in the sample. One μg of pure citreoviridin in 1 ml of ethanol gives 0.120 A at 388 nm, and the recovery of citreoviridin added to the samples ranges from 90 to 95%.

The author applied this chemical method for the determination of citreoviridin in the crude toxin samples and for study of the distribution and excretion of the mycotoxin in rats [21].

(II) Bioassay

After administration of a single dose of citreoviridin near the LD$_{50}$ to mice, characteristic symptoms such as lameness in the posterior extremities, impairment of voluntary movement, tremor, convulsions and respiratory arrest appears. These toxicological and clinical signs are used as an indicator of the neurotoxic mycotoxin of *P. citreo-viride*. The author proposed a standardized method of bioassay for cereal grains invaded by citreoviridin-producing fungi, as follows: The finely ground powder is extracted directly with 2 vol. of ethanol, and the resulting ethanol solution is evaporated to dryness under reduced pressure. The benzene-soluble fraction of the residue is dissolved in olive oil for subcutaneous administration to mice. When the characteristic symptoms described above are observed within several hours of administration, it is highly possible that the extract contains citreoviridin, and the minimum detectable amount ranges between ten and several tens of μg of citreoviridin per mouse.

Distribution of the toxic fungi

Since the discovery by Sakaki [1] that an ethanol extract of naturally moulded rice causes paresis and respiratory arrest in animals, Miyake and his collaborators, Naito and Tsunoda, carried out an extensive microbial survey on domestic and imported rice. It should be stressed that their survey for neurotoxic fungi under the "rice-fungus-toxin" hypothesis was expanded and led to the discovery of the hepatotoxic "Yellow Rice" of *P. islandicum* and the nephrotoxic one of *P. citrinum*. Furthermore, recent research by Tsunoda on a toxic Fusarium sp., which is responsible for the "barley (wheat)-fungus-toxin", also revealed the aetiology of Akakabi (red-mould)-mycotoxicosis in Japan. The efforts of these mycologists thus played an important role in the resolution of mycotoxicosis problems in Japan.

Miyake *et al.* [2] detected *P. citreo-viride* for the first time in rice imported from Formosa in 1937, and in domestic rice harvested in 1937—1939, and they noted that many batches of rice were infested by this fungus under a system of free-marketing. These ungraded batches were withdrawn from commercial dealings when government inspection was introduced. Thereafter, Tsunoda *et al.* [24] and Tsuruta *et al.* [25] examined the fungal pollution of cereal grains imported during 1954—1957 and found that *P. citreo-viride* was detected in 7.4% of the total samples examined and that a high incidence was observed in samples from Thailand, Burma, Italy, Spain and others. These findings suggest that *P. citreo-viride* is distributed in rice throughout the world.

Kurata *et al.* [26] and Miyaki *et al.* [27] also detected the fungus in domestic rice harvested in 1965 and 1967, respectively, although the incidence was very low in comparison with other toxic fungi. Several strains were detected in wheat flour [28], "home-made" Miso (bean-paste) [29] and maize [22].

Besides these mycological surveys, no other (chemical) survey for citreoviridin has been reported. In Japan, rice inspection and the storage of harvested grains under controlled temperature and humidity conditions are preventing heavy fungus pollution. However, the author fears that the fungus, as well as citreoviridin, may be present in cereals harvested in areas where climatic conditions, such as shortage of sunshine, low temperature and high humidity, favour the growth of the fungus.

Toxicology

(1) Acute toxicity

According to Uraguchi [5], the ethanol extract of *P. citreo-viride*-moulded rice typically induces acute poisoning when given i.p., s.c., or p.o. to cats, dogs and other animals. Early symptoms are progressive paralysis of the hind legs, vomiting or convulsions. Respiratory distress appears gradually, and at an advanced stage, cardiovascular disturbance and hypothermia is marked. In the final stage, dyspnoea, gasping, and Cheyne-Stokes respiration is followed by respiratory arrest. These symptoms are marked in higher animals such as cats and dogs.

In mice, the s.c. LD_{50} of the extract fluctuated, depending upon the preparation, from 1 to 10 g/kg with mortality occurring after 65—480 min. Usually no fatal cases are observed 1 day after poisoning, and the LD_{50} values by various routes were in a ratio of 10 (s.c.) : 8 (i.p.) : 30 (p.o.).

In dogs which were dying 1 h after i.p. injection, the heart looked spherical due to the dilatation of the right ventricle, and stagnation in the systemic circulation with unoxygenated, diluted blood in arteries was observed.

Employing electrophysiological techniques, Uraguchi was able to describe the process as follows; (*1*) spinal and medullary depression are responsible for progressive paralysis, (*2*) the toxin selectively inhibits motor neurones or internuncial neurones along the spinal cord and motor nerve cells in the medulla oblongata, (*3*) the attack on the respiratory center is the cause of death, and (*4*) effects on peripheral nerves and striated muscle are not seen.

A study of the toxicity of the purified mycotoxin is now in progress. In cats and dogs the isolated compound induced symptoms identical with those produced by the alcohol extract of mouldy rice [8]. The LD_{50} of citreoviridin in male mice was 11 (s.c.), 7.5 (i.p.) and 29 (p.o.) mg/kg, and in female rats, the s.c. LD_{50} was 3.6 mg/kg [21].

After a single dose near the LD_{50}, the poisoned mice died mostly within 0.5—6 h (s.c.), 0.5—3 h (i.p.) and 1—4 h (p.o.). Acute symptoms developed such as lameness of the posterior extremities, impairment of voluntary movement and tremors, and the mice became unable to stand due to advancing paralysis. Later depression, coma, gasping and convulsions followed by respiratory arrest occurred. On autopsy, the right ventricle was dilated.

Subacute experiments showed that, in mice administered toxin orally in doses of 5, 7.5 and 10 mg/kg/day (0.5 LD_{50}) about half of the male and female mice died. In mice administered the toxin s.c. at a dose of 3 mg/kg for 7 weeks, the death rate was 14/18 in the male and 2/10 in the female. These results show that the cumulative toxicity of citreoviridin is not demonstrable by short-term feeding and that the male is more susceptible than the female to the subacute intoxication by citreoviridin.

Subacute administration induced damage to the central nervous system

References p. 301

in cats. In 20 cats given the crude citreoviridin (20—30% purity) in various doses between 1 and 20 mg/kg (i.v.), 10—30 mg/kg (s.c.) or 12.5—60 mg/kg (p.o.), the following two cases showed interesting symptoms: (*1*) in a female cat (1.6 kg body weight) administered citreoviridin i.v. four times in 2.5 h (a total dose of 15.4 mg/kg) ascending paralysis occurred in the acute stage and was followed by loss of eyesight several days after the administration. Pathological examination after 1 year revealed atrophy of the optic nerve; (*2*) in a female cat (1.1 kg body weight) administered citreoviridin s.c. in two doses of 30 mg/kg and then in one or two doses of 15 mg/kg per week for one month, paroxysmal convulsions occurred after completion of the administration. The clonic and tonic convulsions occurred sporadically and continued for about 10 min at intervals of 1—2 weeks. After this stage, the cat became nervous and abnormal behaviour such as anxiety was found. These symptoms indicate that the neurotoxic action of citreoviridin impairs the central nervous system in cats.

Employing the spectrophotometry and TLC methods described previously, the author investigated absorption, distribution and excretion of citreoviridin in rats [21]. Citreoviridin administered s.c. to female rats disappeared rapidly from the injection site. This indicates that in spite of its lipophilic property the rate of absorption from the injection site is rapid and also explains the rapid onset of the toxicity of citreoviridin. Among the four major organs, the liver showed the highest content of mycotoxin; 3, 1.7 and 0.7% of the total dose were recovered as citreoviridin from the liver 8, 20 and 52 h respectively, after administration.

As for the excretion route, about 3 and 1% of the mycotoxin administered s.c. to rats were detected from the faeces at 21 and 45 h, respectively, and no detectable amount was recovered from the urine. The low recovery of citreoviridin from the tissues and excreta is indicative of *in vivo* metabolic degradation of citreoviridin. In dogs and cats administered the mycotoxin, a small amount of citreoviridin was detected in the vomit.

The influence of vitamins and other nutritional factors upon the development of acute toxicity of the crude toxin was examined. Uraguchi [31] demonstrated that pretreatment of mice with a small dose (0.002 mg) of vitamin B_1 was to some extent effective in reducing the acute intoxication. An excessive dose (2.0 mg) caused acceleration of poisoning when the toxin was given immediately after the administration of vitamin B_1 but an inhibition of poisoning when the toxin administration was deferred. Vitamin C at doses between 0.02—20 mg caused prolongation of the lethal time in poisoned mice.

These findings indicate that the development of acute poisoning depends upon the nutritional condition of the animal.

(II) Chronic toxicity

Sakai *et al.* [32] administered orally to rats various doses of the crude extract of *P. citreo-viride*-moulded rice and DAB, separately or in combination, and observed their body weights, symptoms and death rate for 6 to 16 months. On feeding a large dose (one-sixth of LD_{50} p.o.) over 30% of the rats died in 10 days and with a small dose (one-hundredth of LD_{50} p.o.) 55% of the rats died in 3 months. The body weight curves were depressed in the early stage of intoxication and depressed growth appeared again later in fatal cases. Histologically, atrophy, pleomorphism and other changes of the cells were observed in the liver of rats dying later in the experiment.

Torikai [33] also carried out a long-term feeding experiment with rats and mouldy rice, but no significant pathological changes except anaemia were detected.

As a whole, characteristic features of the chronic toxicity of *P. citreo-viride*-moulded rice or its extract have not been clearly established, and detailed characteristics of the toxicity of purified citreoviridin remain to be clarified.

In summary, cardiac beriberi, which is differentiated from other beriberis and used to prevail mostly in rice-eating countries of Asia in the past, is presumed to be caused by eating ungraded or mouldy rice. From the similarity of the clinical manifestations of acute cardiac beriberi in humans with toxicological features in poisoned animals, citreoviridin from *Penicillium citreo-viride* is suspected of being a pollutant of the mouldy rice and the cause of cardiac beriberi. Introduction of improved techniques of preparation and storage of rice affected the incidence of beriberi. In this respect, ecological studies of fungal metabolism as well as dietary habits of man have contributed much to an understanding of mycotoxicoses.

References

1 J. Sakaki, Z. Tokio-Med. Gesellsch., 5 (1891) 21 (in Japanese).
2 I. Miyake, H. Naito and H. Tsunoda, Beikokuriyo Kenkyujo Hokoku, 1 (1940) 1 (in Japanese).
3 I. Miyake, Nisshin Igaku, 34 (1947) 161 (in Japanese).
4 H. Naito, Studies on Moulds Injurous to Stored Rice, Doct. Thesis (in Japanese).
5 K. Uraguchi, Nisshin Igaku, 34 (1947) 155 (in Japanese).
6 Y. Hirata, J. Chem. Soc. Japan, 68 (1947) 63 and 74 (in Japanese).
7 N. Sakabe, T. Goto and Y. Hirata, Tetrahedron Letters, 27 (1964) 1825.
8 K. Uraguchi, J. Stored Prod. Res., 5 (1969) 227.
9 Y. Ueno, Production of citreoviridin, a neurotoxic mycotoxin of *Penicillium citreo-viride* Biourge, in I.F.H. Purchase (Ed.), Symposium on Mycotoxins in Human Health, MacMillan, London, 1971, pp. 115—132.
10 T. Irisawa, R. Tazawa, T. Sakamoto and N. Fujii, Beriberi, Nanzando Shoten, Tokyo, 1936, pp. 40—41 (in Japanese).
11 H. Wright, Brit. Med. J., 1905 (1905) 1095.
12 Ch. Toyama, Aetiological Study on Beriberi, Ohtaka Press, 1921 (in Japanese).
13 E. Uryu, Keio Igaku, 24 (1944) 229 (in Japanese).
14 T. Fukushima, Nyuji Igaku Zasshi, 12 (1932) 371 (in Japanese).
15 H. Miyake, K. Ota, Y. Kondo, Y. Oda, Y. Tohoda and S. Hayashi, Tokyo J. Med. Sci., 74 (1965) 94 (in Japanese).
16 W. Liao, Studies on Oriental Kakke-diseases, Kariya-shoten, Tokyo, 1936 (in Japanese).

17 T. Ogata, S. Kawakita, S. Suzuki and S. Kagoshima, Nisshin Igaku, 13 (1924) 742 (in Japanese).
18 U. Suzuki and T. Shimamura, Tokyo Kagaku Kaishi, 32 (1911) 4 (in Japanese).
19 C. Funk, J. Physiol., 43 (1911) 395.
20 K. Uraguchi, Pharmacology of Mycotoxins, in H. Raskova (Ed.), International Encyclopedia of Pharmacology and Therapeutics, Section 71, Pergamon, Oxford, 1971, pp. 143—298.
21 Y. Ueno and I. Ueno, Jap. J. Exptl. Med., 42 (1972) 91.
22 D.W. Nagel, P.S. Steyn and D.B. Scott, Phytochemistry, 11 (1972) 627.
23 Y. Ueno, Jap. J. Exptl. Med., 42 (1971) 107.
24 H. Tsunoda, O. Tsuruta and M. Takahashi, Bull. Food Res. Inst., 13 (1958) 29 (in Japanese).
25 O. Tsuruta and H. Tsunoda, Bull. Food Res. Inst., 14 (1956) 32 (in Japanese).
26 H. Kurata, S. Udagawa, M. Ichinohe, Y. Kawasaki, M. Takeda, M. Tazawa, A. Koizumi and H. Tanabe, J. Food Hyg. Soc. Japan, 9 (1968) 1.
27 K. Miyaki, M. Yamazaki, Y. Horie and S. Udagawa, Ann. Rept. Inst. Food Microbiol., Chiba Univ., 23 (1970) 31 (in Japanese).
28 H. Kurata and M. Ichinohe, J. Food Hyg. Sci. Japan, 8 (1967) 237 (in Japanese).
29 Y. Yamazaki, Y. Horie, S. Suzuki, H. Fujimoto, Y. Maebayashi, K. Miyaki and S. Hatakeyama, J. Food Hyg. Sci. Japan, 12 (1971) 370 (in Japanese).
30 K. Uraguchi, F. Sakai and S. Mori, Nisshin Igaku, 42 (1955) 690 (in Japanese).
31 K. Uraguchi, Nisshin Igaku, 35 (1948) 166 (in Japanese).
32 F. Sakai and K. Uraguchi, Nisshin Igaku, 42 (1955) 609 (in Japanese).
33 T. Torikai, Jap. J. Gastroenterol., 41 (1942) 478.

Chapter 14

PENICILLIUM ISLANDICUM (TOXIC YELLOWED RICE) — LUTEOSKYRIN — ISLANDITOXIN -- CYCLOCHLOROTINE

MAKOTO ENOMOTO and IKUKO UENO

Department of Pathology, St. Marianna University School of Medicine, Kawasaki; and Department of Carcinogenesis and Cancer Susceptibility, Institute of Medical Science, University of Tokyo, Tokyo (Japan)

Penicillium islandicum was first isolated by Sopp (1912) from a culture which was found on the Island of Skyr, Norway. C. Thom described this species from a culture, obtained in 1922 from Putterill, Cape Town, South Africa and classified it in the series with *Penicillium funiculosum*. The species is characterized by a more or less tufted, pigmented colony with areas of mixed orange red and dark yellow-green [1]. The species is widely distributed in nature and is classified as a storage fungus or a member of mycoflora of soils.

P. islandicum was a frequent contaminant of rice shipped largely to Japan after World War II from various rice-producing areas of the world. H. Tsunoda, a Japanese mycologist, isolated this fungus in 1948 from yellowed rice* imported from Egypt. When Tsunoda was in Giolo Island during the World War II, a toxicosis causing oedema of legs occurred among Japanese soldiers following the ingestion of rice contaminated with this fungus [3]. To confirm his own observations, he carried out a feeding experiment with mouldy rice inoculated with *P. islandicum*, which he isolated from Egyptian rice. Liver cirrhosis of the postnecrotic type was found in rats fed this rice for around one month. The subsequent extensive studies revealed the hepatotoxic and hepatocarcinogenic activities of the fungal metabolites of *P. islandicum*. Luteoskyrin and cyclochlorotine were identified as the toxins responsible for the hepatic disorders in experimental animals fed on mouldy rice [4—6]. The tumourigenic effect of luteoskyrin on the liver was observed after giving a daily oral dose of luteoskyrin to mice [7].

Outbreaks of disease attributable to the fungus

Acute intoxication of domestic animals or humans by contamination of food with *P. islandicum* has only been reported once — an outbreak of

---

*Yellowed rice is a yellowing of rice grains caused by fungal contamination. It was named by Ichiro Miyake, who, in 1937, detected *Penicillium toxicarium* Miyake (*Penicillium citreo-viride* Biourge), growing on rice of Formosa [2].

poisoning of chicks in Japan. A total of 2891 out of 13 610 chicks died in Nagano prefecture during the autumn of 1963. All of the mice fed with a diet containing this poultry feed died within 16 days after showing inactivity with moderate jaundice. The histopathological findings of the liver, characterized by centrilobular necrosis and subsequent collapse of the stroma with interconnection of necrotic areas, were similar to those reported after *P. islandicum* poisoning [8].

Recent investigation of the mycoflora of rice — domestic and imported in Japan — revealed a relatively low incidence of *P. islandicum* [9]. However, the stability of the taxonomical and mycotoxin-producing properties of this fungus, in addition to its occasional occurrence in human foodstuffs, shows that it still has importance as a possible causative agent of human liver diseases. The experimental evidence that hepatic disorders, including acute liver atrophy, liver cirrhosis and liver tumours, can be induced by metabolites of *P. islandicum* growing on rice grains, presents an important problem for rice-eating peoples in most of the Asiatic countries.

Toxic metabolites of the fungus

Compounds from *P. islandicum* include seven pigments — islandicin, iridoskyrin, catenarin, erythroskyrin, luteoskyrin, skyrin and rubroskyrin. All of the pigments except luteoskyrin, were isolated by Howard and Raistrick [10]. The seventh, a lipophilic and toxic pigment was first isolated by Tatsuno and his collaborators in 1955 from fungus mat of *P. islandicum* Ud*, which was cultured in Czapek medium [11].

Later, it was found that the strains of *P. islandicum* were divided chemotaxonomically into the following three groups.

(1) *The luteoskyrin-producing type strains.* Their colonies are more or less floccose, green-yellow in colour and with brown-red reverse. This group produces the pigments luteoskyrin, islandicin, catenarin, iridoskyrin, dicatenarin and rubroskyrin. The representative strains belonging to this group are NHL 6286, Ud, Ea** and NRRL 1036.

(2) *The skyrin-producing type or flavoskyrin strains.* The colonies are velvety and yellow orange with reverse of the same colour. Both NRRL 1175 and NHL 6244 belong to this group which produces flavoskyrin, chrysophanolic acid and skyrin.

(3) *The erythroskyrin-producing type strains.* The Sa strain, isolated from imported Spanish rice grains and LSHTM.B.B. 233, comprises the representative fungi producing erythroskyrin and rubroskyrin. Their colonies are velvety, and dark-orange in colour with a greenish orange or greenish grey tone. The reverse shows a dark brownish-grey colour [13]. *Mycelia sterilla* was found by Nishikawa to produce rhodomycelin, flavomycelin and purpuromycelin [14]. Later, flavomycelin was confirmed to be identical to luteoskyrin [15] (Table I).

---

*Isolate from imported Californian rice (Tsunoda [12]).
**Isolate from imported Egyptian rice (Tsunoda [12]).

TABLE I

PHYSICOCHEMICAL DATA AND CHEMICAL STRUCTURES OF LUTEOSKYRIN AND RELATED COMPOUNDS [18, 31, 35, 72]

Compounds			mp.	Colour	Colouration with Mg(OAc)$_2$	with H$_2$SO$_4$	PPC R_F(TLC)^a
Luteoskyrin	C$_{30}$H$_{22}$O$_{12}$		281 (decomp)	yellow	—	—	0.52 (0.40)
Rugulosin	C$_{30}$H$_{20}$O$_{10}$		290 (decomp)	yellow	—	red	0.52 (0.40)
Rubroskyrin	C$_{30}$H$_{22}$O$_{12}$		281 (decomp)	brown-red	green	bleu	0.25 (0.10)
Skyrin	C$_{30}$H$_{18}$O$_{10}$	R$_1$=OH, R$_2$=H	>360	yellowish-brown	orange-red	emerald green	0.42 (0.20)
Iridoskyrin	C$_{30}$H$_{18}$O$_{10}$	R$_1$=H, R$_2$=OH	360	orange	reddish-violet	bleu	0.97^b (0.90)
Islandicin (Funiculosin, Rhodomycelin)	C$_{15}$H$_{10}$O$_5$	R$_1$=H, R$_2$=OH	218	red	reddish-violet	reddish-violet	0.97^b
Catenarin	C$_{15}$H$_{10}$O$_6$	R$_1$=R$_2$=OH	246	red	reddish-violet	reddish-violet	0.80
Chrysophanol	C$_{15}$H$_{10}$O$_4$	R$_1$=R$_2$=H	196	yellow	orange-red	red	0.97
Emodin	C$_{15}$H$_{10}$O$_5$	R$_1$=OH, R$_2$=H	257	orange-red	orange-red	red	—
Flavoskyrin	C$_{15}$H$_{12}$O$_5$		208	yellow	—	orange-red ↓ reddish-violet	—
Erythroskyrin	C$_{24}$H$_{31}$O$_6$N		—	yellow-brown	yellow	—	0.63 (0.58)

^{a} ref. 24.

^{b} R_F (TLC: petroleum ether (b.p. 40—50°) as a developer): iridoskyrin, 0.40—0.55 (violet) and islandicin, 0.95 (reddish violet with Mg(OAc)$_2$).

A water-soluble toxic compound was isolated independently by Tatsuno *et al.* [11] and Marumo [16] from culture filtrates of *P. islandicum*. Chemical analysis revealed that the compound was a cyclic peptide including two chlorine atoms in its molecule. This compound was given two different names, cyclochlorotine [17] and islanditoxin [16], by these workers. The interesting fact was that the existence of two kinds of toxic agents in the methanol extract from the fungus mat was presumed from observation of the biphasic effects, which were shown in the survival time and bromosulph-thalein (BSP) excretion of the animals treated with the samples. As far as survival time is concerned the death of the mice occurred either less than 15 h or more than 24 h after injection. The result of the BSP-test also showed a similar biphasic disorder of hepatic function. It was further con-firmed by histopathological findings in the liver, revealing basic differences in hepatic injury. One was the so-called acinus peripheral cytotoxic pattern, characterized by vacuolation of the liver cells and endothelial cells in the perilobular region, caused by the rapid-acting hydrophilic compound. An-other was the so-called acinus central cytotoxic pattern characterized by necrosis of the centrilobular liver cells, induced by the slow-acting lipophilic component, luteoskyrin.

Chemical and physical properties of metabolites, including structure

Luteoskyrin

Luteoskyrin, $C_{30}H_{22}O_{12}$ (m.p. 287° (decomp.)), $Vc=o$ 1623, $Vo-H$ 3378 cm^{-1} (Nujol); λ_{max} 245, 275, 430 nm, is a yellow rectangular crystal with optical rotation of $[\alpha]_D^{25}$ −880° (0.1% in acetone).

The true hydroxyanthraquinone derivatives, including islandicin, cate-narin, skyrin, iridoskyrin and rubroskyrin, react with magnesium acetate to develop characteristic colours. However, luteoskyrin shows no development of colour with this reagent. The structural formula of luteoskyrin was con-firmed recently by nuclear magnetic resonance spectra [18] and an X-ray diffraction study on structure and absolute configuration of D-rugulosin [19]. D-Rugulosin is the dehydroxylated form of luteoskyrin and in reversed relation to L-luteoskyrin stereochemically [20] (Table I). When treated with $Na_2S_2O_4$, luteoskyrin is split into islandicin. Treatment of luteoskyrin with 60% H_2SO_4 yields islandicin and iridoskyrin. Rubroskyrin is shown to yield luteoskyrin when treated with pyridine and chlorobenzene.

Luteoskyrin is known to convert to a brownish red quinoid compound when exposed to the sun. This photochemical product was named lumiluteo-skyrin. $C_{30}H_{20}O_{12}$ (m.p., over 360°), which develops a sky-blue colour with magnesium acetate in butanol [17].

Cyclochlorotine

Cyclochlorotine, white needles of m.p. 251° (decomp.), with optical rotation of $[\alpha]_D^{16}$ −92.9° in ethanol, $C_{25}H_{36}N_5O_8Cl_2$, V_{max} 3400, 1670, 1540, 1535, 1510, 1495, 1460, 695 cm^{-1}; λ_{max} 257 nm (ϵ 2.75), is a cyclic

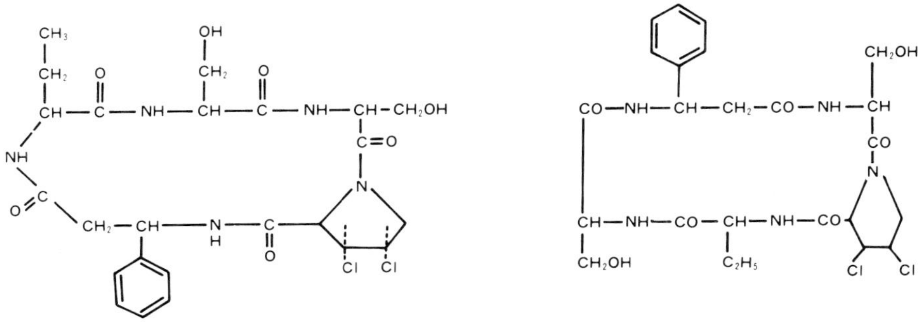

Islanditoxin   Cyclochlorotine

Fig. 1. The chemical structure of cyclochlorotine and islanditoxin.

peptide containing chlorine. Sato and Tatsuno [21] elucidated the amino acid sequence of the peptide as α-pyrrol-carboxyl-L-α-amino-butylyl-L-seryl-β-amino-β-phenylpropinoyl-L-serine amide. Recent investigation with X-ray crystallographic analysis of this chlorine-containing peptide disclosed its structure as shown in Fig. 1 [22]. Cyclochlorotine shows a positive Biuret reaction, but is negative in Sakaguchi reaction, ninhydrin reaction or Millon reaction.

Islanditoxin is a compound isolated by Marumo from a filtrate of *P. islandicum* [16]. Its chemical properties, such as m.p. and type of constituent amino acid, and biological effect in animals seemed to be identical to those of cyclochlorotine. However, the proposed structure by Marumo is different from cyclochlorotine in the number and sequence of the amino acids (Fig. 1).

Biosynthesis of the toxic metabolites

Luteoskyrin

As Gatenbeck [23] suggested, luteoskyrin was presumed to be biosynthesized from acetate-malonate system. This was confirmed by labelling experiments with ^{14}C, revealing incorporation of [^{14}C]malonate, [^{14}C]acetate, [^{14}C]glutamate, [^{14}C]asparate into luteoskyrin [24], rubroskyrin [24] and rugulosin [25]. It was further shown that acetyl-CoA carboxylase acted as a key enzyme for biosynthesis of these pigments [26]. An increased content of citrate at the onset of secondary metabolism will influence the acetyl-CoA carboxylase, leading to a larger CoA pool [27]. Addition of avitin, an inhibitor of acetyl-CoA carboxylase, inhibited the synthesis of pigments, while biotin reversed this inhibition [28].

Cyclochlorotine

Sato and Tatsuno have succeeded in the chemical synthesis of the dehydrochlorinated peptide amide based on its amino acid sequence, which was elucidated by chemical analysis of cyclochlorotine [21]. It was shown that the chemicophysical characteristics such as m.p. R_F values, IR and nuclear magnetic resonance (NMR) spectra and elementary analysis of this synthetic peptide were identical to those of the natural one. The complete biosynthesis of cyclochlorotine will soon become possible because it has recently been disclosed how two chlorine atoms bind in the molecule of cyclic peptide [22].

Environmental and laboratory conditions favouring production of the toxic metabolites

P. islandicum can grow over a wide range of temperatures from below 10° to 45°. The optimum temperature for growth is 30°. However, exposure of cultures for 10 min at 53° will kill the fungus. An optimal temperature for production of toxic metabolites by *P. islandicum* is around 33°. Culture conditions with exposure to a temperature lower than 30° reduced toxin production [29]. For animal feeding experiments, strains of *P. islandicum* were inoculated onto polished rice grains and incubated at 33° for 7—14 days. The cultured mouldy rice was then dried, powdered and mixed with the sterile powder of polished rice in different ratios. Milk casein, McCollum salt and vitamins were added to it. The extensive feeding experiments using this feed disclosed that changes in the liver varying from acute damage to cirrhosis and tumours were produced in mice and rats, depending upon the amount of metabolites produced by this mould [12].

The pH varied during culture from an initial level of pH 6.0 to a low of pH 4.2, followed by an increase to pH 5.8 again. For obtaining the largest production of pigments, a Czapek's nutrient mixture was used: 0.5 g magnesium sulphate, 0—1.0 g ferrous sulphate, 1.0—9.0 g potassium monophosphate, 0.5 g potassium chloride, less than 0.5 g sodium nitrate dissolved in 1 l water [29]. Comparison of toxin yields on rice, barley, wheat and maize demonstrated that elaboration of toxic metabolites was highest on rice, and lowest on maize [30].

Chemical analysis for the toxic metabolites

Luteoskyrin

(1) Detection by paper and thin-layer chromatography

Paper and thin-layer chromatography are useful tools for both separation and identification of luteoskyrin from other naturally occurring hydroxyanthraquinone homologues. The hydroxyanthraquinones were separated by development of one-dimensional paper chromatography (Toyo Roshi No. 3) with the upper layer of a mixture of acetone—petroleum benzine (b.p. 60—70°)—water (5:5:3.5) as solvent [31]. Each pigment was identified by the respective R_F values and the colours developed by spraying with magnesium acetate reagent [32]. The R_F values of the hydroxyqui-

nones were 0.25 (rubroskyrin) and 0.97 (islandicin, iridoskyrin and chryso-
phanol) [31].

A method of thin-layer chromatography using oxalic acid-treated
Kieselgel G as adsorbent was proposed by Fujita [33]. According to the
modified method of Ueno *et al.* [24], the R_F values of the hydroxyanthraquinones on the plate of 0.1 M oxalic acid-treated Kieselgel G with the
solvent of acetone—n—hexane—water mixture (4:2:1) were as follows:
rubroskyrin 0.10; skyrin 0.20; luteoskyrin 0.40; iridoskyrin 0.90 and islandicin 0.95.

For chemical analysis of the toxic pigments produced by *P. islandicum*
invading rice grains, Tatsuno *et al.* [34] devised a method of so-called "one-
grain culture chromatography" as follows: Each rice grain, artificially inoculated, was put on the upper part of a small strip of filter paper in a test tube and
the lower end of the paper was dipped into the water at the bottom of the tube.
After 7 to 14 days' culture at 26° to 29° with the tube sealed with cottonwool,
the rice grain was crushed and the fungal metabolites were extracted with
acetone to chromatograph on the filter paper. Using this simple method, it was
confirmed that *P. islandicum* growing on rice grains produced the same
pigments as in Czapek medium [34].

(2) Extraction and determination

Luteoskyrin extracted from samples with organic solvents, can be assayed spectophotometrically in organic solvents or in buffered aqueous solutions. The visible absorption maxima and their molar extinction coefficients
are as follows [35]: λ_{max} 433 (ϵ 34 300), 448 (ϵ 34 400) in ethanol;
λ_{max} 436 (ϵ 32 300), 457 (ϵ 32 100) in chloroform; λ_{max} 432, 448
(ϵ 35 500) in acetone; λ_{max} 353 (ϵ 22 400), 350 (ϵ 26 300) in 0.1 M potassium phosphate buffer (pH 7.4).

(a) Phenol extraction. Luteoskyrin can be extracted from powdered
mouldy rice with 90% phenol solution after washing with petroleum ether to
remove the rice-bran oil and pigments such as islandicin, iridoskyrin and
erythroskyrin. After steam distillation of the phenol, the pigment is extracted with ether. Taking the 60% recovery into account, the estimated
content of luteoskyrin was reported to be approximately 1 g in 150 g of the
mouldy rice powder tested [12].

(b) Acetone extraction. For chemical assay of the amount of luteoskyrin and other pigments, Ueno and Ishikawa [24] recommended acetone
extraction of the dried mycelium of *P. islandicum* followed by thin-layer
chromatography and spectrophotometric measurement [33].

(c) Ether extraction. Ueno [36] extracted luteoskyrin with ether from
the feed pellet to which luteoskyrin had been added and from the liver of
mice administered the toxin. The recovery rate of luteoskyrin in this method
was higher than that after phenol extraction. 1 g of the pellet is powdered,

suspended in 5% perchloric acid (PCA) solution and luteoskyrin is extracted with ether. In case of the liver, the sample (1 to 10 g) is homogenized with water and PCA added to a final concentration of 5%. After washing the precipitate with n-hexane, luteoskyrin is extracted with ether. Extracted luteoskyrin is assayed by the spectrophotometric method in acetone, directly or after purification by thin-layer-chromatography. As little as 0.5 μg of luteoskyrin can be detected or more than 90% of the added luteoskyrin is recovered from one gram of the pellet containing 50 μg of the toxin. This method should be applicable for the detection of luteoskyrin with high recoveries from the subcellular fractions of liver, from urine and faeces of the animals administered luteoskyrin.

Chlorine-containing peptide (cyclochlorotine)

(1) Detection by thin-layer chromatography

Ishikawa *et al.* [37] devised a method of thin-layer chromatography using Kieselgel G and a mixture of n-butanol—acetone—water (4:1:4) for detection of cyclochlorotine. On spraying with $2 N$ $H_2 SO_4$ and heating the plate at $100°$, the peptide forms a brown spot at R_F 0.7.

(2) Spectrophotometric determination

The chlorine-containing peptide shows a weak UV absorption spectrum due to the benzene ring (λ_{max}, 253, 259, 265). Upon ammonolysis in $2N$ $NH_4 OH$-methanol at $37°$ for 24 h, the peptide is cleft completely to dehydrochlorinated-peptide amide, exhibiting a UV absorption maximum at 268 nm due to the pyrole ring. The optical density at 268 nm is proportional to the concentration of the peptide between 2 μg to 50 μg per ml. Employing this photometric technique, the content of the peptide in the culture filtrate can be determined [37].

(3) Application

A chemical method for fractionation and determination of the peptide produced in the culture filtrate of *P. islandicum* depends on a combination of adsorption on charcoal, gel filtration and spectrophotometry [37,38]. To reduce the volume of the filtrate, the peptide was first adsorbed on charcoal treated with $1 N$ HCl and eluted with methanol or butanol. The crude toxin eluate was then chromatographed on Sephadex LH-20 column with methanol—water (1:1 by vol.). Pure toxin was obtained from the ammonolysis-positive fraction with 91.4% recovery.

Natural occurrence of the toxic metabolites

In spite of the experimental evidence of the hepatotoxic activity of the metabolites produced by *P. islandicum*, the aetiological relationship between the contamination of human food with this mould and human diseases has not yet been elucidated. Recent field investigations on the isolation of mycotoxin-producing fungi contaminating human foodstuffs in Japan showed a relatively low incidence of this fungus in human foodstuffs [9]. Lack of adequate analytical methodology for luteoskyrin or cyclochlorotine contained in food has delayed confirmation of suspected contamination of foodstuffs with the metabolites of *P. islandicum*.

However, it should be emphasized that bioassay using mice revealed that all the tested strains of *P. islandicum* isolated from the foodstuffs in our field and storage investigations were hepatotoxic. A similar method, using liver pathology of mice combined with liver-function tests (including BSP-test and measurement of SGOT and SGPT) was available for screening the toxicity of foodstuffs suspected of being contaminated with the metabolites of *P. islandicum* [39].

Contamination by *P. islandicum* has occasionally been reported, for example as a cause of chick intoxication in Japan [8], as one of the major isolates from various grains, including domestic as well as imported rice in Japan [3,40], from the staple diet "Teff" in Ethiopia [41], from Danish barley [42], and as a prevalent infection in prepared foodstuffs in South Africa [43].

As Uraguchi *et al.* [44] have pointed out, the Japanese are estimated to consume more than 150 kg of rice a year on average. If one in a thousand rice grains is infected with *P. islandicum*, which is capable of producing 6 g of luteoskyrin per kg of mouldy rice [12], the annual intake of luteoskyrin could be approx. 0.9 g. This amounts to about one-tenth of a single oral LD_{50} (220 mg/kg) for a human, if the susceptibility of man to luteoskyrin is identical to that of the mouse.

It is also an important fact that rugulosin, chemically related to luteoskyrin, is known to be produced by a number of regular contaminants of foodstuffs, such as *P. rugulosum*, *P. brunneum*, and *P. tardum* [45].

Biological activity in experimental systems

Luteoskyrin

A lethal dose of luteoskyrin given subcutaneously (s.c.) or orally (p.o.) as an oily suspension produces poisoning which develops rather slowly in animals and causes death mostly from 2 to 4 days after the administration [46,47]. The LD_{50} is 145 (s.c.) and 221 (p.o.) mg/kg in dd mice (Table II). The low slope of the dose—response curve indicates a wide difference in susceptibility among individuals. However, in mice receiving intravenous (i.v.) administration of leuteoskyrin as a neutralized saline solution, the toxic symptoms appeared within a few hours. The i.v. LD_{50} was 6.65 mg/kg in dd mice (Table II) [35]. The dose—response curve showed a steep slope in the latter cases. Slow absorption of luteoskyrin following subcutaneous or oral administration was confirmed by pharmacokinetic studies with ^{3}H-labelled luteoskyrin [48] and by the autopsy finding that luteoskyrin remained for long periods of time at the injection-site or in the gastrointestinal tract of the animal.

References p. 323

TABLE II

DATA ON BIOLOGICAL EFFECTS OF LUTEOSKYRIN

Compounds	Mice LD_{50}(mg/kg)		Bacteria $ID_{50}(\mu g/ml)$			*Tetrahymena pyriformis* $ID_{50}(\mu g/ml)$	Ehrlich ascites tumour ED_{50} $\mu g/kg$
Luteoskyrin	221	(po) in olive oil	*E. coli*, F-11	0.4	Gram (+)	1 — 2.5	1 — 2
	145	(sc) in olive oil	*B. subtilis*	0.13			× 7
	40.8	(ip) in olive oil	*St. aureus*	0.22			days
	40.8	(ip) in saline aq.	*Alcaligenes*				
	6.65	(iv) in saline aq.	*faecalis*	1.00	Gram (—)		
			Serratia				
			marcescens	0.35			
Rugulosin	55	(ip) in olive oil	*E. coli*, F-11	1.5	Gram (+)	5 — 10	4 — 8
	83	(ip) in 0.9% NaCl-CMC solution	*B. subtilis*	0.30			× 7
			St. aureus	0.22			days
			Al. faecalis	4.50	Gram (—)		
			Serratia				
			marcescens	1.60			

The morphological changes of the acutely damaged liver usually appear 24 h after administration of luteoskyrin. The liver is initially enlarged and yellow but later becomes atrophic with minute red dots. The characteristic histological changes of the liver are centrilobular necrosis and fatty degeneration of the liver cells [46,17]. Similar changes are also observed in the tissue culture cells (Chang's liver cells, HeLa cells and rat liver cells) treated with 1 μg/ml of luteoskyrin [49,50].

Effects on liver function

Biochemical tests on the luteoskyrin-poisoned animals showed that the severity of the abnormal liver function was parallel to the extent of the histopathological lesions of the liver.

Serum enzyme activity. An increased serum GPT activity is one of the earliest changes of luteoskyrin poisoning in animals [48,51]. The activity usually begins to increase 24 h after administration. The same pattern was observed in GOT activity, although the latter was somewhat less pronounced than that of GPT, especially with low dosages of the toxin. These enzyme activities reached their highest level 2 days after administration and then gradually returned to the normal range in a week [48]. The histopathological examination revealed that the increase in enzyme activities correlated with the occurrence of centrilobular degeneration and necrosis of the liver cells [51]. Using the [³H]luteoskyrin, the highest concentration of luteoskyrin was detected in the liver 1 day before the appearance of the highest enzyme activities [48]. Thus the increased serum transaminase activity can be a reliable index of acute liver injury produced by luteoskyrin.

The activities of malic dehydrogenase (MDH) and glucose-6-phosphatase (G6Pase) in the serum were also enhanced by the administration of luteoskyrin, although not so markedly as serum transaminase activity. Serum alkaline phosphatase showed no increase even in the seriously poisoned cases with high serum GPT [36].

BSP test. BSP injected i.v. in luteoskyrin poisoned mice was retained longer in the blood than the control, possibly because of impaired liver function with disordered excretion of the dye [39].

Blood sugar level and liver glycogen content. These were reduced in the acutely poisoned mice [51]. However, the extent of the reduction was less than that observed after administration of CCl_4, rubratoxin B or chlorine-containing peptide [36]. Luteoskyrin is not likely to have a direct effect on glucose metabolism.

Lipid contents in the liver. In confirmation of the histopathological findings of diffuse fatty metamorphosis in the liver, an increased content of triglyceride in the liver and lowered concentration in the blood plasma were observed in the mice treated with luteoskyrin [36].

There are a few reports on the elevation of cholesterol ester in the serum of mice and rabbits fed mouldy rice (2 or 5%) or fungal mat (5 mg) for 100 days, respectively [52,53] but no elevation was observed with luteoskyrin [36].

Absorption, distribution and excretion of luteoskyrin

The isotopic experiments using [^{3}H]luteoskyrin revealed that the highest radioactivity in blood was only 0.05% of the dose administered and that in the liver it was lower than 3%, showing a poor absorption of luteoskyrin in general. Nevertheless, the radioactivity in the liver was extremely high when compared with that in the other organs including kidney, heart, spleen, lung, brain and muscle. These results seem to indicate the selective concentration of luteoskyrin in the liver [48]. Luteoskyrin was shown to be excreted *via* bile duct and kidney by chemical identification of luteoskyrin in the faeces and in the urine [36]. The excretory ratio *via* these two routes was 19% in faeces and 6% in urine. [^{3}H]luteoskyrin was excreted rather slowly in the male mice with recovery of 25% of the administered dose within 18 days [48].

Intracellular distribution of luteoskyrin

The [^{3}H]luteoskyrin in the liver was distributed in mitochondrial (30—50% of total), microsomal and soluble fractions (both less than 10%) and in cell debris containing nuclei (near 50%). The highest specific radioactivity was observed in the mitochondrial fraction and the ^{3}H-compound was found to be [^{3}H]luteoskyrin itself [54]. These data may indicate the specific affinity of luteoskyrin with the mitochondrial component.

Factors influencing the toxicity and carcinogenicity of luteoskyrin

The recent toxicological and histopathological studies of the biological activity of luteoskyrin revealed that the hepatotoxicity and carcinogenicity of the compound were influenced by various factors. *Species*: Luteoskyrin is known to cause hepatotoxic effects in mice [7,46,55,56], rats [36,57], rabbits [55] and monkeys [55]. Acute liver lesions seen in these animals are almost the same, characterized by centrilobular liver cell necrosis. Rabbits and mice are more susceptible to luteoskyrin than rats. *Sex*: A marked difference in toxicity was observed between male and female mice. The s.c. LD_{50} in the female was over 2 g/kg, which was more than ten times larger than the value of the male (154 mg/kg) [58]. In the female, no increase of the serum transaminase activity was observed even at the dose effective in the male. Adult male mice castrated before or at weaning were less susceptible to toxicity than the entire males [54].

The serum transaminase activity indicated that the degree of liver damage caused by luteoskyrin is several times greater in male mice than in male rats. Rats also showed a sex difference in susceptibility to luteoskyrin, though not so extensive as that observed in mice [36]. Chemical analysis of the luteoskyrin content of the liver showed higher levels in males, which, however, diminished after castration [59]. The hepatic content of $[^3H]$-luteoskyrin, given i.v. to mice, decreased more rapidly in the female mice. These findings suggest a possible influence of sex hormones on the excretory rate of the compound from the liver [36]. *Age*: The toxicity of luteoskyrin was higher in younger mice. The subcutaneous LD_{50} values (mg/kg) in newborn (within 24 h) dd mice was 7.2 in male and 6.3 in female, or less than one-twentieth of the value for adult male mice (154 mg/kg at 10 week of age) [58]. The luteoskyrin content in the suckling liver was higher than that of the adult liver [54]. *Diet*: The level of protein in diet was found to influence the susceptibility of the mice to the toxic effects of mouldy rice. The death rate of the ddD mice fed a high (34%) protein diet containing 3% mouldy rice was significantly lower than that of the mice fed a low (11%) protein diet containing 3% mouldy rice [52]. A study on dietary composition, predisposing the liver of mice to the action of luteoskyrin, was carried out by using three basal diets provided either 2.25 g rice plus 2.25 g barley, 4.5 g rice or 4.5 g barley daily. Luteoskyrin-induced carcinogenesis was enhanced by feeding a rice diet, either alone or admixed with barley. The fat content of the diets (2.4% for rice and 1.2% for barley) was suggested to be the main factor influencing this result, probably through enhancement of the absorption of fat-soluble luteoskyrin [7]. *Strains*: There is a difference in the susceptibility of mice of different strains to the acute toxic effect of luteoskyrin. When a single dose of luteoskyrin (90 mg/kg corresponding to around 2/3 of a single s.c. LD_{50}) was injected s.c. in mice, acinus-central hepatotoxic changes appeared faster in the C57BL/6 strain than in C3H/He and DDD. The most pronounced cytotoxic lesions 24 h after treatment were seen in C3H/He followed by C57BL/6 and DDD [60].

Chronic effect of luteoskyrin

The chronic effects of luteoskyrin have only been studied in mice, which develop liver tumours after long-term feeding. The oral carcinogenic dose of luteoskyrin in mice (ddS) is around 50 to 100 ppm in the feed [7]. Liver cell tumours of various types, including liver cell adenoma (Hepatoma I), well-differentiated liver cell carcinoma without metastasis (Hepatoma II) and undifferentiated liver cell carcinoma with frequent metastasis (Hepatoma III), were induced usually after 6 months of feeding. Hepatoma I and II developed more frequently than Hepatoma III. Liver cirrhosis was unusual in luteoskyrin-treated mice. However, chronic liver injury, showing advanced pleomorphism of the liver cell nuclei accompanied by giant and bizarre nuclei and a frequent deposit of ceroid-like substance in the Kupffer cells, was seen in the livers of most of the mice dying after 6 months of feeding with or without liver cell tumours. Foci of the infiltrating cells, endothelial cells and the necrotic liver cells were occasionally found in the liver [7,60].

Interim observations on livers of the sacrificed mice and biopsy liver

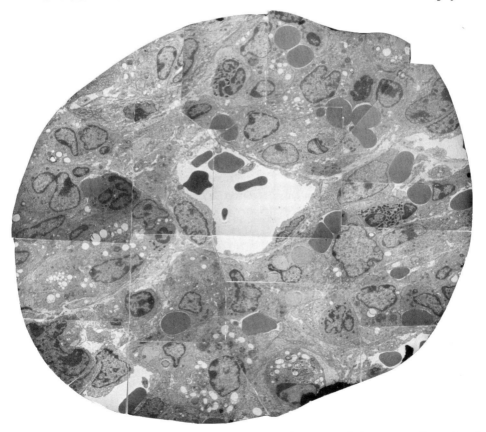

Fig. 2. Electron photomicrograph of a centrilobular area of the mouse liver showing accumulation of the endothelial cells and inflammatory cells as a replacement of the necrotic hepatic cells. The mouse was sacrificed on the 4th day of feeding with 10% *P. islandicum*-infected rice. (Prepared by Takenaka, Inst. of Med. Sci., Univ. of Tokyo, Tokyo.)

Fig. 3. Electron micrograph of a acinus peripheral and portal (lower half of the photograph) regions of the mouse liver showing slight pleomorphism of the liver cells and increased mesenchymal cell reaction. The mouse was sacrificed after 8 weeks of feeding with 5% *P. islandicum*-infected rice. (Prepared by K. Miyata, Dept. of Pathol., St. Marianna Univ., Kawasaki.)

specimens of rats fed metabolites of *P. islandicum* revealed chronic toxic damage, morphologically resembling human "chronic hepatitis". These hepatic lesions, characterized by necrosis of the liver cells and interstitial reactions and proliferation of the liver cells, were observed frequently prior to cirrhosis or tumour (hepatoma) development in mice and rats in the above experiments (Figs. 2 and 3). In the less severely damaged liver only single scattered liver cells were necrotic, resembling the acidophilic body in viral hepatitis. Recently, attention has been focused on the possible role of hepatotoxic mycotoxins in the aetiology of chronic hepatic lesions seen in both animals and humans [60—62].

Cyclochlorotine

Acute effect

Cyclochlorotine is a "periportal toxic agent", causing peripheral damage to the liver lobule. The LD_{50} of cyclochlorotine is 0.33 mg/kg (i.v.), 0.47 mg/kg (s.c.) and 6.55 mg/kg (p.o.). Acute symptoms of cyclochlorotine intoxication were respiratory and circulatory disturbances, followed by convulsions in mice and rats. Dogs administered cyclochlorotine i.v. (1 mg/kg) showed vomiting, cyanosis, palpitation and low skin temperature [63]. Cyclochlorotine acted rapidly, causing death in rats and mice in less than 24 h in most cases. Hematocrit values increased for 2 h after the injection followed by a gradual decline to the initial level. Mice intoxicated with cyclochlorotine developed an initial hyperglycaemia and a subsequent hypoglycaemia, a rapid fall of liver glycogen, decreased succinic dehydrogenase and inhibition of oxidative phosphorylation [64]. A study of distribution of tritium-labelled cyclochlorotine revealed its rapid concentration in the liver (64% of the total labelled compound after 30 min) and excretion from the kidneys (73%) [35].

The liver appeared pale and anaemic 5 min after administration of cyclochlorotine and gradually turned red 30 min later [46, 63,65,66]. Light microscopy showed vacuolation and degeneration of the liver cells followed by the appearance of hyaline droplets and erythrocytes, almost exclusively in the periportal regions of the liver. Extensive haemorrhage occurred throughout the lobules when damage was severe [46]. The endothelial (Kupffer) cells of the sinusoids are also sensitive to the toxic effects of cyclochlorotine, with vacuolation degeneration and nuclear damage observed in these cells [57].

Tissue culture cells showed cytological changes similar to those observed in liver *in vivo* [49]. The parenchymal cells of rat liver were more sensitive to cyclochlorotine than Chang's liver cells or HeLa cells [50].

Chronic effect of cyclochlorotine

Cyclochlorotine is one of the periportal cirrhogenic agents. The mesenchymal reaction proceeds around portal triads with piecemeal necrosis and cellular infiltration (Fig. 4). The simultaneously increasing fibrosis and re-

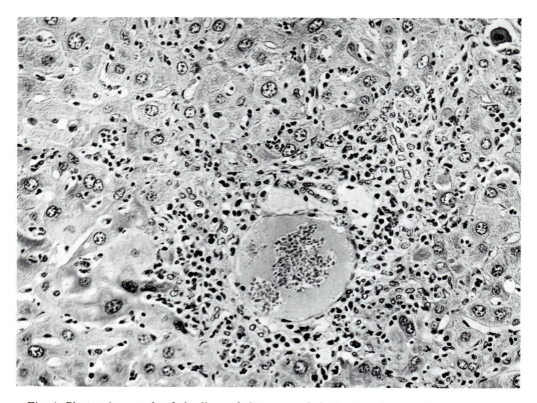

Fig. 4. Photomicrograph of the liver of the mouse fed 20 μg cyclochlorotine per day for 120 days. "Piecemeal necrosis" and degeneration of the liver cells. Haematoxylin and eosin stain. × 200. (From Enomoto and Saito, Acta Pathol. Japon., 23 (1973) 655.)

generation nodules produce cirrhosis of periportal type, though in its final form it cannot be readily distinguished from a non-portal type [60].

Only a few mice fed 60 μg of cyclochlorotine daily for over 200 days developed liver cell tumours. Reticuloendotheliomas, probably originating from the reticuloendothelial cells of the liver, were found occasionally in the liver of the mice given 40 and 60 μg/day of cyclochlorotine [7].

Toxicity of the anthraquinone-type pigments, homologues to luteoskyrin

Rugulosin

Rugulosin has a biological activity similar to luteoskyrin in various systems. It caused acute hepatic lesions of centrilobular toxic type in mice and rats and induced hepatoma in mice. The pathological and functional changes of the liver produced by rugulosin were similar to those produced by luteoskyrin, although the effective dose of the former was somewhat larger than that of the latter. The reported death rates in mice were 1/6 at 1.3 g/kg (p.o.) [55], 10/10 at 2.0 g/kg (p.o.) [55] and 1/10 at 300 mg/kg (s.c.) [46].

The intraperitoneal (i.p.) LD_{50} is 55 mg/kg in olive oil and 83 mg/kg in saline solution [45].

Daily administration of rugulosin at a dose of 0.3 mg per animal for over 500 days produced hepatoma in mice [38]. Rugulosin also has anti-tumour, antibacterial and antifungal activities [45,67].

Islandicin, iridoskyrin, catenarin, skyrin and rubroskyrin

All of the above compounds are metabolites of *P. islandicum* Sopp and are related to luteoskyrin. However, they showed no lethal effect on mice at a subcutaneous dose of 300 mg/kg [46]. Both rugulosin and rubroskyrin inhibited the succinoxidase system of rat liver homogenate, but to a lesser extent than luteoskyrin. Islandicin showed no effect even at a far higher concentration than that at which luteoskyrin exhibited its inhibitory effect [36].

Mode of action of the toxic metabolites

Luteoskyrin

(1) On mitochondrial function

With tricarboxylic acid (TCA) cycle intermediates as substrate, luteo-skyrin (10^{-4} to $10^{-5} M$) inhibited oxygen uptake in the homogenates of rat liver, kidney and heart *in vitro*. However, luteoskyrin inhibited each part of the electron-transferring chain of the succinoxidase system including succinic dehydrogenase and SC-factor and cytochrome c oxidase at a higher concentration. Consequently, the inhibitory action of luteoskyrin on the respiratory chain cannot be attributed to a direct action on the specific site as in the case of antimysin A [68]. In a further experiment conducted with rat liver mitochondria and their digitonin fragments, Ueno revealed that the P:O ratio was diminished by luteoskyrin at a concentration of $10^{-6} M$, which was lower than that required for the inhibition of the respiratory chain. From the detailed investigation of the inhibitory action on the ATP-Pi and ATP-ADP exchange reactions and on the ATPase, Ueno concluded that luteo-skyrin may inhibit the oxidative phosphorylation by two mechanisms; one is an oligomycin-like action on the energy transfer system and another is a dinitrophenol-like uncoupling action on phosphorylation [69].

In connection with the functional impairment of mitochondria Ueno *et al.* [70] have shown that luteoskyrin (10^{-5} to $10^{-7} M$) inhibited both swelling and contraction of rat liver mitochondria *in vitro*. Biochemical demonstration of a decreased P:O ratio, an increased ATPase activity and an accumulation of luteoskyrin in mitochondria in luteoskyrin-poisoned animals, may suggest that luteoskyrin impairs the mitochondrial function *in vivo* as well as *in vitro*. The electron microscope also showed the swollen mitochondria with irregular destruction or lytic changes of the cristae, appearing early in the liver cells of mice: *i.e.*, 24 h after s.c. administration of luteoskyrin [57,71].

(2) On nuclei

Physicochemical studies revealed a luteoskyrin-DNA binding. On the basis of the spectral change of luteoskyrin, it was presumed that luteoskyrin bound *in vitro* with thymus DNA and DNH [72,73] in the presence of magnesium ion to form a complex (DNA—Mg^{2+}—luteo.) with a molar ratio of 4 to 6:1:1 in DNA-P:Mg^{2+}:luteoskyrin. Subsequently, the (DNA—Mg^{2+}—luteo.) complex was isolated by gel filtration on SE sephadex C-25 and measurement of flow dichroism demonstrated the orientation of the planar chromophore of luteoskyrin to be parallel with the axis of the DNA double helix. *In vitro* binding studies showed further that the red shift in visible spectrum of luteoskyrin was caused not only by DNA but also by other nucleophiles including apuric acid, poly U and poly C. These data lead to the conclusion that luteoskyrin may bind to native DNA with a molecular ratio of one pyrimidine base/one magnesium ion/one luteoskyrin [74].

The inhibitory effect of luteoskyrin on DNA-dependent RNA polymerase activity was demonstrated by Ueno *et al.* [75] and Sentenac *et al.* [76] using a preparation of *E. coli* isolated by the method of Chamberlin and Berg [77]. The chelating action of luteoskyrin with Mg^{2+} ion may be one of the factors causing the inhibition of the polymerase reaction [75]. However, reexamination using Mg^{2+} free RNA polymerase prepared by the glycerin method from *Pseudomonas aeruginosa* revealed that luteoskyrin directly inhibited DNA polymerase and chain-initiation reactions by inactivation of the enzyme [78]. The investigation of the effect of luteoskyrin on the synthesis of DNA, RNA and protein in the Ehrlich ascites tumour cells demonstrated that $10^{-6}M$ luteoskyrin selectively inhibited the synthesis of RNA in nuclei [79]. This result contradicts the data of Schachtschabel *et al.* [80], in which gradual suppression of DNA synthesis and cell multiplication was shown as the main effect of luteoskyrin on Ehrlich ascites tumour cells. A transient decrease in DNA of mouse liver cells 24 h after administration of luteoskyrin was confirmed by microspectrophotometry. The oedematous karyoplasm with reduced heterochromatin and nucleolus-associated chromatin were observed microscopically (Fig. 5) [81,82]. These changes had recovered 48 h later. A similar decrease in DNA was also observed after application of aflatoxin B_1 and ^{60}Co irradiation. Some of the liver cell nuclei of luteoskyrin-treated mice showed segregation of nucleolar granular and fibrillar elements, indicating inhibition of the synthesis of rapidly sedimenting nucleolar RNA. It is a well-known and interesting fact that a similar change in the nucleolus is observed following treatment with aflatoxin B_1 and other chemical carcinogens.

(3) On microsomes

Lowered content of microsomal protein and decreased activities of G6Pase and drug-metabolizing enzymes such as aminopyrine-*N*-demethylase and aniline hydroxylase, were observed in the liver of luteoskyrin-poisoned mice [36]. These biochemical alterations were shown to parallel the enhancement of SGPT activity [36]. Impaired transfer of sulphur-containing

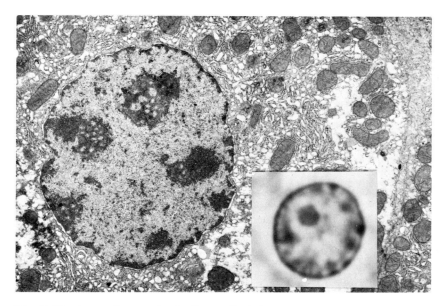

Fig. 5. Electron micrograph of a liver cell of the mouse (DDD strain), sacrificed 24 h after s.c. injection of luteoskyrin (2.5 mg/0.2 ml). Vesicular and cystic dilatation of the endoplasmic reticulum is evident. The nucleus shows a deformed configuration. The inset shows a liver cell nucleus with oedema of karyoplasm and reduction in both heterochromatin and nucleolus-associated chromatin. (From Yokoyama, Japan. J. Clin. Med., 30 (1972) 950. Reproduced by permission.)

amino acids into protein was suggested by Morooka *et al.* based on their observation of the abnormal increase in acid-soluble sulphur content in liver damaged by luteoskyrin [56,83]. Electron microscopy showed vesicular and cystic dilatation of the endoplasmic reticulum of the liver cells after 24 h in luteoskyrin-injected mice (Fig. 5) [57,71,81]. However, no observable effect on the microsomal enzymes was found *in vitro*. Consequently, it is still not certain, whether luteoskyrin acts directly on microsomes or not.

(4) On soluble and lysosomal fractions

Luteoskyrin showed no observable effect on glycolytic activity *in vitro*, but an increased activity of G6P dehydrogenase and acid phosphatase was observed in the early stage of intoxication in mice injected subcutaneously with luteoskyrin [51].

(5) Microbial assay

A mutant F-11, derived from *E. coli* Q-11, is sensitive, especially to luteoskyrin and rugulosin [84]. *Tetrahymena pyriformis* is also sensitive to these hepatotoxic mycotoxins [85]. Luteoskyrin showed antibacterial activity against *B. subtilis*, *S. marcescens* and *S. aureus* (Table II).

Luteoskyrin seems to be a mutagenic agent producing damage to DNA subjected to recombination repair process. This was suggested from the result of a "rec-assay", comparing the lethal action of luteoskyrin in a set of H-17 (rec$^+$) and M-45(rec$^-$) strains of *B. subtilis*. Luteoskyrin was not lethal to the H-17 strain, but lethal to the M-45 strain [86]. Anti-protozoal effects with abnormal cell division were observed in *Tetrahymena pyriformis* treated with luteoskyrin [85].

Cyclochlorotine

The mode of the action of cyclochlorotine on the liver was mainly investigated in acutely poisoned mice. No antimicrobial [84] or antiprotozoal effects [35] have yet been observed with cyclochlorotine.

(1) On carbohydrate metabolism

One of the biochemical characteristics of the poisoned mice was a marked disorder of carbohydrate metabolism. In mice the toxin (0.2 to 0.5 mg/kg) caused an alteration of the blood sugar level, showing an initial elevation followed by a fall to a level lower than the normal one. A rapid reduction of the liver glycogen content was also observed, in parallel with an enhancement of SGOT activity [87]. The biochemical disorders in the liver of the poisoned mice include (*1*) reduced glycogen content in the glycogen particles, (*2*) suppression of both synthesis of glycogen formation from glucose and incorporation of [U-^{14}C]glucose or [2-^{14}C]acetate into glycogen, (*3*) suppressed incorporation of [U-^{14}C]glucose into glycogen and enhanced formation of $^{14}CO_2$ from [^{14}C]glucose in the liver slices, and (*4*) increased D-glucose-6-pyrophosphate: NADP oxidoreductase, decreased uridine-5-pyrophosphate glucose-glycogen glycosyl-transferase in cell-free systems. However, there was (*5*) no change in the activities of anaerobic glycolysis, phosphorylase, D-glucose-1-phosphate phosphohydrolase and D-fructose-1,6-diphosphate 1-phosphohydrolase. From these observations, Ueno *et al.* assumed that the accelerated glycogen catabolism and inhibited glycogen neogenesis were attributable to the change in the activities of both D-glucose-6-phosphate : NADP oxidoreductase and uridine-5-pyrophosphate glucose-glycogen glycosyl-transferase [87].

(2) On oxidative phosphorylation

The liver of cyclochlorotine-poisoned animals showed an inhibition of the respiratory enzyme system and coupled phosphorylation [64,88]. However, this effect was relatively slight as compared with the suppressive effect of cyclochlorotine on the glycogen metabolism. No inhibition of the respiratory enzyme system was observed *in vitro* [89]. Consequently, the observed change in mitochondrial function may reflect a primary impairment of the liver cells in general.

(3) On fat synthesis

Accelerated incorporation of [2-^{14}C]acetate into fat was observed in liver slices of the poisoned mice. This acceleration may be ascribable to an increased formation of NADPH which is essential to fat synthesis, followed by an increase of D-glucose-6-phosphate : NADP oxidoreductase activity [87].

(4) On protein synthesis

The suppressed ^{14}C-labelled amino acid incorporation into the protein and the reduced microsomal protein content were observed in the livers of the poisoned mice. However, cyclochlorotine had no inhibitory action on protein synthetic reaction with cell-free systems *in vitro*. The observed effect of the toxin *in vivo* is possibly a secondary depression of protein synthesis due to disorganization of the endoplasmic reticulum, which was confirmed by the electron microscopic study. Cystic dilatation and a disaggregation of the endoplasmic reticulum were observed in the cytoplasm of the liver cells in the early sample examined; *i.e.* 10 min after i.v. injection of cyclochlorotine. A decrease in the number of granules of the rough endoplasmic reticulum was also seen. Where the damage was more severe, disruption of the cytoplasmic membrane of the liver cells and marked dilatation of the bile canaliculi with loss of their microvilli were seen, followed by the appearance of hyaline droplets [57,64]. In another experiment conducted with mouse liver slices *in vitro*, Yamazoe *et al.* reported a depressed incorporation of [14C]glycine into the protein at a concentration of 4 μg/ml of cyclochlorotine [90].

(5) On vascular permeability

In addition to the biochemical effects of cyclochlorotine on membrane systems as mentioned above, an increased permeability of blood capillaries produced by cyclochlorotine was suggested from the findings of its effects in the liver and skin of the animals. The increased liver weight during the initial anaemic phase of the liver and the decrease in transparency around the portal vein may reflect the marked exudation caused by an increased vascular permeability of the capillaries [91]. The same effect of cyclochlorotine on vascular capillaries was also observed on the skin of the dog injected s.c. with over 100 μg/0.1 ml of this peptide [63,65]. However, unlike histamine or polyvinylpyrrolidone, cyclochlorotine induced necrosis of the epidermis and leucocytic infiltration in addition to the vascular change in the skin [65].

References

1 K.B. Raper and C. Thom, A Manual of Penicillia, Hafner, New York, 1968, p. 623.
2 I. Miyake, H. Naito and H. Tsunoda, Beikokuriyo Kenkyusho Hokoku, 1 (1940) 1, in Japanese.
3 H. Tsunoda, Proc. 1st U.S.—Japan Conf. Toxic Microorganisms, Honolulu, U.S. Dept. of Interior, Washington, D.C., 1968, p. 143.
4 Y. Kobayashi, K. Uraguchi, I. Miyake, F. Sakai, M. Tsukioka, Y. Sakai, T. Sato, M. Miyake, M. Saito, M. Enomoto, T. Shikata and T. Ishiko, Proc. Jap. Acad., 34 (1958) 639.

5 Y. Kobayashi, K. Uraguchi, T. Tatsuno, F. Sakai, M. Tsukioka, Y. Sakai, O. Yonemit-su, T. Sato, M. Miyake, M. Saito, M. Enomoto, T. Shikata and T. Ishiko, Proc. Jap. Acad., 34 (1958) 736.

6 Y. Kobayashi, K. Uraguchi, F. Sakai, T. Tatsuno, M. Tsukioka, Y. Noguchi, H. Tsunoda, M. Miyake, M. Saito, M. Enomoto, T. Shikata and T. Ishiko, Proc. Jap. Acad., 35 (1959) 501.

7 K. Uraguchi, M. Saito, Y. Noguchi, K. Takahashi, M. Enomoto and T. Tatsuno, Food Cosmet. Toxicol., 10 (1972) 193.

8 H. Tsunoda and S. Ito, Shokuryo Kenkyusho Kenkyu Hokoku, 19 (1965) 15, in Japanese.

9 H. Kurata, J. Food Hyg. Soc. Japan, 9 (1968) 431, in Japanese.

10 B.H. Howard and H. Raistrick, Biochem. J., 44 (1949) 227, 56 (1954) 56 and 57 (1954) 221.

11 T. Tatsuno, M. Tsukioka, Y. Sakai, Y. Suzuki and Y. Asami, Pharm. Bull. (Tokyo), 3 (1955) 497.

12 M. Miyake, M. Saito, M. Enomoto, T. Shikata, T. Ishiko, K. Uraguchi, F. Sakai, T. Tatsuno, M. Tsukioka and Y. Sakai, Acta Pathol. Japon., 10 (1960) 75.

13 M. Saigo, K. Ogura and H. Tsunoda, J. Food Hyg. Soc. Japan, 4 (1963) 85, in Japanese.

14 H. Nishikawa, Tohoku J. Agr. Res., 5 (1955) 285.

15 S. Shibata, I. Kitagawa and H. Nishikawa, Pharm. Bull. (Tokyo), 5 (1957) 383.

16 S. Marumo, Bull. Agr. Chem. Soc. Japan, 19 (1955) 262.

17 M. Saito, M. Enomoto, T. Tatsuno and K. Uraguchi, Yellowed rice toxins, in A. Ciegler, S. Kadis and S.J. Ajl (Eds.), Microbial Toxins, Vol. 6, Academic Press, New York, 1971, pp. 299—380.

18 U. Sankawa, S. Seo, N. Kobayashi, Y. Ogihara and S. Shibata, Tetrahedron Letters, (1968) 5557.

19 N. Kobayashi, Y. Iitaka, U. Sankawa, Y. Ogihara and S. Shibata, Tetrahedron Letters, (1968) 6135.

20 S. Shibata, Chem. Brit., 3 (1967) 110.

21 M. Sato and T. Tatsuno, Chem. Pharm. Bull. (Tokyo), 16 (1968) 2182.

22 H. Yoshioka, Nippon Kagakukai Yokoshu, 28 (1973).

23 S. Gatenbeck, Acta Chem. Scand., 16 (1962) 1052.

24 Y. Ueno and I. Ishikawa, Appl. Microbiol., 18 (1969) 406.

25 S. Shibata and T. Ikekawa, Chem. and Ind. (London), (1962) 360.

26 Y. Ueno and T. Tatsuno, Chem. Pharm. Bull. (Tokyo), 17 (1969) 1175.

27 S. Gatenbeck, Z. Physiol. Chem., 353 (1972) 126.

28 Y. Ueno, J. Bacteriol., 97 (1967) 786.

29 H. Tsunoda, K. Uraguchi, T. Tatsuno and M. Saito, Shokubutsu Byori, 20 (1955) 40 in Japanese.

30 H. Tsunoda, M. Saigo and K. Ogura, Tokyo Nogyo Daigaku Nogaku Shuho, 6 (1961) 288, in Japanese.

31 S. Shibata, M. Takido and T. Nakajima, Pharm. Bull. (Tokyo), 3 (1955) 286.

32 S. Shibata, J. Pharm. Soc. Japan., 61 (1941) 320, in Japanese.

33 R. Fujita, Personal communication.

34 T. Tatsuno, H. Wakamatsu, Y. Kanazawa, T. Sato and H. Tsunoda, J. Pharm. Soc. Japon., 77 (1957) 689.

35 K. Uraguchi, Pharmacology of mycotoxins, in H. Raskova (Ed.), International Ency-clopedia of Pharmacology and Therapeutics, Pergamon, Oxford, 1971, pp. 143—298.

36 I. Ueno, Unpublished data.

37 I. Ishikawa, Y. Ueno and J. Tsunoda, J. Biochem., 67 (1970) 753.

38 Y. Ueno, Personal communication.

39 M. Tsukioka, Folia Pharmacol. Japon., 55 (1959) 1367, in Japanese.

40 K. Miyaki, M. Yamazaki, Y. Horie and S. Udagawa, J. Food Hyg. Soc. Japan, 11 (1970) 373.

41 D. Pavlica and I. Samuel, Brit. J. Cancer, 24 (1970) 22.
42 R.B.A. Carnaghan, Vet. Ann., (1966) 84.
43 P.M.D. Martin, G.A. Gilman and P. Keen, The incidence of fungi in foodstuffs and their significance based on a survey in the Eastern Transvaal and Swaziland, in I.F.H. Purchase (Ed.), Mycotoxins in Human Health, MacMillan, London, 1971, pp. 281—290.
44 K. Uraguchi, T. Tatsuno, M. Tsukioka, Y. Sakai, F. Sakai, Y. Kobayashi, M. Saito, M. Enomoto and M. Miyake, Japan. J. Exptl. Med., 31 (1961) 1.
45 Y. Ueno, I. Ueno, N. Sato, Y. Iitoi, M. Saito, M. Enomoto and H. Tsunoda, Japan. J. Exptl. Med., 41 (1971) 177.
46 K. Uraguchi, T. Tatsuno, F. Sakai, M. Tsukioka, Y. Sakai, O. Yonemitsu, H. Ito, M. Miyake, M. Enomoto, T. Shikata and T. Ishiko, Japan. J. Exptl. Med., 31 (1961) 19.
47 K. Uraguchi, I. Ueno, A. Akabori and M. Enomoto, Folia Pharmacol. Japon., 64 (1968) 1, in Japanese.
48 K. Uraguchi, I. Ueno, Y. Ueno and T. Komai, Toxicol. Appl. Pharmacol., 21 (1972) 335.
49 M. Umeda, Acta Pathol. Japon., 14 (1964) 373.
50 M. Umeda, Acta Pathol. Japon., 41 (1971) 195.
51 I. Ueno, K. Tokita and Y. Iitoi, Folia Pharmacol. Japon., 65 (1969) 74, in Japanese.
52 Y. Noguchi, Folia Pharmacol. Japon., 56 (1960) 126, in Japanese.
53 K. Takahashi, Ann. Rept. Co-op. Res. (Med.) Min. Educ. Japan, (1956) 45, in Japanese.
54 I. Ueno, T. Hayashi and Y. Ueno, Japan. J. Pharmacol., 22 (1972) 115.
55 T. Itano, J. Nara Med. Assoc., 10 (1959) 1.
56 N. Morooka, N. Nakano, N. Uchida and A. Takase, Japan. J. Med. Sci. Biol., 20 (1966) 293.
57 M. Saito, Acta Pathol. Japon., 9 (1959) 785.
58 I. Ueno, Folia Pharmacol. Japon., 65 (1969) 6, in Japanese.
59 I. Ueno, T. Yoshida, Y. Murata, Y. Ueno and N. Sato, Japan. J. Pharmacol., 23 (1973) 132.
60 M. Enomoto and M. Saito, Acta Pathol. Japon., 23 (1973) 655.
61 M. Enomoto and M. Saito, Carcinogens produced by fungi, in C.E. Clifton (Ed.), Annual Review of Microbiology, Vol. 26, Annual Reviews, Palo Alto, 1972, pp. 279—312.
62 F.O. Torres, I.F.H. Purchase and J.J. van der Watt, J. Pathol., 102 (1970) 163.
63 Y. Ohshima, K. Takahashi, T. Kashiwagi and Y. Yamada, Sogo Igaku, 16 (1959) 341, in Japanese.
64 T. Hara, Tokyo J. Med. Sci., 72 (1964) 136, in Japanese.
65 T. Hara, Tokyo J. Med. Sci., 72 (1964) 165, in Japanese.
66 H. Itoh, S. Yamasawa, T. Yamaguchi and Y. Horiguchi, Yokohama Med. Bull., 14 (1963) 1.
67 J. Breen, J.C. Dacre, H. Raistrick and G. Smith, Biochem. J., 60 (1955) 618.
68 I. Ueno, Y. Ueno, T. Tatsuno and K. Uraguchi, Japan. J. Exptl. Med., 34 (1964) 135.
69 I. Ueno, Seikagaku, 38 (1966) 741, in Japanese.
70 Y. Ueno, I. Ueno, T. tatsuno and K. Uraguchi, Japan. J. Exptl. Med., 34 (1964) 197.
71 B. Takagi, Electron Microscopy, 8 (1959) 154, in Japanese.
72 Y. Ueno, A. Platel et P. Fromageot, Biochim. Biophys. Acta, 27 (1967) 134.
73 Y. Ueno, I. Ueno, K. Mizumoto and T. Tatsuno, J. Biochem. (Tokyo), 63 (1968) 395.
74 Y. Ueno, I. Ueno and K. Mizumoto, Japan. J. Exptl. Med., 38 (1968) 47.
75 Y. Ueno, I. Ueno and T. Tatsuno, Seikagaku, 38 (1966) 687.
76 A. Sentanac, A. Ruet and P. Fromageot, Bull. Soc. Chim. Biol., 49 (1967) 247.
77 M. Chamberlin and P. Berg, Proc. Natl. Acad. Sci. (U.S.), 48 (1962) 81.
78 K. Mizumoto, Y. Ueno and Y. Kajiro, Seikagaku, 40 (1968) 623, in Japanese.
79 Y. Ueno, I. Ueno, K. Ito and T. Tatsuno, Experientia, 23 (1967) 1001.
80 D.O. Schachtschabel, F. Zilliken, M. Saito and G.E. Florey, Exptl. Cell Res., 57 (1969) 19.

81 T. Yokoyama, Japan. J. Clin. Med., 30 (1972) 950, in Japanese.
82 T. Yokoyama, Arch. Pathol. Anat. Histol., 11 (1972) 133.
83 N. Morooka, N. Nakano, N. Uchida and A. Takase, Japan. J. Med. Sci. Biol., 20 (1967) 501.
84 H. Yamakawa and Y. Ueno, Chem. Pharm. Bull., 18 (1970) 177.
85 Y. Ueno and M. Saheki, Japan. J. Exptl. Med., 38 (1968) 157.
86 T. Kada, Igaku No Ayumi, 84 (1973) 776, in Japanese.
87 Y. Ueno, M. Kaneko, T. Tatsuno, T. Tanaka and K. Uraguchi, Seikagaku, 35 (1963) 224, in Japanese.
88 I. Tanaka, Y. Ueno and K. Uraguchi, Folia Pharmacol. Japon., 56 (1960) 166, in Japanese.
89 I. Tanaka, Y. Ueno and K. Uraguchi, Folia Pharmacol. Japon., 57 (1961) 156, in Japanese.
90 S. Yamazoe, M. Nakao, K. Hayashi, K. Nagano, T. Motegi, S. Uesugi and Kanoh, J. Med. Soc. Gumma Univ., 12 (1963) 73.
91 T. Yamada, Sogo Igaku, 16 (1959) 902, in Japanese.
92 S. Seo, U. Sankawa and S. Shibata, Tetrahedron Letters, (1969) 767.

Chapter 15

TOXINS PRODUCED BY SWEET POTATO ROOTS INFECTED WITH *CERATOCYSTIS FIMBRIATA* AND *FUSARIUM SOLANI*

BENJAMIN J. WILSON and MICHAEL R. BOYD*

*Center in Environmental Toxicology, Department of Biochemistry, and *Department of Pharmacology, School of Medicine, Vanderbilt University, Nashville, Tenn. 37232 (U.S.A.)*

The mould-damaged sweet potato (*Ipomoea batatas*) has been recognized both in Japan and the United States as a cause of fatal respiratory disease in livestock, especially cattle. Research in both countries has clearly shown that such disease outbreaks are caused by poisons produced by the host plant in response to invasion by certain fungi that are not otherwise toxigenic. Therefore, the substances to be discussed in this chapter are not mycotoxins in the usual definition of that word. Although the infecting fungi have a demonstrable stimulatory role in altering sweet potato metabolism, there is no evidence at present either establishing or disproving their further participation as modifiers of possible toxin precursors formed by the host.

The sweet potato is used in several parts of the world, especially in the United States, for human food. It is subject to many microbial diseases that may cause a substantial loss of the total annual crop [1]. Additionally, it may be attacked by nematodes and insects and can sustain biochemical injury as a result of freezing or prolonged storage at low temperatures [2,3]. Culled sweet potatoes are often used as a feed for cattle which usually consume them greedily. The resulting enzootics of bovine lung disease have been documented over several decades both in Japan and the United States [4—6].

A day or so after feeding commences animals lose their appetite and develop dyspnoea which becomes progressively worse until the animals die a few days later. The severe respiratory distress is manifested by a rapid respiratory rate, and sometimes a frothy fluid is noted around the muzzle shortly before death. The afflicted animal often stands with the head and neck extended in a vain affort to facilitate air passage. Moist râles may be heard over the thoracic cage, and in some cases trapped air bubbles may escape from the lungs into the mediastinum and adjacent subcutaneous tissues.

Peckham *et al.* [7] described an outbreak occurring in Tift county, Georgia in 1969 from which 69 adult purebred Herefords died out of a herd

References p. 342

of 275. *Post mortem* examination revealed lesions involving principally the lungs. These organs were heavy and wet and remained expanded when the thorax was opened. Marked congestion of the lungs justified the term "hepatization" which was especially evident in cut sections. The lung surface showed extensive evidence of diffuse emphysema which was confirmed microscopically. Stained histological sections also revealed marked oedema, proliferation of alveolar epithelium, hyaline membrane formation, and areas of congestion and haemorrhage. No other tissues or organs showed significant changes. Experimental feeding studies implicated sweet potatoes contaminated with an isolate of *Fusarium solani* as the source of poisoning. Based on the pathological picture these authors suggested the name *atypical interstitial pneumonia* as being the most appropriate and descriptive of several names formerly applied to this and closely related conditions in cattle. No particular therapeutic measures are effective once the disease is well established.

Similar outbreaks in cattle have been attributed to other food sources including lush pasture, silage, and mouldy hay [8]. In Tasmania (Australia) leaves and twigs of the stinkwood plant, *Zieria arborescens*, are sometimes consumed with production of an oedematous condition of the lungs [9].

Toxic metabolites produced by mould-damaged sweet potatoes

Ipomeamarone. The first toxic metabolite isolated from mould-damaged sweet potatoes was a furanosesquiterpene called ipomeamarone [(+)-*cis*-5-(3-furyl)tetrahydro-2-methyl-2(4-methyl-2-oxopentyl)furan]. Japanese workers identified the poisonous material among numerous other compounds representing products of "abnormal" metabolism of sweet potatoes infected with the black rot fungus, *Ceratocystis fimbriata* [10,11]. The structure (Table I) was proposed by Kubota and Matsuura [12] and confirmed by Kubota by chemical synthesis [13]. The stereochemistry of the compound has been provided by Kubota and Matsuura [13,14] and Hegarty *et al.* [15].

Ipomeamarone, which reportedly imparts a bitter taste to the sweet potato, was shown to be hepatotoxic to mice, with an intraperitoneal (i.p.) LD_{50} of 230 mg/kg [16]. It is interesting to note that the enantiomer of ipomeamarone, ngaione, also occurs naturally and likewise has been shown to be hepatotoxic [17]. Although ipomeamarone presumably occurs as an abnormal metabolite of the sweet potato, the levorotatory ngaione is a normal constituent of the Ngaio tree (*Myoporum laetum*) and other shrubs in Australia and New Zealand. Sheep, cattle, and other livestock may consume the leaves of these plants, often with fatal results. The induced liver disease may result in signs of icterus and photosensitivity.

Ipomeamaronol. In addition to ipomeamarone, a closely related derivative has been obtained recently from mould-infected sweet potatoes. Two groups, Yang *et al.* [18] in the United States, and Kato *et al.* [19] in Japan, simultaneously reported the isolation and characterization of ipomeamaronol, an hydroxylated derivative of ipomeamarone (Table I), from sweet potatoes infected with *F. solani* and *C. fimbriata*, respectively. Although detailed toxicity tests have not yet been carried out, it appears that ipomeamaronol is also hepatotoxic to mice [18].

The lung oedema factor

Neither of the known hepatotoxins accounts for the respiratory pathology which is the principal feature in natural outbreaks of poisoning by damaged sweet potatoes. Wilson *et al.* [20] therefore postulated the existence of another component, termed the "lung oedema (LO) factor" to account for the pulmonary toxicity of such damaged tubers. Wilson *et al.* [21,22] and Boyd and Wilson [23,24] have subsequently shown that two compounds, 4-ipomeanol and ipomeanine, comprise, at least in part, the LO factor. Structures of these compounds, along with their LD_{50} values, are also given in Table I. Both of these substances produce similar pulmonary pathology.

4-Ipomeanol [1-(3-furyl)-4-hydroxy-1-pentanone] was the first recognized component of the LO factor [21,23]. Confirmation of the assigned structure (Table I) has recently been attained by synthesis [25]. The compound was obtained from toxic ether extracts of *F. solani*-infected sweet potatoes. During the isolation procedure, lung oedema, a highly characteristic action of the compound, was utilized to monitor for the presence of the toxin in various fractions. Feeding or injection of the toxic fractions, or pure 4-ipomeanol, produces a striking response in mice characterized by pulmonary oedema and massive pleural effusion (see Fig. 1).

After receiving lethal doses of toxin, mice become increasingly dyspnoeic and die, usually within 24 h. At *post mortem*, clear fluid may drip from the nostrils of the animal when the carcass is held with the head downward. The lungs are sometimes congested and almost always surrounded by 1—2 ml of clear pleural fluid (Fig. 1) which clots on standing. Microscopic sections of the lungs stained with haematoxylin and eosin show intra-alveolar oedema (Fig. 1) and congested blood vessels.

In mice the nature of the pathological response to 4-ipomeanol and the degree of toxicity are similar in both sexes and do not differ greatly with route of administration (Table I) [24]. Likewise the potency of the toxin in several different species tested is similar, although the pattern of pathological response may be somewhat different [24]. For example, in the rat pulmonary oedema and congestion are prominent, but pleural effusion is usually minimal. In the guinea pig at *post mortem* the lungs may be dark red and completely filled with bloody fluid.

4-Ipomeanol is related to ipomeanine [1-(3-furyl)-1,4-pentanedione] (Table I) as a dihydro derivative. The production of significant quantities of the latter compound by sweet potatoes infected with *C. fimbriata* was reported by Japanese investigators several years ago. However, until quite recently no toxicological data for the compound were available. In view of the close structural similarity of 4-ipomeanol and ipomeanine, Wilson and Boyd [22] examined the toxicity of the latter compound. Since only traces were

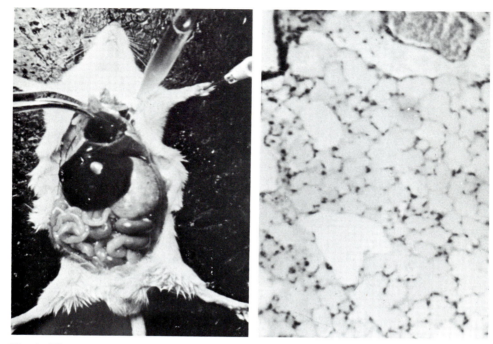

Fig. 1. Photographs showing pathological response to 4-ipomeanol in mice. Left photo shows massive pleural effusion (fluid in pipette). Photo at right is H and E stained section showing prominent intraalveolar oedema of lung.

detected in *F. solani*-infected sweet potatoes, ipomeanine was synthesized by an adaptation of the method of Kubota and Ichikawa [26]. Toxic properties of the synthetic material proved to be quite similar to those of 4-ipomeanol, establishing its role as a potential component of the LO factor.

A compound isomeric with 4-ipomeanol, 1-(3-furyl)-1-hydroxy-4-penta-none, has also been isolated from damaged sweet potatoes [21,23]. In addition to this material, other closely related compounds have been detected in crude extracts. Studies of the physical, chemical, and biological properties of these substances are in progress and will soon be published*.

The respective contributions of 4-ipomeanol and ipomeanine to the lung oedema factor are not yet clear, although 4-ipomeanol predominates in *F. solani*-infected roots in terms of quantities present. However, use of *C. fimbriata* by Japanese workers as the infecting organism reportedly results in significant production of ipomeanine, suggesting that the diketone could contribute significantly to the pulmonary toxicity of black-rotted plants.

Chemical and physical properties of the toxins

The compounds in Table I, with one exception, are colourless oils.

---

* See Note added in proof, p. 344.

TABLE I

STRUCTURES AND TOXICITIES OF POISONOUS METABOLITES OF MOULD-DAM-AGED SWEET POTATOES

Name	Structure	Toxicity (mg/kg)[a]		
		i.p.	i.v.	Oral
Ipomeamarone	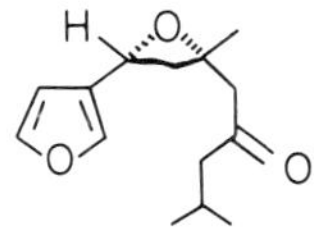	230	b	b
Ipomeamaronol	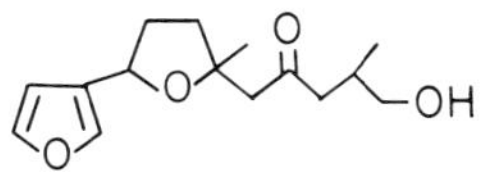	b	b	b
Ipomeanine	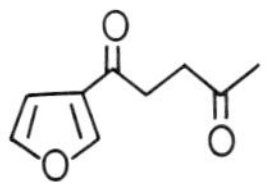	25 ± 1	14 ± 1	26 ± 1
4-Ipomeanol	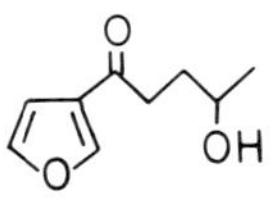	36 ± 4 ♂ 28 ± 2 ♀ 20—40 (est.) (rabbit) 10—20 (est.) (guinea pig)	21 ± 1 18 ± 5 (rat)	38 ± 3

[a] All values given are for *mice*, unless otherwise indicated.
[b] Data not available.

Ipomeanine is a colourless crystalline solid. None of the pure toxins has a characteristic odour. All are stable in neutral media but tend to degrade in the presence of alkali or acid. The hydroxylated compounds are thermally unstable and hence are not distillable. However, protection of the hydroxyl group may be afforded by formation of the trimethylsilyl (TMS) derivatives which have sufficient thermal stability and volatility for gas chromatographic separation and/or purification (see later section this Chapter). Silylation is readily accomplished with a commercially available reagent formulation ("Tri-Sil/BSA," Pierce Chemical Co.; Rockford, Ill., U.S.A.). Semicarbazone derivatives of the keto compounds can be readily prepared (see Table II).

A particularly useful property of the toxins is their reaction with Ehrlich's reagent to yield coloured derivatives. The distinctive colours produced are especially useful in qualitative analysis for the toxins (see later section).

References p. 342

TABLE II

PHYSICAL PROPERTIES AND SPECTRAL DATA FOR SWEET POTATO TOXINS

	Ipomeamarone	Ipomeamaronol	4-Ipomeanol	Ipomeanine
Formula	$C_{15}H_{22}O_3$	$C_{15}H_{22}O_4$	$C_9H_{12}O_3$	$C_9H_{10}O_3$
bp or mp	$122°/2.5$ mm (bp)	a	a	$41-42°$ (mp) $117-120°/$ 2 mm (bp)
Refractive Index	1.4720 (at $32°$)	a	1.4995 (at $25°$)	a
$[\alpha]_D$	+ 32.4 (C 5. 0, EtOH (at $19.5°$)	a	a	—
Derivatives	Semicarbazone, mp $133-134°$; 2,4-dinitro-phenylhydra-zone, mp $148-149°$	3,5-dinitroben-zoate, mp $81°$	Semicarba-zone, mp $137-137.5°$	Semicarbazone, mp $214-216°$; bis -2, 4-dinitro-phenylhydrazone, mp $232-234°$
UV (nm)	λ (EtOH) 211 (ϵ 5650)	λ (MeOH) 212 (ϵ 9100)	λ (MeOH) 211 (shoulder, ϵ 6100), 251 (ϵ 2970)	a
IR (cm^{-1})	ν (neat) 2980, 1715, 1520, 1490, 1390 1165, 1055, 1030, 925, 880	ν ($CHCl_3$) 3400, 1710, 1500, 875	ν (neat) 3380, 3100, 2940, 2900, 1665, 1560, 1505, 1160, 875	ν (CCl_4) 2900, 1720, 1680, 1560, 1510, 1390, 1350, 1150
NMR (ppm)	($CDCl_3$, 100 MHz), δ 0.87 (6 H, d, J = 6.0 Hz), 1.29 (3 H, s), 1.85 (4 H, m), 2.29 (2 H, d, J = 6.2 Hz), 2.62 (2 H, m), 4.84 (1 H, m), 6.29 (1 H, d), 7.29 (1 H, d, J = 1.8 Hz), 7.30 (1 H, s)	($CDCl_3$, 60 MHz), δ 0.87 (3 H, d, J = 6.1 Hz), 1.26 (3 H, s), 2.61 (2 H, s), 2.65 1 H, m), 7.27 (2 H, m)	(CCl_4), δ (TMS) 1.18 (3 H, d), 1.73 (2 H, m), 2.83 (2 H, t), 3.30 (1 H, s), 3.77 (1 H, m), 6.70 (1 H, m) 7.40 (1 H, m), 8.08 (1 H, m)	(CCl_4), δ (TMS) 2.18 (3 H, s), 2.87 (4 H, m), 6.75 (1 H, m), 7.44 (1 H, m), 8.08 (1 H, m)
MS, m/e (%)	250 (molecular ion), 173, 101, 85	266 (molecular ion), 248, 173, 101, 85	168(5) (mo-lecular ion), 153 (2), 150 (6), 124(5) 121(2), 110(47), 95(100)	166 (molecular ion)

TABLE II (continued)

	Ipomeamarone	Ipomeamaronol	4-Ipomeanol	Ipomeanine
Formula	$C_{15}H_{22}O_3$	$C_{15}H_{22}O_4$	$C_9H_{12}O_3$	$C_9H_{10}O_3$
References	11, 18, 19 47	18, 19	21, 25	11,26,b

a Data not available.
b From unpublished data by the authors.

Considerable spectral data are available for the toxins. Table II provides a summary of infrared (IR), ultraviolet (UV), nuclear magnetic resonance (NMR), and mass spectra (MS) as well as some physical properties.

Chemical synthesis of sweet potato toxins

Chemical syntheses of naturally occurring toxins are desirable both for unequivocal confirmation of the proposed chemical structures as well as provision of a source of the compounds in sufficient quantities to allow more extensive toxicological and pharmacological evaluations. With the exception of ipomeamaronol, chemical syntheses of the toxins known to occur in damaged sweet potatoes have been accomplished.

Fig. 2 summarizes the synthetic routes. As is evident from the diagram, the key intermediate in all the syntheses is ethyl 3-furoylacetate (IV). This compound can be easily prepared in 40—60% yield by the Claisen condensation of ethyl acetate with ethyl 3-furoate (III). A simple and convenient method for preparation of the latter compound from the commercially available diethyl 3,4-furandicarboxylate (I) has been reported [25]. More recently this method has been applied commercially, and ethyl-3-furoate has become available from chemical suppliers (Aldrich Chemical Company).

Synthesis of ipomeamarone. Kubota [13] first described a synthetic procedure for ipomeamarone (XIV, Fig. 2). IV was reacted with α-bromo-β-methylcrotonate to give diethyl-4-(3-furoyl)-2-methylbut-1-ene-1,4-dicarboxylate (IX). Hydrolysis and decarboxylation of IX then yielded 4-(3-furoyl)-2-methylbut-1-ene-1-carboxylic acid (X). Formation of the methyl ester of X with diazomethane was followed by reduction of the product with aluminum isopropoxide under Meerwein-Pondorf conditions. The hydroxyester (XI) which resulted was refluxed with sodium hydroxide solution and the product, 5-(3-furyl)tetrahydro-2-methyl-2-furylacetic acid (XII), was obtained. Neutralization of this material with methanolic sodium methoxide, followed by reaction with oxalyl chloride, gave the acid chloride (XIII). Reaction of the chloride with diisobutylcadmium followed by decomposition of the complex with dilute acid gave (±)-*epi*ipomeamarone [*trans*-5-(3-furyl)tetrahydro-2-methyl-2-(4-methyl-2-oxopentyl)furan]. Epimerization of

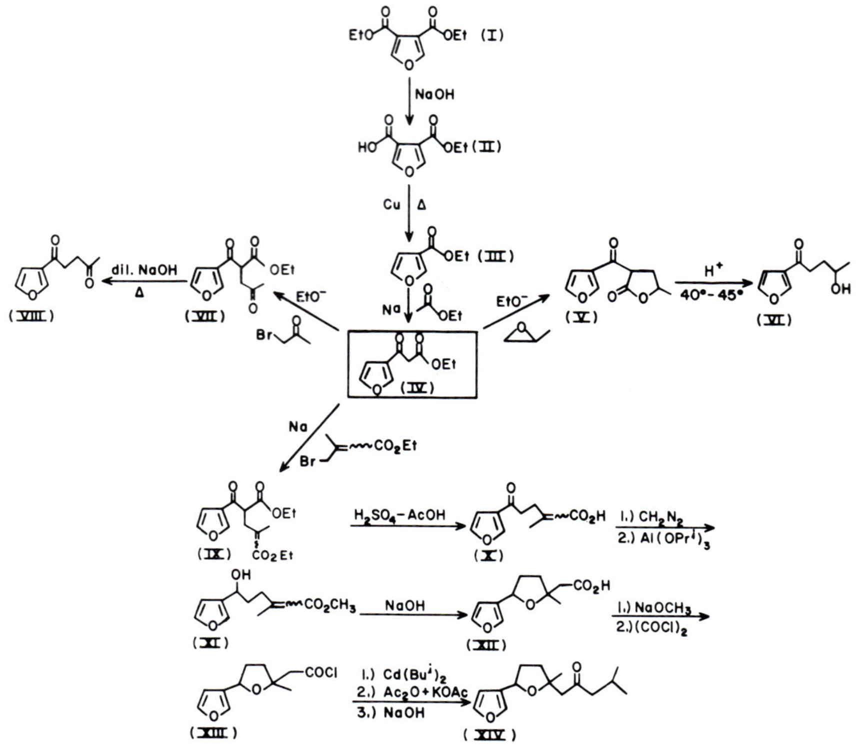

Fig. 2. Chemical syntheses of sweet potato toxins.

this compound was effected by refluxing in a mixture of potassium acetate and acetic anhydride followed by hydrolysis in base to yield ± ipomeamarone (XIV).

Although Kubota's synthesis of ipomeamarone confirmed the structure of the compound, the sequence involves several steps and the overall yield of the synthetic route is quite low. For these reasons the ipomeamarone synthesis is not especially suited for preparation of quantities of the hepatotoxin required for biological studies. Bioproduction (see later section, this chapter) remains, at present, the best method for obtaining the compound. Present research is directed toward devising an alternate synthetic sequence which may be more suitable for preparative-scale production of ipomeamarone.

Chemical synthesis of 4-ipomeanol. Synthesis of the first recognized lung oedemagenic component, 4-ipomeanol, was reported recently [25] and is outlined in Fig. 2. Lactone (V) was prepared by the reaction of IV, in base, with propylene oxide. Decarboxylation of V by gentle heating in dilute acid afforded (±)4-ipomeanol (VI) in about 75% overall yield from IV. The properties and spectra of the synthetic compound were consistent with those of the natural product. The synthesis provides an acceptable source of material for biological studies and is readily adaptable to the preparation of various analogues and modifications of 4-ipomeanol which may aid in probing the biological and pathogenic mechanisms of action of the compound.

Synthesis of ipomeanine. Kubota and Ichikawa [26] provided a synthesis of ipomeanine (VIII). Ethyl 3-furoylacetate (IV) was reacted with bromoacetone to yield product (VII), which was hydrolyzed and decarboxylated in base to give ipomeanine. The overall yield of product from IV was 20—40%.

Biosynthesis

Nearly all information on the biosynthesis of furanoterpenoids in the sweet potato has been derived from Japanese work and relates primarily to ipomeamarone. Moreover, these studies have been confined to sweet potato stimulation with black rot organism *C. fimbriata*. The nature of the stimulus that causes modification of sweet potato metabolism is not clear but appears to have a somewhat specific aetiological basis since only certain fungi are capable of inducing these toxic metabolites.

Infection of live root slices with *C. fimbriata* causes cell necrosis of the surface and stimulates adjacent intact cells to alter their metabolic patterns in favour of furanoterpenoid synthesis. Certain evidence points to this phenomenon as a defensive mechanism in which ipomeamarone exerts an inhibitory action on the invading fungus [27]. Concentrations of the "phytoalexins" decrease in cells progressively further away from the infection site [28]. "Diseased tissue" (layer of cells beneath the infected region) also shows an increase in polyphenolic compounds [29,30] which presumably are antimicrobial in nature. An increase in phospholipids and sterol due to infection has also been noted [31].

Infection with *C. fimbriata* stimulates a marked increase of several host enzymes which progresses over a period of several hours to a few days. These include phenylalanine deaminase [32], tyrosine deaminase [32], and phosphatase [33]. 24 h after infection there also is an increase in phosphate ion uptake by diseased tissue followed by a subsequent decrease. Cut tissue shows a higher than normal ion uptake over a 4-day period [34].

Available evidence suggests that ipomeamarone and related terpenoids in the infected sweet potato arise from alterations of the tricarboxylic acid cycle and fatty acid synthesis [35]. A postulated biosynthetic pathway is given below:

acetate ↔ acetyl CoA ↔ acetoacetyl CoA ↔ β-hydroxy-β-methylglutaryl CoA ↔ mevalonate → → → ipopentenyl pyrophosphate → → → farnesyl pyrophosphate → → → ipomeamarone

Radioactive acetate is apparently incorporated by way of mevalonate and 3-methyl-3-butenyl pyrophosphate. Studies by Oshima and Uritani [36] showed that both $[2\text{-}^{14}C]$ acetate and $[2\text{-}^{14}C]$ mevalonate were incorporated into ipomeamarone, the latter at a lower rate than the former. No dilution effect of mevalonate on incorporation of radioactive acetate was noted, and no such effect of acetate on [2-^{14}C]mevalonate was seen, indicating that the latter probably was not converted to acetate prior to incorporation into ipomeamarone. Uniformly labeled leucine was also incorporated apparently in connection with a proposed pathway of leucine metabolism involving formation of isovaleryl-CoA, dimethylcrotonyl-CoA, and β-methyl gluta-conyl-CoA [37]. [2-^{14}C]Ethanol was incorporated more efficiently and rapidly than acetate, a fact which led to postulation of the presence of an alcohol dehydrogenase which may convert ethanol to acetyl-CoA by way of acetaldehyde. This idea was supported by the demonstration that pyrazole, a potent inhibitor of alcohol dehydrogenase, decreased ethanol incorporation into ipomeamarone [38].

A comparison of radioactive acetate, pyruvate, and citrate as substrates indicated rates of incorporation in that order. The prolonged lag in observed citrate carbon incorporation suggested that ATP-citrate lyase probably does not play a role in acetyl group transport from mitochondria to the surrounding space as occurs in animal tissues [39].

Cell-free systems from normal and diseased sweet potato tissue were utilized in demonstrating the synthesis of 5-phosphomevalonate, mevalonate-5-pyrophosphate, and isopentenyl pyrophosphate. This synthetic ability was greater in diseased than in normal tissues [40]. Similar studies had previously established that a β-hydroxy-β-methylglutarate derivative was generated from acetyl-CoA in response to fungus infection [35]. 2-[2-^{14}C]Acetate was incorporated into geraniol and farnesol in the black-rot damaged sweet potato, and the addition of these terpenoids to infected slices markedly suppressed incorporation of [2-^{14}C]acetate into ipomeamarone as well as the synthesis of lipids other than furanoterpenoids [41]. The effect on acetate incorporation was interpreted as a direct inhibition of the biosynthetic pathway for ipomeamarone as well as a minor dilution effect. [2-^{14}C]farnesol was incorporated into ipomeamarone to the extent of 4.8% and into the "lipid fraction" to the extent of 59.3% [42]. It remains to be determined what role the phosphorylated derivatives of these supposed terpene intermediates play in the biosynthesis of the furanoterpenoid metabolites.

In their studies on ngaione and other related furanoterpenoids from the leaves of the Ngaio tree and other plants, Sutherland and Park [43] have suggested that both cyclopentanoid monoterpenes and furanoid sesquiterpenes could result from an allylic type oxidation of an activated methyl group. Presumably farnesol (XV) would be oxidized to oxo-farnesol (XVI) which would then cyclize to a hemiacetal (XVII) and undergo 1,4-dehydration to form dendrolasin (XVIII) which is a recognized metabolite of certain ants and a woody plant (*Torreya nucifera*), as well as the sweet potato.

(XV)   (XVI)   (XVII)   (XVIII)

Geraniol (XIX), a C_{10} precursor, would similarly give perillene (XX) which could by further allyic methylene oxidation yield perilla ketone (XXI), a compound found in *Perilla frutescens*.

(XIX)   (XX)   (XXI)

Isolation and analysis

Preparative-scale separation of the toxins

It is often desirable to obtain relatively large quantities of the naturally occurring sweet potato toxins, either to permit confirmatory chemical identification or as the final step in bioproduction of quantities of the compounds for biological studies.

The procedure for separation and purification of ipomeamarone used by earlier investigators [12,44] consisted essentially of a distillation of the organic solvent extract of mouldy sweet potatoes. Later, maintaining that distillation procedures alone were inadequate, Akazawa [45] provided a new scheme which additionally included repeated silicagel column chromatography.

Newer methods for the isolation of ipomeamarone and for 4-ipomeanol have been reported in which preparative-scale gas chromatography is utilized [23,46]. Column chromatography is used first to obtain toxin-rich fractions from mould-infected sweet potatoes and to remove many of the non-volatile impurities from the crude extract. Thin-layer chromatography (TLC) is used to monitor column fractions for the various toxins. Prior to gas chromatography (GC) the fraction residues containing 4-ipomeanol are silylated (see earlier section, this chapter). The enriched fractions are subjected to GC using a glass column with silicone UC-W98 as the stationary phase. (Details of column, instrument, conditions, and procedures are available [23,46]). Identification of the toxin peaks in the gas chromatogram can be easily accomplished by injection of small amounts of the authentic materials (*e.g.*, obtained by synthesis) and noting the retention times on the column in use. Finally, trapping of the GC effluent representing the toxin peaks provides the pure compounds. 4-Ipomeanol is, of course, obtained as its silyl derivative. The silyl group may be easily removed by mild acid hydrolysis to permit recovery of pure toxin.

Qualitative and quantitative analysis

At present, the two analytical techniques most suitable for sweet potato toxins are TLC and GC. TLC is most useful in qualitative analysis for the toxins, and a method has been described recently [46]. Commercially available thin-layer sheets (Baker-Flex Silica Gel, IB-F; J.T. Baker Chemical

Co., U.S.A.) are used routinely. The sheets are developed ascendingly in 10% methanol in benzene. The toxins, along with other abnormal metabolites, are detected on the developed plates by spraying with Ehrlich's reagent (5 g of *p*-dimethylaminobenzaldehyde in 20 ml conc. HCl and 80 ml ethanol), followed by gentle warming to develop spot colours. A typical chromatogram is shown in Fig. 3.

A method involving a combination of TLC and colourimetric determination for the quantitative analysis of ipomeamarone in sweet potato tissue has been reported [47]. However, although the investigators did not provide data on the precision or accuracy of their method, another procedure utilizing gas chromatography has more recently appeared [46] and is probably superior in these respects. The GC procedure also provides greater sensitivity.

Prior to GC analysis, the whole sweet potato is weighed and shredded. After further chopping to fine particle size in a food blender, a representative sample is homogenized in 5% methanol in chloroform. The homogenate is filtered through a Buchner funnel containing a spun glass filter. The residue is washed with solvent mixture, the filtrates combined, washed with water, and dried over anhydrous sodium sulphate.

Authentic samples of ipomeamarone (obtained either from synthesis or bioproduction) are used to prepare a calibration curve showing the amount of pure ipomeamarone injected *vs.* detector response (measured either as peak area or peak height). A series of column types have been evaluated for the analysis [46]. A 6 ft. × 1/4 in. glass column containing 10% UC-W98 on 100/120 Chrom Q (Applied Science Laboratories) was found most satisfactory. Details of the equipment and optimum operating conditions are available (see ref. 46).

A calibration plot, covering the range of 50—500 ng ipomeamarone injected, is prepared. The plot has been found to be linear within this range [46]. The sweet potato extract is concentrated or diluted with solvent until GC analysis can be obtained from which the ipomeamarone content of a known volume of material injected may be determined from the calibration plot. With the total volume of the extract and the weight of potato tissue known, the average ipomeamarone content of the sweet potato tissue can be calculated. The average percent recovery for the extraction procedure is reported to be in the range of 91—107%.

Very recently a more convenient and improved GC analytical method* for ipomeamarone has been devised in which an internal standard is utilized. In addition, the method provides simultaneous quantitative analysis for the lung oedemagenic toxins. Fig. 3 shows a typical gas chromatogram of a mixture of the toxins along with a correlation of the GC peaks with the TLC spots.

---

* Details of this analytical procedure will soon be published by the authors of this chapter.

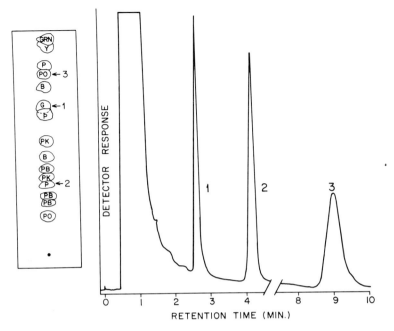

Fig. 3. Typical TLC of crude extract from mouldy sweet potatoes developed ascendingly in 10% methanol in benzene and sprayed with Ehrlich's reagent; spot colours coded as follows: GRN, green; G, gray; Y, yellow; PK, pink; PB, brownish purple; P, purple; PO, pinkish orange turning dark gray; and B, blue. Right—correlation of toxin peaks in GC with corresponding TLC spots (1, ipomeanine; 2, 4-ipomeanol silyl ether; 3, ipomeama-rone). GC obtained using a $10' \times 1/4''$ column containing 12% OV-101 on Chrom Q; column temp. 190°; He flow 90 ml/min; flame ionization detection; sample, 5 μl of 10 μM solution of toxins in acetone.

Natural occurrence and environmental conditions for bioproduction

No extended analytical surveys have been carried out that would indicate the incidence and levels of toxic furanoterpenoids in marketed sweet potatoes. Freshly harvested sweet potato tubers, with absolutely no areas of blemish, have been found consistently negative for the Ehrlich-positive compounds. However, older, slightly blemished or discoloured tubers offered for sale in supermarkets often may contain readily detectable quantities of both ipomeamarone, 4-ipomeanol, and several other Ehrlich-positive metabolites. In one limited selection of such blemished specimens the quantity of ipomeamarone varied from 9 to 950 mg per sweet potato [46].

Japanese workers have claimed that insect damage by the weevil *Cylas formicarius elegantulus* can lead to ipomeamarone accumulation [44]. However, since fungus invasion is a likely sequel to this type of mechanical damage, and since no fungus isolations were attempted in this study, the exact role of the insect in this phenomenon is open to question. Chemical stimulation by mercuric chloride and iodoacetic acid has also been claimed [48].

Infection with Ceratocystis fimbriata. In general, the procedures used by Japanese workers with *C. fimbriata* involve growing the organism in shake cultures using white potato-dextrose medium to obtain a high concentration of spores and other fungal elements. The culture is used to inoculate the surface of live sweet potato slices. Infection is allowed to progress for a few hours to a few days in a humid environment at about 30°. The infected slices are solvent-extracted and ipomeamarone is purified using one of several methods which have been described including steam distillation [12], and column [45] and thin-layer chromatography [39]. High levels of ipomeamarone (up to 20 mg per gram of tuber [47]) have been attained in this manner. Ipomeanine reportedly accompanies ipomeamarone along with undetermined quantities of β-furoic acid, batatic acid, and many other unidentified compounds representing altered metabolism of the sweet potato [11]. Japanese workers emphasize the enhanced responsiveness of the disease-resistant varieties of sweet potato, such as Norin I, in toxin production [49].

Infection with Fusarium solani. Among several different fungi used in attempts to stimulate 4-ipomeanol biosynthesis only members of the genus Fusarium proved useful for this purpose [20]. *Fusarium solani (javanicum)*, a recognized sweet potato pathogen, appears to be the best member of this genus for providing both hepatotoxins and lung oedemagenic toxins. However, only traces of ipomeanine have been detected in extracts which contain good yields of ipomeamarone and 4-ipomeanol. Routinely the fungus is grown as a shake culture in sweet potato slurry. Washed sweet potatoes are decontaminated with glutaraldehyde, sliced (1/4 inch thick), and then coated with sterile 15% mannitol solution followed by application of the sweet potato culture slurry. The slurry also contains 10 μg streptomycin per ml as an inhibitor of bacterial contaminants. The inoculated slices are spread out several layers thick in teflon-lined aluminum trays and covered with aluminum foil. The trays are placed in an incubator at 22° for 6 days. As the fungus invasion advances the slices develop a pungent scent similar to that of geraniol or closely related terpenes. Extraction followed by TLC of the extract reveals the presence of several Ehrlich-positive compounds including ipomeamarone and 4-ipomeanol. These may be separated and purified by silicagel column chromatography followed by preparative GC as described earlier.

The presence of spores of pathogenic fungi in the soil and on the surfaces of sweet potatoes poses a threat to the plant, and a fungal invasion may commence as the result of various types of mechanical injury permitting spore entry into the fleshy root. The exposed surface of cut slices tends to dry rapidly, a phenomenon that discourages or even inhibits infection completely. The reported antifungal property of ipomeamarone and possibly other phytoalexins may further tend to retard infection, but at best this can only be considered a delaying action on the part of the host against further invasion.

In addition to its role as stimulator of abnormal metabolism the fungus conceivably may alter some of the metabolites formed by the sweet potato, thereby playing an additional role in bioproduction of the observed spectrum of products.

Biological activities of the toxins in experimental systems and a consideration of the mode of action

Little information is currently available regarding biological activity of the toxins in experimental systems. As mentioned earlier, ipomeamarone reportedly has antimicrobial properties [27]. It has been found to inhibit the germination of spores and mycelial growth of several species of common fungi, including certain sweet potato pathogens [50]. Uritani and Akazawa [51] found that ipomeamarone in low concentration prevented absorption of inorganic phosphate from the culture medium and conversion of inorganic P to acid-soluble organic P and insoluble P by *C. fimbriata*. With the exception of the referenced work dealing with pathology induced by the administration of ipomeamarone, no data are available regarding the *in vivo* or *in vitro* effects of the compound in mammalian species.

With regard to 4-ipomeanol, recent work* has uncovered certain pharmacological actions of this substance. With i.v. doses of 3—12 mg/kg in the anesthetized dog, 4-ipomeanol produces a marked, but transient hypotensive response. At concentrations around 10^{-3} M the compound produces a complete but reversible neuromuscular blockade in the rat phrenic nerve-diaphragm preparation. At this concentration it also effects a reversible blockade of prostaglandin E_2-induced contraction of the rat stomach muscle. The relationship, if any, of these properties of 4-ipomeanol to its mode of toxic action is unknown.

The lethal effects of 4-ipomeanol in experimental animals are not prevented by pretreatment with a variety of compounds. Pretreatment with maximally tolerated doses of 5-hydroxytryptamine antagonists in mice or of antihistamines in guinea pigs did not modify the toxicity of 4-ipomeanol in these species. Pretreatment of mice with various autonomic agonists and antagonists such as atropine, propranolol, phenoxybenzamine, isoproterenol, and epinephrine, was without significant effect.

Numerous compounds are known which are capable of producing pulmonary oedema in animals. The subject has been reviewed by Visscher *et al.* [52], by Aviado [53], and by Louisada [54]. In general, the mechanism(s) of action of oedemagenic substances are not well understood. Such is also the case for 4-ipomeanol. It is interesting, however, to compare the effects of 4-ipomeanol with those of α-naphthylthiourea (ANTU), a compound which has been used as a rat poison [55].

LD_{50} values for ANTU are similar to those of 4-ipomeanol. Administration of toxic portions of the compound to experimental animals is followed

---

* Details of these studies will soon be published by the authors of this chapter.

References p. 342

by a sequence of events similar to that described earlier for 4-ipomeanol. In the mouse, for example, following a latent period of 4—8 h, overt signs of pulmonary oedema are observed, and the animal becomes progressively anoxic. The sequence terminates in a brief convulsion and death. At *post mortem*, only the lungs show striking pathology, with pulmonary oedema and massive pulmonary effusion grossly evident.

Another common property of 4-ipomeanol and ANTU is the production of tolerance to normally lethal doses of the compounds by pretreatment with a series of sublethal doses. McClosky and Smith [55] found that tolerance to the acute toxicity of ANTU can be induced in rats by administration of increasing, but sublethal, doses of ANTU at 3- to 4-day intervals. No pleural effusions were found in "tolerant" rats dying from large doses of the poison. Similarly, mice pretreated for 1 week with 10 mg/kg/day (i.p.) of 4-ipomeanol were completely protected from a normally lethal dose of 70 mg/kg of the compound [24]. SKF-525A (50 mg/kg) given an hour before challenge with toxin did not modify the induced tolerance. By gradually increasing daily pretreatment doses over a period of several days the LD_{50} of 4-ipomeanol was increased several fold. Pretreatment with phenobarbital or 3-methylcholanthrene induced some tolerance to challenge with the toxin, although not so effectively as did pretreatment with the toxin itself.

The phenomenon of tolerance development is common among many lung-oedemagenic agents. This subject has been reviewed by Stokinger [56] and, more recently, by Fairchild [57]. It will be of interest to determine if 4-ipomeanol induces tolerance by a mechanism similar to that for ANTU and other similar lung-toxic substances.

Acknowledgment

This work was supported by Research Grant 5 RO1 ES 00569-08 and Center in Environmental Toxicology Grant ES 00267 to Vanderbilt University from the U.S. Public Health Service. M.R. Boyd also acknowledges support by the Vivian B. Allen Foundation.

References

1 C.E. Steinbauer and L.J. Kushman, Sweet Potato Culture and Diseases, Agricultural Handbook No. 388, Agricultural Research Service, U.S. Department of Agriculture, Washington, D.C., 1971, pp. 1—3.
2 A.H. Dempsey, L.J. Kushman and J.E. Love, in Thirty Years of Cooperative Sweet Potato Research 1939—1969, Southern Cooperative Series Bulletin No. 159, Sept. 1970, pp. 36—38.
3 I. Uritani, H. Hyodo and M. Kuwano, Agr. Biol. Chem. (Japan), 35 (1971) 1248.
4 S. Abo and S. Nomura, J. Vet. Hyg. Assoc., 10 (1942) 17.
5 A.A. Hansen, North Am. Vet., 9 (1928) 31.
6 W. Monlux, J. Fitte, G. Kendrick and H. Dubuisson, Southwest. Vet., 6 (1953) 267.
7 J.C. Peckham, F.E. Mitchell, O.E. Jones and B. Doupnik Jr., J. Am. Vet. Med. Assoc., 160 (1972) 169.

8   C.L. Vickers, W.T. Carll, B.W. Bierer, J.B. Thomas and H.D. Valentine, J. Am. Vet. Med. Assoc., 137 (1960) 507.
9   B.L. Munday, Aust. Vet. J., 44 (1968) 501.
10 M. Hiura, Rept. Gifu Agr. Coll., 50 (1943) 1.
11 T. Kubota, Tetrahedron, 4 (1958) 68.
12 T. Kubota and T. Matsuura, J. Inst. Polytech. Osaka City Univ., 2 (1952) 82, 94, 104; 4 (1953) 104, 108, 112.
13 T. Kubota and T. Matsuura, J. Chem. Soc., (1958) 3667.
14 T. Kubota and T. Matsuura, Chem. Ind., (1956) 521.
15 B.F. Hegarty, J.R. Kelly, R.J. Park and M.D. Sutherland, Aust. J. Chem., 23 (1970) 107.
16 T. Taira and Y. Fukagawa, Nippon Nagei Kagaku Kaishi, 32 (1958) 513.
17 F.A. Denz and W.G. Hanger, J. Pathol. Bacteriol., 81 (1961) 91.
18 D.T.C. Yang, B.J. Wilson and T.M. Harris, Phytochemistry, 10 (1971) 1653.
19 N. Kato, H. Imaseki, N. Nakashima and I. Uritani, Tetrahedron Letters, No. 13, (1971) 843.
20 B.J. Wilson, D.T.C. Yang and M.R. Boyd, Nature, 227 (1970) 521.
21 B.J. Wilson, M.R. Boyd, T.M. Harris and D.T.C. Yang, Nature, 231 (1971) 52.
22 B.J. Wilson and M.R. Boyd, Fifth Intern. Cong. Pharmacology, San Francisco, California, July 23—28, 1972 (Abstract No. 1517).
23 M.R. Boyd and B.J. Wilson, J. Agric. Food Chem., 20 (1972) 428.
24 M.R. Boyd and B.J. Wilson, Fifth Intern. Cong. Pharmacology, San Francisco, California, July 23—28, 1972 (Abstract No. 164).
25 M.R. Boyd, B.J. Wilson and T.M. Harris, Nature, 236 (1972) 158.
26 T. Kubota and N. Ichikawa, Chem. and Ind., (1954) 902.
27 S. Yamashita, J. Soc. Brewing, Japan, 9 (1946) 88.
28 H. Imaseki, S. Takei and I. Uritani, Plant and Cell Physiol., 5 (1964) 119.
29 M. Kojima, T. Minamikawa, H. Hyodo and I. Uritani, Plant and Cell Physiol., 10 (1969) 471.
30 M. Kojima, Agr. Biol. Chem., 35 (1971) 632.
31 H. Imaseki, S. Takei and I. Uritani, Plant and Cell Physiol., 5 (1964) 119.
32 T. Minamikawa and I. Uritani, Arch. Biochem. Biophys., 108 (1964) 573.
33 K. Uehara, S. Fujimoto and T. Taniguchi, J. Biochem., 70 (1971) 183.
34 H. Imaseki, K. Akutsu and I. Uritani, Plant and Cell Physiol., 8 (1967) 753.
35 I. Oguni, K. Oshima, H. Imaseki and I. Uritani, Agr. Biol. Chem., 33 (1969) 50.
36 K. Oshima and I. Uritani, Agr. Biol. Chem., 32 (1968) 1146.
37 K. Oshima-Oba, I. Sugiura and I. Uritani, Agr. Biol. Chem., 33 (1969) 586.
38 I. Oguni and I. Uritani, Agr. Biol. Chem., 35 (1971) 357.
39 K. Oba, H. Shibata and I. Uritani, Plant and Cell Physiol., 11 (1970) 507.
40 K. Oshima Oba and I. Uritani, Plant and Cell Physiol., 10 (1969) 827.
41 K. Oshima and I. Uritani, J. Biochem., 63 (1968) 617.
42 I. Oguni and I. Uritani, Agr. Biol. Chem., 34 (1970) 156.
43 M.D. Sutherland and R.J. Park, Sesquiterpenes and their biogenesis in *Myoporum deserti* A. Cunn, in J.B. Pridham (Ed.), Terpenoids in Plants, Chapter 9, Academic Press, London, 1967, p. 147.
44 T. Akazawa, I. Uritani and H. Kubota, Arch. Biochem. Biophys., 88 (1960) 150.
45 T. Akazawa, Arch. Biochem. Biophys., 90 (1960) 82.
46 M.R. Boyd and B.J. Wilson, J. Agric. Food Chem., 19 (1971) 547.
47 T. Akazawa and K. Wada, Plant Physiol., 36 (1961) 39.
48 I. Uritani, M. Uritani and H. Yamada, Phytopathology, 50 (1960) 30.
49 I. Uritani, The role of plant phenolics in disease resistance and immunity, in G. Johnson and E.A. Geissman (Eds.), Symposium on Biochemistry of Plant Phenolic Substances, Colorado State University, Fort Collins, Colo., 1961, p. 98.
50 F. Nonaka and K. Yasmi, Saga Daigaku Nogaku Iho, No. 22 (1966) 39; Chem. Abstr., 65 (1966) 11009.

51 I. Uritani and T. Akazawa, Science, 121 (1955) 216.
52 M.B. Visscher, F.J. Haddy and G. Stephens, Pharmacol. Rev., 8 (1956) 389.
53 D.M. Aviado, Acute pulmonary edema, in The Lung Circulation, Vol. II, Chapter 18, Pergamon, New York, 1965.
54 A.A. Louisada, Pulmonary Edema in Man and Animals, Warren Green, St. Louis, Mo., 1970.
55 W.T. McClosky and M.I. Smith, Public Health Reports, 60 (1945) 1101.
56 H.E. Stokinger, Arch. Environ. Health, 10 (1965) 719.
57 E.J. Fairchild, Arch. Environ. Health, 14 (1967) 111.

Note added in proof

Significant new data are now available regarding both the chemistry and biological mechanisms of action of the lung-oedemagenic toxins (see Boyd *et al.*, Biochim. Biophys. Acta, 337 (1974) 184; Boyd *et al.*, Toxicol. Appl. Pharmacol., (1974) in press; Boyd *et al.*, Federation Proc., 33 (1974) 234).

Chapter 16

OCHRATOXIN A AND RELATED METABOLITES

J. HARWIG

Food Research Laboratories, Health Protection Branch, Department of National Health and Welfare, Tunney's Pasture, Ottawa, Ontario, K1A OL2 (Canada)

The discovery of the aflatoxins led to increased awareness of the potential role of fungal toxins in animal and human disease. Whereas the aflatoxins were discovered as a result of attempts to determine the aetiology of a recognized outbreak of disease, the ochratoxins were discovered in laboratory screening of a large number of fungal cultures for toxicity by South African workers. They attributed the toxicity of a strain of *Aspergillus ochraceus* Wilhelm, reported by Scott in 1965, to ochratoxin A, the main toxic component in culture extracts [1,2]. During isolation and chemical characterization of this toxin, Van der Merwe *et al.* [3], and Steyn and Holzapfel [4] isolated also the methyl and ethyl derivatives of ochratoxin A and of the non-toxic dechloro derivative, ochratoxin B. The work of Theron *et al.* [5], Purchase and Theron [6], and that of others [7,8] characterized ochratoxin A as a potent nephrotoxin and hepatotoxin. Work described by Lai *et al.* [9,10], and by Hesseltine *et al.* [11], provided evidence that other members of the *A. ochraceus* group produced ochratoxins whereas Van Walbeek *et al.* [12,13] reported ochratoxin production by a strain of *Penicillium viridicatum* Westling.

As methods of analysis for ochratoxin A came into use in different laboratories, reports appeared on the natural occurrence of the compound. Low levels were reported in low grade U.S. corn (maize) by Shotwell *et al.* [14] and in mouldy Canadian wheat by Scott *et al.* [15]. Evidence for the natural occurrence of ochratoxin A in some other commodities soon followed. The rapid and successful isolation and chemical characterization of the ochratoxins and recognition of their presence in naturally contaminated substrates can be attributed largely to their relative chemical stability and their property of fluorescence in ultraviolet (UV) light.

Further data on the occurrence of ochratoxin A in animal and human foods and on its toxic effects are required for an evaluation of its role in disease. Recent reports on the occurrences of two nephrotoxins, citrinin and ochratoxin A, in Canadian and Danish grains in association with strains of *P. viridicatum* producing these two toxins have provided a clue that they may be the factors responsible for a recognized disease of unknown aetiology in swine [16—19].

A review with an emphasis on the chemistry of the ochratoxins was recently published by Steyn [20].

Outbreaks of disease of attributable to *P. viridicatum* and *A. ochraceus*

A disease in pigs involving kidney degeneration and referred to as mould nephrosis or mycotoxic nephropathy had long been recognized in Danish slaughter houses. In some years, the disease occurred in up to 7% of the pigs slaughtered. Krogh and Hasselager [19] cited Danish work published in 1928 in which clinical and patho-anatomical symptoms of this disease could be reproduced by feeding pigs naturally mouldy rye that was suspected of being toxic. The disease was later induced in pigs and in rats by feeding them naturally mouldy barley. Among the fungi isolated from the barley, *P. viridicatum* strain 67B was the most toxic to the rats and induced kidney damage in rats and pigs resembling the natural mould nephrosis in pigs [19]. This strain was subsequently shown to produce the nephrotoxins citrinin and oxalic acid in a semi-synthetic medium [21,22]. Citrinin-induced but not oxalate-induced kidney damage in pigs was comparable to the naturally occurring kidney degeneration. Citrinin, perhaps in combination with other synergistically acting compounds, was therefore suggested to be the toxin responsible for the naturally occurring mycotoxicosis. Strain 67B also produced ochratoxin A (Krogh, personal communication). Both ochratoxin A and citrinin were detected in Danish barley and oats fed to swine that on slaughtering had been recognized to suffer from mycotoxic nephropathy [17]. Small amounts of ochratoxin A were present in the kidneys and other tissues of these swine [18]. The two toxins were also detected in naturally mouldy Canadian grains from which ochratoxin A and citrinin-producing strains of *P. viridicatum* could be isolated [15,16]. An ochratoxin A-producing strain grown on barley and fed to rats produced severe kidney lesions [23].

Although the data are not sufficient to provide a complete picture, present evidence indicates a cause and effect relationship between the production by *P. viridicatum* of ochratoxin A and citrinin in feeds and mould nephrosis in swine. Experiments with pigs fed diets containing non-mouldy grains and purified ochratoxin A and citrinin at levels encountered in naturally contaminated grains are needed for a fuller understanding of the aetiology of mould nephrosis.

Nephrotoxicity associated with consumption of naturally mouldy barley was also recognized in Ireland [24]. The clinical signs and histopathological findings with respect to kidney damage closely corresponded with those described by the Danish workers. Fungi isolated from this barley which grew on agar media were *Aspergillus fumigatus*, *Absidia ramosa*, and *P. cyclopium*. The barley was not examined for ochratoxin A.

There is little evidence that *A. ochraceus* may be the cause of recognized outbreaks of animal disease. The mould was implicated in bovine abortion in Wisconsin [25]. A strain of *A. ochraceus* was isolated from mouldy alfalfa-grass hay that made up part of the diet of aborting cows. When the hay was withheld, the abortions ceased. The isolate of *A. ochraceus* produced ochratoxin A in sterile moist corn and, when this was fed as part of a diet to pregnant rats, caused foetal death and resorption. These effects were also observed in rats intubated with pure ochratoxin A. It was not reported whether the mouldy hay contained ochratoxin A.

Ochratoxin-producing fungi

As reviewed by Hesseltine *et al.* [11], members of the taxonomic entity referred to as the *A. ochraceus* group were found in all parts of the world in soil, insects, foods such as rice, oats, wheat flour, chili and Capsicum pepper, and in diseased plants and decaying vegetation. Being osmophilic, they were frequently found in dried and salted fish products in the Orient. *A. ochraceus* was isolated from South African cereals and legume crops and was found in sorghum and soybean seeds from commercial storage in the U.S.A., but seldom in more than 2—3% of surface-disinfected kernels [1,26]. The mould was also present in black and red pepper [27].

A large proportion of *A. ochraceus* isolates from foods and feeds was toxigenic. 17 of 37 strains grown on a mixture of sterilized corn and rice were lethal to rats [28]. Ochratoxin A-producing strains of *A. ochraceus* were isolated from sorghum, hops, peanuts, pecan nuts, mouldy Japanese rice, black pepper, adzuki beans, and ground pepper [2,12,29—33].

P. viridicatum is common in soil and vegetation [34]. It was one of the Penicillium spp. causing "blue eye" of corn, a condition that may develop over winter in cribbed corn [35,36]. Among other Penicillium spp., *P. viridicatum* was consistently associated with stored dent corn but occurred infrequently and only at low levels in corn from the field. In contrast to some other Penicillium spp., *P. viridicatum* is therefore mainly a storage fungus. The mould grew and competed well with other fungi in dent corn adjusted to a moisture content of 22% and incubated at temperatures between 4° and 30° [37]. The mould develops under conditions of relatively low temperatures. Conidia germinated at 16° within 7, 21, and 26 days at relative humidities of, respectively, 86, 83, and 81%, and grew on agar media over a temperature range of −2° to 36° [38].

A number of reports indicate the potential toxicity of *P. viridicatum*. Corn in which the fungus was prevalent killed horses and pigs in Argentina [39]. Extracts of cultures of *P. viridicatum* showed the presence of toxins, as concluded from positive rabbit skin reactions [40]. American workers showed that corn cultures of *P. viridicatum* mixed with the diet of mice, miniature pigs, guinea pigs, and rats caused damage to the liver and kidneys [41—43]. Diets containing 10—15% of rice cultures caused 50% mortality in mice [44]. In rats, these diets produced only a mild kidney alteration [43] unlike the *P. viridicatum*-induced kidney changes described by Danish workers. Toxicity of the isolates of the American workers was associated with mycelial mats grown in liquid culture media under stationary conditions.

Culture filtrates after 2 weeks of growth at 23° were not toxic [45]. Other isolates of *P. viridicatum* produced, under similar conditions, large amounts of ochratoxin A in culture filtrates [13,15].

The hepatic lesions induced in mice by the *P. viridicatum* isolates of the American workers resembled the hepatic lesions in sheep suffering from facial eczema, a photosensitivity disease caused by ingestion of sporidesmins. Mice fed toxic rice cultures mixed with the diet had pronounced hepatic lesions and, when exposed to sunlight, developed a phototoxic syndrome involving gross and microscopic changes in their ears and face [46].

Liver lesions induced in mice by *P. viridicatum* isolates of the American workers were identical to those produced by *P. ochraceum* cultures. These lesions did not resemble those induced by ochratoxin A and ochratoxin A was not detected in the cultures of *P. ochraceum* [47]. A recent report indicates that mice given low levels of rice cultures of these *P. viridicatum* isolates in their diet had an increased incidence of pulmonary tumours [48]. The toxic and carcinogenic factors in these Penicillia have yet to be isolated.

A Penicillium sp. from ham identified as *P. viridicatum* was the first species outside the *A. ochraceus* group to be reported to produce ochratoxin A [12,13]. Ochratoxin A- and ochratoxin A and citrinin-producing strains of *P. viridicatum* and *P. palitans* were present in mouldy Canadian grains in which ochratoxin A and citrinin had been detected [16]. These strains were readily isolated from kernels incubated on moistened filter paper at 10—12°.

TABLE I

FUNGI PRODUCING OCHRATOXIN A

Species	Number of isolates producing Ochr. A/ number of isolates tested	References
A. ochraceus	> 1/5, 4/14	2, 32
A. ochraceus	8/17, 6/11	50, 11
A. ochraceus	2/33, 2/58	33, 31
A. ochraceus	3/3	30
A. ochraceus group	4/34	9
A. alliaceus	2/2, 7/7	50, 11
A. melleus	4/5, 1/3	50, 11
A. ostianus	1/1, 2/3	50, 11
A. petrakii	1/3	11
A. sclerotiorum	2/4, 4/9	50, 11
A. sulphureus	1/1, 1/1	50, 11
P. viridicatum	a	12, 13, 15, 16
P. viridicatum	a	49
P. cyclopium [b]	a	49
P. commune	a	49
P. palitans	a	16
P. purpurescens	a	49
P. variabile	a	49

[a] Ratio not indicated.
[b] Identity of fungus not definitively established.

At higher temperatures, other species emerged from the material [15]. In a survey of the mycotoxin-producing potential of Penicillia isolated from mould-fermented sausages, 17 of 422 isolates produced ochratoxin A. 12 of the 17 isolates were *P. viridicatum* and the rest included 3 or 4 new species [49].

Table I lists the fungi known to produce ochratoxin A and compares the number of toxigenic isolates with the total number examined. Some of the Aspergillus spp. may have lost their original ability to produce the toxin during maintenance of cultures by periodic transfer. Doupnik and Peckham tested 10 strains of *A. ochraceus* from peanuts and all 10 were initially toxic to chicks. After subculturing, however, several of these isolates had lost their toxigenicity [29].

Toxic metabolites of A. ochraceus and P. viridicatum

The structures of the ochratoxins are presented in Fig. 1. These were proposed on the basis of their physical and chemical properties including infrared, UV, nuclear magnetic resonance (NMR), and optical rotatory dispersion spectra [2—4]. The toxins were found to be composed of a dihydroisocoumarin moiety linked over its 7-carboxy group to L-β-phenylalanine. The dihydroisocoumarin closely resembles mellein (3-methyl-8-hydroxy-3,4-dihydroisocoumarin) and 4-hydroxymellein, known metabolites of some *A. ochraceus* strains and possible precursors of ochratoxin A [51].

Ochratoxin A is a colourless crystalline compound with a molecular formula $C_{20}H_{18}ClNO_6$. When crystallized from benzene, it contained 1 mole of benzene of crystallization and then had a melting point (m.p.) of *ca.* 90°. Crystallized from xylene after removal of benzene, it contained no xylene of crystallization and melted at 169°. Evidence for the presence of carboxyl, a secondary amide, and a lactone group in the molecule was derived from its infrared spectrum. The presence of a free phenolic hydroxyl group was indicated by a red colour reaction with ethanolic ferric chloride. Treatment of ochratoxin A with methanol-HCl yielded a methyl ester which still showed a red colour reaction with ethanolic ferric chloride. Subsequent methylation of the hydroxyl group with diazomethane yielded the *O*-methyl ester [3]. Ochratoxin A had $[\alpha]_D$ — 118° (*c* 1.1. in chloroform) [3], and λ_{max}, 213 nm and 332 nm (ϵ 36 800 and 6400, respectively) [20]. On acid

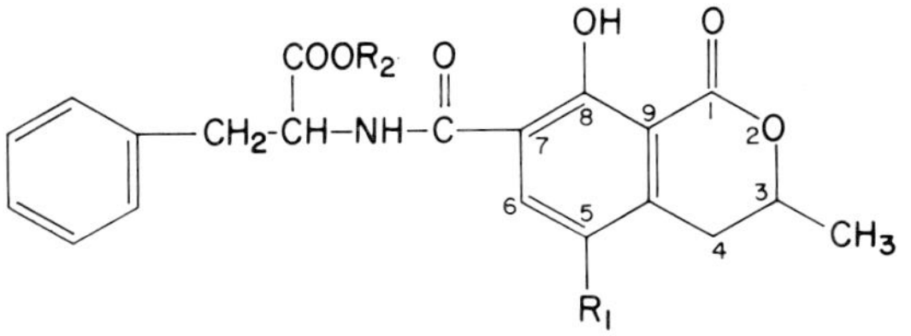

Fig. 1. Structure of the ochratoxins; ochratoxin A: R_1 = Cl, R_2 = H; ochratoxin B: R_1 = H; R_2 = H; ochratoxin C: R_1 = Cl, R_2 = C_2H_5; methyl ester of ochratoxin A: R_1 = Cl, R_2 = CH_3; methyl or ethyl ester of ochratoxin B: R_1 = H; R_2 = CH_3 or C_2H_5.

hydrolysis, it yielded L-β-phenylalanine and an optically active lactone acid $C_{11}H_9O_5Cl$, formulated as 7-carboxy-5-chloro-3,4-dihydro-8-hydroxy-3-methylisocoumarin (ochratoxin α) [3]. The proposed stereochemical configuration of ochratoxin A was proven by direct synthesis [52]. An alternative chemical synthesis for ochratoxin A was described by Roberts and Woollven [53].

Ochratoxin B, $C_{20}H_{19}NO_6$, is the dechloro analogue of ochratoxin A and is also a colourless crystalline compound. Crystallized from acidic aqueous methanol, it had an m.p. 221°, $[\alpha]_D - 35°$ (c 0.15), and λ_{max} 218 and 318 nm (ϵ 37 200 and 6900, respectively) [3].

Ochratoxin C is the amorphous ethyl ester of ochratoxin A with the molecular formula $C_{22}H_{22}ClNO_6$ and had $[\alpha]_D - 100°$ (c 1.2) and λ_{max} 213, 331, and 378 nm (ϵ 32 700, 4100, and 2050, respectively) [4].

Another three neutral compounds related to ochratoxin A, B, and C were isolated from cultures of *A. ochraceus* strain K-804 [4]. Ochratoxin A methyl ester, $C_{21}H_{20}ClNO_6$ had spectral properties virtually identical to those of ochratoxin C, could not be crystallized, and had $[\alpha]_D - 78°$ (c 0.027). Ochratoxin B ethyl ester, $C_{22}H_{23}NO_6$, was crystallized from ether and had m.p. 102–103°, $[\alpha]_D - 49°$ (c 0.04), and λ_{max} 218, 318, and 364 nm (ϵ 32 000, 5200, and 1250, respectively). The methyl ester of ochratoxin B, $C_{21}H_{21}NO_6$, had m.p. 134–135° when crystallized from benzene, $[\alpha]_D - 62°$ (c 0.02), and was spectrally similar to ochratoxin B ethyl ester. Identification of these esters was readily verified by esterification of ochratoxin A and B to the corresponding methyl and ethyl esters; mild alkaline hydrolysis of ochratoxin A and B esters yielded the corresponding toxins.

In addition to the ochratoxins, penicillic acid and hydroxyaspergillic acid, identified by m.p. and spectral properties, were present in the corn meal cultures of *A. ochraceus* strain K-804 [4]. Many members of the *A. ochraceus* group produced penicillic acid and ochratoxin A simultaneously [33,50]. Secalonic acid was present in extracts of ochratoxin A-producing strains of *A. ochraceus* that had been isolated from mouldy rice [54].

Ochratoxin A and B were also found to be metabolites of *P. viridicatum* [12,13,15,16]. Many of the strains from mouldy grains produced the known fungal metabolite citrinin (4,6-dihydro-8-hydroxy-3,4,5-trimethyl-6-oxo-3H-2-benzopyran-7-carboxylic acid) in addition to ochratoxin A [15,16,21,22]. Two other compounds present in cultures of *P. viridicatum* strain ATCC 18411 were 7-carboxy-3,4-dihydro-8-hydroxy-3-methylisocoumarin and the non-toxic 4-hydroxyochratoxin A, $C_{20}H_{18}ClNO_7$. The latter was a colourless crystalline compound with m.p. 216–218° (from benzene) and showed λ_{max} 213 and 334 nm (ϵ 32 500 and 6400, respectively) [55]. This compound was identical to ochratoxin D isolated from this culture by Scott *et al.* [56] (Scott, personal communication).

Biosynthesis of ochratoxin A

By analogy with other structurally related fungal metabolites, the bio-

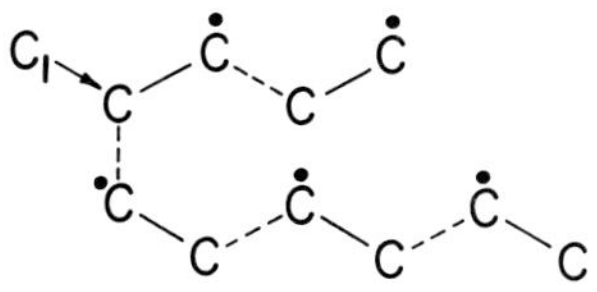

Fig. 2. Proposed scheme for the biosynthesis of ochratoxin A involving head-to-tail condensation of five acetate units and introduction of one C_1-unit (*cf.* Steyn *et al.*, [57].

synthesis of the dihydroisocoumarin moiety of the ochratoxins may involve head-to-tail condensation of five acetate units and incorporation of a C_1-unit at position 7 with subsequent oxidation to a carboxyl group (Fig. 2). The phenylalanine portion of ochratoxin A may be derived through the shikimic acid pathway [57].

Some observations were consistent with this suggestion. In short-term experiments with resting cultures of *A. ochraceus*, radioactivity supplied in the form of [1-^{14}C]acetate or formate was incorporated exclusively into the heterocyclic moiety of ochratoxin A and [1-^{14}C]phenylalanine was introduced solely into the amino acid part of the molecule [58]. By Kuhn-Roth degradation of the heterocyclic moiety, carbon-3 and the methyl carbon at position 3 could be obtained as acetic acid. All the activity, exactly 20% of the total activity of the dihydroisocoumarin moiety, was located in carbon-3. Isolation of the carboxyl group at position 7 by a modified Schmidt decarboxylation reaction involving protection of the phenolic hydroxyl group gave inactive carbon dioxide. This indicates that the carboxyl carbon is not derived from a carboxyl group of acetate. Evidence that this carbon is derived from the C_1-pool was obtained from the observation that cultures incubated with [^{14}C-H_3]methionine yielded radioactive ochratoxin A. The dihydroisocoumarin resulting from acid hydrolysis of this fraction was radioactive and, upon modified Schmidt decarboxylation, yielded labelled carbon dioxide.

In other experiments, involving addition of radioactive precursors at the time of inoculation and an 8-day incubation period, almost all of the incorporated labelled phenylalanine was present in the phenylalanine moiety of ochratoxin A [59]. [2-^{14}C]Acetate was introduced mostly into the dihydroisocoumarin moiety but a significant portion of the radioactivity was present in the phenylalanine moiety, probably resulting from randomization of the label after the long period of incubation. Most of the radioactivity from [2-^{14}C]acetate was incorporated into carbons 9,7 and 5 (see Fig. 1 for numbering system) and little in 1, 8, 6, 3, the methyl carbon at position 3 and the carboxyl carbon at position 7. This pattern of labelling was interpreted to be consistent with acetate-derived synthesis of only a portion of the dihydroisocoumarin moiety.

Another biosynthetic scheme [60] involves condensation of acetate and malonate units and the introduction of a C_1-unit at carbon-7. Addition of [2-^{14}C]malonate to cultures of *A. ochraceus* yielded labelled ochratoxin A with the activity located in the dihydroisocoumarin moiety. Increasing inhibition of ochratoxin production with increasing concentration of ethionine was interpreted to support the derivation of the carboxyl group from compounds of the C_1-pool such as methionine. This was confirmed by the finding that ^{13}C-enriched ochratoxin A was obtained from cultures by adding [13C]formate to the medium. Methylation of the dihydroisocoumarin resulting from hydrolysis yielded the chloroform soluble methoxyisocoumarin carboxylate methyl ester that was suitable for NMR spectroscopy. This indicates that ^{13}C was incorporated solely into the carboxyl group of the dihydroisocoumarin.

An enzyme referred to as ochratoxin A synthetase and capable of coupling phenylalanine to the heterocyclic moiety through an amide bond was active in cell free extracts of *A. ochraceus* containing adenosine triphosphate and magnesium chloride [58].

It is as yet unknown at what stage during the formation of ochratoxin A the chlorine atom is introduced into the molecule. The highest incorporation of $Na^{36}Cl$ into the toxin by *A. ochraceus* cultures occurred when the salt was added on the second or third day of incubation [61].

Laboratory and environmental conditions favouring ochratoxin A production

Synthetic and semi-synthetic media. Glutamic acid or proline [62] and ammonium nitrate [10] were suitable N sources for the production of ochratoxin A in synthetic media by *A. ochraceus* strains or other members of the *A. ochraceus* group. Sucrose [10,62] and glucose [10] were favourable C sources. Levels of ochratoxin A in these media rose to 100—200 mg/l of medium. Addition of 1% phenylalanine to a basal salts medium supplemented with 2% yeast extract greatly stimulated ochratoxin A production [31].

Glutamic acid and proline could not be replaced to induce ochratoxin A synthesis by any one of the 25 amino acids and 9 peptides tested [63]. Derivatives of glutamic acid and proline either did not support toxin formation or the compound was produced at much lower levels. These derivatives added along with glutamic acid or proline had an inhibitory effect on the production of the toxin. The inhibition could be removed by increasing the concentration of glutamic acid.

Addition of the combined trace elements iron, zinc, copper, boron, manganese, and molybdenum increased the synthesis of ochratoxin A by some members of the *A. ochraceus* group [10].

Raulin-Thom and YES (2% yeast extract and 15% sucrose) medium were used for screening members of the *A. ochraceus* group for production of ochratoxin A and penicillic acid [50]. Another convenient liquid medium for ochratoxin A production by *A. ochraceus* consisted of 4% sucrose and 2% yeast extract. Under these conditions, no ochratoxin B was formed. However, on increasing the concentration of sucrose, cultures also produced ochratoxin B [64]. In 5-gallon carboys containing 4 l of 4% sucrose and 2% yeast extract incubated as stationary cultures, maximum amounts of ochratoxin A were formed 7—8 days after inoculation. Yields of pure ochratoxin A from culture filtrates ranged from 39—65 mg/l [65].

YES medium was used also for ochratoxin A production by *P. viridicatum* [13,66]. For production of relatively large amounts of ochratoxin A, stationary cultures of *P. viridicatum* were grown in 2.8 l Fernbach flasks containing about 200 ml YES incubated at 25—28° in the absence of direct light [13].

Sterile food substrates. Ochratoxin A was produced by *A. ochraceus* or other members of the *A. ochraceus* group on moistened and sterilized wheat, soybean, pearled wheat, cracked corn, pecans, polished rice, wheat bran, and shredded wheat [9,11,30,67,68]. Optimal temperature for toxin production by *A. ochraceus* was 28° [67]. Production was much lower at 15 and 37°. Production of ochratoxin A in cracked corn and wheat was higher at 28° than at 20 and 10° but penicillic acid, produced simultaneously, reached higher levels in substrates incubated at 10° [50]. The rate of production of penicillic acid at this temperature was slower than at 28° but at the latter temperature a sharp decline occurred after about two weeks of incubation. Levels of penicillic acid were several orders of magnitude higher than those of ochratoxin A.

Requirements for maximum production of ochratoxin A by *A. ochraceus* ATCC 18642 on moistened and autoclaved shredded wheat included addition of 40—70 ml water to 100 g of substrate and an incubation period of 19—21 days at about 22° [68]. Under these conditions, about 2.5 g/kg substrate was produced. Ochratoxin B was also present in the extracts. Another method giving high yields of ochratoxin A involved the use of moistened and autoclaved pearled wheat and cracked corn in shaken flasks incubated at 28° for 5 days [11,69]. Wheat gave higher yields of ochratoxin A than corn for 12 of the 14 better producing strains of the *A. ochraceus* group [11]. The highest ochratoxin A-producing strain, *A. ochraceus* NRRL 3174, produced an average of 1.3 g/kg wheat and 0.9 g/kg corn. Ochratoxin B was produced in quantities of 0.2 g/kg wheat and 0.05 g/kg corn. The highest ochratoxin B-producing culture was *A. ochraceus* NRRL 3519 which produced 1.0 g ochratoxin A and 0.9 g ochratoxin B/kg wheat. *P. viridicatum* NRRL 3712 grown in shaken flasks for 12 days produced an average of 1.6 g ochratoxin A and 0.1 g ochratoxin B/kg wheat at 20° [69].

The natural production of ochratoxin A is likely to occur in the presence of several interacting factors and some of these, *e.g.* competition among different species, are excluded in laboratory studies involving moistened and autoclaved substrates. The relevance of laboratory conditions to natural conditions favouring toxin production is therefore not always clear. Growth of *A. ochraceus* in natural substrates has not been reported to be associated with naturally occurring ochratoxin A and nothing is known about environmental factors favouring production of the toxin by this species. It may, however, be speculated that production occurs under conditions that favour growth of *A. ochraceus* in natural substrates. *A. ochraceus* became predomi-

nant in corn stored for several weeks at 25° and 18.5% moisture [70].
Likewise, in inoculated sound wheat incubated at 20—25° at a moisture
content of 16% and above, A. ochraceus grew rapidly, mainly in the embryo.
It had, however, difficulty invading kernels in which A. glaucus had already
established itself [71]. As A. glaucus is one of the first invaders of grains
stored in the U.S.A., this observation may explain the low levels of occurrence of A. ochraceus [26]. It may similarly be suspected that toxin production by P. viridicatum occurs in corn stored at 22% moisture over a range of
4—30°. The mould grew well under these conditions in the presence of other
Penicillium spp. [37].

Isolation of the ochratoxins

Ochratoxins were extracted from fungal cultures with methanol—
chloroform (1:1, v/v) [3], methanol—chloroform preceded by extraction
with hexane [72], or hot chloroform [56,73]. The acidic ochratoxins,
present in concentrated extracts or in the precipitate resulting from addition
of hexane to chloroform extracts, were transferred into sodium bicarbonate
[3,72,73]. The acidified bicarbonate solution was then extracted with chloroform. Ochratoxin A and B eluted from acidic silica with benzene—chloroform (3:1, v/v) were separated by ion exchange chromatography on a Dowex
resin column (formate form) [3]. Separation was also achieved by elution
from a silica gel column with benzene—acetic acid as gradient solvent [73].
A third method involved the use of a Sephadex LH-20 column from which
ochratoxin A was eluted with methanol after washing the column with chloroform [72]. Dissolved in chloroform after concentration, this fraction was
rechromatographed on a Sephadex column and ochratoxin A was eluted
with chloroform-methanol.

Concentrated chloroform extracts of P. viridicatum cultures were directly added to a column of silica gel containing 5% oxalic acid [56]. Ochratoxin A was eluted from the column with chloroform—acetone (99:1 and
98:2, v/v) and ochratoxin B with chloroform—acetone (95:5, v/v) after
washing with benzene and chloroform. Further purification required extraction of ochratoxin A containing column fractions with sodium bicarbonate
and transfer into chloroform.

Another method of isolating ochratoxin A involved passing culture filtrates of A. ochraceus directly through a Dowex resin column (formate
form) [65]. Ochratoxin A was eluted with a methanolic solution of formic
acid and transferred into chloroform. The toxin was then eluted from a silica
gel column with benzene—acetic acid. Ochratoxin A was transferred into
bicarbonate and the acidified bicarbonate solution extracted with chloroform. For final purification, ochratoxin A was crystallized from benzene
[3,56,65,72,73] and ochratoxin B from acidic aqueous methanol [3] or
methanol [73].

After transfer of the acidic ochratoxins from concentrated culture extracts into bicarbonate, the remaining neutral fraction was dissolved in 95%
aqueous methanol and extracted with hexane for removal of lipids [4]. The

concentrated methanol extract was separated on preparative silica gel plates with benzene—acetic acid (25:1, v/v) as mobile phase to give two fluorescent bands. A light green fluorescent band was separated by chromatography on formamide-impregnated Whatman No. 1 filter paper with hexane as mobile phase to yield the dull green fluorescent ochratoxin C and ochratoxin A methyl ester. A light blue fluorescent band treated similarly yielded the light blue fluorescent ochratoxin B ethyl ester and ochratoxin B methyl ester. Ethyl and methyl esters of ochratoxin A and B were also obtained by esterification of the purified toxins with boron trifluoride—ethanol or —methanol [73].

Analysis for ochratoxins

The first method described for analysis of ochratoxins involved Soxhlet extraction of mouldy material with chloroform—methanol (1:1, v/v), transfer of the acidic toxins into sodium bicarbonate, and re-extraction of the acidified bicarbonate solution with chloroform [74]. The residue obtained after removal of the solvent was spotted on silica gel thin layer plates; separation was achieved with benzene—acetic acid (3:1, v/v). A second mobile phase used was benzene—methanol—acetic acid (12:2:1, v/v/v). The ochratoxin A content of the mouldy material was estimated by visual comparison of the fluorescent spots of the sample with those of a standard solution while viewed under UV light. Ochratoxin A fluoresced green, ochratoxin B blue green, and ochratoxin C pale green. For confirmation of the presence of ochratoxins, plates were sprayed with 0.1 N NaOH which made all three toxins fluoresce bright blue.

A method that could be used in conjuction with analysis for aflatoxins was suitable for analysis of whole wheat flour, corn meal, rice cereal, and barley cereal [75]. The detection limit was about 25 µg/kg and recoveries ranged from 80—100%. Extraction was performed in a blender with aqueous methanol and hexane. The methanol extract was applied to a Celite column for partitioning. After extraction of the column with hexane, ochratoxin A was eluted with chloroform—hexane (1:1, v/v). The toxin was separated by TLC with toluene—ethyl acetate— 90% formic acid (5:4:1, v/v/v) as mobile phase and amounts were estimated visually by intensity of fluorescence. Exposure of thin-layer plates to ammonia fumes or spraying with a 25% solution of triethylamine in methanol changed the fluorescence of ochratoxin A in UV light to bright blue. A second solvent system, benzene—methanol—acetic acid (24:2:1, v/v/v) was used for confirmation of the presence of ochratoxin A. Cleaning of glassware with a 5% sodium hypochlorite solution was recommended for destruction of the toxin before the usual washing procedures.

Samples were analysed for ochratoxins, zearalenone, and aflatoxins simultaneously by a method involving water—chloroform extraction [76]. The chloroform extract was added to a silica gel column which was then washed with hexane and benzene. After use of appropriate eluents for removal of zearalenone and aflatoxins, the ochratoxins were eluted with benzene—acetic acid (9:1, v/v); the concentrated eluate was chromatographed on thin-layer plates for identification and estimation. Eppley [76] pointed out that confirmation of the identity of ochratoxin A, in addition to similarity in TLC behaviour to authentic ochratoxin A with different solvent systems, could be obtained by transferring the acidic toxins into bicarbonate, followed by acidification and re-extraction with chloroform. Benzene—acetic acid (9:1, v/v) was the solvent system used for separation and quantitation of the ochratoxins. The method was applicable to a wide variety of different commodities but substances interfering with ochratoxin A estimation were present in barley, brazil nuts, green coffee, cotton seed, and Capsicum pepper.

A method of analysis satisfactory for detection of ochratoxins, aflatoxins, and sterigmatocystin in corn and sorghum but not in peanuts involved extraction of ground samples with chloroform—methanol—hexane (8:2:1, v/v/v) [77]. If ochratoxin A could be detected on TLC plates sprayed with 20% KOH and exposed to UV light, indicating levels above 20 μg/kg original material, the concentrated extract was applied to a column containing silica gel in petroleum ether. After removal of sterigmatocystin and aflatoxins with the appropriate eluents, ochratoxin A was eluted with benzene—acetic acid (9:1, v/v). Fractions containing ochratoxin A were concentrated and levels estimated visually by TLC.

Rice cereal, oatmeal, and corn meal were examined for ochratoxin A by blending with hexane and extracting the hexane with sodium bicarbonate [72]. The residues were then blended with chloroform—methanol. The chloroform—methanol extract was concentrated and extracted with the bicarbonate solution used previously for the extraction of hexane. Following acidification of the bicarbonate solution, ochratoxin A was extracted with chloroform. Amounts of ochratoxin A on silica gel plates were estimated by spectrophotofluorometry. There was a linear relationship between concentration of ochratoxin A and fluorescence intensity in the range of 1—10 ng/spot. An 85% recovery of the toxin added to the cereal products was obtained with an overall precision of 3—4%. The lowest concentration that could be detected reliably was 10—50 μg/kg [72]. The excitation maximum on silica gel plates was at 340 nm for ochratoxin A and at 325 nm for ochratoxin B while the emission maximum for both toxins was at 475 nm. Selection of a suitable combination of primary and secondary filters for spectrophotofluorodensitometric measurements should be based on these characteristics [78]. Exposure of the TLC plates to ammonia fumes before densitometry increased sensitivity of the ochratoxin A analysis. Although ammoniated ochratoxin A was stable for at least 3 h, the fluorescence of ammoniated ochratoxin B and C rapidly decreased on storage of TLC plates in the dark [79].

A method for the simultaneous detection of ochratoxins, aflatoxins, zearalenone, sterigmatocystin, and patulin involved extraction of powdered samples with acetonitrile—KCl solution [80]. Lipids were removed by extraction with isooctane. The mycotoxins were transferred into chloroform and the residue dissolved in benzene—acetonitrile (98:2, v/v) after removal of the chloroform. Benzene—methanol—acetic acid (90:5:5, v/v/v) or hexane-acetone-acetic acid (90:10:5, v/v/v) were used as the mobile phase for TLC. Standard solutions of ochratoxin A were prepared in benzene—acetic acid (99:1, v/v) and their concentration was determined spectrophotometrically. For confirmation of identity of ochratoxins, plates were examined alternately under long- and short-wave UV light. The bright fluorescence of ochratoxin A and C was retained under both types of UV light while the faint fluorescence of ochratoxin B and its ethyl ester under long-wave UV light changed to bright blue under short-wave UV light. This multitoxin method was applied to corn, barley, oats, and wheat in which the lowest detectable levels of ochratoxins ranged from 45—100 μg/kg.

For the simultaneous analysis for ochratoxins and citrinin in mouldy grains, the above method was modified to include acidification of the acetonitrile—KCl used for extraction [16]. Toluene—ethyl acetate—90% formic acid (6:3:1, v/v/v) was the mobile phase used for TLC estimation of ochratoxin A. When citrinin interfered, plates were first exposed to ammonia fumes which deleted the UV light-induced yellow fluorescence of citrinin but changed the fluorescence of ochratoxin A to bright blue. Citrinin estimates were performed without delay after extraction and concentration with ether—methanol—water—90% formic acid (95:4:1:1, v/v/v/v) as mobile phase. Confirmation of the presence of ochratoxin A was obtained by the use of another solvent system, benzene—methanol—acetic acid (24:2:1, v/v/v), by solubility in sodium bicarbonate, and formation of the fluorescent methyl ester. The presence of citrinin was confirmed by the disappearance of its yellow fluorescence after exposure to ammonia fumes and appearance of green fluorescence after spraying with boron trifluoride—methanol.

Application of a modified "Best Foods" method allowed detection of ochratoxin A in mouldy grains [16]. In a recent abstract, a sensitive method was described for the detection of ochratoxin A in barley [81].

Penicillic acid and ochratoxin A in extracts of mouldy grains were successfully separated and quantitated on silica gel plates with benzene—acetic acid (9:1, v/v) as mobile phase [50]. Intensity of fluorescence of ammoniated ochratoxin A and penicillic acid was compared with that of standards by spectrophotofluorodensitometric methods. The ammoniated derivative of penicillic acid fluoresced with a linear response between 1 and 9 μg and those of ochratocin A, B, C and the ethyl ester of ochratoxin B between 10 and 50 ng. Identity of penicillic acid was confirmed by its spectral properties and the excitation spectrum of its derivative formed by reaction with phenylhydrazine in ammonia. Identity of ochratoxin A was confirmed by examination of excitation and emission spectra before and after exposure to ammonia fumes and by TLC behaviour of the methyl ester. Spraying TLC plates with aluminum chloride turned the green fluorescence of ochratoxin A under UV light to deep blue [49].

Solvent systems used for TLC of extracts containing ochratoxin A and B and the corresponding R_F values were tabulated by Steyn [20]. A TLC method involving plates coated with silica gel containing oxalic acid and chloroform—methylisobutylketone (4:1, v/v) as solvent system provided good separation of ochratoxin A and B and nine other toxins [82].

Analytical problems resulting from instability of ochratoxin A in extracts and standard solutions may occur rarely. Ethanolic solutions of ochratoxin A stored at refrigerator temperature did not substantially decrease in concentration after one year of storage [72]. Added to cereal products and stored at 4° and 28°, levels of the toxin did not change for at least one week and autoclaving samples for up to 3 h failed to completely destroy the toxin [67]. In aqueous solution buffered at pH 7.5, however, ochratoxin A was unstable in daylight [83].

Natural occurrence of ochratoxin A

Data on the natural occurrence of ochratoxin A are summarized in Table II. It was first detected in one sample of low-grade U.S. corn of 164 samples assayed. Culture plates of the positive sample showed a large number of Penicillium and some Fusarium colonies [14]. This survey of U.S. corn from commercial markets was extended to a total of 283 samples, but no further positive samples were detected [84]. In another survey of 293 samples of U.S. export cargo corn, the presence of ochratoxin A was confirmed in 3 samples of medium grade. 2 of the positive samples also contained traces of ochratoxin B. *A. ochraceus* was not observed in these samples [85]. A survey of 1971 U.S. barley indicated the presence of ochratoxin A at low levels in 18 of 127 samples examined [86]. There was no correlation between the amount of the toxins in samples and their apparent quality. Canadian workers reported ochratoxin A in mouldy wheat. This had been used as a component of feed for cattle that suffered mortality and was therefore analyzed for mycotoxins [15]. This work was extended to another 29 samples of "heated" grains from Saskatchewan, 18 of which were positive for ochratoxin A. 13 of these also contained citrinin. Levels of ochratoxin A encountered in these heated grains were very much higher than in the U.S. survey of grains from commercial markets. The highest levels encountered were 27 ppm ochratoxin A and 80 ppm citrinin. The presence of ochratoxin A was also confirmed in dried white beans destined for human consumption and in a sample of mouldy peanuts [16].

Levels comparable to those reported in Canadian heated grains were found in Danish barley and oats used as swine feed. These grains were examined for citrinin and ochratoxin A as swine that had consumed these feeds were, on slaughtering, recognized to suffer from mould nephrosis. Ochratoxin A was detected in 19 samples, 3 of which also contained citrinin. The highest concentrations of ochratoxin A and citrinin were, respectively, 27.5 and 2 ppm [17]. Ochratoxin A residues were detected in the kidneys of 18 of 19 swine from farms where ochratoxin A-containing feed had been used. The highest residue was 67 ppb. Most but not all of these kidneys showed fibrosis. Liver and adipose tissue showed no pathological changes but

TABLE II

NATURAL OCCURRENCE OF OCHRATOXIN A

Sample description	Number of positive samples	Levels	Location	Chemical confirmation	References
Commercial corn	1/283	110–150 ppb	U.S.A.	Bicarbonate solubility, methyl ester preparation	14, 84
Export cargo corn	3/293	83–166 ppb	U.S.A.	Same as above	85
Commercial barley	18/127	trace-38 ppb	U.S.A.	Methyl ester preparation	86
Samples of mouldy wheat from the same pile	4/4	20–100 ppb	Canada	Ammonia treatment, 2 solvent systems, bicarbonate solubility, methyl ester preparation	15
Heated[a] grains	18/29	0.03–27 ppm	Canada	Same	16
Mixed feeds	1/3	0.53 ppm	Canada	Same	16
Dried white beans	3/?	0.02–2.1 ppm	Canada	Same	16
Mouldy peanuts	1/1	4.9 ppm	Canada	Same	16
Barley and oats associated with mould nephrosis of swine	19/?	up to 27.5 ppm	Denmark	Derivative formation, NMR spectroscopy	17
Tissues of swine consuming above grains					18
Kidneys	18/19	up to 67 ppb	Denmark	Same	
Livers	7/8		Denmark	Same	
Adipose tissue	8/8		Denmark	Same	

a Elevated temperatures caused by rapid microbial growth in wet grains.

References p. 364

the toxin was detected in 7 of 8 livers and in all 8 samples of adipose tissue [18]. These findings indicate that human exposure to ochratoxin A may occur not only through consumption of mouldy plant materials but also through consumption of animal products.

Biological activity of ochratoxins

Lower organisms, tissue culture cells, and isolated mitochondria
    In the search for bioassay systems for confirmation of chemical analysis or for systems for the screening of large numbers of fungal cultures for toxicity, various lower organisms were tested for sensitivity to ochratoxin A. Table III indicates that zebra fish and brine shrimp larvae were sensitive to ochratoxin A. Most of the bacteria tested had little sensitivity. Growth of actinomycetes was inhibited at lower concentrations than that of bacteria [90]. The high sensitivity of a strain of *Bacillus cereus* var. *mycoides* to ochratoxins was used for confirming the presence of ochratoxin A and B and for quantitation of these toxins [91]. The test involved application of varying amounts of toxin to paper discs and placing these on agar seeded with *B. cereus* var. *mycoides*. After an 18-h incubation period at 35°, the diameter of the zone of inhibition was measured. With this procedure 1.5 μg ochratoxin A per disc and 3.0 μg ochratoxin B per disc could be detected. Partial

TABLE III

BIOLOGICAL ACTIVITY OF OCHRATOXIN A TO LOWER ORGANISMS AND TISSUE CULTURE CELLS

Organism	Activity	References
Zebra fish larvae	LC_{50}: 1.7 μg/ml (72-h exposure)	87
Brine shrimp larvae	LC_{50}: 10 μg/ml (16-h exposure)	88
Brine shrimp larvae	20% mortality at 2 μg/ml (24-h exposure)	89
Gram-positive and -negative bacteria	No inhibition at 500 μg/ml	90
Some strains of Proteus and and Bacillus	Minimum inhibitory dosage 250—500 μg/ml	90
Actinomyces spp.	Minimum inhibitory dosage 100—500 μg/ml	90
Bacillus cereus var. mycoides	Minimum inhibitory dosage 1.5 μg/disc	91
HeLa cells	Cytotoxicity at 1—32 μg/ml	33, 90
Monkey epithelial cells	Mitotic abnormalities at 0.1—3.2 μg/ml	92

purification of extracts of moulds and mouldy foods is probably required before application of this test.

Tissue culture cells showed cytotoxic effects at low levels of ochratoxin A. Abnormalities of mitotic processes were apparent at early stages of growth [33,92].

In isolated rat liver mitochondria, adenosine diphosphate-stimulated respiration was inhibited by 34% in $2.1 \cdot 10^{-7}$ M ochratoxin A and by 21% in $1.0 \cdot 1.0^{-7}$ M of ochratoxin α. This is the only system tested in which the dihydroisocoumarin was more toxic than the parent compound. Isolated mitochondria did not metabolize ochratoxin A to the heterocyclic moiety. Inhibition by the two compounds was released by 2,4-dinitrophenol, indicating that the site of inhibition along the chain of reactions leading to adenosine triphosphate formation was beyond the site of action of the uncoupling agent [93].

Ducklings

Although initial studies on ochratoxin A suggested that the oral LD_{50} was 25 μg/50 g duckling [2], subsequent, more definitive work established the LD_{50} at 150 μg per duckling [94]. Ochratoxin B and C were initially reported non-toxic [3], but a later publication reported LD_{50} values of 135—170 μg per duckling for ochratoxin C and the methyl ester of ochratoxin A. Ochratoxin B and its derivatives were non-toxic [4]. 30-day-old ducklings intubated with 100 μg of ochratoxin A developed liver changes characterized by swelling and structural damage of the mitochondria, morphological changes in the endoplasmic reticulum, and fatty infiltration of hepatocytes [5].

Rats

At comparable doses of ochratoxin A, liver lesions in rats were more severe than in ducklings [5]. Weanling male rats (about 60 g each) intubated with 100 μg ochratoxin A developed focal areas of liver cell necrosis and hyaline degeneration. Ultrastructural changes consisted of proliferation of the smooth endoplasmic reticulum with formation of masses of smooth membranes within the liver cell cytoplasm referred to as "fingerprints". These corresponded to the areas of hyaline degeneration observed by light microscopy. In later work, attention was focused on change in the kidney as well as in the liver of rats [6]. Ochratoxin A intubated at levels near the LD_{50} value of 20—22 mg/kg induced necrosis of the renal tubules and periportal cells of the liver. Glycogen accumulation was observed in the cytoplasm of liver cells.

In a feeding study lasting 90 days rats were given a semi-synthetic diet containing 0, 0.2, 1.0, or 5.0 ppm ochratoxin A. The principal pathological changes involved the distal convoluted tubules of the kidney and were observed at levels as low as 0.2 ppm [8].

Ochratoxin A caused foetal death and resorption in pregnant rats intubated with doses of 6.25—25.00 mg/kg. Ochratoxin α had no effect [25].

Ochratoxin A may not readily cross the placenta. Although injected into the blood stream of pregnant sheep at levels sufficiently high to cause maternal death within 24 h, only traces of the toxin appeared in the foetal blood [8].

The 4-hydroxy derivative of ochratoxin A had no effects in rats when injected intraperitoneally at 40 mg/kg [55].

Rainbow trout

Ochratoxin A injected intraperitoneally into 6-month-old rainbow trout had an LD_{50} of 4.67 mg/kg, as determined during a 10-day experimental period [7]. No deaths occurred with ochratoxin B, ochratoxin α, and the lactone moiety of ochratoxin B at the doses used, *i.e.* 66.7 mg/kg, 28 mg/kg, and 26.7 mg/kg, respectively. Pathological changes produced by ochratoxin A involved degenerative changes in the hepatic parenchymal cells, including nuclear swelling and cytoplasmic and nuclear lipid vacuolation, and severe kidney damage. The latter consisted of necrosis of tubules, glomeruli, and haemopoietic tissue. Although 66.7 mg/kg ochratoxin B did not cause mortality, this dosage induced pathological changes similar to those induced by much lower doses of ochratoxin A.

Chicks and hens

Day-old chicks resembled day-old ducklings in their sensitivity to ochratoxin A and C. The LD_{50} values for these two toxins were 166 and 216 μg per chick respectively, as calculated from first week mortalities following single dose, *per os* administration [95]. Gross effects in poisoned birds included diarrhoea, and haemorrhage in the liver, kidney, and proventriculus.

Other studies on day-old chicks given ochratoxin A orally gave LD_{50} values of 116—135 μg per chick. Ochratoxin B was much less toxic than ochratoxin A as no mortalities occurred in chicks given single doses of 800 μg. The LD_{50} for this compound was estimated at 1890 μg per chick [96]. An oral dose of 1000 μg killed 2 of 10 day-old chicks [97]. Sublethal doses of ochratoxin A produced dose-related growth depression. Birds dying after administration of the toxins had visceral gout and acute nephrosis; hepatic lesions occurred less often than renal lesions and included mild, diffuse vacuolation of hepatocytes, and foci of necrosis. Surviving and dead birds showed suppression of haemopoiesis in the bone marrow and depletion of lymphoid elements from the spleen and bursa of Fabricius. The birds also suffered enteritis [96].

Diets of young laying hens, which were mixed with sterile wheat cultures of *A. sulphureus* so that they contained 1, 2, or 4 ppm of ochratoxin A, severely affected performance of the birds [98]. Laying hens fed diets at 2 or 4 ppm levels for 6 months showed 18 and 64% mortality respectively. All birds receiving feed containing the toxin exhibited morbidity, evidence of stress, and depression of body weight. Sexual maturity was delayed and egg production and feed efficiency were reduced. Fertile eggs laid by hens fed the ochratoxin A-containing diets had lower hatchability than eggs from control birds and growth rate of chicks from such eggs was lower in the initial two weeks of life but not later. The effects were more marked at the 4 ppm level than at the 2 and 1 ppm levels.

Growth depression was also noted in male broiler chicks fed diets containing ochratoxin A at 0.5, 1, 2, 4, or 8 ppm [99]. In these birds there was high mortality in the groups receiving 4 and 8 ppm, a dose-dependent increase in relative weight of the crop, pancreas, liver, and gizzard, a significant regression of the bursa of Fabricius, and an increase in the glycogen content of the liver.

The lower toxicity of ochratoxin α was confirmed by the chick embryo test. Whereas the LD_{50} for ochratoxin A was 16.96 μg per egg, as determined 48 h after injection, 100 μg per egg of the dihydroisocoumarin killed only 3 of 10 embryos in 96 h [100]. Gedek reported an LD_{50} of 10 μg of ochratoxin A per egg [101].

Dogs
Ochratoxin A was strongly nephrotoxic to dogs [102].

Carcinogenic potential of ochratoxin A

Compared to control rats, no increased tumour incidence was noted in rats given low levels of ochratoxin A orally or subcutaneously [103]. Tumours were not observed in mice given twice weekly subcutaneous injections of 10 μg of ochratoxin A for 36 weeks and allowed to live up to 81 weeks of age [104]. In rainbow trout, ochratoxin A alone did not produce tumours in 8 months but produced hepatomas when fed at 20 ppb with the cocarcinogen sterculic acid [105].

Mode of action
An understanding of the mode of action of ochratoxin A requires a knowledge of its metabolic fate in biological systems. Injected into rats at a single dose of 10 mg/kg, ochratoxin A was metabolized to ochratoxin α and a green fluorescent compound assumed to be 4-hydroxy-ochratoxin A from its R_F value on thin-layer plates [55,106]. These two compounds and ochratoxin A appeared in the urine and faeces of the rats. Ochratoxin A was hydrolysed *in vitro* to ochratoxin α and phenylalanine by the proteolytic enzymes carboxypeptidase A and α-chymotrypsin from bovine pancreas, ochratoxin A having a much greater affinity for carboxypeptidase A than for α-chymotrypsin [83,100,107].

The hydrolysis of N-carbobenzoxyglycyl-L-phenylalanine, a substrate used for measuring carboxypeptidase A activity, was inhibited by ochratoxin A but not by ochratoxin α. Ochratoxin A is a competitive inhibitor of carboxypeptidase A. The toxin had a considerably higher affinity for the enzyme than the substrate N-carbobenzoxyglycyl-L-phenylalanine [108].

The observed accumulation of glycogen in rat liver could have resulted from an interaction of ochratoxin A with nucleic acids or with the enzymes of carbohydrate metabolism [109]. No evidence was found for interaction with nucleic acids but, in *in vitro* studies, the toxin exerted an inhibitory effect on the phosphorylase system, an enzyme complex catalyzing the breakdown of glycogen to phosphorylated glucose.

In vivo hydrolysis of ochratoxin A may occur by proteolytic enzymes in the liver [83]. Yamazaki *et al.* [100] could not detect hydrolytic activity in rat liver homogenates or microsomal fractions but Doster and Sinnhuber [110] obtained rat tissue extracts of liver and small and large intestine that were active in hydrolyzing ochratoxin A and B. As the *in vitro* rates of hydrolysis of ochratoxin B by carboxypeptidase A and tissue extracts were much greater than those of ochratoxin A, it was suggested that the difference in toxicity of ochratoxin A and B is partly due to the different rates of hydrolysis to the corresponding non-toxic dihydroisocoumarins.

Chu *et al.* [97] provided evidence that the chlorine atom may play an indirect role in determining toxicity of the ochratoxins by affecting the dissociation of the phenolic hydroxyl group of the dihydroisocoumarins. Modification of the phenolic hydroxyl group by preparation of *O*-methylated ochratoxin C led to loss of toxicity. This indicated the importance of the hydroxyl group in the toxicity of the ochratoxins. From shifts in the UV absorption maxima of alkaline solutions of ochratoxins and their absence on addition of acid, it was concluded that these maxima resulted from the ionization of the phenolic hydroxyl group [73]. Dissociation of this group may be necessary for toxicity as there was a correlation between the dissociation constants of the phenolic hydroxyl group of the various derivatives and their toxicity. The acid dissociation constants of the hydroxyl groups of ochratoxin A, C, B, and ochratoxin α are respectively 7.05, 7.10, 7.14, 7.95, and 11.0; ochratoxin A and C with a pK near neutral pH are the most toxic and ochratoxin B and α have little or no toxicity [97]. Binding of ochratoxin A with bovine serum albumin but not with γ-globulin was demonstrated by Sephadex gel filtration, equilibrium dialysis, and spectrophotometric and spectrofluorometric analysis [111]. This suggests that the toxic effects of the ochratoxins may result from *in vivo* interaction of the phenolic hydroxyl group with proteins and enzymes [97,111].

References

1 De B. Scott, Mycopathol. Mycol. Appl., 25 (1965) 213.
2 K.J. van der Merwe, P.S. Steyn, L. Fourie, De B. Scott and J.J. Theron, Nature, 205 (1965) 1112.
3 K.J. van der Merwe, P.S. Steyn and L. Fourie, J. Chem. Soc., (1965) 7083.
4 P.S. Steyn and C.W. Holzapfel, J. South Afr. Chem. Inst., 20 (1967) 186.
5 J.J. Theron, K.J. van der Merwe, N. Liebenberg, H.J.B. Joubert and W. Nel, J. Pathol. Bacteriol., 91 (1966) 521.
6 I.F.H. Purchase and J.J. Theron, Food Cosmet. Toxicol., 6 (1968) 479.
7 R.C. Doster, R.O. Sinnhuber and J.H. Wales, Food Cosmet. Toxicol., 10 (1972) 85.
8 I.C. Munro, P.M. Scott, C.A. Moodie and R.F. Willes, J. Am. Vet. Med. Assoc., 163 (1973) 1269.
9 M. Lai, G. Semeniuk and C.W. Hesseltine, Phytopathology, 58 (1968) 1056 (Abstr.).

10 M. Lai, G. Semeniuk and C.W. Hesseltine, Appl. Microbiol., 19 (1970) 542.
11 C.W. Hesseltine, E.E. Vandegraft, D.I. Fennell, M.L. Smith and O.L. Shotwell, Mycologia, 64 (1972) 539.
12 W. van Walbeek, P.M. Scott and F.S. Thatcher, Can. J. Microbiol., 14 (1968) 131.
13 W. van Walbeek, P.M. Scott, J. Harwig and J.W. Lawrence, Can. J. Microbiol., 15 (1969) 1281.
14 O.L. Shotwell, C.W. Hesseltine and M.L. Goulden, Appl. Microbiol., 17 (1969) 765.
15 P.M. Scott, W. van Walbeek, J. Harwig and D.I. Fennell, Can. J. Plant Sci., 50 (1970) 583.
16 P.M. Scott, W. van Walbeek, B. Kennedy and D. Anyeti, J. Agr. Food Chem., 20 (1972) 1103.
17 P. Krogh, Symp. Intern. Union of Pure and Applied Chemistry, Aug. 21—22, 1972 Kungälv, Sweden (Abstr.).
18 B. Hald and P. Krogh, Symp. Intern. Union of Pure and Applied Chemistry, Aug. 21—22, 1972, Kungälv, Sweden (Abstr.).
19 P. Krogh and E. Hasselager, Royal Veterinary and Agricultural College Yearbook 1968, pp. 198—214.
20 P.S. Steyn, Ochratoxin and other dihydroisocoumarins, in A. Ciegler, S. Kadis, and S.J. Ajl (Eds.), Microbial Toxins, Vol. VI, Academic Press, New York, 1971, pp. 179—205.
21 P. Friis, E. Hasselager and P. Krogh, Acta Pathol. Microbiol. Scand., 77 (1969) 559.
22 P. Krogh, E. Hasselager and P. Friis, Acta Pathol. Microbiol. Scand. Section B, 78 (1970) 401.
23 W. van Walbeek, C.A. Moodie, P.M. Scott, J. Harwig and H.C. Grice, Toxicol. Appl. Pharmacol., 20 (1971) 439.
24 H.G. Buckley, Irish Vet. J., 25 (1971) 194.
25 P.E. Still, A.W. Macklin, W.E. Ribelin and E.B. Smalley, Nature, 234 (1971) 563.
26 C.M. Christensen, Critical Reviews in Environmental Control, 2 (1971) 57.
27 C.M. Christensen, H.A. Fanse, G.H. Nelson, F. Bates and C.J. Mirocha, Appl. Microbiol., 15 (1967) 622.
28 C.M. Christensen, G.H. Nelson, C.J. Mirocha and F. Bates, Cancer Res., 28 (1968) 2293.
29 B. Doupnik Jr. and J.C. Peckham, Appl. Microbiol., 19 (1970) 594.
30 B. Doupnik Jr. and D.K. Bell, Appl. Microbiol., 21 (1971) 1104.
31 M. Yamazaki, Y. Maebayashi and K. Miyaki, Appl. Microbiol., 20 (1970) 452.
32 S. Nesheim, J. Assoc. Offic. Anal. Chem., 50 (1967) 370.
33 S. Natori, F. Sasaki, H. Kurata, S. Udagawa, M. Ichinoe, M. Saito and M. Umeda, Chem. Pharm. Bull., 18 (1970) 2259.
34 K.B. Raper and C. Thom, A Manual of the Penicillia, Hafner, New York, 1968, pp. 489—490.
35 B. Koehler, J. Agr. Res., 56 (1938) 291.
36 G. Semeniuk and H.J. Barr, Ann. Rept. Iowa Corn Res. Inst., 8 (1943) 55.
37 P.B. Mislevic and J. Tuite, Mycologia, 62 (1970) 67.
38 P.B. Mislevic and J. Tuite, Mycologia, 62 (1970) 75.
39 J.B. Marchionatto, Rev. Fac. Agron. Vet. Buenos Aires, 9 (1942) 159.
40 A.Z. Joffe, Bull. Res. Counc. Israel, 9D (1960) 101.
41 W.W. Carlton, J. Tuite and P. Mislevic, Toxicol. Appl. Pharmacol., 13 (1968) 372.
42 W.W. Carlton and J. Tuite, Pathol. Vet., 7 (1970) 68.
43 W.W. Carlton and J. Tuite, Toxicol. Appl. Pharmacol., 16 (1970) 345.
44 I.T. Budiarso, W.W. Carlton and J. Tuite, Toxicol. Appl. Pharmacol., 20 (1971) 357.
45 I.T. Budiarso, W.W. Carlton and J. Tuite, Toxicol. Appl. Pharmacol., 20 (1971) 194.
46 I.T. Budiarso, W.W. Carlton and J. Tuite, Path. Vet., 7 (1970) 531.
47 W.W. Carlton, J. Tuite and R.W. Caldwell, Toxicol. Appl. Pharmacol., 21 (1972) 130.
48 G.M. Zwicker, W.W. Carlton and J. Tuite, Lab. Invest., 26 (1972) 497 (Abstr.).

49 A. Ciegler, D.J. Fennell, H.-J. Mintzlaff and L. Leistner, Naturwissenschaften, 59 (1972) 365.
50 A. Ciegler, Can J. Microbiol., 18 (1972) 631.
51 R.J. Cole, J.H. Moore, N.D. Davis, J.W. Kirksey and U.L. Diener, Agr. Food Chem., 19 (1971) 909.
52 P.S. Steyn and C.W. Holzapfel, Tetrahedron, 23 (1967) 4449.
53 J.C. Roberts and P. Woollven, J. Chem. Soc., Section C (1970) 278.
54 K. Miyaki, Y. Maebayashi and M. Yamazaki, Ann. Rept. Inst. Food Microbiol., Chiba Univ., 23 (1970) 41.
55 R.D. Hutchison, P.S. Steun and D.L. Thompson, Tetrahedron Letters, 43 (1971) 4033.
56 P.M. Scott, B. Kennedy and W. van Walbeek, J. Assoc. Offic. Anal. Chem., 54 (1971) 1445.
57 P.S. Steyn, C.W. Holzapfel and N.P. Ferreira, Phytochemistry, 9 (1970) 1977.
58 N.P. Ferreira and M.J. Pitout, J. South Afr. Chem. Inst., 22 (1969) S1.
59 J.W. Searcy, N.D. Davis and U.L. Diener, Appl. Microbiol., 18 (1969) 622.
60 M. Yamazaki, Y. Maebayashi and K. Miyaki, Tetrahedron Letters, 25 (1971) 2301.
61 R.D. Wei, F.M. Strong and E.B. Smalley, Appl. Microbiol., 22 (1971) 276.
62 N.P. Ferreira, Recent advances in research on ochratoxin, Part 2. Microbiological aspects, in R.I. Mateles and G.N. Wogan (Eds.), Biochemistry of Some Foodborne Microbial Toxins, M.I.T., Cambridge, Mass., 1967, pp. 157—168.
63 N.P. Ferreira, Antonie van Leeuwenhoek J. Microbiol. Serol., 34 (1968) 433.
64 N.D.Davis, J.W. Searcy and U.L. Diener, Appl. Microbiol., 17 (1969) 742.
65 N.D. Davis, G.A. Sansing, T.V. Ellenburg and U.L. Diener, Appl. Microbiol., 23 (1972) 433.
66 P.M. Scott, J.W. Lawrence and W. van Walbeek, Appl. Microbiol., 20 (1970) 839.
67 H.L. Trenk, M.E. Butz and F.S. Chu, Appl. Microbiol., 21 (1971) 1032.
68 A.F. Schindler and S. Nesheim, J. Assoc. Offic. Anal. Chem., 53 (1970) 89.
69 C.W. Hesseltine, Biotechnol. Bioeng., 14 (1972) 517.
70 L.C. Lopez and C.M. Christensen, Phytopathology, 57 (1967) 588.
71 C.M. Christensen, Cereal Chem., 39 (1962) 100.
72 F.S. Chu and M.E. Butz, J. Assoc. Offic. Anal. Chem., 53 (1970) 1253.
73 S. Nesheim, J. Assoc. Offic. Anal. Chem., 52 (1969) 975.
74 P.S. Steyn and K.J. van der Merwe, Nature, 211 (1966) 418.
75 P.M. Scott and T.B. Hand, J. Assoc. Offic. Anal. Chem., 50 (1967) 366.
76 R.M. Eppley, J. Assoc. Offic. Anal. Chem., 51 (1968) 74.
77 L.J. Vorster, Analyst, 94 (1969) 136.
78 F.S. Chu, J. Assoc. Offic. Anal. Chem., 53 (1970) 696.
79 H.L. Trenk and F.S. Chu, J. Assoc. Offic. Anal. Chem., 54 (1971) 1307.
80 L. Stoloff, S. Nesheim, L. Yin, J.V. Rodricks, M. Stack and A.D. Campbell, J. Assoc. Offic. Anal. Chem., 54 (1971) 91.
81 S. Nesheim, N. Hardin, O. Francis and W. Langham, 85th Annual Meeting, Assoc. Offic. Anal. Chem., Oct. 11—14, 1971 (Abstr.).
82 P.S. Steyn, J. Chromatog., 45 (1969) 473.
83 M.J. Pitout, Biochem. Pharmacol., 18 (1969) 485.
84 O.L. Shotwell, C.W. Hesseltine, M.L. Goulden and E.E. Vandegraft, Cereal Chem., 47 (1970) 700.
85 O.L. Shotwell, C.W. Hesseltine, E.E. Vandegraft and M.L. Goulden, Cereal Science Today, 16 (1971) 266.
86 S. Nesheim, 85th Annual Meeting, Assoc. Offic. Anal. Chem., Washington, D.C., Oct. 11—14, 1971 (Abstr.).
87 Z.H. Abedi and P.M. Scott, J. Assoc. Offic. Anal. Chem., 52 (1969) 963.
88 J. Harwig and P.M. Scott, Appl. Microbiol., 21 (1971) 1011.
89 R.F. Brown, J. Am. Oil Chem. Soc., 46 (1969) 119.
90 T. Arai and M. Otomo, Ann. Rept. Inst. Food Microbiol., Chiba Univ., 22 (1969) 81.

91 D. Broce, R.M. Grodner, R.L. Killebrew and F.L. Bonner, J. Assoc. Offic. Anal. Chem., 53 (1970) 616.
92 J.C. Engebrecht and I.F.H. Purchase, South Afr. Med. J., 43 (1969) 524.
93 J.H. Moore and B. Truelove, Science, 168 (1970) 1102.
94 I.F.H. Purchase and W. Nel, Recent advances in research on ochratoxin, Part 1, Toxicological Aspects, in R.I. Mateles and G.N. Wogan (Eds.), Biochemistry of Some Foodborne Microbial Toxins, M.I.T., Cambridge, Mass., 1967, pp. 153—156.
95 F.S. Chu and C.C. Chang, J. Assoc. Offic. Anal. Chem., 54 (1971) 1032.
96 J.C. Peckham, B. Doupnik Jr. and O.H. Jones Jr., Appl. Microbiol., 21 (1971) 492.
97 F.S. Chu, I. Noh and C.C. Chang, Life Sci., 11 (1972) 503.
98 H. Choudhuri, C.W. Carlson and G. Semeniuk, Poultry Sci., 50 (1971) 1855.
99 T.L. Tucker and P.B. Hamilton, Poultry Sci., 50 (1971) 1637 (Abstr.).
100 M. Yamazaki, S. Suzuki, Y. Sakakibara and K. Miyaki, Jap. J. Med. Sci. Biol., 24 (1971) 245.
101 B. Gedek, Zentralbl. Vet. Med. B., 19 (1972) 15.
102 G.M. Szczech, W.W. Carlton and J. Tuite, Lab. Invest., 26 (1972) 492 (Abstr.).
103 I.F.H. Purchase and J.J. van der Watt, Food Cosmet. Toxicol., 9 (1971) 681.
104 F. Dickens and H.B. Waynforth, Survey of compounds which have been tested for carcinogenic activity, 1968—1969 Volume, U.S. Department of Health, Education, and Welfare, p. 630.
105 R.C. Doster, R.O. Sinnhuber, J.H. Wales and D.J. Lee, Federation Proc., 30 (1971) 578 (Abstr.).
106 W. Nel and I.F.H. Purchase, J. South Afr. Chem. Inst., 21 (1968) 87.
107 M.J. Pitout, Biochem. Pharmacol., 18 (1969) 1829.
108 M.J. Pitout and W. Nel, Biochem. Pharmacol., 18 (1969) 1837.
109 M.J. Pitout, Toxicol. Appl. Pharmacol., 13 (1968) 299.
110 R.C. Doster and R.O. Sinnhuber, Food Cosmet. Toxicol., 10 (1972) 389.
111 F.S. Chu, Arch. Biochem. Biophys., 147 (1971) 359.

Chapter 17

STERIGMATOCYSTIN

J.J. VAN DER WATT

Division of Toxicology, National Research Institute for Nutritional Diseases, South African Medical Research Council, P.O. Box 70, Tiervlei, C.P. 7503 (Republic of South Africa)

Studies on the metabolic products of *Aspergillus versicolor* (Vuill.) Tiraboschi initially resulted in the isolation of sterigmatocystin and subsequently the closely related compounds 5-methoxysterigmatocystin, demethylsterigmatocystin, dihydrosterigmatocystin and dihydrodemethylsterigmatocystin.

These investigations did not result from a study of the aetiology of a health problem, as was the case with the aflatoxins, but formed part of a systematic survey of fungal products.

Sterigmatocystin was the first known naturally occurring compound to contain the dihydrofurobenzofuran system and after elucidation of its structure a large number of similar compounds have now been isolated from moulds or moulded material. The distinctive feature of this series is that all these compounds contain either the unusual 7,8-dihydrofurano (2,3-b) furan (DHFF) or the more fully reduced 2, 3, 7, 8-tetrahydrofuro (2, 3-b) furan (THFF) [1]. The group of furofuran mould metabolites can be divided into 3 subgroups [1], *viz.*

(*i*) the aflatoxin group in which a substituted coumarin is fused to the 4,5-ring positions of DHFF or THFF;

(*ii*) the sterigmatocystin group in which a substituted xanthone is fused to the 4,5-ring positions of DHFF, and

(*iii*) the mould pigments in which a substituted anthraquinone is fused to the 4,5-ring positions of DHFF or THFF.

From this subdivision of the furofuran compounds Rodricks [1] postulated that, due to the extreme toxicity and carcinogenicity of aflatoxin B_1, sterigmatocystin and perhaps other compounds of this group may be important naturally occurring toxic agents. This assumption was shown to be correct in the case of sterigmatocystin, demethylsterigmatocystin and some other mould metabolites or analogues in subgroup (*ii*) [2,3]. Sterigmatocystin was found to be carcinogenic to rats on subcutaneous injection [2], and on oral administration [4,5], and both acutely and chronically toxic to subhuman primates [6] while its toxicity to zebra fish larvae has been reported to be equal to that of aflatoxin B_1 [7].

The above information which has been accumulated over the 23 years since sterigmatocystin was first isolated [8], now strongly suggests that the DHFF compounds, in which a substituted xanthone is fused to the 4,5-ring positions, may be a potential hazard to the health of man and domestic animals. Evidence of food contamination by these compounds is gradually accumulating [9,10].

Toxic metabolites

The major toxic metabolite of *Aspergillus versicolor*, *A. nidulans* and an undescribed species of Bipolaris (I.M.I. 115076) is sterigmatocystin while 5-methoxysterigmatocystin and demethylsterigmatocystin have also been isolated. Sterigmatocystin was first isolated from *Aspergillus versicolor* (Vuillemin) Tiraboschi in 1954 [8] but the structural proof was only completed some years later [11]. Subsequent intensive investigations resulted in the isolation and characterisation of 5-methoxysterigmatocystin [12], demethylsterigmatocystin [13], dihydrosterigmatocystin [14] and dihydromethylsterigmatocystin [14]. In addition two other sterigmatocystin derivatives, *O*-methylsterigmatocystin [15] and aspertoxin [16] have been isolated from aflatoxin producing cultures of *A. flavus*, while a third, dihydro-*O*-methylsterigmatocystin was isolated from a non-aflatoxin producing strain of *A. flavus* [17].

The isolation of these latter three sterigmatocystin-related compounds from *A. flavus* cultures provides strong support for the hypothesis that sterigmatocystin and the aflatoxins share a biogenetic pathway or that aflatoxin B_1 may be derived from a DHFF-xanthone related to sterigmatocystin [1,18].

Chemical and physical properties of the toxic metabolites

Two different numbering systems exist for the chemical structure of sterigmatocystin (I). The original numbers proposed by Roberts and co-workers and depicted in formula A (Fig. 1) continue to be used by Rodricks *et al.* [27] and Elsworthy and his colleagues [13], but have been changed by others to that illustrated in formula B [28,29].

In Table I the most important physical properties of the eight naturally occurring toxic metabolites of the DHFF group are listed. It has been suggested that nuclear magnetic resonance spectroscopy provides the best physical evidence for the presence of the dihydro furofurano-structure [1]. Sterigmatocystin is a chemically labile compound undergoing numerous reactions. Hot ethanolic potassium hydroxide converts sterigmatocystin (I) into an optically inactive compound, isosterigmatocystin (X) [11,30] (Fig. 2). Dihydrosterigmatocystin (VI) remains unchanged under the above reaction conditions [11]. Sterigmatocystin is methylated by both methylsulphate and methyliodide but remains unchanged after treatment with diazomethane in ether methanol solution [30]. Diazomethane does, however, methylate the 3-phenolic group of isosterigmatocystin [11]. Catalytic hydrogenation of sterigmatocystin in glacial acetic acid over palladised charcoal gives dihydrosterigmatocystin (VI) [30]. Oxidation of the latter compound by the Ellis persulphate reaction gives the 5-hydroxy-derivative from which the

TABLE I

PHYSICAL PROPERTIES OF THE NATURALLY OCCURRING TOXIC METABOLITES

	Structure	mp. °C	(α) 20 A	Formula	Molecular weight	Source	References
Sterigmatocystin	I	248	-398° (CHCl$_3$)	$C_{18}H_{12}O_6$	344	A. versicolor A. nidulans A. regulosus Bipolaris	31, 36, 37
5-Methoxysterigmatocystin	II	223	-360° (CHCl$_3$)	$C_{19}H_{14}O_7$	354	A. versicolor	12, 38
Demethylsterigmatocystin	III	255	-483° (CHCl$_3$)	$C_{18}H_{12}O_7$	330	A. versicolor	13
Dihydrosterigmatocystin	VI	230	-311.7° (CHCl$_3$)	$C_{18}H_{14}O_6$	326	A. versicolor	14
Dihydrodemethylsterigmatocystin	VII	202	-376.6° (CHCl$_3$)	$C_{17}H_{12}O_6$	?12	A. versicolor	14
Dihydro-O-methylsterigmatocystin	VIII	283	-243° (CHCl$_3$)	$C_{19}H_{16}O_6$	340	A. flavus	17, 35
O-Methylsterigmatocystin	IV	265	—	$C_{19}H_{14}O_6$	338	A. flavus	15, 34
Aspertoxin	V	327	-140° (DMF)	$C_{19}H_{14}O_7$	354	A. flavus	16, 28, 29

References p. 381

Fig. 1 (structure A)

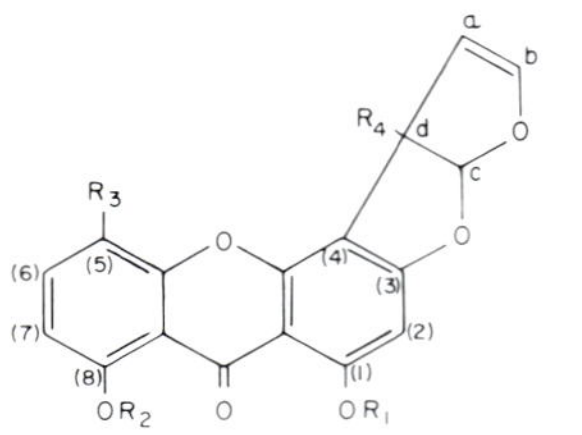

		R_1	R_2	R_3	R_4
I	STERIGMATOCYSTIN	CH_3	H	H	H
II	5-METHOXYSTERIGMATOCYSTIN	CH_3	H	CH_3O	H
III	DEMETHYLSTERIGMATOCYSTIN	H	H	H	H
IV	O-METHYLSTERIGMATOCYSTIN	CH_3	CH_3	H	H
V	ASPERTOXIN	CH_3	CH_3	H	OH

Fig. 1 (structure B)

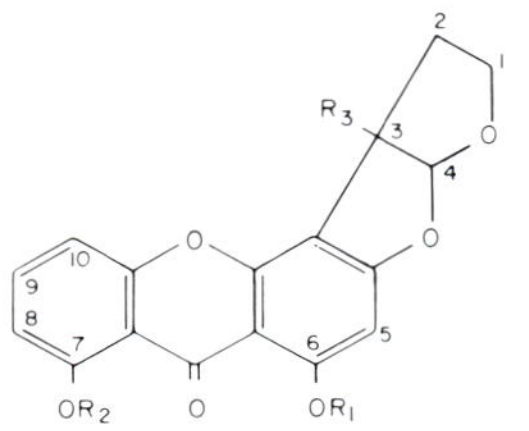

		R_1	R_2	R_3
VI	DIHYDROSTERIGMATOCYSTIN	CH_3	H	H
VII	DIHYDRODEMETHYLSTERIGMATOCYSTIN	H	H	H
VIII	DIHYDRO-O-METHYLSTERIGMATOCYSTIN	CH_3	CH_3	H
IX	DIHYDROASPERTOXIN	CH_3	CH_3	OH

Fig. 1. Chemical structures of the toxic metabolites.

5-O-methyl ether is prepared by monomethylation with dimethylsulphate and potassium carbonate in acetone. The dihydro-5-methoxysterigmatocystin produced in this manner is identical with the product prepared by hydrogenation of the naturally occurring 5-methoxysterigmatocystin (II) [12].

Oxidation of sterigmatocystin with potassium permanganate in refluxing acetone gives γ-resorcyclic acid. 3,8-Dihydroxy-1-methoxyxanthone-4-carboxylic acid is obtained when a smaller proportion of oxidant is used [30].

Methanol and ethanol in acid solution add to the vinyl ether system of sterigmatocystin to give dihydromethoxysterigmatocystin (XI) and dihydroethoxysterigmatocystin (XII), respectively [11]. No stereochemical assignment of the components has yet been made. Isomeric α-β-acetoxydihydrosterigmatocystin (XIII, XIV) prepared by refluxing a glacial acetic acid solution of sterigmatocystin were separated on preparative TLC. The stereochemical assignment here was based on the chemical shift of the acetoxy group [31].

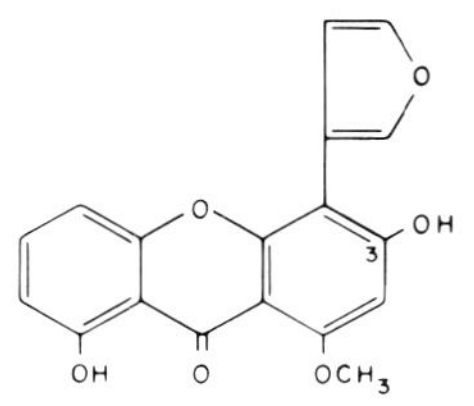

X   ISOSTERIGMATOCYSTIN

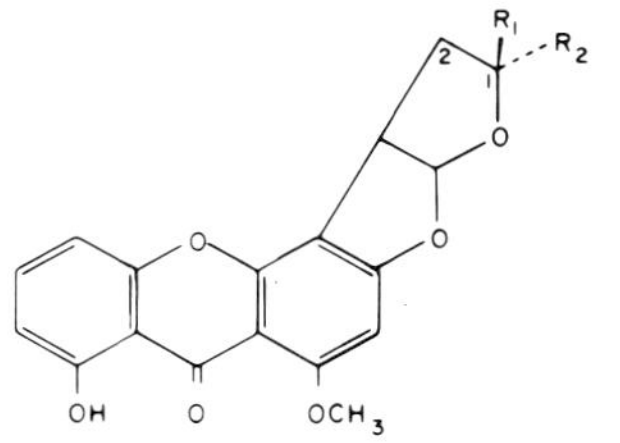

		R_1	R_2
XI	DIHYDROMETHOXYSTERIGMATOCYSTIN	OCH_3	H
XII	DIHYDROETHOXYSTERIGMATOCYSTIN	OC_2H_5	H
XIII	α - ACETOXYDIHYDROSTERIGMATOCYSTIN	H	$COCH_3$
XIV	β - ACETOXYDIHYDROSTERIGMATOCYSTIN	$COCH_3$	H

Fig. 2. Chemical structures of the derivatives of sterigmatocystin.

Acetylation of sterigmatocystin under mild conditions with acetic anhydride in pyridine gives O-acetylsterigmatocystin while acetylation under more rigorous conditions gives O-acetyldihydro-acetoxysterigmatocystin. Aluminium chloride in chlorobenzene converts sterigmatocystin to 1,3,8-trihydroxyxanthone and bromination gives a compound with the formula $C_{18}H_{11}O_7Br_3$ [30,32].

Both (±) O-methylsterigmatocystin (IV) [33] and (±) dihydro-O-methyl-sterigmatocystin (VIII) [34] have been synthesized by different routes. The conversion of sterigmatocystin into dihydroaspertoxin (IX) has been reported [29] and the absolute configuration of sterigmatocystin has been shown to be the same as that of aflatoxin B_1 [13].

Biosynthesis

As early as 1964 it was suggested that aflatoxins may be derived from sterigmatocystin or that they share a common biogenic precursor [18]. The proposed pathway by which aflatoxin was produced from sterigmatocystin consisted of cleavage of the phenol-containing ring and the lactone-containing ring in the xanthone moiety of the molecule followed by aldol condensation, dehydration and ring closure [18].

Following on the above hypothesis, Thomas [19] proposed a pathway by which sterigmatocystin could be derived from averufin, an anthraquinone pigment which is also a metabolite of *A. versicolor*, with the subsequent conversion of the sterigmatocystin thus obtained into aflatoxin. However, these hypotheses were not substantiated by the results of feeding experiments in which ^{14}C-labelled sterigmatocystin, derived from [1-^{14}C]acetate feedings of *A. versicolor* was not incorporated into aflatoxin B[1] by *A. flavus* [18]. *A. versicolor*, when fed [1-^{14}C]acetate, produces labelled sterigmatocystin in which the radioactive distribution supports the acetate malonate hypothesis for the origin of the xanthone ring system [20]. The distribution of radioactivity was similar to that observed in aflatoxins but the carbon atoms of the furano-system were significantly less active than those of the xanthone system indicating a biosynthesis from the acetate-derived chains rather than from a single chain [21]. The four-carbon unit of the bisfuran system appears to be derived by head-to-tail linkage of two acetate units; the C—C bond joining this moiety to the xanthone ring is apparently formed between two carbon atoms, both of methyl group origin from acetate. Biogenetically sterigmatocystin is thus apparently derived from two separate preformed ketide units.

Aspertoxin, on the other hand, is almost certainly derived from *O*-methyl-sterigmatocystin rather than *vice versa* because the additional hydroxy group present in the bisfuran portion of the metabolite is attached to an acetate group. In cultures of *A. parasiticus* dihydro-5-hydroxysterigmatocystin specifically labelled with ^{14}C in the *O*-methyl group is a biogenetic precursor of both aflatoxins B[2] and G[2] [13].

The most recent results on biosynthetic studies indicate that ^{14}C-sterigmatocystin isolated from cultures of *A. versicolor* supplemented with [1-^{14}C]acetate was efficiently converted to aflatoxin B[1] by the resting mycelium of *A. parasiticus*. The results obtained in this study tend to indicate a biosynthetic pathway leading from 5-hydroxysterigmatocystin to sterigmatocystin and then to aflatoxin B[1] [61].

Environmental and laboratory conditions favouring production of the toxic metabolites

Due to the sparsity of information on the natural occurrence of sterigmatocystin and related toxic DHFF compounds, no information is currently available on the environmental conditions favouring the production of these metabolites in products destined for human or animal consumption.

In respect of laboratory conditions favouring the production of sterigmatocystin, Rabie *et al.* [22] reported that yields of up to 12 g of sterigmatocystin per kg of maize were obtained with *A. versicolor* after cultivation for 21 days at 27° on a solid medium consisting of whole maize supplemented with Soytone. For optimal production of sterigmatocystin various nitrogen sources were tested in liquid media and it was found that peptone was far superior to either nitrate or ammonia nitrogen. Similarly the best carbon source was shown to be glucose supplemented with either galactose or lactose. Corn steep liquor drastically reduced yields in both liquid and solid media while extracts of soya beans (Soytone) lowered the yields in liquid media but on addition to whole maize, appreciably higher yields were obtained. Optimum results were obtained when 20 g Soytone was added to 1 kg of maize.

The size of the culture vessels used in the above experiments was shown to have a very marked influence on the ability of the mould under test to produce sterigmatocystin. The highest production of toxin was obtained in 250-ml flasks, while yields decreased with an increase in the size of the culture vessels. The sterigmatocystin producing moulds used in these investigations consisted of various isolates of *A. versicolor*, *A. nidulans*, *A. ustus*, a new sterigmatocystin-producing strain, and *Bipolaris sorokiniana*. An isolate of *A. versicolor* gave the highest yields of toxin both in liquid and on solid media.

Analysis for toxic metabolites

The currently most important known members of the DHFF group, are sterigmatocystin and demethylsterigmatocystin, two carcinogens. No standard, routine method has yet been described for the detection of the latter compound in food products while four methods have been reported for the analysis and chemical confirmation of sterigmatocystin [23—26].

In the first method [23] the fluorescence intensity of a standard spot of sterigmatocystin on a TLC plate is compared with that of a spot of the sterigmatocystin-containing sample. This method is relatively insensitive as the smallest amount of sterigmatocystin detectable by fluorescence on thin-layer chromatograms is ± 500 ng; even at much higher levels detection is difficult due to the dull brick-red colour of the fluorescence [26]. The second method [24] involves extraction of the sample with chloroform—methanol (87:13) for 5 h in a Soxhlet apparatus or 1 min in a high speed blender. After considerable cleanup involving liquid-liquid partition, the sterigmatocystin is converted to its monoacetate by treatment with acetic anhydride-pyridine at 110°. The acetylated derivative is highly fluorescent and can be detected at levels of ± 2.5 ng as a blue fluorescent spot on TLC plates at a R_F less than the R_F of sterigmatocystin. This method of analysis, as used by the original investigators, is also unsatisfactory as the fluorescence intensity of the acetate fades very rapidly. It was found that a 50% reduction in fluorescence intensity of the acetate occurred after 22 min.

The third method for analysis [25] forms part of a screening method for aflatoxins, ochratoxins and sterigmatocystin. The quantitative step for this procedure is based on the conversion of sterigmatocystin to a more fluorescent derivative by spraying a developed TLC plate with a 20% KOH solution and estimating the sterigmatocystin by comparison with a standard. This method was found to be reliable for sterigmatocystin [26], although visual comparison must be used because, in the time required to measure fluorescence with a fluorodensitometer, the fluorescence will have faded. The smallest quantity of sterigmatocystin detectable by this method was 200 µg/kg.

References p. 381

The most sensitive method for the detection of sterigmatocystin in grain samples is that of Stack and Rodricks [26]. This method consists of initial extraction of a 50 g finely ground sample with acetonitrile—water (9:1) containing KCl which is followed by partition of the extract first against hexane and then against $CHCl_3$ and, if required, silica gel column chromatography. This sequence provides an extract in which the sterigmatocystin can be estimated by fluorescence intensity comparison on TLC plates with a pure sterigmatocystin standard. The important improvement in this method is the enhancement of visualization of the sterigmatocystin on the TLC plates by an $AlCl_3$ spray reagent. After preparative TLC the identity of sterigmatocystin is confirmed by formation of an acetate derivative and of a derivative formed by treating the suspected extract with aqueous acid. In this manner sterigmatocystin could be detected and estimated in samples at levels as low as 30 $\mu g/kg$ with an overall recovery, under experimental conditions, of 90%.

A multimycotoxin detection method for aflatoxins B_1, B_2, G_1 and G_2, ochratoxins A and B and their ethyl esters, zearalenone, sterigmatocystin and patulin has also been described [27]. This method is based on selective extraction with acetonitrile—water, defatting with iso-octane and removing water-soluble components by transfer of the mycotoxins to chloroform. The detection limit for sterigmatocystin achieved by this method is 60 $\mu g/kg$ which is low enough to be useful in a routine screening procedure.

Natural occurrence

Moulds of the *A. versicolor* group are of ubiquitous distribution and have been found on a wide variety of decaying animal and vegetable products [39]. Sterigmatocystin producing strains of *A. nidulans* (Eidam) Wint. were isolated from maize and groundnuts while an undescribed species of Bipolaris (I.M.I. 115076) isolated from animal fodder was also shown to be capable of producing sterigmatocystin [23].

From the above it may be assumed that this mycotoxin could be an important mould metabolite contaminating cereal and grain products. This assumption has been verified by the detection of sterigmatocystin in grain [10] and coffee-beans [9] as well as in food samples obtained from Moçambique during a survey of mycotoxin contamination of stored and prepared foods from an area with the world's highest liver cancer incidence [40].

Biological activity in experimental systems

During an investigation of the toxicity of 59 species of moulds recovered from domestic cereal and legume crops Scott [41] found *A. nidulans* to be one of 22 species to be acutely toxic to ducklings, mice and rats while the *A. versicolor* strain under test was non-toxic. Further investigations indicated that, in addition to *A. versicolor*, strains of *A. nidulans* and a Bipolaris species produced sterigmatocystin and that this metabolite was both acutely toxic and a potential carcinogen [2,6,23].

Mice have been shown to be very resistant to the toxic effects of sterig-matocystin. In this species the LD_{50} value is in excess of 800 mg/kg [42,43]. Detailed studies of the acute toxicity of sterigmatocystin in Wistar rats indicated that the 10-day LD_{50} value was 166 mg/kg in males and 120 mg/kg in females. After intraperitoneal (i.p.) administration the LD_{50}s in males were between 60 and 65 mg/kg depending on whether dimethylforma-mide or wheat-germ oil was used as a vehicle. The principal *postmortem* alterations consisted of hepatic and renal necrosis. The site of necrosis within the hepatic lobule differed according to the route of administration of sterigmatocystin, being centrilobular after oral dosing and periportal after i.p. injection. In addition, an extensive zone of fatty changes was located along the periphery of the necrotic areas when the toxin was administered in oil. Extensive haemorrhage within the renal parenchyma and necrosis of tubules and glomeruli of the kidneys occurred at high doses (144 mg/kg) while marked degenerative alterations and necrosis were evident at lower doses (10—100 mg/kg). In the kidneys, the tissue damage was independent of the route of administration or the vehicle employed [44]. Studies extend-ing over a period of 16 weeks showed that the livers of Wistar rats exposed to a diet containing 100 ppm of sterigmatocystin underwent degenerative and regenerative alterations culminating in a parenchyma consisting mainly of regenerative nodules [45]. During the acute and subacute hepatotoxic phase preceding regenerative hyperplasia, the histopathological picture of disturbed lobular arrangement, ballooning degeneration, intracytoplasmic hyaline bodies, nuclear and cellular pleomorphism and bile duct reduplication close-ly resembled the changes that were seen in some human cases of hepatitis during an epidemic of this disease in Moçambique [46,47]. The culmination of the initial investigations of the biological effects of sterigmatocystin in rats was the evidence of its hepatocarcinogenicity in males and females [4,5]. From these studies it was also concluded that aflatoxin was no more than ten times more potent as a carcinogen than sterigmatocystin and that sterigmatocystin was a hepatocarcinogen and not carcinogenic due to cirrho-genic properties. The potency of this latter mycotoxin was further evidenced by its ability to induce a series of skin lesions, including squamous cell carcinomas, in rats after application to the shaved skin on the backs of the experimental animals [48]. These observations indicate that, in contrast to aflatoxin, which is incapable of producing neoplastic skin lesions, sterigmato-cystin may be a proximal carcinogen.

Evaluation of the influence of dietary protein levels on sterigmatocystin-induced hepatocellular neoplasia in rats revealed that 63% of experimental animals developed hepatic malignancy after 52 weeks on diets containing more than 12% assimilable protein in contrast to animals receiving 4% assimi-lable protein in which no neoplasms were observed after the same period [49]. In rats receiving sterigmatocystin [50], the specific activity of liver nuclear acid DNAase increases gradually to reach a maximum value (200% of the control value) at the stage when the entire hepatic parenchyma consists of foci of regenerative nodules. The specific activity then declines dramatically to the control value and remains there, while the histological alterations in the liver progress to neoplasia. This enhancement of acid DNAase activity in progressively altered hepatic tissue may have considerable diagnostic value as similar results have been obtained with aflatoxin B_1 and an aminoazodye [51].

Studies on male vervet monkeys (*Cercopithecus aethiops*) indicated that these non-human primates were also susceptible to the toxic effects of sterigmatocystin [52] and that the i.p. 10-day LD_{50} value in animals weighing between 0.5 and 5.5 kg was 32 mg/kg. On *postmortem* examination it was found that the animals exposed to the higher doses of the toxin (150 mg/kg) were icteric, and petechia could be observed on all the serosal surfaces and in parenchymal organs. In the lower dose groups, some macroscopic signs of liver and kidney damage were also observed. Microscopic examination of kidney sections revealed degeneration of the glomeruli with oedema in Bowman's space. Hyaline droplet degeneration was also present in the proximal and distal convoluted tubules. In the medullary rays fatty changes and necrosis of the tubular epithelium was prominent. The liver sections from animals in the lowest dose groups showed predominantly centrilobular changes consisting of small foci of necrosis, fatty vacuolation of hepatocytes and ballooning degeneration. Macrophages, plasma cells and round cells infiltrated the necrotic areas.

Both the liver and kidney lesions increased in severity with an increase in the dose of toxin and at the highest doses employed the livers of the monkeys were enlarged, yellow and friable; petechiae and haemorrhages were present in the kidneys, gastro-intestinal tract, epicardium and on all the serosal surfaces.

Histological examination revealed that there were diffuse haemorrhages throughout the renal parenchyma with hyalinisation of the glomeruli or fragmentation of the capillary loops. The epithelial cells of the nephrons showed hyaline degeneration, fatty changes and necrosis. The livers showed diffuse fatty changes, extensive single cell and central necrosis with haemorrhage. Intracellular bile stasis was present in most of the surviving parenchymal cells. Zenker's hyaline degeneration was present in the myocardium.

Oral dosing of monkeys with sterigmatocystin at a rate of 20 mg/kg every 14 days for a period of 4 to 6 months resulted in chronic hepatitis [6]. This condition was characterised by portal fibrosis with fibrous elements extending a short way into the parenchyma from the enlarged portal tracts. The lobular structure at this stage remained intact but the portal tracts were infiltrated with round cells and plasma cells. With continued dosing of the toxin this phase was followed by chronic aggressive hepatitis. Single-cell necrosis of the hepatocytes and progressive growth of the fibrous septa led to disruption of the lobular architecture. Bands of reticulin fibres commencing in the enlarged portal tracts extended between the foci of hepatocytes. Kupffer cell prominence and bile duct epithelial proliferation could also be observed.

Simultaneously focal reactive hyperplasia of the liver cells commenced and after 12 months' dosage large hyperplastic nodules, up to 8 mm in diameter, were observed throughout the entire organ in biopsy specimens. These hyperplastic foci and the grossly disturbed intrahyperplastic parenchyma contained hepatocytes of varying dimensions with pleomorphic nuclei. The cytoplasm of these cells was hyperchromatic and the chromatin could be seen as basophilic masses situated mainly at the periphery of the nucleus.

Animals dying at this stage had essentially the same histological alterations in the liver, but the entire organ was in addition affected by marked fatty changes of the hepatocytes within hyperplastic nodules and the surrounding parenchyma.

After 24 months' dosage the livers of the monkeys were grossly deformed with irregularly circumscribed nodular masses up to 2 cm in diameter protruding above a coarsely granular liver surface (Fig. 3). On the cut surface, these nodular masses were yellowish-white in colour with irregular green-yellow bile stained areas. Histological examination of these livers revealed a variable pathological process consisting of active chronic aggressive hepatitis, macronodular transformation, slight steatosis, some induction of liver cells, rosette formation and cholestasis. Marked hepatocellular pleomorphism and karyomegaly were also present. Hepatoma formation commenced within the macronodules and gradually enlarged to invade the surrounding compressed tissues [43].

The biological effects of sterigmatocystin and its analogues have been studied in primary monkey kidney epithelial cells [53,54]. From these studies it was concluded that the toxic effects of this mycotoxin were mani-

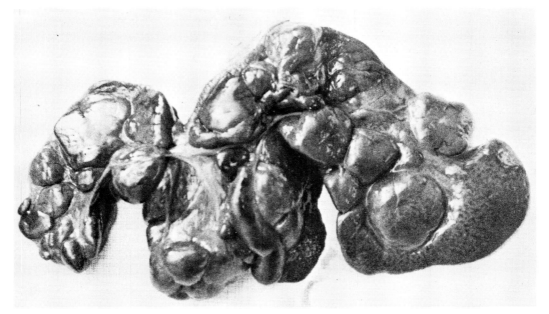

Fig. 3. Liver of a monkey after 24 months exposure to sterigmatocystin. Irregular, circumscribed, nodular masses can be seen protruding above the surface of the liver.

fested by a loss of chromatin from the nucleus and progressive cellular degeneration. Other specific cytological alterations included mitotic inhibition, nucleolar fragmentation and the formation of unidentified eosinophilic intranuclear masses. Ultrastructural evaluation showed that there was a redistribution of nucleolar components similar to the nucleolar segregation produced by inhibitors of DNA-dependent RNA synthesis. Sterigmatocystin, O-methylsterigmatocystin and demethylsterigmatocystin inhibit the uptake of [^{3}H]uridine most potently while dihydrosterigmatocystin, O-methyldihydrosterigmatocystin and demethyldihydrosterigmatocystin were less active in this respect. These investigations indicated that the toxicity of sterigmatocystin and its analogues were dependent on their chemical structures in that an unsaturated $\Delta^{1,2}$-furobenzofuran system and the position of methoxy and hydroxy groups on the xanthone ring of the analogues affect the cytoxicity in cultures of monkey kidney epithelial cells.

Investigations of the effect of sterigmatocystin on the activity of acid DNAase in a similar *in vitro* system indicated that a marked increase in the activity of this enzyme occurred after 2 h exposure to the toxin but after 24 h exposure the activity of acid DNAase was lower than that of the control cells. In contrast, alkaline DNAse activity was increased after both 2 and 24 h. Continuous exposure of monkey kidney epithelial cells for 10 days gave rise to a new population of cells which morphologically resembled endothelial cells. These studies thus indicated that the repair of DNA damaged by exposure to sterigmatocystin was incomplete [55].

The other DHFF compounds that have been shown to be biologically active include demethylsterigmatocystin, a carcinogen [3], O-methylsterigmatocystin, a substance of low toxicity to mice and ducklings, and aspertoxin, which has been found to be severely toxic to developing chick embryos [15,16].

Mode of action

The definite sequence of biochemical events leading up to morphological alterations characteristic of neoplasia is still unknown but there are certain biochemical and morphological responses of tissues which can be associated with the neoplastic process. The mode of action of aflatoxin B_1 for instance, can be ascribed to the ability of this compound to bind to DNA [57]. This binding results in inhibition of transcription of DNA by RNA-polymerase causing a disturbed RNA synthesis which in turn causes the typical nucleolar lesions observed in cells after exposure to toxic levels of aflatoxin B_1 [58].

Sterigmatocystin similarly causes a segregation of nucleolar components in the nuclei of cultured monkey kidney epithelial cells [53]. In addition, sterigmatocystin severely inhibits the incorporation of [^{3}H]thymidine indicating an inhibition of DNA synthesis [54]. From the effects of sterigmatocystin on nucleolar morphology, inhibition of mitosis and impaired DNA synthesis it is evident that this mycotoxin has a profound effect on the nucleus. The evidence that sterigmatocystin damages DNA directly was the effect of low levels of toxin on the incorporation of [³H]thymidine [54] by cell cultures; a marked increase was shown in the number of lightly labelled nuclei without a reduction in the number of heavily labelled nuclei [56]. An increase in unscheduled incorporation of [³H]thymidine indicates repair synthesis of damaged DNA [59].

Furthermore, it has been shown that exposure of cells to sterigmatocystin results in an increased activity of the nuclear DNA-repair enzyme acid and alkaline deoxyribonuclease. These observations confirm that DNA is damaged by sterigmatocystin [56].

It is highly unlikely that the aforementioned effects are caused by a metabolite of this mycotoxin, as the major metabolite obtained after exposure of a vervet monkey to sterigmatocystin was an unchanged sterigmatocystin conjugate [60].

References

1 J.V. Rodricks, J. Agr. Food. Chem., 17 (1969) 457.
2 F. Dickens, H.E.H. Jones and H.B. Waynforth, Brit. J. Cancer, 20 (1966) 134.
3 J.S.E. Holker, as quoted by J.C. Engelbrecht and B. Altenkirk, J. Natl. Cancer Inst., 48 (1972) 1647.
4 I.F.H. Purchase and J.J. van der Watt, Food Cosmet. Toxicol., 6 (1968) 555.
5 I.F.H. Purchase and J.J. van der Watt, Food Cosmet. Toxicol., 8 (1970) 289.
6 J.J. van der Watt and I.F.H. Purchase, Mycotoxins in Human Health, MacMillan, London, 1971, p. 209.
7 Z.H. Abedi, as quoted by J.V. Rodricks, J. Agr. Food Chem., 17 (1969) 457.
8 Y. Hatsuda and S. Kuyama, J. Agr. Chem. Soc. (Japan), 28 (1954) 989.
9 I.F.H. Purchase and M.E. Pretorius, J. Assoc. Off. Anal. Chem., 56 (1972) 225.
10 P.M. Scott, IUPAC International Symposium on Control of Mycotoxins, Göteborg, August 21—22, 1972, p. 20 (Abstr.).
11 E. Bullock, J.C. Roberts and J.G. Underwood, J. Chem. Soc., (1962) 4179.
12 J.S.E. Holker and S.A. Kagal, Chem. Commun., (1968) 1574.
13 G.C. Elsworthy, J.S.E. Holker, J.M. McKeown, J.B. Robinson and L.J. Mulheirn, Chem. Commun., (1970) 1069.
14 Y. Hatsuda, T. Hamasaki, M. Ishida, K. Matsui and S. Hara, Agr. Biol. Chem., 36 (1972) 521.
15 H.J. Burkhardt and J. Forgacs, Tetrahedron, 24 (1968) 717.
16 J.V. Rodricks, K.R. Henery-Logan, A.D. Campbell, L. Stoloff and M.J. Verrett, Nature, 217 (1968) 668.
17 R.J. Cole and J.W. Kirksey, Tetrahedron, 35 (1970) 3109.
18 J.S.E. Holker and J.G. Underwood, Chem. Ind., (1964) 1865.
19 R. Thomas, Biosynthetic pathways involving ring cleavage, in Z. Vanek and Z. Hostalek (Eds.), Biogenesis of Antibiotic Substances, Academic Press, New York, 1965, pp. 160—161.
20 J.S.E. Holker and L.J. Mulheirn, Chem. Commun., (1968) 1576.
21 M. Biollaz, G. Büchi and G. Milne, J. Am. Chem. Soc., 92 (1970) 1035.
22 C.J. Rabie, A. Lübben and M. Steyn, Proc. Congr. of the South African Society for Plant Pathology and Microbiology, Salisbury, January, 9—12, 1973, p. 12.
23 C.W. Holzapfel, I.F.H. Purchase, P.S. Steyn and L. Gouws, S. Afr. Med. J., 40 (1966) 1100.
24 L.J. Vorster and I.F.H. Purchase, Analyst. 94 (1969) 694.
25 L.J. Vorster, Analyst. 94 (1969) 136.

26 M. Stack and J.V. Rodricks, J. Assoc. Off. Anal. Chem., 54 (1971) 86.
27 L. Stoloff, S. Nesheim, L. Yin, J.V. Rodricks, M. Stack and A.D. Campbell, J. Assoc. Off. Anal. Chem., 54 (1971) 91.
28 J.V. Rodricks, E. Lustig, A.D. Campbell and L. Stoloff, Tetrahedron, 25 (1968) 2975.
29 A.C. Waiss, M. Wiley, D.R. Black and R.E. Lundin, Tetrahedron, 21 (1968) 3207.
30 R.D. Hutchison and C.W. Holzapfel, Tetrahedron, 27 (1971) 425.
31 J.E. Davies, D. Kirkaldy and J.C. Roberts, J. Chem. Soc., (1960) 2169.
32 J.C. Engelbrecht and B. Altenkirk, J. Natl. Cancer Inst., 48 (1972) 1647.
33 J.H. Birkinshaw and I.M.M. Hammady, Biochem. J., 65 (1957) 162.
34 M.J. Rance and J.C. Roberts, Tetrahedron, 32 (1970) 2799.
35 M.J. Rance and J.C. Roberts, Tetrahedron, 4 (1969) 277.
36 P.J. Aucamp and C.W. Holzapfel, J. Afr. Chem. Inst., 23 (1970) 40.
37 M. Ishida, T. Hamasaki and Y. Hatsuda, Agr. Biol. Chem., 36 (1972) 1847.
38 E. Bullock, D. Kirkaldy, J.C. Roberts and J.G. Underwood, J. Chem. Soc., (1963) 829.
39 Thom and Raper, A Manual of Aspergilli, Williams and Wilkins, Baltimore, Md., 1945, p. 183.
40 Joint South African-Moçambique Liver Cancer Research Group, Unpublished observations, 1972.
41 B. de Scott, Mycopathol. Mycol. Appl., 24 (1965) 213.
42 E.B. Lillehøj and A. Ciegler, Mycopathol. Mycol. Appl., 35 (1968) 373.
43 J.J. van der Watt, Unpublished observations, 1972.
44 I.F.H. Purchase and J.J. van der Watt, Food Cosmet. Toxicol., 7 (1969) 135.
45 J.J. van der Watt and I.F.H. Purchase, S. Afr. Med. J., 44 (1970) 159.
46 J.J. van der Watt, F.O. Torres and I.F.H. Purchase, Proc. Xth Intern. Cancer Congress, Houston, Texas, 1970, p. 80.
47 F.O. Torres, I.F.H. Purchase and J.J. van der Watt, J. Pathol., 102 (1970) 163.
48 I.F.H. Purchase and J.J. van der Watt, Tox. Appl. Pharm., 26 (1973) 274.
49 J.J. van der Watt and I.F.H. Purchase, S. Afr. Med. J., 46 (1972) 1570.
50 P. Kempff, M.J. Pitout and J.J. van der Watt, Biochem. Pharmacol., 22 (1973) 2490.
51 M.J. Pitout, J.J. van der Watt and P. Kempff, Chem.-Biol. Interactions, 6 (1973) 227.
52 J.J. van der Watt and I.F.H. Purchase, Brit. J. Exptl. Pathol., 51 (1970) 183.
53 J.C. Engelbrecht, S. Afr. Med. J., 44 (1970) 153.
54 J.C. Engelbrecht and B. Altenkirk, J. Natl. Cancer Inst., 48 (1972) 1647.
55 J.C. Engelbrecht and M.J. Pitout, Proc. Soc. Physiology and Medical Biochemistry, Scientific Meeting, Pretoria, Sept. 29, 1972.
56 J.C. Engelbrecht, M.J. Pitout and P. Kempff, Personal communication, 1973.
57 A.M.Q. King and B.H. Nicholson, Biochem. J., 114 (1969) 679.
58 C. Frayssinit, Intern. J. Cancer, 6 (1970) 74.
59 H.F. Stich, R.H.C. San and Y. Kawazoe, Nature, 229 (1971) 416.
60 P.G. Thiel and M. Steyn, Biochem. Pharmacol., 22 (1973) 3267.
61 D.P.H. Hsieh, M.T. Lin and R.C. Yao, Biochem. Biophys. Res. Commun., 52 (1973) 992.

Chapter 18

PATULIN

P.M. SCOTT

Health Protection Branch, Health and Welfare Canada, Ottawa, Ontario, K1A OL2 (Canada)

The antibiotic isolated by Birkinshaw *et al.* [1] in 1943 from *Penicillium patulum* and subsequently in the same laboratory from *P. expansum* [2] was named patulin. However, priority for the isolation of the crystalline compound goes to Chain and co-workers [3], who had previously isolated claviformin, later shown to be identical with patulin [4], from *P. claviforme*. Crude preparations of the antibiotic had also been isolated from fungi by other workers prior to 1943 [5—8]. For a detailed review of the early history of patulin see the article by Abraham and Florey [9]. Patulin has since been isolated, under a variety of names, from *Aspergillus clavatus* (clavacin [10], clavatin [11]), *A. giganteus* [12], *A. terreus* [13], *Penicillium expansum* (expansine [14], not to be confused with a phenolase inhibitor named expansin [15]; mycoin $C_{(3)}$ [16,17]), Penicillium spp. (penicidin [18]), *P. melinii* [9], *P. equinum* [19], *P. novae-zeelandiae* [19], *P. divergens* [20], *P. urticae* [13] (= *P. patulum*), *P. griseofulvum* [21], *P. leucopus* (leucopin) [22], *P. cyclopium* [23], *P. lapidosum* [24], and *Byssochlamys nivea* (= Gymnoascus sp.) [25,26]. It was supplied for clinical trials by a British corporation under the brand name "tercinin" [27].

Interest in the antibiotic properties of patulin has now given way to concern over its toxicity to animals and plants coupled with its presence in the environment under certain circumstances.

Mycotoxicoses associated with patulin-producing fungi

No outbreak of disease in animals, associated with the ingestion of feed from which a patulin-producing fungus was isolated, can be definitely attributed to patulin, and other factors (including possibly other mycotoxins) might have been responsible. The toxin itself has not been identified in a toxic feed nor have the effects of feeding pure patulin to large farm animals been studied.

Ukai *et al.* [28] obtained patulin from a *Penicillium* sp., originally referred to as *P. velutinum*, *P. maltum* or the Hori-Yamamoto strain [29,30] but later identified as *P. urticae* [31], that had been isolated from a malt feed responsible for the deaths of over 100 dairy cattle in Japan. Feeding of malt grains inoculated with the fungus to mice and a mature bull caused nervous symptoms, cerebral haemorrhage and death [32]. Intoxication of cattle in Germany by germinated malt containing toxic *Aspergillus clavatus* might also have been due to patulin [33,34]. In France, there have been other disease outbreaks in cattle associated with the presence of this toxic fungus in wheat [35,36]; pulmonary oedema and congestion were found on autopsy of the dead cows in one of these cases [35]. Forgacs *et al.* [37] were unsuccessful in an attempt to isolate patulin from a toxinogenic *A. clavatus* obtained from pelletted feed that caused bovine hyperkeratosis.

Physical and chemical properties of patulin

Patulin is a colourless, crystalline compound with a melting point of about 110° (the highest obtained was 112° for a sample recrystallised from ethyl acetate [38]) and a molecular weight of 154. It sublimes in high vacuum at 70–100° [3,39,40]. Patulin is a neutral substance, soluble in water, alcohols, acetone, ethyl acetate and chloroform, less soluble in ether and benzene, and insoluble in petroleum ether. It can be recrystallised preferably from benzene. Katzman *et al.* [39] found that repeated recrystallisation changed patulin into an amorphous, insoluble product, but other workers, *e.g.* Nauta *et al.* [14], did not encounter this problem.

The infrared (IR) absorption spectrum of patulin has bands at 3390, 1768, and 1745 (shoulder) cm^{-1} in nujol mull [41]; 3360 (broad), 1765 and 1740 cm^{-1} in hexachlorobutadiene mull (first band) and KBr disc [42]; and 3580, 3340 (broad), 1782 and 1755 cm^{-1} in $CHCl_3$ solution [43], confirming the presence of an alcoholic OH group and α,β-unsaturated five-membered lactone ring in the patulin molecule (Fig. 1). The 1745 (1740) cm^{-1} band has been accounted for by the possible small contribution of other structures [41,42]. However, the structure 4-hydroxy-4*H*-furo[3,2-*c*]pyran-2(6*H*)-one (Fig. 1) proposed by Woodward and Singh [44] must be correct [45,46], and is confirmed by the 100 MHz nuclear magnetic resonance (NMR) spectrum (in $CDCl_3$): δ = 5.97 (3P, OC*H*(OH), = C*H*, = C*H*; complex), 4.73 (1P, C*H*O; doublet of doublets, A part of ABX system, J_{AB} = 17 cps), 4.40 (1P, C*H*O; doublet of doublets, B part of ABX system, J_{AB} = 17 cps) and 3.46 (1P, O*H*; doublet, J = 5 cps) [43]. NMR spectroscopy may be useful in detecting the presence of impurities in standard preparations of crystalline patulin. Values reported for the extinction coefficients (ϵ) at about 276 nm in the ultraviolet (UV) absorption spectrum of patulin are given in Table I. Irregularities in using UV spectroscopy to monitor patulin purification were observed by Katzman *et al.* [39].

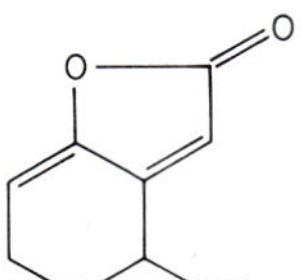

Fig. 1. Chemical structure of patulin.

TABLE I

ULTRAVIOLET ABSORPTION SPECTRUM OF PATULIN

Solvent	λ_{max} (nm)	ϵ	Reference
Alcohol	276	14 540	39
0.005 N alcoholic HCl	276	15 120	39
?	276.5	16 600	11
Ethanol	277	19 950	40
?	275	16 600	47
?	276	16 600	44
Ethanol	275	15 340	43
Ethanol	275	14 540	43
Methanol	275	12 880	48
Water	276	13 350	49
0.1 N HCl	277	15 000	50

Patulin is optically inactive, in spite of the asymmetric carbon atom in the structure (Fig. 1). As a hemiacetal it would be expected to racemise rapidly in solution after production by the fungus.

The mass spectrum of patulin shows a prominent parent ion at m/e 154 [51]. Assignment of structures to seven of the principal fragment ions in the mass spectrum of deuterated patulin was made by Scott and Yalpani [52]. The mass spectrum of patulin trimethylsilyl ether obtained by coupled gas chromatography—mass spectrometry and after subtraction of background is shown in Fig. 2 [53].

Patulin forms other derivatives expected for a secondary alcohol, for example a monoacetate, cinnamate and benzoate [1,14]. It reacts as a sim-

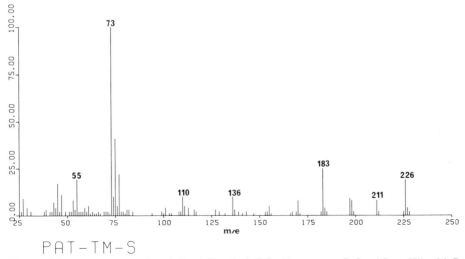

Fig. 2. Mass spectrum of patulin trimethylsilyl ether; recorded with a Hitachi Perkin-Elmer RMS-4 mass spectrometer at 80 eV, following gas chromatography on a column of 0.25% SE-30 on glass beads at 100—150°.

References p. 399

ple carbonyl compound after ring opening at the hemiacetal function and forms a phenylhydrazone, semicarbazone, oxime and similar derivatives [1,11,14]; this functionality has proved most useful for the detection of patulin on thin-layer chromatography (TLC) (see p. 389). Patulin reduces warm Fehlings solution and decolourises potassium permanganate solution [1,11,14]. Reduction of patulin with sodium borohydride yields ascladiol, also a mycotoxin produced by *Aspergillus clavatus* [54], while hydrogenation gives desoxypatulinic acid [40] or other products depending on the catalyst and conditions [1,11,14].

Patulin is unstable in the presence of alkali [1,3,14,55,56]. Although its antibiotic properties are not affected by cold dilute acid (pH 2) [56] and it was recovered unchanged after treatment with methanolic H_2SO_4 for 24 h [57], it is decomposed by boiling in 2 N H_2SO_4 for 6 h [1]. Tetrahydro-γ-pyrone 2-carboxylic acid was isolated as a reaction product. However, other experiments have indicated retention of activity under the following acidic conditions at $100°$: 0.1 N HCl for 30 min [3] and pH 2 for 15 min [56]. Jefferys [58] found patulin to be stable, as an antibiotic, for several weeks in buffers of pH 3.3—6.3 but it was slowly inactivated at pH 6.8. Stansfeld *et al.* [59] found patulin to be stable in buffer pH 6.0 for several months. Some decomposition occurs in distilled water; the solution becomes yellow and acidic after a few days at room temperature [1,60,61], and UV spectroscopy showed that patulin was completely destroyed after 3 months [39]. While Katzman *et al.* [39] found no change in the UV spectrum of patulin in alcohol after this time, Tanenbaum and Bassett [49] reported that patulin formed "pre-patulin" after standing for several days in 95% ethanol and Pohland and Allen [62] noted decomposition of patulin in methanol. Patulin is stable in chloroform (a preferred solvent for TLC standards), benzene and methylene chloride [62], but not as a dry thin film [63].

Patulin is appreciably stable in apple and grape juices but not in orange juice [62,64]. About 50% of added patulin survives heating in apple juice at $80°$ for 10—20 minutes [64]. Sulphur dioxide, sometimes used as a preservative for fruit juices, causes rapid disappearance of patulin [62]. Patulin is unstable in wheat flour, ground sorghum, and ground wet (but not *dry*) corn [62,64,65]. Vitamin B_1 completely inhibited the activity of patulin in dilute aqueous solution [65] and fresh garden soil (pH 7) also destroyed patulin, although it was stable in some more acidic soils [58].

Since the initial observations by Atkinson and Stanley [66] and Cavallito and Bailey [67] that patulin could be inactivated by sulphydryl containing compounds, this reaction has received considerable investigation. Although there is disagreement as to which thiol is the most active, cysteine, thioglycollic acid, glutathione and dimercaptopropanol have all been shown to react with patulin [66—70], at a pH as low as 2.3 [64]. Patulin also reacts with sulphydryl groups in tomato sprouts [69] and rat liver [71]. Although no pure reaction products have been isolated, disappearance of patulin in the presence of thiols has been observed using as assay systems bacteria [16,66—68,72], plants [69], germinating seeds [73], phagocytic cells [74], mito-static action [70], animals [75] and UV spectroscopy [60,64], while disappearance of the sulphydryl group in the presence of patulin has been shown chemically with nitroprusside [66,68], bis(3-carboxy-4-nitrophenyl) disulphide [76] and other reagents [69,71,77,78]. Patulin was thought by Andraud et al. [79] to inhibit urease by reacting with sulphydryl groups in the enzyme. However, Singh [60] did not regard this hypothesis of enzyme inhibition as tenable. He showed that reaction of patulin with cysteine was very slow and that glyceraldehyde-3-phosphate dehydrogenase, which contains sulphydryl groups in its active centre, did not react with patulin. Daniel and co-workers [80] found that a mixture of patulin and cysteine, while less cytotoxic, possessed the same antimitotic power as patulin alone.

The inhibitory effect of human and animal serum on the activity of patulin has been reviewed by Abraham and Florey [9] and by de Rosnay [16]. Some conflicting results have been obtained by different investigators. Fresh rabbit serum appears to be the most active [39,81]. The protein in horse serum and human plasma responsible for the inactivation of patulin is the albumin fraction [16]. Asparagine also neutralises the antibiotic [16].

Patulin reduces the toxicity of tetanus toxin to mice [82].

Biosynthesis

Bu'Lock and Ryan [83] and Tanenbaum and Bassett [84] showed that in *Penicillium patulum* cultures patulin could be derived from ^{14}C-labelled 6-methylsalicyclic acid, which in turn could be formed, together with patulin, from labelled acetate or glucose [84]. In a kinetic pulse-labelling study, in which radiolabelled acetate and pertinent secondary metabolites of *P. urticae* (*P. patulum*) were fed to submerged fermentor cultures of this fungus (a non-pigmented mutant) in modified Czapek-Dox medium, a preferred route for patulin biosynthesis from 6-methylsalicylic acid was found to exist (Fig. 3) [50]. Acetate, 6-methylsalicylic acid, *m*-cresol, *m*-hydroxybenzyl alcohol, *m*-hydroxybenzaldehyde and gentisaldehyde were readily

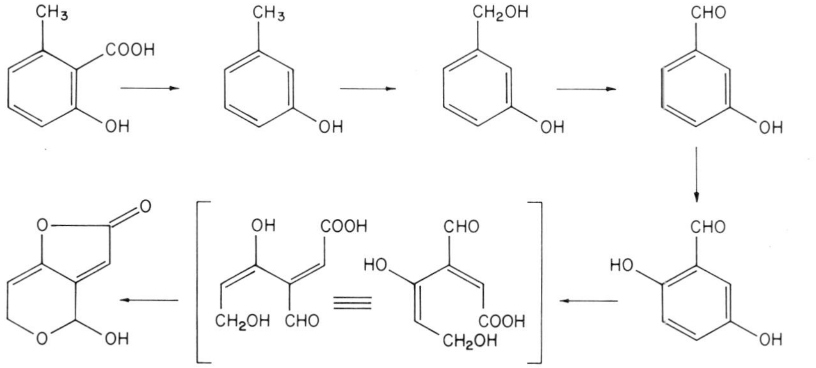

Fig. 3. Pathway of patulin biosynthesis from 6-methylsalicylic acid in *Penicillium urticae* NRRL 2159A.

converted into patulin. Toluquinol and gentisyl alcohol, previously shown to incorporate deuterium from added deutero-*m*-cresol [52], were not converted into patulin nor was patulin transformed further. An enzyme which converts acetyl coenzyme A and malonyl coenzyme A into 6-methylsalicylic acid in the presence of NADPH has been prepared from *P. patulum* [85,86]. Of the enzymes required for conversion of 6-methylsalicylic acid into patulin, only 6-methylsalicylic acid decarboxylase [87] and *m*-hydroxybenzyl alcohol dehydrogenase [88] have been detected or characterised. The immediate precursor of patulin (Fig. 3) has been denoted "pre-patulin" by Bassett and Tanenbaum [89], who claim to have isolated it as a red amorphous compound from *P. patulum* grown with added calcium carbonate [49]. However, the structure of this material has not been proved.

Substrates favouring patulin production

The two synthetic media most commonly used for patulin production by various Penicillium and Aspergillus spp. have been Czapek-Dox and Raulin-Thom or modifications thereof [1,2,13,22,38,39,48—50,65,90]. Barta and Mečiř [20] found that *Penicillium divergens* grown on Czapek-Dox medium showed no antibacterial activity but produced a low yield of patulin on modified Raulin-Thom solution. Lochhead *et al.* [38], using a modified Czapek-Dox solution as control medium, found that glucose was the best carbon source for patulin production by a Penicillium sp. and, contrary to the observations of Katzman *et al.* [39] with *Aspergillus clavatus*, that addition of yeast extract or cornsteep liquor depressed patulin production. Maltose, fructose or galactose can be used instead of glucose [21,91,92]. Lochhead *et al.* [38] also showed that better yields of patulin were obtained in stationary (surface) culture than in shaken (submerged) culture and that 20—25° was a more favourable temperature than 30°; maximum production of patulin was at 8—12 days of incubation. Darken and Sjolander [93] obtained a high yield of patulin (2.1 g/l) from a strain of *A. clavatus* grown in shaker flasks containing a glucose-nitrate medium plus 0.1% cornsteep liquor and 1% potassium hydrogen phthalate, while 2.35 g patulin/l was obtained in 86-h tank fermentations using a slightly different medium and a strain of *P. urticae*.

Norstadt and McCalla [94,95] found potato dextrose broth much superior to Raulin-Thom for patulin production by their strains of *Penicillium urticae*. The maximum yield was 2.7 g/l after 14 days at 25° (a range of temperatures from 5° to 35° was studied). Other semi-synthetic media that have been shown to support patulin production include yeast extract-sucrose [96] and Saboraud [29,92,97] liquid media.

Patulin has been isolated from culture media by adsorption on activated charcoal, followed by elution with a polar solvent such as acetone or ethyl acetate [28,38,39], or by direct extraction with ethyl acetate [1,38,95], usually after concentration of the medium. A comparison of methods showed the latter procedure to be the most efficient [38]. Further purification of the patulin has been achieved using column chromatography on alumina [1,95], silica gel [10,96], or ion exchange resin (permutit) [39].

Natural substrates that have been shown to allow production of patulin by Penicillium spp. are barley, malt, rice [92], wheat straw [94], soil containing root bark or leaves of apple trees or root residues from other fruit trees (sour cherry, peach, pear and quince) [98,99], and apples [90,100]. Maximum amounts of patulin were detected in apples 13—14 days after inoculation with *P. expansum* [90]. An antibiotic assumed to be patulin was produced by *P. patulum* on fresh lawn mowings, bracken, Timothy grass, sugar-beet pulp and fresh wheat straw, and by *A. clavatus* on the last three substrates [101].

Analysis for patulin

Patulin was originally assayed by its antibacterial action using either a serial dilution method [39] or by measuring zones of inhibition in the cylinder-plate method [38,65] with *Escherichia coli* or *Staphylococcus aureus*, for example, as test organisms. Atkinson *et al.* [18] devised a procedure for the estimation of patulin (penicidin) in fungal cultures on the basis of its antiluminescent activity with luminescent cultures of *Photobacterium fischeri*.

Paper chromatography (PC) (Table II) has been studied extensively as a means of identifying patulin in fungal cultures. Thin-layer chromatography (TLC) was first used in 1965 (Table III) and is now an important technique in the analysis of patulin in foods. Patulin can be detected on the chromatogram by bioautography using, for example, *Bacillus subtilis* [106,107] or *Escherichia coli* [116]. Yamamoto [102] used 4% phenylhydrazine as a spray reagent, followed by heating the paper chromatogram at 100° for 5—10 min, whereupon patulin was detected as an orange yellow band (detection limit 1—2 μg). This reagent, as the hydrochloride, has subsequently been found to be useful for the detection of patulin by TLC [51,64] with a detection limit of 0.02—0.05 μg patulin per spot [63]. Several other spray reagents, some of them not too sensitive, have been described (colours, detection limits, and type of chromatography are given in parentheses): anthrone in acidic ethyl acetate (red, PC) [21], aniline oxalate solution (yellow, PC) [21], naphthoresorcinol in chloroform (green, 10 μg PC) [21], o-phenylenediamine solution (brown, 3—5 μg PC) [21]; aqueous dioxan solutions of diazotized sulfanilic acid salt, followed by ammonia (middle chrome, PC) [104], diazotized 4-benzoylamino-2,5-dimethoxyaniline Zn salt followed by ammonia (terracotta, PC) [104], diazotized o-dianisidine Zn salt followed by ammonia (venetian red, PC) [104], and p-nitrobenzenediazonium fluoroborate followed by ammonia (gold, PC) [104]; diazotized p-nitroaniline solution (brown orange, TLC) [50]; aqueous $KMnO_4$ (white on pink, TLC, PC) [50,104]; benzidine solution, then 120°/5 min or 80°/15 min (yellow brown, 3 μg PC) [99,103]; ammonia (yellow, tan UV fluorescence, PC) [105]; o-dianisidine in acetic acid (brown, PC, 0.2 μg

TABLE II

PAPER CHROMATOGRAPHY OF PATULIN

Paper	Solvent system (A, ascending; D, descending)	R_F value	Reference
Toyo No. 50	Phenol—water (1:99, 5:95), A	0.84	102
Whatman No. 4	Butanol—acetic acid, D	1.0	21
?	Butanol—acetic acid—water (4:1:5)	0.77	103
?	Butanol—acetone—water (2:7:1)	0.90	103
Whatman No. 1 or No. 3MM	n-Butanol—0.5 N NH$_4$ OH—ethanol (7:2:1), A	0.80	49
Whatman No. 1 or No. 3MM	Benzene—propionic acid—water (2:2:1), A	0.66	49
Whatman No. 1 or No. 3MM	n-Butanol—acetic acid—water (4:1:1), A	0.80	49
Whatman No. 1	Ethyl methyl ketone—acetone— water—formic acid (80:4:12:2), A	0.83	104
Whatman No. 1	Ethyl methyl ketone—water —diethylamine (921:77:2), A	0.00	104
Whatman No. 1	Methyl isobutyl ketone—formic acid— -water (1000:4:96) (2 phases), A	0.78	104
Whatman No. 1	Chloroform—methanol—water—formic acid (1000:100:96:4) (2 phases), A	0.70	104
Whatman No. 1	Benzene—ethyl methyl ketone—water -formic acid (900:100:98:2) (2 phases), A	0.46	104
Whatman No. 1	Benzene—formic acid—water (1000:2:98) (2 phases), A	0.13	104
Schleicher and Schüll 2043b	Butanol—acetic acid—water (4:1:1), A	0.70	99
Schleicher and Schüll 2043b	Isopropanol—ethanol—water (4:1:1), D	0.91	99
Schleicher and Schüll 2043b	Butanol—KCl—water (2.5:10:100), A	0.83	99
Schleicher and Schüll 2043b	Acetic acid—water (15:85), D	0.83	99
Schleicher and Schüll 2043b	Ethyl acetate—formic acid—water (10:2:3), A	0.99	99
?	Ethanol—water (4:1)	0.76	105
Whatman No. 1	Water, A	0.94	106, 107
Whatman No. 1	n-Butanol satd. with water, A	0.78	106, 107
Whatman No. 1	Ethyl acetate satd. with water, A	0.91	106, 107

TABLE II (continued)

Paper	Solvent system (A, ascending; D, descending)	R_F value	Reference
Whatman No. 1	Benzene satd. with water, A	0.94	106, 107
Whatman No. 1	Isoamyl acetate—methanol—formic acid—water (4:2:1:3, bottom layer), A	0.91	106, 107
Whatman No. 1	n-Butanol—methanol—water (4:1:5, bottom layer), A	0.84	106, 107
Whatman No. 1	Methanol—n-hexane (6:4, bottom layer), A	0.74	106, 107
Whatman No. 1	Benzene—cyclohexanone—0.15 M pH 7.4 phosphate buffer (5:35:60, bottom layer), A	0.84	106, 107
Whatman No. 1 or 3 MM	Benzene—acetic acid—water (136:72:3) or Acetic acid—water (2:98)	?	108

TLC) [49,117]; 2,4-dinitrophenylhydrazine hydrochloride solution then 60°/5—10 min (yellow, PC) [99]; 1% methanolic diphenylbor(in)ic acid, β-aminoethyl ester, then 105°/15 min (blue UV fluorescence, 0.1 µg TLC) [112]; acidic ethanolic p-anisaldehyde then 130°/8—20 min (red, 0.1 µg TLC) [110]; and 0.5% aqueous 3-methyl-2-benzothiazolinone hydrazone hydrochloride then 130°/15 min (yellow UV fluorescence, 0.01 µg TLC) [63]. This latter reagent appears to be the most sensitive so far discovered. Patulin has also been reported detectable as a dark spot under short wavelength UV light [108], but quenching of UV fluorescence on Mallinckrodt 7GF silica gel layers incorporating a 254 nm inorganic phosphor is considerably more useful as the detection limit is then 0.04 µg patulin per spot [114].

Other colourimetric methods reported for the detection or determination of patulin were not originally applied as spray reagents for PC or TLC. Laubie [118] obtained a blue-green colour changing to rose on heating, when several mg of patulin were treated with veratrole in dilute HCl. Broom et al. [119] found that patulin could be determined in blood extracts after formation of a red colour with alcoholic ammonia and ethyl cyanoacetate. In the author's laboratory it has been found that about 1 µg patulin could be detected on TLC using these latter reagents as sprays and heating [120], but the veratrole reagent was not useful for TLC.

Gas chromatography (GC) of underivatised patulin was reported by Scott et al. [51] who used a glass column packed with 5% OV-210 on 60—80 mesh Diatoport S at a column temperature of 135°. The secondary alcohol group of patulin readily lends itself to the formation of derivatives. Pohland et al. [121] formed the trimethylsilyl ether with N,O-bis-trimethylsilyl acetamide (BSA) in ethyl acetate, the acetate with acetic anhydride and pyridine in tetrahydrofuran, and the chloroacetate with chloroacetic anhy-

TABLE III

THIN-LAYER CHROMATOGRAPHY OF PATULIN

Adsorbent	Solvent system	R_F value	Reference
Whatman CC41 cellulose	Benzene—acetic acid—water (136:72:3) or acetic acid—water (2:98)	?	108
Silica gel	Chloroform—acetic acid (85:15)	?	109
Silica gel (0.25 mm)	Chloroform—acetic acid (9:1)	0.32	50
Silica gel (0.25 mm)	Ethyl acetate—petr. ether (60—70°)—acetic acid (60:90:2)	0.20	50
Adsorbosil 5 (silica gel) (0.25 mm)	Ether	0.77	64
Adsorbosil 5 (0.25 mm)	Ether	0.70	51
Adsorbosil 5 (0.25 mm)	Toluene—ethyl acetate—formic acid (6:3:2)	0.52	51
Adsorbosil 5 (0.3 mm)	Toluene—ethyl acetate—formic acid (6:3:1)	0.41	110
Adsorbosil 5 (0.3 mm)	Benzene—methanol—acetic acid (24:2:1)	0.21	110
Adsorbosil 5 or Silica gel G (0.25—0.3 mm)	Toluene—ethyl acetate—formic acid (5:4:1)	0.5	63
Adsorbosil 5 or Silica gel G (0.25—0.3 mm)	Chloroform—acetone (9:1)	0.5	63
Adsorbosil 5 or Silica gel G (0.25—0.3 mm)	Chloroform—methanol (95:5)	0.4	63
Adsorbosil 5 or Silica gel G (0.25—0.3 mm)	n-Hexane—ether (1:3)	0.4	63
Silica gel G	Hexane—ether (1:3)	0.47	111
Silica gel G	Benzene—tetrahydrofuran (8:2)	0.58	111
MN Silica gel G-HR (0.25 mm)	Toluene—chloroform—acetone (45:35:25)	0.50—0.65	112
EM Silica gel 5763 (0.25 mm)	Chloroform—methanol (3:1)	0.65	113
EM Silica gel 5763 (0.25 mm)	Chloroform—methanol (85:15)	0.51	113
EM Silica gel 5763 (0.25 mm)	Chloroform—methanol (95:5)	0.28	113
EM Silica gel 5763 (0.25 mm)	Chloroform—methanol—water (95:5:satd.)	0.02	113
EM Silica gel 5763 (0.25 mm)	Chloroform—methanol (1:3)	0.67	113
Silica gel 7GF (0.25 mm)	Benzene—methanol—acetic acid (90:5:5)	0.34	114

TABLE III (continued)

Adsorbent	Solvent system	R_F value	Reference
SilicAR-7GF (0.25—0.5 mm) (two-dimensional TLC)	Chloroform—acetone (9:1)	0.5—0.8	115
	Chloroform—acetic acid—ether (17:1:3)	0.5—0.8	115
Eastman Chromagram sheets, silica gel 6060 (ITLC)	Methanol	0.7	116
Eastman Chromagram sheets, silica gel 6060 (ITLC)	Chloroform—methanol (9:1)	0.65	116
Eastman Chromagram sheets, silica gel 6050 (ITLC)	Chloroform	0.3	116
Eastman Chromagram sheets, silica gel 6060 (ITLC)	Benzene—methanol (99:1)	0.25	116
Eastman Chromagram sheets, silica gel 6060 (ITLC)	Chloroform—benzene—methanol (50:49:1)	0.16	116
MN Polygram SIL N-HR silica gel (ITLC)	Ethanol—water (4:1)	0.71	117
MN Polygram SIL N-HR silica gel (ITLC)	Toluene—ethyl acetate—formic acid (6:3:1)	0.37	117
MN Polygram SIL N-HR silica gel (ITLC)	Benzene—methanol—acetic acid (24:2:1)	0.13	117
MN Polygram SIL N-HR silica gel (ITLC)	Benzene—propionic acid—water (2:2:1)	0.64	117
MN Polygram SIL N-HR silica gel (ITLC)	Chloroform	0.04	117
MN Polygram SIL N-HR silica gel (ITLC)	Chloroform—methanol (1:1)	0.71	117
MN Polygram SIL N-HR silica gel (ITLC)	Methanol	0.66	117

References p. 399

dride and pyridine in tetrahydrofuran. The electron capture detector allowed detection of 0.012 μg of patulin by GC of the latter derivative on a column of 3% JXR on Gas Chrom W at 130°, but the authors felt that the 4-fold sensitivity improvement compared to the flame ionization detector was insufficient. The trimethylsilyl ether was chromatographed on a column packed with 1% SE-30 on Gas Chrom Q at 110°, but was not the preferred derivative under their conditions owing to incomplete reaction, instability of the derivative even under refrigeration, tailing of the GC peak, and coating of the detector [121]. Pero *et al.* [111], however, found that trimethylsilyl patulin prepared by the complete reaction of patulin with BSA and trimethyl-chlorosilane in pyridine (3:1:9) was stable for 2 weeks at 5° and had good peak characteristics on 3% OV-17, Dexsil 300 and OV-25 phases using temperature programming; the quantitative detection limit was 0.025 μg. GC of patulin and its trimethylsilyl derivative has also been reported by Suzuki *et al.* [122].

Overall quantitative or semi-quantitative methods for analysis of patulin in foods have only recently become available. The preferred solvent system for the extraction of patulin from apple juice is ethyl acetate; three extractions are necessary for complete transfer of patulin. Scott and Somers [64] carried out a partial cleanup of the extract by column chromatography on silica gel followed by TLC using phenylhydrazine hydrochloride as spray reagent. The detection limit of the method was 100–300 μg patulin/l. This method has recently been modified by improving the column chromatography and using the more sensitive spray reagent 3-methyl-2-benzothiazolinone hydrazone hydrochloride, which results in lowering the detection limit for patulin in clear apple juice (from a freshly opened can) to about 20 μg/l [63]. The method described by Pohland *et al.* [121], which included no column chromatographic cleanup step and used GC for the final determination, was sensitive to only 700 μg patulin/l.

Patulin can be extracted from grains or flours with ethyl acetate–water (10:1) [64], ethyl acetate–water (7:1) followed by ethyl acetate [122], acetonitrile–hexane (4:1) [114], and acetonitrile–4% KCl solution (9:1) [115]. The most sensitive overall method appears to be that of Pohland and Allen [114], which includes a preparative TLC cleanup step followed by analytical TLC on phosphor-containing silica gel. The lower limit of detection of patulin in corn by this method was about 40 μg/kg. Column chromatographic procedures for the cleanup of grain extracts were used by Scott and Somers [64] and Suzuki *et al.* [122], while Stoloff *et al.* [115] used solvent partition. Meats have been analysed by a method involving extraction with acetonitrile–hexane (20:9), cleanup of the acetonitrile phase on a Celite 545 column and TLC [112]; use of diphenylbor(in)ic acid as spray reagent gave an overall detection limit of 500 μg patulin/kg. To date, no patulin has been found to occur naturally in grains or meat.

There is currently a lack of suitable tests for the confirmation of patulin at low levels in extracts of apple juice or other foods. Mass spectroscopy, either of patulin [51] or its trimethylsilyl ether (Fig. 2) [53], after GC purification, appears to be the best technique for positive identification and deserves further application. Recent attempts to find suitable simple bioassay systems for patulin have not resulted in a sufficiently sensitive test. The LC_{50} value for zebra fish (*Brachydanio rerio*) larvae was 18.0 μg patulin/ml [123]. Brine shrimp (*Artemia salina*) larvae and the crustacean *Cyclops fuscus* were also less sensitive to patulin than to several other mycotoxins [124—126]; 100 μg patulin/ml was necessary to cause about 60% mortality [125] and 100% mortality [126], respectively, in these two organisms. 50 μg patulin/ml was required for effective inhibition of leucine uptake by rabbit reticulocytes, a test used by Ueno *et al.* [127] for scirpene toxins. Chick embryos were investigated by Stansfeld *et al.* [59] in 1944, but embryos survived inoculation with 0.25 ml patulin solution containing less than 625 μg patulin/ml. Thus the well-known antibacterial assays may find new application for the confirmation of patulin in food extracts. Some species, for example, *Bacillus subtilis*, *Escherichia coli*, *Staphylococcus aureus*, and *Agrobacterium tumefaciens* can be inhibited by as little as 2—5 μg pure patulin/ml [9,60,128,129]. Other sensitive micro-organisms include paramecia (killed by 1 μg/ml) [39] and two other protozoa *Trypanosoma equiperdon* and *Glaucoma piriformis* (inhibited by 0.1 μg/ml and 2 μg/ml respectively) [129]; and the fungi *Pythium debaryanum* (completely inhibited by 2 μg/ml) [2], *Pythium ultimum* (inhibited by 1.2 μg/ml) [2,9] and other Pythium species [2]. In addition, patulin is cytotoxic to HeLa cells and other cells in culture at a concentration of 1—3.2 μg/ml [129—131].

Natural occurrence

Penicillium expansum is a common storage rot organism in apples, and patulin was first found in sap from natural rots caused by this fungus by Brian *et al.* [100], who identified it by its antimicrobial spectrum. Subsequently, Walker [108] also detected patulin in apples infected with *P. expansum* using paper chromatography and ultraviolet spectroscopy. Quantitative data were obtained using TLC by Harwig *et al.* [90] who found up to 17.7 mg patulin per rotten apple (240 μg/ml in the sap) and noted that no fungi other than *P. expansum* grew from tissue of the 28 apples that contained patulin. Thus contamination of apple juice and other apple products with patulin appeared to be a possibility if unsound apples were included in the starting material. The occurrence of patulin in commercial apple juice at a level of 1 mg/l has recently been demonstrated and confirmed by mass spectrometry [51], lower concentrations (20—120 μg/l) have been found in five other samples on the evidence of four TLC solvent systems [63]. Experiments have shown that patulin disappears from apple juice fermented by Saccharomyces spp. and its presence in apple cider thus appears to be unlikely [132]. The significance of the presence of patulin in apple juice as a possible human health hazard cannot be evaluated on the present toxicological data (see p. 396).

The occurrence of 0.1—0.3 ppm patulin in spontaneously mouldy baked goods, including bread, has been reported [133]. The evidence for this is restricted to one TLC solvent system and it appears that the whole extract (in 1 ml chloroform) must have been spotted on the TLC plate to give sufficient patulin to be detectable by the spray reagent used [117].

Patulin has been found in soil (1.5 ppm) and wheat straw residues (75, 40 ppm) from subtilled plots in Nebraska [134]. Two-dimensional PC was used to detect and determine the patulin, which was confirmed by extracting sufficiently large samples to obtain IR and UV spectra. It had previously been found that a strain of *Penicillium urticae* that produced patulin was one of the more potent phytotoxic fungi isolated from stubble-mulched plots [105,135].

The natural occurrence of patulin in eggs by transmission from chicken feed appears to be unlikely; hens dosed with 200 mg [^{14}C] patulin over 4 days yielded eggs containing ^{14}C activity but these contained no patulin detectable by an *Escherichia coli* bioassay [136].

Biological activity of patulin in experimental systems

(A) Toxicity to animals

Most studies on the acute toxicity of patulin have been carried out with mice. LD_{50} values found by various routes of administration were 15 mg/kg [119] and 8—10 mg/kg [39] (subcutaneous (s.c.) injection), 25 mg/kg [119] and 15.6 mg/kg [92] (intravenous (i.v.) injection), 30 mg/kg [137], 15 mg/kg [119] and 15 mg/kg [75] (intraperitoneal (i.p.) injection), and 35 mg/kg orally [119]. In general it appears that patulin is more toxic by the s.c. and i.p. routes than by i.v. injection [16,59,119] and is least toxic when given orally [39,119]. Mice receiving a lethal dose of patulin by s.c. injection developed marked oedema at the site of injection [59,119]. Death of the animal may be preceded by convulsions [119,138]. The most striking features on autopsy are oedematous and haemorrhagic lungs [119,138]. Histologically, congestion occurs in the lungs, kidneys, liver and spleen [16,17, 119]. While de Rosnay *et al.* [16] observed no change in the heart and brain on autopsy of mice that received lethal doses of patulin by s.c. or i.v. injection, Hopkins [138] found oedema, congestion of the vessels and a round-cell infiltration on examination of the brain.

Data on the acute toxicity of patulin to other animal species is more limited. Katzman *et al.* [39] reported that the LD_{50} (s.c. injection) in rats was about 25 mg/kg while Broom *et al.* [119] found 15 mg/kg by the same route and between 25 and 50 mg/kg by i.v. injection. Baron *et al.* found that 25 mg/kg (s.c.) caused 45% mortality and 20 mg/kg (i.p.) caused 55% mortality. Generally the same effects were observed in rats as in mice [39,119]. The oral LD_{50} for patulin in rats has not been determined. Freerksen and Bönicke [139] found three 50-mg doses were lethal when given orally as was 25 mg injected directly into the stomach or intestine.

Patulin had a marked antidiuretic effect in rats when administered either by injection [39,119] or by stomach tube [119].

Other animals found to be affected by patulin include the rabbit, cat, dog, and tropical fish (guppies) [39,119,138—140]. The oral LD_{50} in chicks is 170 mg/kg [141]. Patulin caused a rise in blood pressure, followed by a fall, when injected i.v. in anaesthetised cats (4—8 mg dose) or non-anaesthetised rabbits (0.5—5.0 mg/kg dose) and the rate of respiration was increased [119,140].

The reaction product of patulin and glutathione is non-toxic to mice, chick embryos and rabbit skin [75].

Although patulin could be detected in the stomach contents of rats dosed orally 3 times with 100 mg patulin within 45 min, it was not found in the blood, urine, intestinal secretions, or lymph [139]. With half this dose in rats or a larger dose in rabbits, patulin was not detectable in any of these fluids [139]. However, Broom et al. [119] found 5% of patulin injected into the rabbit was recovered in the urine within 48 h, although again it was not detectable in the blood. Patulin was in fact unstable in animal blood in vitro [119,139] and also in tissues, particularly liver and kidney, although it was not inactivated by heart tissue [139].

There are some contradictory results concerning the sub-acute toxicity of patulin. Broom et al. [119] noted evidence of cumulative toxic action (pulmonary oedema and death of half the animals) when 0.5 mg (25 mg/kg body weight) was given orally to mice every day for 2 weeks. However, Freerksen and Bönicke [139] found that mice tolerated 0.6 mg (0.5% aqueous solution) given orally every 3 h for 8 days and no deaths occurred. Much lower doses (0.5 mg/kg body weight) given daily by i.p. injection produced no microscopic lesions in mice after 11 days [119]. Mice tolerated 0.03 mg per animal per day for 16 weeks in the sub-acute study of Lembke and Hahn [142], but daily i.p. injections of 0.1 mg patulin per mouse markedly reduced blood lymphocyte counts after 3—4 weeks and mice often died from abscesses in different organs [143]. De Rosnay et al. [16] observed liver necrosis and renal and pulmonary congestion on subcutaneous injection of the same dose for 26 days. Liver lesions have also been found in chicks given daily oral doses of 0.2 mg patulin for 6 weeks [144].

The evidence for patulin being labelled a carcinogen comes from the studies of Dickens and Jones [78], who injected 0.2 mg of patulin in arachis oil s.c. twice weekly into male rats for 61—64 weeks. Sarcomas were produced at the site of injection in 6 rats among 8 animals surviving after 64—69 weeks. One fibrosarcoma (out of 3 tested) was capable of continued growth on subcutaneous implantation into young female rats. The question of the validity of subcutaneous injection as a test for carcinogenicity has been reviewed by Grasso [145] and Ciegler et al. [146]. There is need for long-term oral feeding studies in animals to answer questions as to whether patulin could be harmful when ingested by humans.

(B) Effects on humans

During a clinical trial to determine whether patulin could be effective in treatment of the common cold, aqueous solutions of patulin (1:20 000 to

1:5000 dilution) were sprayed into the patient's nose and throat, or (1:10 000 dilution) were snuffed up from the hand, instilled into the anterior nares with a pipette or gargled [138]. Treatment was every 2 or 4 h and usually continued for 24—48 h. No ill effects were observed and, indeed the treatment appeared to be beneficial with regard to curing colds. Stansfeld *et al.* [59], in addition to concluding that patulin had no effects on colds or conjunctivitis, recorded that a 1:10 000 dilution of patulin produced some irritation in the eyes of patients with the latter disease. Patulin (0.3 mg) causes oedema in rabbit skin [75] and this condition was produced in the skin of a human subject after three applications of ointment containing 1% patulin [147]. Intense tissue swelling, in addition to epidermal and dermal injury, was also observed within 48 h in a patch test carried out with patulin [148]. A keloid later resulted at the site of the reaction.

Patulin (mycoin) was irritating to the stomach and not tolerated well by invalids when given orally as tablets [16]. I.v. perfusion of 100 mg patulin in 500 ml solution was, however, well tolerated [16]! Other reports have stated that patulin produced nausea or vomiting in humans [139,149].

(C) Other biological activity (see also p. 396)

The phytotoxic activity of patulin has been reviewed by Singh [60, 129] who refers to the inhibition of seed germination and plant wilting caused by patulin. For example, Timonin [65] found that concentrations of patulin greater than 50 μg/ml reduced the germination of wheat seed. More recently, Norstadt and McCalla [150], concerned about crop yield problems in stubble-mulched plots, observed that patulin added to soil or sand also caused a reduction in leaf length of the wheat, and at maturity the root and straw weights were lowered. Biochemical effects produced by patulin in plant systems include inhibition of the release of scopoletin from sterile oat roots [151], inhibition of induced nitrate reductase activity in cauliflower and white mustard [152], and increased rate of pigment exosmosis from red beet root discs [153,154]. Some algal species have been adversely affected by treatment with patulin [154].

The antifungal activity of patulin was discovered early in its history and indeed was noted in cultures of *Penicillium expansum*, from which patulin (expansine) was later isolated, by Van Luijk in 1938 [5]. Patulin inhibits the growth of many species of fungi, including plant pathogens [9,60,65] — for example it is fungicidal and fungistatic to *Cerostomella ulmi* [155], the cause of Dutch elm disease, and *Ustilago tritici* (loose smut of wheat) [65] — and dermatophytes [9,60,65,156], for example Trichophyton spp. and Epidermophyton spp. As mentioned previously (see p. 398), Pythium species appear to be the fungi most susceptible to patulin [2]. Inhibition of respiration by patulin was observed in detailed studies with *P. aphanidermatum*, *Claviceps purpurea* and *Alternaria solani* [60,61].

Patulin caused mutation in a strain of the yeast *Saccharomyces cerevisiae* [157] when added at a concentration as low as 10 μg/ml at the exponential phase of growth. Toxicity was also apparent at the same concentration range that produced mutation. Patulin has been investigated as a possible antifermentative agent for wine [158]. Myrchink [24] did not find patulin inhibitory to *Torulopsis pulcherima* or a Schizosaccharomyces sp.

Patulin has a broad antibacterial spectrum and is active against both Gram-positive and Gram-negative organisms [9,16,60,129,139,159]. In fact it is more active than penicillin against the latter [138]. It was thus unfortunate that patulin could not find commercial application as an antibiotic on account of its toxicity (see p. **397**). De Rosnay *et al.* [16] cited the successful treatment of bovine brucellosis and some good results with certain human diseases. It might even have found use as a preservative for milk and saké in Japan [160,161].

Patulin is lethal to, or inhibits growth of, all protozoa that have been tested at concentrations varying from 0.1 to 200 μg/ml [39,129]. Several bacterial viruses have also been inactivated by the antibiotic [129].

(D) Cytological effects and mode of action

For detailed information on the cytological and biochemical effects of patulin, the reviews by Singh [60,129] and Dustin [162] and the recent work by Kawasaki *et al.* [163] should be consulted. Mouse fibroblasts (L cell line) are highly sensitive, multiplication being inhibited by < 1 μg/ml patulin [131]. Patulin specifically affects dividing cells, *i.e.* it is a mitotic poison [162]. Of particular interest are the reports by Vollmar [164] on the inhibition of malignant cells by patulin, by Sentein [165] that patulin caused chromosome breakage in salamander eggs during mitosis, and by Withers [166] that in human leucocyte cultures a high percentage (12%) of polyploid cells were produced by 0.54 μg/ml patulin. Patulin induced single and double strand breaks in HeLa cell DNA at a concentration of 32 μg/ml [167].

Evidence that patulin inhibits aerobic respiration, in addition to that found from studies with micro-organisms [60,61], has been obtained using guinea pig brain and kidney tissue [137]. The exact site where patulin acts on respiration is not clear, although it possibly occurs in the respiratory reactions prior to the terminal electron transport chain [129].

References

1 J.H. Birkinshaw, S.E. Michael, A. Bracken and H. Raistrick, Lancet, 245 (1943) 625.
2 W.K. Anslow, H. Raistrick and G. Smith, J. Soc. Chem. Ind., 62 (1943) 236.
3 E. Chain, H.W. Florey and M.A. Jennings, Brit. J. Exptl. Pathol., 23 (1942) 202.
4 E. Chain, H.W. Florey, M.A. Jennings, D. Crowfoot and B. Low, Lancet, 246 (1944) 112.
5 A. van Luijk, Meded. Phytopathol. Lab. Scholten, 14 (1938) 43; cited in ref. 9.
6 N. Atkinson, Aust. J. Exptl. Biol. Med. Sci., 20 (1942) 287.
7 S.A. Waksman, E.S. Horning and E.L. Spencer, Science, 96 (1942) 202.
8 B.P. Wiesner, Nature, 149 (1942) 356.

9 E.P. Abraham and H.W. Florey, Substances produced by Fungi Imperfecti and Ascomycetes, in H.W. Florey, E. Chain, N.G. Heatley, M.A. Jennings, A.G. Sanders, E.P.
   Abraham and M.E. Florey (Eds.), Antibiotics, Vol. I, Oxford University Press,
   London, 1949, pp. 273—355.
10 I.R. Hooper, H.W. Anderson, P. Skell and H.E. Carter, Science, 99 (1944) 16.
11 F. Bergel, A.L. Morrison, A.R. Moss and H. Rinderknecht, J. Chem. Soc. (Lond.),
   (1944) 415.
12 H.W. Florey, M.A. Jennings and F.J. Philpot, Nature, 153 (1944) 139.
13 J. Kent and N.G. Heatley, Nature, 156 (1945) 295.
14 W.T. Nauta, H.K. Oosterhuis, A.C. van der Linden, P. van Duyn and J.W. Dienske,
   Rec. Trav. Chim., 65 (1946) 865.
15 J.R.L. Walker, Nature, 227 (1970) 298.
16 C.D. de Rosnay, C. Martin-Dupont and R. Jensen, J. Méd. Bord., 129 (1952) 189.
17 F. Baron, A. Buzas, G. Clément and C. Dufour, Bull. Soc. Chim. Fr., (1951) 526.
18 N. Atkinson, R.A.W. Sheppard, N.F. Stanley and P. Melvin, Aust. J. Exptl. Biol. Med.
   Sci., 22 (1944) 223.
19 H.S. Burton and B.F. Pausacker, unpublished results (1947); cited in ref. 9.
20 J. Barta and R. Mečiř, Experientia, 4 (1948) 277.
21 P. Simonart and R. de Lathouwer, Zentralbl. Bakteriol. Parasitenkd. Abt. II, 110
   (1956—57) 107.
22 H. Umezawa, Y. Mizuhara, K. Uekane and M. Hagihara, J. Penicillin (Japan), 1 (1947)
   6; Chem. Abstr., 41 (1947) 6918f.
23 O.M. Efimenko and P.A. Yakimov, Tr. Leningr. Khim.-Farm. Inst., (1960) 88; Chem.
   Abstr., 55 (1961) 21470f.
24 T.G. Myrchink, Antibiotiki, 12 (1967) 762.
25 E.O. Karow and J.W. Foster, Science, 99 (1944) 265.
26 H.H. Kuehn, Mycologia, 50 (1958) 417.
27 Anon., Lancet, 245 (1943) 641.
28 T. Ukai, Y. Yamamoto and T. Yamamoto, J. Pharm. Soc. Jap., 74 (1954) 450.
29 M. Hori and T. Yamamoto, J. Pharm. Soc. Jap., 73 (1953) 1097.
30 M. Hori, T. Yamamoto, A. Ozawa, Y. Matsuki, A. Hamaguchi and H. Soraoka, Jap. J.
   Bacteriol., 9 (1954) 1105.
31 T. Yamamoto, J. Pharm. Soc. Jap., 74 (1954) 797.
32 T. Yamamoto, J. Pharm. Soc. Jap., 74 (1954) 810.
33 J. Schultz, R. Motz, M. Schäfer and W. Baumgart, Monatsh. Vet.-Med., 24 (1969) 14.
34 J. Schultz, Monatsh. Vet-Med., 23 (1968) 598.
35 C. Moreau and M. Moreau, Compt. Rend. Séances Acad. Agric. Fr., 46 (1960) 441.
36 J. Jacquet, P. Boutibonnes and J.-P. Cicile, Bull. Acad. Vét. Fr., 36 (1963) 199.
37 J. Forgacs, W.T. Carll, A.S. Herring and B.G. Mahlandt, Am. J. Hyg., 60 (1954) 15.
38 A.G. Lochhead, F.E. Chase and G.B. Landerkin, Can. J. Res., 24E (1946) 1.
39 P.A. Katzman, E.E. Hays, C.K. Cain, J.J. van Wyk, F.J. Reithel, S.A. Thayer, E.A.
   Doisy, W.L. Gaby, C.J. Carroll, R.D. Muir, L.R. Jones and N.J. Wade, J. Biol. Chem.,
   154 (1944) 475.
40 B.G. Engel, W. Brzeski and P.A. Plattner, Helv. Chim. Acta, 32 (1949) 1166.
41 J.F. Grove, J. Chem. Soc. (Lond.), (1951) p. 883.
42 F. Lalau-Kéraly, P. Nivière and P. Tronche, Compt. Rend. Hebd. Séances Acad. Sci.,
   Paris, 261 (1965) 4028.
43 P.M. Scott, unpublished results, (1971).
44 R.B. Woodward and G. Singh, J. Am. Chem. Soc., 71 (1949) 758.
45 R.B. Woodward and G. Singh, Experientia, 6 (1950) 238.
46 M.M. Shemyakin and A.S. Khokhlov, Dokl. Akad. Nauk SSSR, 75 (1950) 47.
47 H.J. Dauben Jr. and F.L. Weisenborn, J. Am. Chem. Soc., 71 (1949) 3853.
48 Y.-C. Pei, Y. Wang and W.-Y. Huang, K'o Hsüeh T'ung Pao, (1957) 588; Chem.
   Abstr., 53 (1959) 14216e.
49 S.W. Tanenbaum and E.W. Bassett, Biochim. Biophys. Acta, 28 (1958) 21.

50 P.I. Forrester and G.M. Gaucher, Biochemistry, 11 (1972) 1102.
51 P.M. Scott, W.F. Miles, P. Toft and J.G. Dubé, J. Agric. Food Chem., 20 (1972) 450.
52 A.I. Scott and M. Yalpani, Chem. Commun., (1967) p. 945.
53 P.M. Scott and W.F. Miles, unpublished results, (1973).
54 T. Suzuki, M. Takeda and H. Tanabe, Chem. Pharm. Bull. (Tokyo), 19 (1971) 1786.
55 T. Yamamoto, J. Pharm. Soc. Jap., 76 (1956) 1419.
56 N.G. Heatley and F.J. Philpot, J. Gen. Microbiol., 1 (1947) 232.
57 B.G. Engel, W. Brzeski and P.A. Plattner, Helv. Chim. Acta, 32 (1949) 1752.
58 E.G. Jefferys, J. Gen. Microbiol., 7 (1952) 295.
59 J.M. Stansfeld, A.E. Francis and C.H. Stuart-Harris, Lancet, 247 (1944) 370.
60 J. Singh, Mechanism of Antifungal Action of Patulin, Ph. D. Thesis, University of
    Illinois, Urbana, Ill., 1966.
61 D. Gottlieb and J. Singh, Riv. Patol. Veg., 4 (1964) 455.
62 A.E. Pohland and R. Allen, J. Assoc. Off. Anal. Chem., 53 (1970) 688.
63 P.M. Scott and B.P.C. Kennedy, J. Assoc. Off. Anal. Chem., 56 (1973) 813.
64 P.M. Scott and E. Somers, J. Agric. Food Chem., 16 (1968) 483.
65 M.I. Timonin, Sci. Agric., 26 (1946) 358.
66 N. Atkinson and N.F. Stanley, Aust. J. Exptl. Biol. Med. Sci., 21 (1943) 255.
67 C.J. Cavallito and J.H. Bailey, Science, 100 (1944) 390.
68 W.B. Geiger and J.E. Conn, J. Am. Chem. Soc., 67 (1945) 112.
69 G. Miescher, Phytopathol. Z., 16 (1950) 369.
70 E.G. Rondanelli, P. Gorini, E. Strosselli and D. Pecorari, Haematologica (Pavia), 42
    (1957) 1427.
71 G. Andraud, R. Cuvelier, P. Tronche and J. Couquelet, Compt. Rend. Séances Soc.
    Biol. Fil., 158 (1964) 2341.
72 H. Rinderknecht, J.L. Ward, F. Bergel and A.L. Morrison, Biochem. J., 41 (1947)
    463.
73 F. Bustinza Lachiondo and A.C. Lopez, An. Jardin Bot. Madrid, 7 (1947) 177; Biol.
    Abstr., 25 (1951) 15188.
74 A. Delaunay, P. Daniel, C. de Roquefeuil and M. Henon, Ann. Inst. Pasteur (Paris), 88
    (1955) 699.
75 K. Hofmann, H.-J. Mintzlaff, I. Alperden and L. Leistner, Fleischwirtschaft, 51
    (1971) 1534, 1539.
76 F. Dickens and J. Cooke, Brit. J. Cancer, 19 (1965) 404.
77 I. Goodman and R.B. Hiatt, Biochem. Pharmacol., 13 (1964) 871.
78 F. Dickens and H.E.H. Jones, Brit. J. Cancer, 15 (1961) 85.
79 G. Andraud, P. Tronche, J. Couquelet and M. Dorel, Compt. Rend. Séances Soc. Biol.
    Fil., 159 (1965) 686.
80 P. Daniel, E. Lasfargues and A. Delaunay, Compt. Rend. Séances Soc. Biol. Fil., 149
    (1955) 18.
81 H.A. Krebs, Biochem. J., 38 (1944) xxix.
82 E. Neter, J. Infect. Dis., 76 (1945) 20.
83 J.D. Bu'Lock and A.J. Ryan, Proc. Chem. Soc., (1958) p. 222.
84 S.W. Tanenbaum and E.W. Bassett, J. Biol. Chem., 234 (1959) 1861.
85 P. Dimroth, H. Walter and F. Lynen, European J. Biochem., 13 (1970) 98.
86 A.I. Scott, G.T. Phillips and U. Kircheis, Bio-org. Chem., 1 (1971) 380.
87 R.J. Light, Biochim. Biophys. Acta, 191 (1969) 430.
88 P.I. Forrester and G.M. Gaucher, Biochemistry, 11 (1972) 1108.
89 E.W. Bassett and S.W. Tanenbaum, Experientia, 14 (1958) 38.
90 J. Harwig, Y-K. Chen, B.P.C. Kennedy and P.M. Scott, Can. Inst. Food Sci. Technol.
    J., 6 (1973) 22.
91 E.W. Bassett and S.W. Tanenbaum, Biochim. Biophys. Acta, 28 (1958) 247.
92 T. Yamamoto, J. Pharm. Soc. Jap., 74 (1954) 806.
93 M.A. Darken and N.O. Sjolander, Antibiot. Chemother., 1 (1951) 573.
94 F.A. Norstadt and T.M. McCalla, Plant Soil, 34 (1971) 97.

 95 F.A. Norstadt and T.M. McCalla, Appl. Microbiol., 17 (1969) 193.
 96 P.M. Scott, B. Kennedy and W. van Walbeek, Experientia, 28 (1972) 1252.
 97 G. Cantini, J.C. Scurti, G. di Modica and S. Tira, Atti Acad. Sci. Torino, Cl. Sci. Fis. Mat. Natur., 104 (1970) 171; Microbiol. Abstr. A, 5 (1970) 102.
 98 H. Börner, Phytopathol. Z., 49 (1963) 1.
 99 H. Börner, Phytopathol. Z., 48 (1963) 370.
100 P.W. Brian, G.W. Elson and D. Lowe, Nature, 178 (1956) 263.
101 E. Grossbard, J. Gen. Microbiol., 6 (1952) 295.
102 T. Yamamoto, J. Pharm. Soc. Jap., 76 (1956) 1375.
103 H. Opel, Naturwissenschaften, 44 (1957) 306.
104 L. Reio, J. Chromatogr., 1 (1958) 338.
105 F.A. Norstadt and T.M. McCalla, Science, 140 (1963) 410.
106 V. Betina, J. Chromatog., 15 (1964) 379.
107 V. Betina, Sb. Prac., (1964) 33.
108 J.R.L. Walker, Phytochemistry, 8 (1969) 561.
109 J.D. Bu'Lock, D. Hamilton, M.A. Hulme, A.J. Powell, H.M. Smalley, D. Shepherd and G.N. Smith, Can. J. Microbiol., 11 (1965) 765.
110 P.M. Scott, J.W. Lawrence and W. van Walbeek, Appl. Microbiol., 20 (1970) 839.
111 R.W. Pero, D. Harvan, R.G. Owens and J.P. Snow, J. Chromatog., 65 (1972) 501.
112 F. Tauchmann, L. Tóth and L. Leistner, Fleischwirtschaft, 51 (1971) 1079.
113 A.I. Schepartz, R.A. Fleischman and J.H. Cisle, J. Chromatog., 69 (1972) 411.
114 A.E. Pohland and R. Allen, J. Assoc. Off. Anal. Chem., 53 (1970) 686.
115 L. Stoloff, S. Nesheim, L. Yin, J.V. Rodricks, M. Stack and A.D. Campbell, J. Assoc. Off. Anal. Chem., 54 (1971) 91.
116 A. Aszalos, S. Davis and D. Frost, J. Chromatog., 37 (1968) 487.
117 J. Reiss, Chromatographia, 4 (1971) 576.
118 H. Laubie, Bull. Soc. Pharm. Bord., 92 (1955) 213.
119 W.A. Broom, E. Bülbring, C.J. Chapman, J.W.F. Hampton, A.M. Thomson, J. Ungar, R. Wien and G. Woolfe, Brit. J. Exptl. Pathol., 25 (1944) 195.
120 P.M. Scott, unpublished results (1967).
121 A.E. Pohland, K. Sanders and C.W. Thorpe, J. Assoc. Off. Anal. Chem., 53 (1970) 692.
122 T. Suzuki, M. Takeda and H. Tanabe, Shokuhin Eiseigaku Zasshi, 12 (1971) 489; Chem. Abstr., 77 (1972) 3886u.
123 Z.H. Abedi and P.M. Scott, J. Assoc. Off. Anal. Chem., 52 (1969) 963.
124 J. Harwig and P.M. Scott, Appl. Microbiol., 21 (1971) 1011.
125 J. Reiss, Zentralbl. Bakteriol. Hyg. I Abt. Orig. B., 155 (1972) 531.
126 J. Reiss, J. Assoc. Off. Anal. Chem., 55 (1972) 895.
127 Y. Ueno, M. Hosoya and Y. Ishikawa, J. Biochem. (Tokyo), 66 (1969) 419.
128 H. Yamakawa and Y. Ueno, Chem. Pharm. Bull. (Tokyo), 18 (1970) 177.
129 J. Singh, Patulin, in D. Gottlieb and P.D. Shaw (Eds.), Antibiotics, Vol. I, Springer, Berlin, 1967, pp. 621—630.
130 M. Umeda, Jap. J. Exptl. Med., 41 (1971) 195.
131 D. Perlman, N.A. Guiffre, P.W. Jackson and F.E. Giardinello, Proc. Soc. Exptl. Biol. Med., 102 (1959) 290.
132 J. Harwig, P.M. Scott, B.P.C. Kennedy and Y-K. Chen, Can. Inst. Food Sci. Technol. J., 6 (1973) 45.
133 J. Reiss, Naturwissenschaften, 59 (1972) 37.
134 F.A. Norstadt and T.M. McCalla, Soil Sci., 107 (1969) 188.
135 T.M. McCalla, W.D. Guenzi and F.A. Norstadt, Z. Allg. Mikrobiol., 3 (1963) 202.
136 J. Lovett, Ann. Meeting Am. Soc. Microbiol., Apr. 23—28, 1972, Philadelphia, Pa., p. 17 (Abstr.).
137 G. Andraud, P. Tronche and J. Couquelet, Ann. Biol. Clin., 22 (1964) 1067.
138 W.A. Hopkins, Lancet, 245 (1943) 631.

139 E. Freerksen and R. Bönicke, Z. Hyg., 132 (1951) 274.

140 A. Schweitzer, Exptl. Med. Surg., 4 (1946) 289.

141 J. Lovett, Poultry Sci., 51 (1972) 2097.

142 A. Lembke and B. Hahn, Kiel. Milchwirtsch. Forschungsber., 6 (1954) 41, 219; Chem. Abstr., 49 (1955) 16239i.

143 W. Bollag, Experientia, 5 (1949) 447.

144 H.-J. Mintzlaff and W. Christ, Fleischwirtschaft, 51 (1971) 1802.

145 P. Grasso, Chem. Brit., 6 (1970) 17.

146 A. Ciegler, R.W. Detroy and E.B. Lillehoj, Patulin, penicillic acid, and other carcinogenic lactones, in A. Ciegler, S. Kadis and S.J. Ajl (Eds.), Microbial Toxins, Vol. VI, Academic Press, New York, 1971, pp. 409—434.

147 M.A. Jennings, unpublished results, 1944 (quoted in ref. 9).

148 J.E. Dalton, Arch. Dermatol. Syphilol., 65 (1952) 53.

149 K. Walker and B.P. Wiesner, Lancet, 246 (1944) 294.

150 F.A. Norstadt and T.M. McCalla, Soil Sci., 111 (1971) 236.

151 P. Martin, Arch. Mikrobiol., 29 (1958) 154.

152 M.M.R.K. Afridi, Hind. Antibiot. Bull., 5 (1962) 51.

153 G. Andraud, J. Couquelet, M. Dorel and P. Tronche, Compt. Rend. Séances Soc. Biol. Fil., 160 (1966) 325.

154 J. Meyer, R. Sartory, J. Malgras and J. Touillier, Bull. Assoc. Diplômés Microbiol. Fac. Pharm. Nancy, 46 (1952) 30; Chem. Abstr., 48 (1954) 2990g.

155 S.A. Waksman and E. Bugie, Proc. Soc. Exptl. Biol. Med., 54 (1943) 79.

156 K. Loewenthal and J.A. Tolmach, J. Invest. Dermatol., 8 (1947) 357.

157 V.W. Mayer and M.S. Legator, J. Agric. Food Chem., 17 (1969) 454.

158 P.G. Garoglio and C. Stella, Riv. Vitic. Enol. (Conegliano), 10 (1957) 195; Chem. Abstr., 52 (1958) 19008d.

159 T. Korzybski, A. Kowszyk-Gindifer and W. Kuryłowicz, Antibiotics. Origin, Nature and Properties, Vol. II, Pergamon, New York, 1967, pp. 1223—1230.

160 Y. Inomoto and W. Hashida, J. Ferment. Technol., 30 (1952) 287.

161 Y. Inomoto, W. Hashida and A. Yamamoto, J. Ferment. Technol., 29 (1951) 204.

162 P. Dustin Jr., Pharmacol. Rev., 15 (1963) 449.

163 I. Kawasaki, T. Oki, M. Umeda and M. Saito, Jap. J. Exptl. Med., 42 (1972) 327.

164 H. Vollmar, Z. Hyg. Infektionskr., 127 (1947) 316.

165 P. Sentein, Compt. Rend. Séances Soc. Biol. Fil., 149 (1955) 1621.

166 R.F.J. Withers, Symposium on the Mutational Process, Mechanism of Mutation and Inducing Factors, Prague, 1965, pp. 359—364.

167 M. Umeda, T. Yamamoto and M. Saito, Jap. J. Exptl. Med., 42 (1972) 527.

Chapter 19

MALTORYZINE

HIROSHI IIZUKA

Institute of Applied Microbiology, University of Tokyo, Tokyo (Japan)

In November, 1954 the author and coworkers investigated cases of milk cow poisoning which had occurred in T Pasture in the vicinity of Tokyo and in S Pasture in Chiba Prefecture. From their findings they deduced that the feed used must have been responsible for the poisoning. Therefore, they examined the feed for its toxic effect in rabbits, mice and cows, and conducted chemical, microbial, pharmacological and pathological tests on it. As a result of these tests, they drew the conclusion that a toxic substance, which is produced when a particular mould, Aspergillus sp. IAM 2950 grows on malt rootlets, was present in the feed and was responsible for these cases of poisoning [1]. This toxic substance was extracted and isolated as phenol-like pale yellowish-brown needle-like crystals (Fig. 1) (LD_{50}: 3 mg/kg, m.p.: 68.5--69°, specific rotary activity: $[\alpha]_D^{25} = 0$; molecular formula: $C_{11}H_{14}O_4$). From the chemical characteristics of these crystals, they found that the toxin was a new substance which they named "maltoryzine" [2—5].

Fig. 1. Crystal of maltoryzine.

References p. 417

Up to that time toxic substances such as hordenine and maltoxine present in malt rootlets themselves were considered to be responsible for poisoning by malt rootlet feed. As the result of the authors' experiments, however, it was clear that toxicity of the feed is caused by another mechanism. A particular strain of Aspergillus present in normal malt rootlets grows rapidly under the influence of particular environmental factors and forms maltoryzine as a secondary metabolic product by utilizing the constituents of malt rootlet feed as precursors.

After the authors' clarification of this cause for poisoning, the method of handling malt rootlet feed was improved. Since then, no case of poisoning by the feed has been seen in Japan.

Formation of maltoryzine

Maltoryzine is formed in cultures of Aspergillus sp. IAM 2950 in media mixed with malt rootlet extract. So far, the formation of maltoryzine has not been observed in cultures of this strain in various synthetic or natural media used for the culture of microorganisms, nor in cultures of strains such as *A. oryzae* or *A. sojae* used in the Japanese fermentation industry when grown in malt rootlet extract media or in mixtures of synthetic or natural media often used in the culture of microorganisms.

Some experimental methods for the production of maltoryzine are described below.

(1) Preparation of malt rootlet extract

Fresh and normal malt rootlets available in a brewery or a malt house were immersed in distilled water (100 ml/10 g of malt rootlets at 80—90°). After 20 min immersion, the malt rootlets were filtered off with cotton cloth. The filtrate obtained was used as malt rootlet extract.

(2) Composition of malt rootlet extract media

These media can be prepared by adding an equal volume of malt rootlet extract to any of the synthetic media used in the culture of microorganisms, for example Czapek's medium, bouillon media and koji extract media. It is advisable, however, to use synthetic media which are as simple in composition as possible so that isolation of maltoryzine is easier. In this experiment, a mixture of Czapek's medium having the following composition and an equal volume of malt rootlet extract was used as medium.

$NaNO_3$	3.0 g	$FeSO_4 \cdot 7 H_2O$	0.01 g
K_2HPO_4	1.0 g	Sucrose	30.0 g
$MgSO_4 \cdot 7 H_2O$	0.5 g	H_2O	1000 ml
KCl	0.5 g		

This medium will hereafter be called malt-rootlet—Czapek's medium.

(3) Seed cultures

100 ml of malt-rootlet—Czapek's medium (pH 6.0) were put in a 500-ml shaking flask and sterilized at 120° for 15 min. Then, 1 ml of a spore suspension prepared by culturing Aspergillus sp. IAM 2950 in a test tube was added to the medium. The mixture was incubated at 30° for 72 h with shaking at 125 rev./min.

(4) Tank culture

100 l of malt-rootlet—Czapek's medium were put in a 200-l stainless steel culture tank and sterilized at 120° for 20 min. After cooling, 2 l of the seed culture of Aspergillus sp. IAM 2950 were inoculated into the medium. During the process of culture, the culture tank was kept turning at 200 rev./min while sterile air (30°) was continuously introduced into the tank at a rate of 100 l/min. Soybean oil of fine quality was added as defoaming agent when necessary. Samples of the cultured liquid were collected at various times during culture for measurement of toxicity and other properties.

(5) Toxicity test

The changes in toxicity during the culturing period and the behaviour of the toxin in the course of its extraction, purification and isolation were examined by the following methods. The samples collected during the process of culture were sterilized by Seitz filtration and injected into the abdominal cavities of male mice ranging from 10 g to 15 g in body weight. The toxicity of each sample is expressed as the death rate of the test animals to which it had been administered.

(6) Toxicity observed in the case of tank culture

The toxicity test was conducted with 1.0, 0.6 and 0.4 ml of each sample collected after 1.5, 2.0, 2.5, 3.0 and 3.5 days' culture. The 1-ml dose of the 1.5-day sample (pH 6.8) killed 50% of the test animals within 20 h of injection. The same amount of the 2-day sample (pH 6.9) killed 100% of the animals within 12 h of administration, while in the case of the 2.5-day cultured sample (pH 7.0) both the 1.6 and 0.6 ml doses killed 100% within 2 h. The 1.0-ml dose of the 3-day sample (pH 6.8) killed 100% of mice within 24 h of injection. However, no toxicity was observed in the case of the 3.5-day sample (pH 6.5). Thus, the peak in toxicity was observed at about 2.5 days.

When shaking cultures of the fungus were prepared in glass flasks under the same conditions as those used in the case of seed culture, none of the 1-,

References p. 417

2-, 3- and 4-day samples killed mice with doses smaller than 1.0 ml. How-ever, the 1.0-ml dose of the 5-day sample killed 100% of the test animals within 45 min and the 0.75-ml dose about 50% within 24 h of injection. In the case of the 6- and 7-day samples, their 1-ml dose killed 100% of mice within 25—40 min of administration. After 8 days' culture, the toxicity of the cultured liquid decreased.

The stainless steel of which the culture tank was made was found to have exerted no effect either on the growth of the strain or on the develop-ment of toxicity.

Extraction, purification and isolation of maltoryzine [3,4]

The toxic principle was isolated from concentrated tank-cultured liquid by extraction and purification using mortality after administration to mice as an indicator of toxicity.

(1) Concentration of cultured liquid

The tank-cultured liquid was filtered in a filter press for 72 h. The filtrate was concentrated to 1/10 of its original volume under reduced pres-sure. During the process of concentration, n-butanol was added when neces-sary. The concentrated liquid was allowed to stand at 5° for 18 h. The resulting turbid liquid was centrifuged for 30 min at 3000 rev./min. The precipitate was found to be toxic. The supernatant was not toxic when administered to mice in 0.2-, 0.1- and 0.05-ml doses (corresponding to about 2 ml, 1 ml and 0.5 ml, respectively, of the cultured liquid before concentra-tion).

The precipitate was freeze-dried at —40°. The yield of this precipitate was about 0.8 g/100 l. About 90% of the toxicity of the cultured liquid was found in this precipitate, which had an LD_{50} of 30 mg/kg.

(2) Solubility of the precipitate

The solubility of the precipitate in 19 different solvents was examined. It showed a solubility of approx. 90% in 2 N NaOH and absolute ethanol. In 2 N NaOH, however, it became gradually coloured and lost its toxicity in 2—3 h. It had solubility values ranging from 50% to 60% in benzene, metha-nol and chloroform, but was sparingly soluble in CCl_4, CS_2, pyridine and acetone and insoluble in 5% $NaHCO_3$, 2 N HCl, ligroin, chlorobenzene, ni-trobenzene, isoamyl alcohol, ethyl acetate and butyl acetate.

(3) Toxicity of ethanol extract

The precipitate (50 mg) was stirred in 100 ml of absolute ethanol for

6 h at 60°. The mixture was centrifuged for 10 min at 3000 rev./min and the supernatant was filtered. After distilling off ethanol from the filtrate, the ethanol-soluble and ethanol-insoluble fractions were freeze-dried. The insoluble fraction showed no toxicity at all when it was administered to mice in 50-mg and 20-mg doses. All of the test animals died when 10 mg and 5.0 mg of the soluble fraction were injected into them. About 50% of the animals died after a dose of 2.5 mg.

(4) Toxicity of alkaline ether extract

50 mg of the ethanol-soluble fraction was dissolved in about 100 ml of an NaOH solution (pH 9.0). The solution was extracted 3 times with an equivalent volume of absolute ether. The combined ether extracts were dehydrated with anhydrous sodium sulphate, filtered and evaporated to dryness. Pale yellowish brown crude crystals were formed.

The ether-soluble fraction killed mice in 20-, 10- and 5-mg doses, indicating that the toxin was not decomposed by alkali. On recrystallization in benzene and water, pale brownish yellow crystals formed. The toxin is phenolic because it showed a brown colour with the $FeCl_3$ in methanol.

(5) Toxicity of acid ether extract

The alkaline water layer was adjusted to pH 2.5 and then extracted with ether in Soxhlet's extractor for 3 consecutive days. The ether extract was dehydrated with anhydrous sodium sulphate and evaporated to dryness when a yellowish brown residue was obtained. This ether-insoluble fraction did not show toxicity in 30-mg and 20-mg doses.

(6) Paper chromatography

The toxic crystals were examined by ascending paper chromatography. The following colour reagents were used:

(i) 0.1% $FeCl_3$ (in methanol); (ii) 0.1% alcoholic methyl red solution; (iii) sucrose solution (2 g of sucrose was dissolved in a mixture of 10 ml. conc. HCl and 90 ml absolute ethanol); (iv) diazotized sulfanilic acid. The sample (0.02—0.03 ml) (about 100% in MeOH) was placed on the filter paper (Toyo No. 50, 2 × 40 cm) at a point 3 cm from its end. The chromatogram was developed for about 26 cm at room temperature (15 ± 2°), and air-dried and then sprayed with the colour reagents. The solvents and R_F values are as follows:

TABLE I

Solvents	R_F values[a]
n-BuOH saturated with water	0.95
n-BuOH saturated with 5 N NH$_4$OH	0.85
Benzene—HAc—H$_2$O (2 : 2 : 1)	0.95
Benzene—HAc—H$_2$O (1 : 1 : 2)	0.94
n-BuOH saturated with ammonium carbonate buffer (1.5 N NH, 1.5 N (NH$_4$)$_2$CO$_3$)	0.95
n-BuOH—benzene—buffer (80 : 5 : 15)	0.97
n-BuOH—benzene—buffer (40 : 11 : 19)	0.94
Acetone	0.98

[a] All R_F values were obtained from FeCl$_3$-positive spots.

The sample was FeCl$_3$-positive in acid, neutral and alkaline developers showing always only a single brown spot. It showed a pink spot with methyl red but was negative with sucrose and diazotized sulphanilic acid.

Filter press
Mycelium   Filtrate
                Concentrated *in vacuo*
                Centrifuge
Upper   Precipitate
                Lyophilization
                Extracted with hot abs. ethanol
Ethanol insol. Solvent layer
                Dried *in vacuo*
                Adjusted to pH 9.0 with 2 N NaOH
                Extracted with abs. ethyl ether
Solvent layer   aq. layer
                Acidified to pH 2.0 with 2 N HCl
                Extd. continuously with abs. ether
aq. layer   Solvent layer
                Dried *in vacuo*
                Cellulose column partition chromatography
                (solvent: BuOH saturated with water)
                FeCl$_3$ positive fraction
                Recrystallization
                Maltoryzine (Yield : *ca.* 300 mg/100 l cultured medium)

Fig. 2. Fractionation procedure of maltoryzine. Medium cultured for 2.5 days in a 100-l tank.

(7) Cellulose column partition chromatography [6]

A column (12 mm) was packed with pure cellulose powder to a height of 370 mm by the wetting method. A solution of 100 mg of the ethanol-soluble fraction in 1.5 ml of butanol saturated with water was chromatographed on a cellulose column with butanol saturated with water as the mobile phase. Fractions (0.5 ml) were collected and Nos. 23—38 showed a brown colour with $FeCl_3$ and a pink colour with methyl red. Their UV absorption spectra showed peaks at 220, 280 and 320 nm. On repeated recrystallization in benzene or chloroform fraction No. 34 gave pale brownish yellow needles (Fig. 1). Their yield was 300 mg/100 l of cultured medium. The methods of extraction and purification described above are summarized in Fig. 2.

Toxicity of maltoryzine

Mice to which maltoryzine had been administered developed muscular paralysis and showed swollen yellow livers. Shown below is an example of its toxicity to mice.

TABLE II

Time to death	Dose (mg)	Death rate
14 min	0.05	4/4
5.4 h	0.03	4/4
18 h	0.015	1/4

The LD_{50} is thus 3 mg/kg.

Physical and chemical properties

Melting point: 68.5°—69.0° (decomp.).
Molecular weight: 217.2 (Rast method).
Elementary quantitative analysis: Negative in all reactions for the detection of halogen (Beilstein's method), sulphur (Na melting method), and nitrogen (Na melting method).
Elementary analytical values: C: 62.87 H: 6.78.
Molecular formula: $C_{11}H_{12}O_4$.
Colour reactions: *(1)* Negative to the ninhydrin reaction and hence without an amino group; *(2)* Negative to Schiff's reaction and hence without an aldehyde group; *(3)* shows a brown colour with $FeCl_3$ and red with diazo compounds, indicating a phenolic compound; *(4)* shows a pink colour with methyl red, indicating a neutral or weakly acid compound; *(5)* maltoryzine decolourized permanganic acid at room temperature, indicating the presence of an aliphatic double bond.

Formation of 2,4-dinitrophenylhydrazone: A solution of 0.5 g of 2,4-dinitrophenylhydrazine in 100 ml of $2N$ HCl was added to a methanol solution of maltoryzine. The resulting reddish brown crystals of 2,4-dinitrophenylhydrazone were recrystallized from water, and had a melting point of 105.5°—109°. Since Schiff's reaction was positive, a ketone group is present.

Methylation [7—10]: Acetone (40 ml) was added to a mixture of maltoryzine 0.5 g, $(CH_3)_2SO_4$: 15 ml, K_2CO_3 : 8.4 g and the whole mixture was heated on a hot water bath for 15 h. The K_2CO_3 and acetone were removed. The residue was heated with 17.5 ml of 5% NaOH on a hot water bath for 2 h, acidified with HCl and extracted with ether. The residue from the ether extract was recrystallized in water. The analytical values of the final crystals were: C: 67.10 H: 7.89 $(OCH_3)_3$: 36.9. This result indicates the presence of 3 OH groups.

Paper electrolysis [11]: Maltoryzine showed no decomposition in a phosphate buffer at pH 7.4 or 5.0. At pH values over 7.4, however, the substance decomposed immediately.

Optical rotation: $[\alpha_D^{25}] = 0$ (C = 2.8, $CHCl_3$). Hence, no asymmetric carbon atoms or intramolecular dissymmetry is present.

Colour reactions of phenolic substances: Since maltoryzine was supposed to be a phenol derivative with 3 OH groups, 11 different colour tests were conducted on it and pyrogallol, phloroglucinol and hydroxyhydroquinone as control samples. As indicators phosphomolybdic acid, phosphomolybdic acid in NH_3, Guarschi's indicator, vanillin, fluorescence, Liebermann's indicator, iodine, piner lumber, fluorone, $FeCl_3$ in MeOH and methyl red were used. Maltoryzine was found to have the same colour reactions as hydroxyhydroquinone. Hence, the 3 OH groups are in the same positions as those of hydroxyhydroquinone.

UV absorption spectrum [12—15]: A solution of maltoryzine in methanol and of acetophenone in the same solvent were measured for their molecular extinction coefficients with a Shimazu spectrophotometer (QB-50). The spectrum of this substance is shown in Fig. 3 while the molecular extinction coefficients are compared with those of acetophenone in Table III. The UV

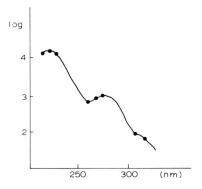

Fig. 3. UV absorption spectrum of maltoryzine in methanol.

TABLE III

UV-ABSORPTION SPECTRA OF MALTORYZINE AND ACETOPHENONE IN METH-
ANOL

Absorption zone	K-band		B-band		R-band	
Substance	λ_{max}	log ϵ	λ_{max}	log ϵ	λ_{max}	log ϵ
Acetophenone	220	4.2	280	3.0	320	1.7
Maltoryzine	220	4.1	280	3.1	320	2.1

absorption spectra of these two substances showed 3 peaks, one with a
strong band at 220 nm (log ϵ 3.1), another with a shoulder band at 280 nm
(log ϵ 3.1) and the other with a weak band at 320 nm (log ϵ 2.1). Rasmussen
et al. [12] suggest that the presence of a benzyl containing compound is
indicated by the absorption band at 280 nm (β-band) and the presence of
β-unsaturated ketone from those at 220 nm (K-band) and 320 nm (R-band).
Hence, the value of the molecular extinction coefficients of this substance
indicates the presence of α- and β-phenyl unsaturated ketone.

IR absorption spectrum [16] : The IR absorption spectrum of this substance
was measured with an IR spectrophotometer (Koken DS 301). The absorption spectrum had distinct absorption bands at 3300 cm^{-1} and 1700 cm^{-1}
and conjugated double bond absorption bands at 1600 cm^{-1} and 1500 cm^{-1}
(Fig. 4). From the elementary composition and colour reactions of this substance, the distinct absorption band at 3300 cm^{-1} indicates the presence of
OH groups: that at 1700 cm^{-1} indicates the presence of a ketone group if the
fact that the substance was negative to Schiff's reaction and that the IR
absorption band in the neighbourhood of this wave length disappeared after

Maltoryzine

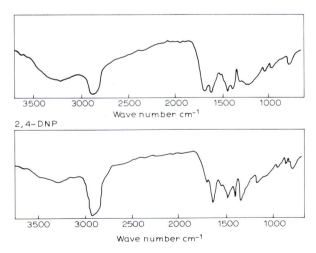

2,4-DNP

Fig. 4. IR absorption spectra of maltoryzine and its 2,4-DNP as nujol mull.

References p. 417

treatment with 2,4-dinitrophenylhydrazone is taken into consideration. From the conjugated double bond absorption bands at 1600 cm^{-1} and 1500 cm^{-1} and UV absorption spectrum, the presence of a phenyl group can be deduced.

Gibb's reaction [17—19]: This substance was dissolved in a buffer solution (M/10 NaOH or M/10 H_3BO_3 adjusted to pH 9.4). Gibb's reagent was added to this solution for its colour reaction but the result was negative. Stiller *et al.* [20], Jacob *et al.* [21], Calam *et al.* [22] and Smith *et al.* [23] published reports on the determination of the positions of OH groups. The fact that Gibb's reaction was negative indicates that the positions directly opposite to the OH groups have been substituted or that the OH groups have lost their activity forming a chelating ring as stated by Tomita *et al.* [18]. Hence, the OH groups are presumed to be in positions 2, 3 and 6 or 2, 4 and 5 in maltoryzine. From the physical and chemical properties stated above, maltoryzine is assumed to have the following structure:

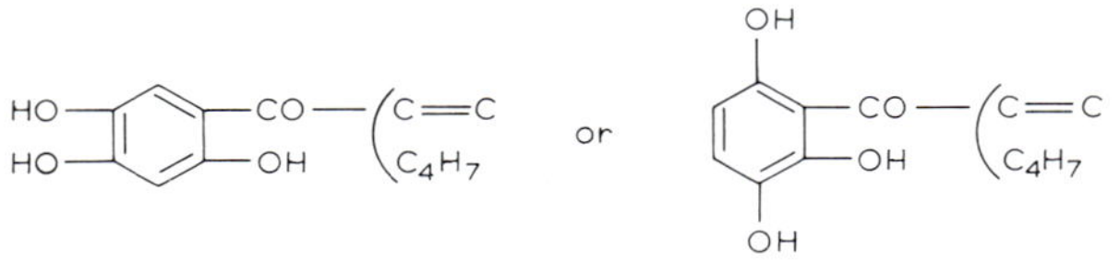

Decomposition of maltoryzine

With the object of clarifying its chemical structure, maltoryzine was subjected to various decomposing reactions. Maltoryzine has various physical and chemical properties but from the fact that it decolourizes $KMnO_4$ rapidly at normal temperature and that its spectrum has an IR absorption band around 1600 cm^{-1} it was presumed to have an aliphatic double bond in its side chain. Thus, maltoryzine was ozonized and the resulting ozonide was subjected to ketonic decomposition and then to acid decomposition. As the result of this experiment, the ozonide was proved to be 2, 3, 6-trimethoxy-phenyl-β-ketonic acid. Hence, it was found that the side chain of maltoryzine has the structure γCO-CH$_2$ · CH=CH · CH$_3$. The experiment is described below.

(1) Ozonization

200 mg of maltoryzine methylate were dissolved in 50 ml of $CHCl_3$. Sufficient ozone (60—70 V, 1.3 vol. %) was introduced into the solution gently in dry ice. After completion of the reaction, $CHCl_3$ was distilled off from the reaction solution under reduced pressure. The residue was acidified with sulphuric acid and then steam-distilled to separate it into the volatile and non-volatile components.

The volatile component was dissolved in a 5% $NaHCO_3$ and subjected to ascending paper chromatography [24]. The R_F values shown by the volatile component in 4 different developers agreed with those of acetic acid.

Further, the chromatogram of an equivalent mixture of the volatile component and acetic acid showed only one spot.

The non-volatile component, on the other hand, was acidified with HCl and then ether-extracted in Soxhlet's extractor for 3 consecutive days. The ether extract was dehydrated with Na_2SO_4 and evaporated to dryness. In this manner, 20 mg of pale yellow plates m.p. 159—160° (decomp.) were obtained. By examining their properties, these crystals were found to be of 2,4-dinitrophenylhydrazone.

From the analytical results for C, H and methoxyl groups, the non-volatile component obtained on ozonization was found to have the molecular formula $C_9H_5O_3(OCH_3)_3$. On paper electrophoresis [11], on the other hand, it moved 1.8 cm toward the positive pole from the starting point in 1h and 3.5 cm in 2 h. It was thus assumed to be acidic. It is proposed that on ozonization maltoryzine had its aliphatic double bond split and produced 1 mole acetic acid and an acid substance, $C_9H_5O_3(OCH_3)_3$.

(2) Ketonic decomposition

20 mm of the ether-soluble ozonization product were kept in a 50-ml egg-apple-shaped flask fitted with a Dimroth-condenser with 20 ml of 1 N H_2SO_4 at 120° for 1—2 min. Carbon dioxide which had been formed during the reaction was absorbed into 1/10 N $Ba(OH)_2$. After precipitates had been removed, the $Ba(OH)_2$ solution was titrated with 1/2 N HCl. The amount of the acid (16.9 ml) consumed for this purpose corresponded to 1 mole of CO_2. Hence, on ketonic decomposition the ozonization product of maltoryzine produced 1 mole of CO_2.

The ether-soluble part of the reaction solution in the flask formed 2,4-dinitrophenylhydrazone (m.p. 110—112°). The ketonic decomposition product of maltoryzine is considered, therefore, to be a carbonyl compound.

(3) Acid decomposition

A solution of 500 mg ozonide in 50 ml 1 N ethyl alcoholic KOH was heated to 75 ± 20° on a water bath for 15 min. After cooling, the reaction solution was acidified with H_2SO_4 and divided into the volatile and non-volatile parts by steam distillation.

The volatile part was subjected to ascending paper chromatography [24] with 4 different developers. The R_F value obtained for each of the developers agreed with that of acetic acid for the same developer. The spot was eluted in 5% $NaHCO_3$ and was shown to be acetic acid. The non-volatile part was acidified with HCl and then extracted with ether for 2 consecutive days. The ether-soluble part obtained was dissolved in 5% $NaHCO_3$, again acidified and extracted with ether for 24 consecutive hours. After the extract had been dehydrated with Glauber's salt, ether was distilled off from it. The residue was recrystallized in water. In this manner 5 mg of white needle-like crystals having a melting point of 148° (decomp.) were obtained.

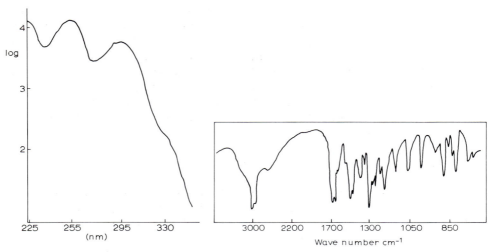

Fig. 5. UV absorption spectrum of the hydrolysate in methanol.

Fig. 6. IR absorption spectrum of the hydrolysate as nujol mull.

From the results of their elementary analysis, their substance was found to have a molecular formula of $C_{10}H_{12}O_5$. Their UV spectrum in methanol is shown in Fig. 5 and their IR absorption spectrum in Fig. 6. From these results, the non-volatile part obtained on acid decomposition is supposed to be an acid compound with the molecular formula $C_7H_3O_2(OCH_3)_3$, that is, trimethoxybenzoic acid.

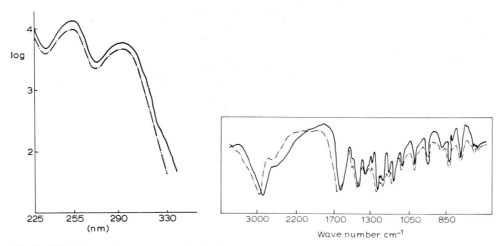

Fig. 7. UV absorption spectra of the hydrolysate and 2,3,6-trimethoxybenzoic acid in methanol. ——, hydrolysate; — · — ·, syn. 2,3,6-trimethoxybenzoic acid.

Fig. 8. IR absorption spectra of the hydrolysate and 2,3,6-trimethoxybenzoic acid as nujol mull. ——, hydrolysate; — · — ·, syn. 2,3,6-trimethoxybenzoic acid.

On the basis of the above-mentioned results of Gibb's reaction and the phenolic colour reaction, the 3 phenolic OH groups are supposed to have the positions 2, 3 and 6 or 2, 4 and 5. Further, given the published values for the melting points of trimethoxybenzoic acids, this acid substance, $C_7H_3O_2(OCH_3)_3$, agrees with 2,3,6-trimethoxybenzoic acid. The authors then synthesized this compound [24—28]. As a result, they confirmed that the acid substance and synthetic 2,3,6-trimethoxybenzoic acid are one and the same, melting at the same temperature and also showing the same UV (Fig. 7) and IR absorptions (Fig. 8).

(4) Structure of side chain of maltoryzine

As was stated above, the ozonide of maltoryzine methylate forms CO_2 and a carbonyl compound on ketonic decomposition and acetic acid and 2,3,6-trimethoxybenzoic acid on acid decomposition. This being so, the ozonization product of maltoryzine is considered to be 2,3,6-trimethoxy-benzoic acid and it is thus reasonable to suppose that the side chain of maltoryzine has the structure $-CO-CH_2-CH=CH-CH_3$. If the presence of 3 phenolic OH groups is taken into consideration, this supposition is compatible with the molecular formula assumed on the basis of analytical values.

Chemical structure of maltoryzine

Maltoryzine methylate ($C_{14}H_{20}O_4$) produces 1 M acetic acid and $C_9H_5O_3(OCH_3)_3$, an acid substance, on ozonization. This acid substance produced 1 M CO_2 and a carbonyl compound on ketonic decomposition. The carbonyl compound produced 1 M acetic acid and 2,3,6-trimethoxyben-zoic acid on acid decomposition. This acid substance was thus proved to be 2,3,6-trimethoxyphenyl-β-ketonic acid.

The above experimental results prove that the side chain of maltoryzine has the structure shown above and hence the chemical structure of maltory-zine is 1-(4-pentenyl)-2,3,6-trihydroxybenzene.

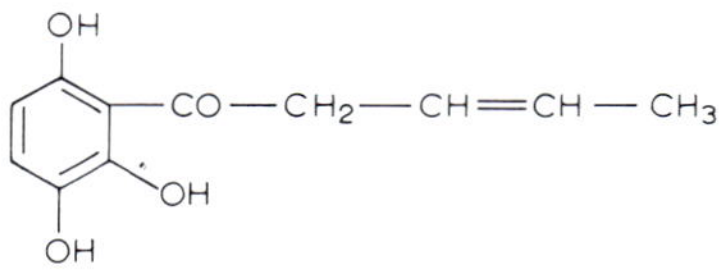

OH

$-CO-CH_2-CH=CH-CH_3$

OH

OH

Accordingly it is a new substance different either from hordenine [28], clavatine [29] or from maltoxine [30].

References

1 Y. Ohkubo, N. Urakawa, T. Hayama, Y. Seto, T. Miura, Y. Kano, S. Motoyoshi, S. Yamamoto, K. Ishida, H. Iizuka and M. Iida, Jap. J. Vet. Sci., 17(4) (1955) 145.

2 H. Iizuka and M. Iida, Nature, 196 No. 4855 (1962) 681.

3 H. Iizuka and M. Iida, Ann. Meeting Agric. Chem. Soc. Japan, 1956, p. 1 (Abstract).

4 M. Iida and H. Iizuka, Ann. Meeting Agric. Chem. Soc. Japan, 1957, p. 87 (Abstract).

5 H. Iizuka and M. Iida, J. Gen. Appl. Microbiol., 4(2) (1958) 133 (Abstract).

6 S. Marumo and K. Miyao, Bull. Agric. Chem. Soc. Japan, 19 (1955) 262.

7 P.W. Clutterbuck, W. Koerber and H. Raistrick, Biochem. J., 31 (1937) 1089.

8 H. Gilman and J.R. Thirtle, J. Am. Chem. Soc., 66 (1944) 895.

9 S. Rajagoham and T.R. Seshadri, Proc. Ind. Acad. Soc., 30A (1949) 289.

10 Z. Horii and Y. Komiyama, J. Pharm. Soc. Japan, 72 (1952) 1520.

11 H.E. Wade and D.M. Morgan, Biochem. J., 56 (1954) 41.

12 R.S. Rasmussen, D.O. Tunicliff and R.R. Brattain, J. Am. Chem. Soc., 71 (1949) 1068.

13 A.E. Gilman and E.S. Sterm, An Introduction to Electronic Absorption Spectroscopy in Organic Chemistry, 1954, p. 124.

14 E.O. Stecher and A. Clements, J. Am. Chem. Soc., 76 (1954) 503.

15 F.M. Dean and D.R. Randell, J. Chem. Soc., (1959) 1071.

16 L.J. Bellamy, The Infrared Spectra of Complex Molecules, 1954, p. 54 and 1958, p. 65.

17 H.D. Gibbs, J. Biol. Chem., 72 (1927) 649.

18 M. Tomita and S. Kamio, J. Chem. Soc. Japan, 61 (1940) 449.

19 F. Feigl, J. Japan. Chem., 13 (1959) 65.

20 E.T. Stiller and J.C. Keresztesy, J. Am. Chem. Soc., 61 (1930) 1237.

21 A. Jacob and A.R. Todd, J. Chem. Soc., (1940) 651.

22 C.T. Calam and P.W. Clutterbuck, Biochem. J., 41 (1947) 458.

23 J.N. Smith, B. Spencer and R.T. Williams, Biochem. J., 47 (1950) 289.

24 R.L. Reid and M. Lederer, Biochem. J., 50 (1952) 60.

25 W. Baker and N.C. Brown, J. Chem. Soc., (1948) 2303.

26 L.E. Smith and F.B. LaForge, J. Am. Chem. Soc., 53 (1931) 3072.

27 M. Ikawa, J. Chem. Soc. Japan, 62 (1941) 1052.

28 S. Kirkwood, J. Am. Chem. Soc., 72 (1950) 2522.

29 S. Ukai, Y. Yamamoto and T. Yamamoto, J. Pharm. Soc. Japan, 74 (1954) 450.

30 N. Urakawa, Am. J. Physiol., 190 (1960) 5.

Chapter 20

MYCOTOXIC NEPHROPATHY

PALLE KROGH

Institute of Hygiene and Microbiology, Royal Veterinary and Agricultural University, Copenhagen (Denmark)

Kidney lesions associated with the use of mouldy feed have been known in animal husbandry for several decades. In 1928 a report on a particular nephropathy in swine was published, which included observations of field cases of swine nephropathy associated with mouldy feed as well as successful experimental reproduction of porcine nephropathy by feeding a batch of mouldy rye [1]. Among the symptoms of porcine nephropathy, polydypsia and polyuria were the most characteristic and polyuria associated with the feeding of mouldy grain or hay is a well-known syndrome in horses [2]. Thus mycotoxic nephropathy seems to occur naturally in at least two species of domestic animals, and may perhaps exist in other species as well. At present most information obtained concerns the disease in swine, including items such as clinical and biochemical aspects, pathology, kidney function, epidemiology, and aetiology. Therefore this description will be based upon mycotoxic *porcine* nephropathy.

Mycotoxic porcine nephropathy

Clinical aspects

As was mentioned previously Larsen observed that polydypsia and diuresis (polyuria) are characteristic symptoms of the disease and perhaps the only observable symptom under field conditions [1]. During the renewed interest in mycotoxic porcine nephropathy since 1960 several swine feeding experiments were carried out, using nephrotoxic batches of barley collected from farms where the disease was prevalent. The results indicated that, besides polyuria, growth depression was a characteristic symptom [3]. When the nephrotoxic component was removed from the diet, the growth curve of the pigs returned to a normal shape, but the polyuria continued [3].

Pathology

In cases of porcine nephropathy, induced by nephrotoxic grains, ochratoxin-contaminated barley, or crystalline citrinin, the alteration is found in the kidneys only and lesions are absent in all other organs and tissues [4—6].

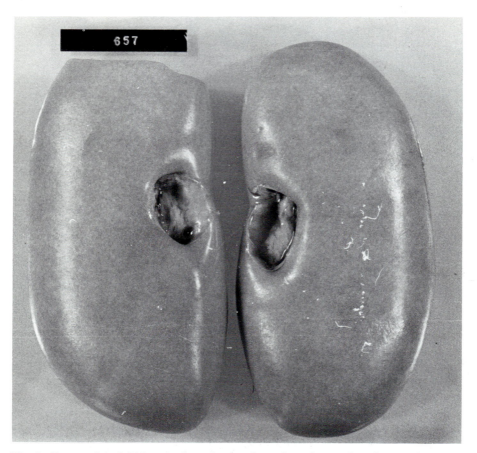

Fig. 1. Decapsulated kidneys of a pig slaughtered at 6 months of age. The colour has changed from the normal red-brown to grey. The surface has a slightly mottled appearance. Magnification: 0.75 ×. (Photo by F. Elling)

Macroscopically the kidneys are enlarged, and the colour has changed from the normal red-brown to grey (Fig. 1). The cut surface shows a diffusely distributed fibrosis in the cortex which in advanced cases includes small cysts. Microscopically the alterations consist of degeneration of the proximal tubules and formation of connective tissue in the interstitium (Fig. 2). The degeneration of the proximal tubules includes loss of brush border, atrophy of epithelial cells, and thickening of the basement membrane (Fig. 3). The interstitial fibrous tissue is diffusely distributed in the cortex, and in markedly indurated kidneys the fibrous tissue surrounding the Bowman capsules continues into the adjacent interstitial fibrous tissue. In advanced cases alteration also takes place in the glomeruli, in the form of sclerotic tufts and atrophied glomeruli. Lesions are not found in the medulla and the pelvis. It must be emphasized that the observed lesions develop in the course of the normal lifetime of bacon pigs, i.e. 6—8 months.

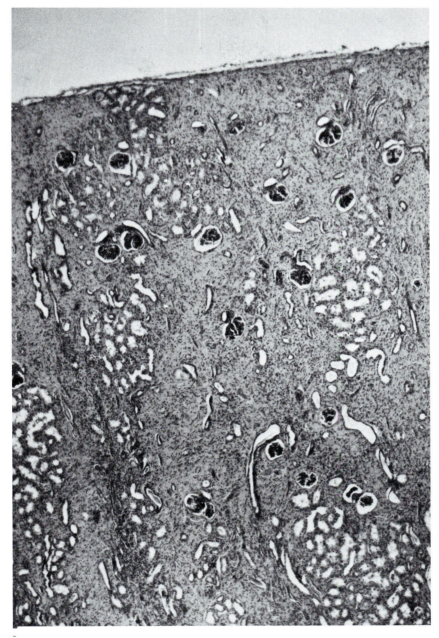

Fig. 2. Histological section of cortex of the kidney. Note the extensive interstitial fibrosis, the tubular dilatation, and the atrophy of tubules. H and E, 50 ×. (Photo by F. Elling)

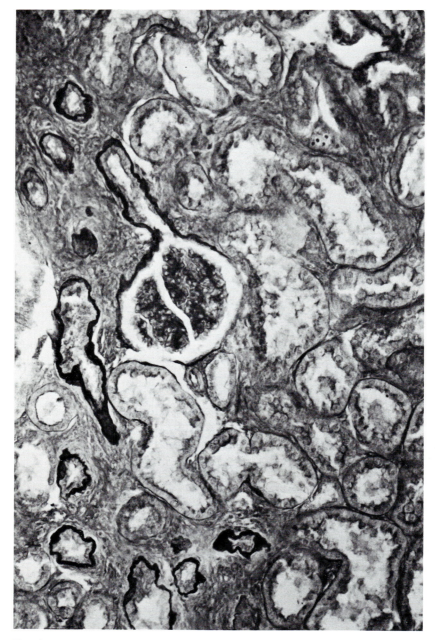

Fig. 3. Same section as Fig. 2. Glomerulus and proximal convoluted tubules surrounded by collagen deposits. Note the thickened and blurred basement membrane. PAS, 200 ×. (Photo by F. Elling)

Kidney function

A large-scale feeding experiment with pigs has recently been performed using barley naturally contaminated with ochratoxin A [6]. The pigs of a body weight of 20 kg to 90 kg (slaughter weight), were fed at three levels of ochratoxin A: 200, 1000 and 4000 ppb. Among the parameters studied were various aspects of kidney function. A decrease of glomerular filtration was observed, as measured by renal clearance of inulin, as well as a low renal clearance of *para*-aminohippuric acid (PAH) indicating a reduced renal blood flow. The capacity of tubular reabsorption decreased, as measured by Tm_{PAH}. Blood urea was increased but a state of uraemia was not reached. In the urine glucose was detected as well as small amounts of protein.

Quantitative investigations of selected enzymes in the kidney and urine showed that leucine aminopeptidase (LAP) is excreted in the urine, and a decrease of LAP in the renal cortex was observed. In the renal cortex LAP is concentrated in the brush border of the proximal tubules, so the loss of brush border observed by histological investigations corresponds very well with the concomitant excretion of LAP into the urine.

Based upon these studies it may be concluded that the action of nephrotoxic mycotoxins (ochratoxin A) takes place in the renal tubular system, especially in the proximal tubules [7].

Epidemiology

Since its discovery in 1928, mycotoxic porcine nephropathy has been observed regularly at meat inspection in Danish slaughterhouses. Most cases of mycotoxic nephropathy are uncomplicated, resulting in condemnation only of the kidneys, whereas the remaining part of the slaughtered pig is accepted for consumption. Only in a few cases is the the kidney function inhibited to such an extent that a state of uraemia is developed. In these cases the whole carcass is condemned at meat inspection.

In 1970 a Central Registration of mycotoxic porcine nephropathy was established through the Danish Veterinary Service, covering all slaughterhouses in Denmark. Based upon the data obtained through this registration the incidence rate has been calculated for 1971 and 1972 (Table I) [7]. A considerable district-to-district variation of the incidence rate has been encountered. Thus incidence rates reaching 1000—1500 per 100 000 were observed in several districts. The variation (district-to-district and year-to-year)

TABLE I

INCIDENCE RATE OF MYCOTOXIC PORCINE NEPHROPATHY IN DENMARK

Year	Cases per 100 000 pigs slaughtered
1971	67
1972	27

References p. 427

may be due to climatic variation, but apparently local habits of harvesting and storage of cereals are also contributory factors. Outside Denmark mycotoxic porcine nephropathy has been encountered in Ireland [8]. The disease has also been observed in Norway and Sweden [9].

Aetiology

Investigations of the aetiology of mycotoxic porcine nephropathy have been conducted in Denmark since 1966. The first step in these investigations was the isolation of a nephrotoxic strain of *Penicillium viridicatum* Westling from a batch of barley associated with porcine nephropathy [10]. The nephrotoxicity of the fungal organism was demonstrated by feeding sterilized barley inoculated with the fungus to rats and pigs which subsequently developed nephropathy. Chemical isolation and characterization of nephrotoxic compounds produced by this strain of *P. viridicatum* West. (No. 400, formerly No. 67 B) revealed the strain to be a producer of citrinin and ochratoxin A, as well as oxalic acid [11,4]. Separate feeding experiments in pigs with sodium oxalate and crystalline citrinin demonstrated that nephropathy similar to the disease observed in field cases could be produced by citrinin, whereas only a slight interstitial fibrosis in the kidneys was produced by oxalate, even at a high dose of 1 g sodium oxalate per kg body weight [4]. The experimental observations thus suggested both citrinin and ochratoxin A as possible causes of naturally occurring porcine nephropathy [4,6]. Based upon this assumption a survey of cereals associated with field cases of mycotoxic porcine nephropathy was carried out [12]. Through the Central Registration 35 pig farms where cases of nephropathy had occurred recently, from three districts with a high incidence rate were traced. Samples of cereals used as pig feed (mainly barley) were collected and analysed. Ochratoxin A was detected in 58% of the samples, with a mean concentration of 2.97 ppm. Citrinin was found in 9% of the samples, always together with ochratoxin A. The mean concentration of citrinin was 1.05 ppm.

Thus porcine nephropathy seems to be caused by ochratoxin A and citrinin, and ochratoxin A is apparently the more important of the two mycotoxins, as it occurs in higher concentration and with higher frequency. However, other nephrotoxic mycotoxins, so far unknown, may also be involved.

Ochratoxin A has now been detected in samples of cereals used as animal feed from four Scandinavian countries: Denmark, Finland, Norway, and Sweden, as revealed by surveys of cereals conducted by the Scandinavian Mycotoxin Project [13,38].

Ochratoxin

Mycological, chemical, and toxicological aspects of ochratoxin A are dealt with in detail in Chapter 16.

Citrinin

Citrinin is a secondary metabolite of microscopic fungi, included in several

TABLE II

PRODUCERS OF CITRININ

Genus: Penicillium Link		
Monoverticillata		
	P. lividum series	*P. lividum* West.
	P. implicatum series	*P. implicatum* Biourge
	P. decumbens series	*P. fellutanum* Biourge
		P. citreo-viride Biourge
	The Ramigena series	*P. velutinum* van Beyma
Asymmetrica-Divaricata		
	P. canescens series	*P. canescens* Sopp
		P. jenseni Zalecki
Asymmetrica-Velutina		
	P. citrinum series	*P. citrinum* Thom
		P. steckii Zalecki
	P. chrysogenum series	*P. notatum* West.
Asymmetrica-Fasciculata		
	P. viridicatum series	*P. viridicatum* West.
		P. palitans West.
	P. expansum series	*P. expansum* Link
	P. claviforme series	*P. claviforme* Bainier
Genus: Aspergillus Micheli		
	Aspergillus flavipes group	*A. niveus* Blochwitz
	Aspergillus terreus group	*A. terreus* Thom

species belonging to the fungal genera Penicillium Link and Aspergillus Micheli. Citrinin was first isolated from *P. citrinum* Thom [14]. The citrinin-producing capacity of this species has later been confirmed by several investigators [15—20]. However, several other species of Penicillium have later been observed as producers of citrinin, as well as two species of Aspergillus (Table II) [11,21—29].

It appears that *P. viridicatum* is the main producer of citrinin in cereals, at least in temperate regions of the world [11,25]. Citrinin has been found as a natural contaminant of wheat, rye, barley, and oats in Canada, and of barley and oats in Denmark, as previously mentioned [25,12].

It is interesting to note that the species *P. viridicatum* and *P. palitans* are producers of citrinin as well as ochratoxin, and that these two mycotoxins may even be isolated from the same strain [24,25].

Citrinin in the crystalline form is a yellow compound, with the following physicochemical properties: m.p. 170—171°, $[\alpha]_D^{22}$ —27.7°, $[\alpha]_{578}^{22}$ —31.5°, $[\alpha]_{546}^{21}$ —46.1° (c. 0.66 in abs. ethanol), mol. wt. 259 [11]. The spectral

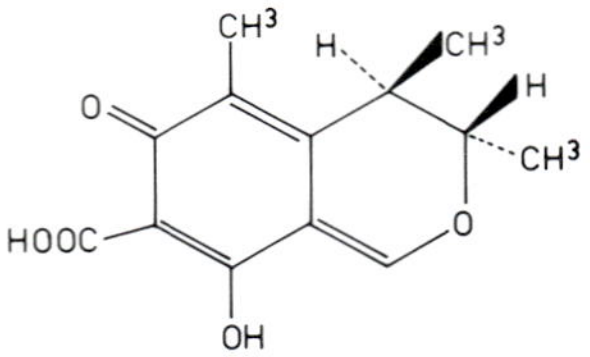

Fig. 4. Citrinin.

data of citrinin have recently been reviewed [35]. Single-crystal X-ray diffraction study of citrinin has shown that it is a *p*-quinone methide (Fig. 4) in
which intramolecular repulsion between the methyl groups has resulted in a
marked distortion of the molecule [36].

The citrinin skeleton is formed *via* the polyacetate pathway and the
extra-skeletal carbons arise from C_1 donors [37].

Citrinin was part of the "oo-hen-mai" problem ("yellowed rice") in post-
war Japan, and in this connection a method for detection was developed,
consisting of colourimetric determination employing a ferric chloride complex of citrinin [30]. A method for detection of citrinin in rice based upon
fluorometry was developed by Kawashira *et al.* [31]. Scott *et al.* reported a
TLC method for screening of fungal extracts including citrinin. With the
solvent system used citrinin appears as a streak [32]. Later the same group
developed a modified method for determination of citrinin in cereals (Method 2 A), with less tendency to streaking; however the precision is not completely satisfactory (63% recovery at 0.50 ppm level) [25]. Recently a more
precise TLC method for detection of citrinin in barley has been developed
employing extraction of acidified grain with chloroform, followed by den-
sitometric TLC quantitation [33].

The kidneys are the primary target organ for perorally dosed citrinin, as
demonstrated experimentally in the rat [11,16,18], rabbit [34], guinea pig
[34], and pig [4].

Public health aspects of nephrotoxic mycotoxins

Ochratoxin A has been found as a natural contaminant of several natural
products (see p. 358), and citrinin has been detected in wheat, rye, barley,
and oats [12,25]. In connection with the aforementioned survey of cereals
associated with field outbreaks of porcine nephropathy in Denmark the
investigations were extended to elucidate possible residues of mycotoxins in
meat. A pig farm using ochratoxin A-contaminated feed was traced, and
samples of liver, kidney and adipose tissue were collected during a 2-month
period, when bacon pigs were delivered at a slaughterhouse. The samples
were analysed for residues of ochratoxin A, which was detected in almost all
samples, with the highest concentration being 67 ppb [39]. In connection
with the aforementioned swine feeding experiment using ochratoxin A-
contaminated feed, samples of kidney, liver, muscle tissue, and adipose tissue
were analysed for residues, and ochratoxin A was detected in all 4 types of

tissue [6]. As lesions are only found in the kidneys, and all other organs and the carcass can pass the meat inspection, except in the very few cases where the kidney dysfunction has resulted in uraemia, demanding total condemnation, ochratoxin A-contaminated foodstuffs of animal origin will reach the human food channel. However, the concentration of ochratoxin A in the meat is much lower than in the feed used for animal production.

Thus humans may be exposed to nephrotoxins in two ways: By consuming foodstuffs directly contaminated by mycotoxin-producing fungi or by consuming residue-containing meat of slaughter animals fed mycotoxin-contaminated feed.

References

1 S. Larsen, Maanedsskr. Dyrl., 40, (1928), 259, 289.
2 R. Manninger and J. Mocsy, Spezielle Pathologie und Therapie der Haustiere, Fischer, Jena, 1954, p. 669.
3 A. Madsen, B. Laursen and H.P. Mortensen, Forsøgslab. Årbog, (1965) 106.
4 P. Friis, E. Hasselager and P. Krogh, Acta Path. Microbiol. Scand., 77 (1969) 559.
5 H.E. Nielsen and E. Hasselager, Forsøgslab. Årbog, (1965) 91.
6 P. Krogh, N.H. Axelsen, F. Elling, N. Gyrd-Hansen, B. Hald, J. Hyldgaard-Jensen, A.E. Larsen, A. Madsen, H.P. Mortensen, T. Møller, O.K. Petersen, U. Ravnskov, M. Rostgaard and O. Aalund, Acta Pathol. Microbiol. Scand., 1974, Section A Suppl.
7 P. Krogh, (in preparation), 1974.
8 H.G. Buckley, Irish Vet. J., 25(10) (1971) 194.
9 T. Møller, Personal communication, 1972.
10 P. Krogh and E. Hasselager, Roy. Vet. Agric. Col. Yearbook (1968) 198.
11 P. Krogh, E. Hasselager and P. Friis, Acta Pathol. Microbiol. Scand. Section B, 78(4) (1970) 401.
12 P. Krogh, B. Hald and J. Pedersen, Acta Pathol. Microbiol. Scand. Section B, 81 (1973) 689.
13 P. Krogh, Personal communication, 1973.
14 A.C. Hetherington and H. Raistrick, Phil. Trans. Roy. Soc. Ser. B., 220 B (1931) 269.
15 C. Verona and P. Gambogi, Phytopathol. Z., 19(4) (1952) 423.
16 F. Sakai, Nihon Yakuri-gaku Zasshi 51(5) (1955) 431.
17 S. Udagawa, Y. Hashimoto and S. Hirayama, Eisei-shikenjo Hokoku, 74 (1965) 299.
18 J. Nagai, M. Hayashi and K. Mizobe, Fukuoka Igaku Zasshi 48(2) (1957) 311.
19 T.G. Mirchink, V.S. Blagoveshchenskij and V.A. Fedorov, Mikrobiologiya, 35(2) (1966) 263.
20 T.G. Mirchink, V.S. Blagoveshchenskij and V.A. Fedorov, Mikrobiologiya, 36(6) (1967) 1036.
21 A.V. Pollock, Nature, 160(4062) (1947) 331.
22 V. Betina, P. Nemec, M. Kutkova, J. Balan and S. Kovac, Chem. Zvesti, 18 (1964) 128.
23 A. Jabbar and A. Rahim, J. Pharm. Sci., 51(6) (1962) 595.
24 P.M. Scott, W. Walbeek, J. Harwig and D.I. Fennell, Can. J. Plant Sci., 50(5) (1970) 583.
25 P.M. Scott, W.Walbeek, B. Kennedy and D. Anyeti, Agr. Food Chem., 20(6) (1972) 1103.
26 H.J. Mintzlaff, A. Ciegler and L. Leistner, Z. Lebensmittelunters. Forsch., 150(3) (1972) 133.
27 V.I. Bilaj, Mikroskopicheskie Griby-Produtsenty Antibiotikov, Izdat. Akad. Nauk Ukr. SSR, Kiev, 1961, pp. 24—25.

28 M.I. Timonin and J.W. Rouatt, Can. J. Publ. Health, 35 (1944) 80.
29 H. Raistrick and G. Smith, Biochem, J., 29 (1935) 606.
30 T. Taira and S. Yamatodani, Penishirin 1(5) (1947) 275.
31 I. Kawashiro, H. Tanabe, H. Takeuchi and Ch. Nishimura, Eisei-shikenjo Hokoku, 73 (1955) 191.
32 P.M. Scott, J.W. Lawrence and W. Walbeek, Appl. Microbiol., 20(5) (1970) 839.
33 B. Hald and P. Krogh, J.A.O.A.C. 56 (1973) 1440.
34 A.M. Ambrose and F. DeEds, Proc. Soc. Exptl. Biol. Med., 59 (1945) 289.
35 W.C. Neely, S.P. Ellis, N.D. Davis and U.L. Diener, J. Assoc. Off. Anal. Chem., 55(5) (1972) 1122.
36 O.R. Rodig, M. Shiro and Q. Fernando, J. Chem. Soc., 23 (1971) 1553.
37 I.T. Glover, Ph. D. Thesis, University of Virginia, 1964.
38 P. Krogh, B. Hald, P. Englund, L. Rurqvist and O. Swahn, Acta Pathol. Microbiol. Scand. Section B, 82(2) (1974) 301.
39 P. Krogh, Proc. 2nd Intern. Congr. Plant Pathol., Minneapolis, September 1973.

SUBJECT INDEX